Deutsche Forschungsgemeinschaft

Carcinogenic and Anticarcinogenic Factors in Food

Symposium

Editors:
Gerhard Eisenbrand (Chairman), Anthony David Dayan, Peter Stefan Elias, Werner Grunow and Josef Schlatter
Scientific Secretariate:
Monika Hofer und Eric Fabian

Senate Commission on Food Safety
SKLM

WILEY-VCH

DFG

Deutsche Forschungsgemeinschaft
Kennedyallee 40, D-53175 Bonn, Federal Republic of Germany
Postal address: D-53175 Bonn
Phone: ++49/228/885-1
Telefax: ++49/228/885-2777
E-Mail: (X.400): S = postmaster, P = dfg, A = d400, C = de
E-Mail (Internet RFC 822): postmaster@dfg.de
Internet: http://www.dfg.de

Library of Congress Card No.: applied for

A catalogue record for this book is available from the British Library.

Die Deutsche Bibliothek – CIP Cataloguing-in-Publication Data
A catalogue record for this publication is available from Die Deutsche Bibliothek

ISBN 3-527-27144-9

Printed on acid-free and chlorine-free paper.

Cover Design and Typography: Dieter Hüsken
Composition: ProSatz Unger, D-69469 Weinheim
Printing: betz-druck gmbh, D-64291 Darmstadt
Bookbindung: J. Schäffer GmbH & Co. KG, D-67269 Grünstadt

Printed in the Federal Republic of Germany

Contents

Preface

The DFG Symposium "Carcinogenic/Anticarcinogenic Factors in Food: Novel Concepts?" was the third in a series of symposia organised by the Senate Commission on Food Safety (SKLM). It was held in Kaiserslautern from 4th to 7th October 1998. Scientists from all over the world joined the members of the Senate Commission to take a critical look at the current state of knowledge on this topic. In their role as an advisory body, the SKLM evaluated results, drew conclusions and formulated recommendations which are published in this volume together with the individual contributions of the speakers and the poster contributions. This symposium volume not only provides an overview of our current knowledge on this topic but also points to gaps in our knowledge, draws conclusions and formulates recommendations from the point of view of the Senate Commission. Thus the SKLM is making available an up-to-date, scientifically substantiated source of information for a wide range of readers and users.

The SKLM wishes to thank all those involved in creating this volume, especially the speakers, and also the rising young scientists who made an important contribution to the symposium with their posters. Special thanks are due to the members of the editorial committee who formulated the main conclusions and recommendations as a team, in particular the chairmen and rapporteurs Prof. Dayan, Prof. Elias, Prof. Grunow and Dr. Schlatter.

We would also like to thank the scientific secretariate of the SKLM, especially Dr. Hofer and Dr. Fabian, for their contribution to the publication of this volume. This comprised not only the planning and preparation of the symposium but also the work which had to be completed before the manuscript could be submitted to the DFG editorial department and the publisher.

Our cordial thanks are also extended to the program director of the DFG, Dr. Hans Hasso Lindner, for the efficient and the very helpful manner in which he has supported the work of the Senate Commission. The SKLM is particularly indebted to the Deutsche Forschungsgemeinschaft which made this symposium possible through its support.

Prof. Dr. Gerhard Eisenbrand
Chairman of the Senate Commission
on Food Safety

I Hauptschlussfolgerungen und Empfehlungen

1 Einleitung

Die Ergebnisse vieler epidemiologischer Studien lassen erkennen, dass das Risiko des Menschen, an bestimmten Krebsarten zu erkranken, von seiner Ernährung beeinflusst wird. Derzeitige Schätzungen gehen davon aus, dass etwa ein Drittel aller Krebserkrankungen in der westlichen Welt einen Zusammenhang mit der Ernährung zeigt.

Für die Untersuchung und Erkennung von Krebsrisikofaktoren ist nicht allein die Kenntnis von krebsauslösenden (carcinogenen) bzw. krebshemmenden (protektiven) Stoffen von Bedeutung. Erheblichen Einfluss auf das Krebsrisiko hat auch der persönliche Lebensstil. Besonders deutlich wird dies am Beispiel Tabakkonsum bzw. Tabakrauchen, wo klare Dosis/Zeit/Häufigkeitsbeziehungen vor allem für Tumoren der Atemwege, aber auch anderer Organe sowohl tierexperimentell als auch für den Menschen belegt sind. Ähnliches gilt für den Alkoholkonsum, auch hier zeigt die epidemiologische Erkenntnislage einen klaren Zusammenhang zwischen Tumorhäufigkeit des oberen Gastrointestinaltrakts sowie der Leber und erhöhtem Alkoholkonsum. Ein weiterer bedeutsamer Lebensstilfaktor ist die Energieaufnahme mit der Nahrung. Tierexperimentelle Erkenntnisse und epidemiologische Studien lassen keinen Zweifel, dass überkalorische Ernährung einen besonders gewichtigen Einfluss auf das Krebsrisiko hat und dass verminderte Zufuhr an Nahrungsenergie dieses drastisch senken kann. Schließlich werden auch Ausmaß und Stärke körperlicher Tätigkeit, beispielsweise durch Sport, als günstige Einflussfaktoren diskutiert.

Hauptgegenstand der Beiträge des vorliegenden Symposiums war die kritische Erfassung und Bewertung der Bedeutung von krebsauslösenden bzw. krebshemmenden Faktoren in der Nahrung. Dabei stand die wissenschaftliche Diskussion von Wirkmechanismen fördernder bzw. hemmender Kausalfaktoren ebenso im Vordergrund wie die Frage nach neuen Sichtweisen bzw. neuen Konzeptansätzen zum Zentralthema „Krebs und Ernährung".

2 Krebs und Ernährung: Ein multifaktorieller Zusammenhang

Nahrungsmittel sind in der Regel komplexe Mischungen aus Nährstoffen und Begleitstoffen unterschiedlichster Art. Manche Bestandteile gelten aufgrund epidemiologischer oder tierexperimenteller Befunde als Risikofaktoren für die Krebsentstehung, andere besitzen krebshemmende Eigenschaften. Da sie mit der Nahrung in wechselnden Mengen aufgenommen werden und sich ihre Wirkungen zum Teil gegenseitig aufheben, bestehen keine einfachen und leicht zu durchschauenden Beziehungen zwischen Ernährung und Tumorhäufigkeit.

Planung und Interpretation von tierexperimentellen bzw. Humanstudien, die die Beziehung zwischen Ernährung und Krebshäufigkeit zum Gegenstand haben, erfordern aus den genannten Gründen besondere Sorgfalt.

Die Entwicklung einer klinisch manifesten Krebserkrankung verläuft in der Regel über viele Jahre, während denen sich Gesamternährungsverhalten, Exposition und Lebensstil des betroffenen Individuums ändern können. Außerdem können individuelle Entwicklung und Alterung Körperprozesse zusätzlich beeinflussen. Untersuchungen über den Zusammenhang zwischen Krebserkrankung und Ernährung des Menschen müssen solche möglichen Veränderungen bei der Prüfung der Rolle von Einzelfaktoren berücksichtigen. Darüber hinaus werden Stoffwechselwege und -lage (Homoeostase) und andere biologische Prozesse, wie z.B. die Reparatur von DNA-Schäden, durch individuell unterschiedliche genetische, aber auch phenotypische Faktoren beeinflusst (individuelle Susceptibilität).

Migrationsstudien an Bevölkerungsgruppen können Anhaltspunkte dafür liefern, in welcher Phase der Carcinogenese ernährungsabhängige Faktoren modifizierend wirken. So sprechen Effekte, die relativ schnell nach der Migration einsetzen, wie dies beim Dickdarmkrebs beobachtet wurde, eher für eine Wirkung auf späte Phasen der Carcinogenese. Im Gegensatz hierzu scheinen beim Magenkrebs eher in frühen Phasen einwirkende ernährungsabhängige Effekte von Bedeutung zu sein.

Die Wirkung eines vermuteten protektiven Mechanismus kann unabhängig vom Anwendungsort oder dem untersuchten Gewebe auftreten. Beispielsweise kann sich eine veränderte metabolische Aktivierung oder Detoxifizierung eines mit der Nahrung aufgenommenen Carcinogens in der Leber auf das Auftreten von Tumoren in der Blase oder der Brust auswirken. Daher ist für jeden Einzelfaktor die Frage des Wirkorts und des Wirkmechanismus zu prüfen.

3 Carcinogene Faktoren

Nach dem derzeitigen Kenntnisstand gelten vor allem Überernährung und hoher Fett- und Fleischverzehr als ernährungsbedingte Risikofaktoren. Die mit dem Körpergewicht direkt korrelierte Kalorienaufnahme beeinflusst physiologische, metabolische und molekulare Vorgänge, die sich u. a. auf die Zellproliferation auswirken. In tierexperimentellen Studien konnte gezeigt werden, dass neben physiologischen Parametern, wie Körpertemperatur und motorische Aktivität, in Abhängigkeit von der Kalorienzufuhr auch die Expression von Fremdstoff-metabolisierenden Enzymen und die Menge an endogen gebildeten Radikalen beeinflusst werden. Auch wirkt sich eine reduzierte Kalorienzufuhr günstig auf DNA-Reparaturprozesse aus, erhöht die Expression von Tumorsuppressorgenen und verstärkt Apoptoseprozesse.

Es kann heute als gesichert gelten, dass der wirksamste Weg zur Verringerung der Krebsinzidenz beim Menschen über die Nahrung darin besteht, Überernährung zu vermeiden, einen unangemessen hohen Anteil an Fett- und Fleischverzehr abzubauen und täglich reichlich Obst und Gemüse zu verzehren, sich also möglichst ausgeglichen zu ernähren.

Viele epidemiologische Studien zeigen, dass Alkohol als Risikofaktor für Tumoren des oberen Verdauungskanals und der Leber anzusehen ist. Der alkoholbedingte Anteil am Risiko für Dickdarm- und Brustkrebs erscheint zwar als verhältnismäßig gering, ist aber angesichts der großen Häufigkeit dieser Tumoren nicht unbedeutend. Tierversuche stützen die Annahme, dass Alkohol vor allem als Co-Carcinogen oder Tumor-Promotor anzusehen ist.

Als carcinogene Faktoren sind auch einige in Lebensmitteln auftretende Toxine erkannt worden. So besteht ein Zusammenhang zwischen der Häufigkeit an Leberzellkrebs zu erkranken und einer nahrungsbedingten Exposition gegenüber dem Schimmelpilzgift Aflatoxin B_1. Dies gilt für tropische Länder, in denen Lebensmittel häufig mit Schimmelpilzen befallen und stark mit Aflatoxinen kontaminiert sind und in denen primärer Leberkrebs wesentlich häufiger vorkommt als in Europa. Gleichzeitig spielen hierbei Infektionen mit dem Hepatitis-B-Virus eine bedeutende Rolle. Für ein weiteres Mykotoxin, das Ochratoxin A, existieren Hinweise auf einen Zusammenhang von ernährungsbedingter Aufnahme über Lebensmittel und der Häufigkeit von Nierentumoren in bestimmten Gebieten des Balkans.

Epidemiologische Hinweise auf Humancarcinogenität liegen auch für anorganisches Arsen vor. Ein potentieller Zusammenhang zwischen alimentärer Arsen-Exposition und der Häufigkeit von Tumoren der Haut, Leber, Lunge, Niere und Blase ist beschrieben. Diese Befunde stammen aus Gebieten, bei denen eine extrem hohe Exposition einhergeht mit Mangelernährung, wie dies für einige Gebiete Taiwans beobachtet wurde.

Beispiele für in Lebensmitteln vorkommende Stoffe, bei denen bisher keine klaren Hinweise aus der Epidemiologie vorliegen, die sich aber im Tierversuch als potente Carcinogene erwiesen haben, sind heterocyclische aromatische Amine, *N*-Nitrosoverbindungen, polycyclische aromatische Kohlenwasserstoffe und Ethylcarbamat.

Zusätzlich zur Exposition des Menschen mit exogenen Risikofaktoren gilt nach heutigem Kenntnisstand die Exposition mit endogen gebildeten carcinogenen Faktoren als wesentliche Einflussgröße. So könnte z. B. die endogene Synthese von N-Nitrosoverbindungen (NOC) aus Vorläufern im Organismus möglicherweise eine Hauptquelle der humanen Exposition mit NOC darstellen. Dabei ist zu berücksichtigen, dass die endogene Nitrosierung nicht notwendigerweise zu stabilen, analytisch fassbaren NOC führen muss, sondern u. U. zu direkt mutagen wirkenden Nitrosierungsprodukten führen kann. Die Vorstufen für eine endogene Nitrosierung finden sich im Wesentlichen in Lebensmitteln. Die endogene Bildung von NO aus Arginin hat vermutlich zusätzlich erhebliche Bedeutung.

Weiterhin können Stickoxid- und Sauerstoff-Radikalreaktionen auf vielfache Weise gentoxische Schäden bzw. Mutationen auslösen. Als Folge von oxidativem Stress können DNA-Basen mit Produkten der Lipidperoxidation reagieren und so exocyclische DNA-Addukte, wie z. B. Etheno-DNA-Addukte, gebildet werden. Diese endogene oxidative DNA-Schädigung gilt als weiterer wesentlicher Risikofaktor für Genmutationen und deren Konsequenzen in Säugerzellen.

Da für gentoxische Carcinogene im Prinzip kein Schwellenwert für gesundheitliche Unbedenklichkeit festgelegt werden kann, sollten Präventions- bzw. Schutzmaßnahmen eine möglichst weitgehende Verringerung der Exposition gegenüber Stoffen und Prozessen, die ein Krebsrisiko für den Menschen darstellen, zum Ziel haben (Minimierungskonzept).

4 Anticarcinogene Faktoren

Es kann heute als gesichert gelten, dass ein hoher Verzehr an Obst und Gemüse mit einer Verringerung der Krebsrate in vielen Geweben assoziiert ist. Offensichtlich überwiegt bei Personen mit reichlichem Verzehr an Obst und Gemüse die Exposition mit krebshemmenden Substanzen in der Regel jene mit krebsfördernden bei weitem. Für viele Pflanzeninhaltsstoffe unterschiedlicher Stoffgruppen ist eine protektive Wirkung experimentell gezeigt worden. Allerdings ist die Evidenz am Menschen in den meisten Fällen sehr viel weniger überzeugend. Aus experimentellen Studien liegen Hinweise vor, dass protektive Wirkungen von *Allium*-Arten, wie z. B. Zwiebeln und Knoblauch, auf Thiolverbindungen, u. a. Diallylsulfid, zurückgeführt werden können. Weitere Studien ergaben Hinweise auf Isothiocyanate und Indol-3-carbinol als protektive Substanzen aus Brokkoli sowie Cruciferen-Arten. Aus der Epidemiologie liegen Hinweise auf eine Erniedrigung des Tumorrisikos durch Carotinoide, die Vitamine C und E sowie Selen vor.

Zur Schutzwirkung kurzkettiger Fettsäuren, wie Butyrat, auf den Gastrointestinaltrakt liegen eine Reihe von experimentellen Hinweisen vor. Kurzkettige Fettsäuren entstehen u. a. durch die Fermentation bestimmter Polysaccharide in der Darmflora. Stärke und Fructooligosaccharide haben sich als gute Substrate für Butyrat-produzierende Mikroorganismen erwiesen.

Weiterhin liegen Hinweise dafür vor, dass regelmäßige körperliche Tätigkeit einen präventiven Einfluss auf die Entstehung von Dickdarmkrebs, möglicherweise auch von Brustkrebs hat.

Zur ernährungsabhängigen Krebsprävention können zahlreiche biologische und metabolische Prozesse beitragen. So können beispielsweise protektive Mechanismen zu einer verringerten Resorption oder einer verringerten metabolischen Aktivierung von carcinogenen Substanzen führen. Es können aber auch inaktivierende Stoffwechselwege induziert oder die Ausscheidung von Carcinogenen bzw. deren Metaboliten gefördert werden. Unter Umständen können auch biologische Prozesse der Krebsentstehung dadurch gehemmt werden, dass beispielsweise Zellen in geringerem Ausmaß oder gar nicht initiiert werden, Reparaturprozesse verstärkt werden oder auf andere Weise die Entwicklung einer Neoplasie gebremst wird. Dies kann zum Beispiel auch durch Verringerung der Mitoserate sowie durch Induktion von Differenzierung und/oder von programmiertem Zelltod (Apoptose) geschehen. Ebenso sind vielfältige Beeinflussungen der zellulären und humoralen Immunantwort denkbar. Viele Untersuchungsergebnisse stammen allerdings aus *in vitro*-Versuchen bzw. Tierversuchen, und ihre Relevanz für den Menschen ist in vielen Fällen noch ungeklärt bzw. ungesichert.

In der Regel erlaubt es deshalb der gegenwärtig noch vorherrschende Mangel an gesicherten Erkenntnissen nicht, fundierte Empfehlungen darüber zu geben, welche spezifischen Ergänzungen der Nahrung vor einer Krebserkrankung schützen könnten.

5 Erkenntnislücken

Das Verständnis der Mechanismen, wie über die Ernährung die Entstehung von Krebs verhindert oder vermindert werden kann, ist bislang noch ungenügend.

So bestehen Erkenntnislücken in Bezug auf die Aufnahme, Bioverfügbarkeit, und den Stoffwechsel vieler potentiell protektiver Lebensmittelinhaltsstoffe beim Menschen. Ebenso fehlen ausreichende Daten zur Exposition mit potentiellen Humancarcinogenen aber auch mit potentiellen Schutzstoffen über die Muttermilch.

Auch die Protektivmechanismen, die als Folge einer verminderten Kalorienzufuhr wirksam werden können, sind noch nicht genügend aufgeklärt.

Die Frage der Übertragbarkeit von Ergebnissen aus experimentellen Modellen auf den Menschen ist in vielen Fällen nicht ausreichend untersucht. In Verbindung mit epidemiologischen Studien bzw. Migrationsstudien ist die Entwicklung und Validierung neuartiger Biomarker bzw. neuer molekularer Techniken notwendig, um die Aussagekraft epidemiologischer Studien zu verbessern (molekulare Epidemiologie).

Erkenntnislücken bestehen auch bezüglich der Aussagekraft von Humanstudien an definierten Gruppen mit erhöhtem Krebsrisiko als Modell für die Gesamtbevölkerung. Ein konkretes Beispiel ist die Beobachtung eines krebsfördernden Effekts einer β-Carotin-Supplementierung bei Rauchern und Asbestexponierten. Inwieweit dieser anscheinend paradoxe Effekt auf die Gesamtbevölkerung übertragbar ist, ist gegenwärtig noch offen.

6 Empfehlungen zum Forschungsbedarf

Die Exposition eines Individuums gegenüber einem Nahrungsinhaltsstoff ist häufig schwierig zu ermitteln. Sie erfordert entweder vielfache Nahrungsanalysen über längere Zeit oder aber die Erfassung des Stoffes bzw. seiner wirksamen Metaboliten im Organismus. Dies kann z. B. anhand von Blutplasmaspiegeln des Stoffs oder seiner Metaboliten bzw. der Ausscheidungsprodukte im Urin geschehen. Angemessene analytische Methoden sind hierzu einzusetzen bzw. zu entwickeln, zu standardisieren und zu validieren. Auch besteht ein großer Bedarf an Entwicklung und Einsatz geeigneter Biomarker, z. B. um die Exposition mit krebsfördernden/krebshemmenden Stoffen zuverlässig zu messen. Biomarker für Ernährung können eine indirekte Ermittlung von Verzehrmengen eines Lebensmittels, u. U. aber auch die Aufnahme noch nicht identifizierter Schutzfaktoren erfassen.

Weiter besteht Bedarf an der Entwicklung von Biomarkern für biologische Messgrößen, die von Lebensmittelinhaltsstoffen beeinflusst werden. Beispiele hierfür sind die Erfassung von Veränderungen zirkulierender Enzyme, die Erfassung spezifischer Addukte an Biopolymeren oder die Erfassung von Indikatorläsionen bzw. von Reaktionsprodukten aus oxidativen Radikalkettenreaktionen oder eine expositionsinduzierte Veränderung eines zellulären Rezeptormoleküls. Diese neuen molekularen Methoden der Epidemiologie haben ein großes Zukunftspotential.

Als Biomarker für eine Exposition mit gentoxischen Agentien sollten verstärkt DNA-Addukte in Zellen und Geweben, DNA-Basenaddukte im Urin oder Hämoglobinaddukte in Blutzellen herangezogen werden. Als Biomarker zur Bestimmung individuell unterschiedlicher Susceptibilitäten sollte die Erfassung von DNA-Reparaturvorgängen, von Enzyminduktion und Polymorphismen oder von Mutationsspektren bestimmter Onkogene bzw. Tumorsuppressorgene vorangetrieben werden. Klinisch können Surrogatmarkerenzyme wie das Carcinoembryonale Antigen (CEA) oder das Prostata-spezifische Antigen (PSA) eingesetzt werden, um beispielsweise Entstehung und Wachstum eines Tumors zu verfolgen – vorausgesetzt die Beziehung zwischen dem gemessenen Faktor und dem Krankheitsprozess ist bekannt. Da die erwähnten molekularen Techniken die Aussagekraft experimenteller und epidemiologischer Studien ganz wesentlich verbessern können, verdient ihre Entwicklung besonders intensive Forschungsarbeit und Forschungsförderung. Dabei ist es essentiell, validierte Methoden zu entwickeln, um zuverlässige Aussagen zu erhalten. Dies gilt in besonderem Maße, wenn es darum geht, die Relevanz eines Faktors oder Mechanismus, der bisher nur *in vitro* bzw. im Tierversuch untersucht wurde, für die Situation am Menschen zu klären. Intensive Forschung ist weiter erforderlich, um die Rolle der Mikroflora für Struktur und Funktion des Gastrointestinaltrakts und den Stoffwechsel von Nahrungsinhaltsstoffen aufzuklären.

II Main Conclusions and Recommendations

1 Introduction

The results of many epidemiological studies show that the risk of humans becoming afflicted by certain types of cancer is influenced by their diet. Current estimates assume that about a third of all cases of cancer in the western world are related to nutrition.

Knowledge of carcinogenic or anticarcinogenic (protective) substances alone is insufficient to investigate and recognise cancer risk factors. Personal lifestyle also has a considerable influence on the risk of cancer. This is especially evident in the case of tobacco consumption or smoking, for which a clear relationship has been demonstrated both in animal studies and in man between dose, time and frequency of smoking and the incidence of tumours, particularly of the respiratory tract, and also of other organs. The same is true of alcohol consumption. In this case, epidemiological findings also show a clear relationship between elevated alcohol consumption and the incidence of tumours of the upper gastrointestinal tract and the liver. Another significant lifestyle factor is the energy of the ingested food. Evidence from animal studies and epidemiological studies leave no doubt that a high-calorie diet has an especially important influence on the risk of cancer and that a reduced intake of nutritional energy can substantially reduce this risk. Finally, the extent and intensity of physical exercise, such as sporting activities, is also under discussion as a beneficial factor.

The main subject of the contributions to this symposium was the critical consideration and evaluation of the importance of cancer-inducing and cancer-inhibiting factors in food. However, equal emphasis was placed on scientific discussion of the mode of action of cancer causing or inhibiting factors and on the question of new aspects and new conceptual approaches to the central theme of "Cancer and Nutrition".

2 Cancer and Food: A Multifactorial Relationship

Food is generally a complex mixture of nutrients and accompanying substances of various kinds. Epidemiological findings and the results of animal studies have indicated that some ingredients are cancer risk factors, whereas others have anticarcinogenic properties. As they are ingested in varying amounts and as their effects may to some extent counteract each other, there is no simple and easily recognisable relationship between nutrition and the incidence of tumors.

For these reasons studies in animals and humans to investigate the relationship between food and the incidence of cancer require extremely careful planning and interpretation.

A clinically manifest cancer usually develops over a period of many years, during which the entire eating habits, exposure and lifestyle of the affected person may change. In addition, individual development and ageing can influence the processes in the body. Investigations of the relationship between cancer and food in humans must take such possible changes into account when evaluating the role of individual factors. Moreover, metabolic pathways, the metabolic state (homeostasis) and other biological processes, such as the repair of DNA damage, are influenced by individual genetic and phenotypic factors (individual susceptibility).

Migration studies on population groups can provide indications of the stage of carcinogenesis at which nutritional factors exert a modifying effect. Thus, effects with a relatively rapid onset after migration, such as have been observed in the case of large bowel cancer, point rather to an effect on the late phases of carcinogenesis. In contrast, nutrition-related effects seem to be of greater importance in early stages of stomach cancer.

The effect of a suspected protective mechanism can be independent of the site of application or the tissue examined. For instance, a modified metabolic activation or detoxification by the liver of a carcinogen ingested with food can have an effect on the incidence of tumors in the bladder or in the breast. Hence the site specificity needs to be examined when considering the possible influence of each individual factor.

3 Carcinogenic Factors

The current state of knowledge indicates that over-eating, in particular, the high consumption of fat and meat represent food-related risk factors. Calorie intake, which is directly correlated with body weight, influences physiological, metabolic and molecular processes, which may effect cell proliferation. Animal studies have shown that the intake of calories not only influences physiological parameters, such as body temperature and physical activity, but also the expression of enzymes which metabolise foreign substances and endogenously formed free radicals. A reduced intake of calories has a beneficial effect on DNA repair processes, increases the expression of tumor suppressor genes and intensifies the processes of apoptosis.

There is now firm evidence for the fact that the most effective way of reducing the incidence of cancer in humans via nutritional measures is to avoid eating an excessive proportion of fat, to reduce the intake of meat and to consume plenty of fruit and vegetables every day, i. e. to ensure a diet as balanced as possible.

Many epidemiological studies show that alcohol is a risk factor for tumors of the upper digestive tract and the liver. Although the risk of cancer of the large intestine and of breast cancer posed by alcohol seems relatively low, it cannot be regarded as insignificant in view of the high incidence of these tumors. Animal studies lend support to the assumption that alcohol in particular can be considered as a co-carcinogen or tumor promoter.

Some toxins which occur in food have also been recognised as carcinogenic factors. A relationship has been established between the frequency of cancer of the liver and nutritional exposure to aflatoxin B_1 formed by moulds. This applies especially to tropical countries, where food is frequently infested with moulds and is highly contaminated with aflatoxins, and where primary liver cancer is considerably more frequent than in Europe. Hepatitis B virus infections also play a significant role in this context. In the case of another mycotoxin, ochratoxin A, there are indications of a correlation between the intake of this substance with food and the incidence of tumors of the kidney in certain areas of the Balkans. Epidemiological evidence for human carcinogenicity has also been reported for inorganic arsenic. A potential connection between alimentary exposure to arsenic and the incidence of tumors of the skin, liver, lung, kidney and bladder has been described. These findings have been reported from areas in which there is an extremely high exposure to these substances coupled with nutritional deficiencies, as in certain areas of Taiwan.

Further examples of substances found in food which have shown no clear epidemiological evidence of carcinogenicity so far, but which have proved to be potent carcinogens in animal studies, are heterocyclic aromatic amines, *N*-nitroso compounds, polycyclic aromatic hydrocarbons and ethyl carbamate.

In addition to the exposure of humans to exogenous risk factors, research has shown that exposure to endogenously formed carcinogenic factors can also be of significant influence. For example, the endogenous synthesis of *N*-nitroso compounds (NOC) from precursors in the organism may represent the main source of human exposure to NOC. It must be taken into account that the endogenous nitrosation does not necessarily lead to the formation of stable, analytically detectable NOC, but it still may result in the formation of nitrosation products with a direct mutagenic effect. The precursors for endogenous nitrosation are primarily found in food. The endogenous formation of NO from arginine is presumably also of considerable significance.

Further, nitrogen monoxide and oxygen radical reactions are capable of causing toxic damage to the genes or induce mutations in many different ways. As a consequence of oxidative stress, DNA bases can react with products of lipid peroxidation to form exocyclic DNA adducts, e.g. etheno-DNA adducts. This endogenous oxidative DNA damage is believed to be a further important risk factor for genetic mutations and their consequences in mammalian cells.

Since in principle a threshold level for safety to health cannot be determined for genotoxic carcinogens, the introduction of preventive and protective measures to minimise the exposure of humans to substances and processes which pose a risk of cancer should be our first priority (minimisation concept).

4 Anticarcinogenic Factors

It can now be regarded as an established fact that a higher consumption of fruit and vegetables is associated with a reduced cancer rate in many tissues. As a rule, people who eat plenty of fruit and vegetables must have a greater exposure to cancer-inhibiting substances than to those substances which promote the disease. Experiments have shown that many constituents of plants, which belong to different chemical classes have a protective effect.

In many cases however the evidence gained from human studies is much less convincing. Experimental studies indicate that the protective effects of *Allium species*, e.g. onions and garlic, can be attributed to their content of thiol compounds, including diallyl sulphide. Additional studies provided evidence that isothiocyanates and indole-3-carbinol in broccoli and *Cruciferae* species possess protective properties. Epidemiological studies indicate that carotenoids, vitamins C and E and selenium reduce the risk of tumors.

A series of investigations has shown that short-chain fatty acids, such as butyric acid, appear to exert a protective effect on the gastrointestinal tract. Short-chain fatty acids may be formed by fermentation of certain polysaccharides in the intestinal flora. Starch and fructooligosaccharides have proved to be good substrates for butyrate-producing micro-organisms.

There are also indications that regular physical exercise also has a preventive influence on the development of colon cancer and possibly also of breast cancer. Many biological and metabolic processes can contribute to nutrition-dependent prevention of cancer. For example, protective mechanisms may reduce the absorption of carcinogenic substances or decrease their metabolic activation, inactivating metabolic pathways can be induced and the excretion of carcinogens or their metabolites can be stimulated. Under certain circumstances, biological processes leading to cancer can be inhibited because, e.g. cells are initiated only to a lesser extent or not at all or repair processes are intensified or the development of a neoplasia is slowed down in some other way. This could be achieved by lowering the mitotic rate, and also by the induction of differentiation and/or by programmed cell death (apoptosis). Similarly, a multitude of factors which influence the cellular and humoral immune response are conceivable. However, many results have been obtained in *in vitro* investigations or from animal studies, and their relevance to humans has not yet been clarified or verified in many cases.

As a rule, therefore, the current lack of verified information does not permit conclusive recommendations about which specific supplements should be added to the diet to protect against cancer.

5 Gaps in our Knowledge

Knowledge of the means to prevent the onset of cancer or to reduce its incidence by nutritional measures is still inadequate.

There are still gaps in our knowledge regarding the absorption, bioavailability and metabolism of many potential protective substances in the food we eat. Similarly, data on exposure to potential human carcinogens, and also on potential protective substances supplied in human milk are still insufficient.

The protective mechanisms which result from a reduced supply of calories to the body have not yet been sufficiently clarified.

In many cases the question of the relevance to humans of the results from experimental models has not been sufficiently investigated. The development and validation of new biomarkers or new molecular techniques is necessary in the case of population and migration studies in order to permit more reliable conclusions to be drawn from epidemiological investigations (molecular epidemiology).

Gaps in knowledge also exist with regard to the possible applicability of human studies of people with a high risk of cancer as models for the whole population. One specific example is the observation of a cancer promoting effect as a result of β-carotene supplementation of the diet of smokers and people exposed to asbestos. It is not clear whether this seemingly paradoxical effect is applicable to the entire population.

6 Recommendations for Further Research

The exposure of an individual to a food constituent is often difficult to determine. It requires repeated analysis of the food over a lenghty period or the determination of the substance or the effective metabolites in the organism. This could be achieved by analysing levels of the substance or its metabolites in blood plasma or by analysing its excretion products in urine. Appropriate analytical methods need to be used or developed, standardised and validated. There is also an urgent need for the development and use of suitable biomarkers, e.g. to measure reliably the exposure of individuals to carcinogens and anticarcinogens. Biomarkers of nutrition could be used as an indirect determination of the consumption of a certain foodstuff, and also of the intake of yet unidentified preventive factors.

There is also the need for discovering biomarkers representing biological parameters which are influenced by the food constituents. Examples include the detection of changes in circulating enzymes, the measurement of specific adducts on biopolymers, the assay of indicator lesions or reaction products of oxidative radical chain reactions, or an exposure-induced change in a cellular receptor molecule. These new molecular methods in epidemiology will have a great potential in the future.

DNA adducts in cells and tissues, DNA base adducts in urine or haemoglobin adducts in blood cells should be increasingly used as biomarkers to indicate exposure to agents which are genotoxic. Progress must be made in discovering biomarkers of differences in individual susceptibilities, e.g. by the determination of DNA repair processes, enzyme induction and polymorphisms or mutation spectra of certain oncogenes or tumor suppressor genes. Surrogate marker enzymes such as carcinoembryonal antigen (CEA) or prostate-specific antigen (PSA) can be used clinically to monitor the genesis and growth of a tumor, provided the relationship between the measured factor and the course of the disease is known. As the molecular techniques mentioned above can significantly improve our ability to draw conclusions from experimental and epidemiological studies, their development deserves intensive research and financial support. However, it is essential to devise validated methods in order to be able to draw reliable conclusions. This is especially important when clarifying the relevance to the human situation of a factor or mechanism which has only been investigated *in vitro* or in animal studies.

Intensive research is also required to elucidate the role of the intestinal flora in the structure and function of the gastrointestinal tract and in the metabolism of food constituents.

III Contributions

A General Session

1 Foods, Phytochemicals, and Metabolism: Anticarcinogens and Carcinogens

John D. Potter and Julie A. Ross

1.1 Abstract

There is now extensive evidence that diet influences the risk of cancer. Present in foods are substances that may increase or decrease risk. Further, there are specific behaviors that influence human metabolism in ways that are likely to increase risk (e.g., obesity) and decrease risk (e.g., physical activity). However, this picture is even more complicated than it initially seems because some agents and behaviors may influence risk of cancers differently depending on age, sex, and the specific cancers of interest. Some examples from our own work include the evidence that plant foods, which are generally protective against a variety of adult cancers, may, nontheless, increase the risk of infant leukemia; and obesity, which is a risk factor for postmenopausal breast cancer in women and colon cancer in men, is associated with reduced risk of premenopausal breast cancer and appears not to influence risk of colorectal neoplasia in older women.

1.2 Introduction

A relationship between aspects of diet and human cancer has been the focus of research since the 1930s, and particularly in the last twenty years. Even though much remains to be understood about mechanisms, it is now possible to draw consistent conclusions regarding the role of diet, obesity, exercise, and alcohol in the etiology of cancer, and to make public health recommendations on the basis of those conclusions [1]. Nonetheless, not only do many mechanisms remain to be elucidated, but there are also some important paradoxes which, if properly understood, will have implications for etiology and prevention. For instance, much has been made of the possible contradictory implications for the prevention of heart disease and cancer that follow from the apparent relationships with alcohol. Some of the enthusiasm for alcohol as a preventive may be misplaced [2]. Nonetheless, there are important lessons to be learned from contradictions – real and apparent. Here, we discuss two interesting observations. The first is the different relationship that intake of vegetables has with the majority of adult cancers in contrast to the relationship with infant leukemia. The second is the associations that exist between obesity and risk of cancer, which vary by age, sex, and cancer site.

1.3 Vegetables, fruit, and cancer

1.3.1 Adult cancers

Over 230 studies have presented data that show, in aggregate, that there is a reduced risk of most epithelial cancers in adults in association with higher intakes of vegetables or fruits or both. As examples, there are 24 studies of lung cancer that have reported on vegetables and fruit. Of these, 23 show statistically significant lower risks in association with higher consumption of one or more classes of vegetables and fruits. The story is similar for stomach cancer (30 studies out of 35 show statistically significant lower risk), esophagus (19 of 23), and mouth and pharynx (14 of 16). Even for breast cancer, there is good evidence, with 12 of 22 studies showing lower risk associated with higher consumption. For greater detail and more extensive summaries of the evidence, the reader is referred to Steinmetz and Potter [3], Steinmetz and Potter [4], and WCRF Panel [1].

There is also a great deal of animal evidence that points towards possible mechanisms for this overall protection. There is a long list of potentially protective compounds that may contribute a reduced risk. For detail, the reader is referred to Steinmetz and Potter [5], WCRF Panel [1], and Potter and Steinmetz [6]. A partial list is presented as Table 1.1, and some possible mechanisms are shown in Table 1.2.

Table 1.1: Some potentially anticarcinogenic substances in plant foods.

Allium compounds – found in onions, garlic, etc.
diallyl sulfide
allyl methyl trisulfide
Carotenoids – widespread in many colored vegetables and fruit
α-carotene
β-carotene
cryptoxanthin
lutein
lycopene
others
Dietary fiber – found in vegetables and fruit as well as grains, nuts, and seeds
Dithiolthiones – found in cruciferous vegetables (cabbage family)
Flavonoids – found in a variety of vegetables and fruits
quercetin
kaempferol
Folic acid – found particularly in greens
Isoflavones – found in legumes, particularly soy
genestein
biochanin A
others
Isothiocyanates – found in cruciferous vegetables
sulphorophane
others
Lignans – derived from fiber by colonic bacteria
Selenium – widespread in plant foods; levels depend on soil concentrations
Vitamin C (ascorbate) – widespread in vegetables and fruit
Vitamin E (tocopherols) – found in grains, nuts, seeds, oils

Table 1.2: Some possible anticarcinogenic mechanisms of substance in vegetables and fruit.

Antioxidant effects – relevant both to membrane lipids and DNA
Effects on cell differentiation
Induce enzymes that detoxify carcinogens
Block formation of carcinogens, e. g., nitrosamines
Alter estrogen metabolism – both agonist and antagonist effects
Alter colonic milieu (including bacterial flora, bile acid composition, pH, faecal bulk)
Preserve integrity of intracellular matrices
Effects on DNA methylation
Effects on nucleotide pool
Increase apoptosis of cancer cells
Decrease proliferation
Inhibition of topoisomerase II

1.3.2 Infant leukemia

In children treated with epipodophyllotoxins for a primary childhood neoplasm, second primary leukemias (very commonly AML), occur relatively frequently. These often display a specific acquired translocation involving the long arm of chromosome 11 [7, 8]. These 11q23 lesions are now known to involve the *MLL*-1 (also called *ALL*-1, *HTRX*, or *HRX*) gene, which is a homologue of the *Drosophila trithorax* gene [9, 10].

In 1994, we drew attention to the fact that this same gene frequently shows similar abnormalities in primary leukemias in infants with no known exposures to epipodophyllotoxins. We postulated that the relevant exposure must be *in utero*, and, as the chemotherapeutic agent is a known topoisomerase II inhibitor, hypothesized that the relevant agent(s) in the infants could also have topoisomerase II inhibitory activity [11]. There are a variety of substances in specific foods and specific medications with such activity. Our hypothesis is shown graphically in Figure 1.1. The reader is referred to the original paper for more detail. We subsequently tested this hypothesis in a case-control study of infant leukemia [12]. The results are summarized in Table 1.3. What this shows is that maternal exposure to foods known to contain compounds with potential topoisomerase II activity (beans, fresh vegetables, canned vegetables, fruit, soy,

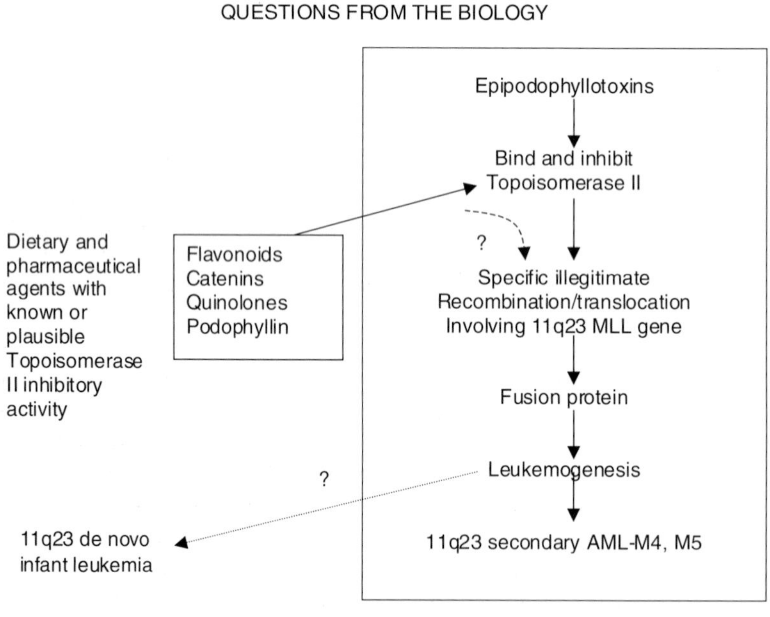

Figure 1.1: Known and hypothesized relationships among epipodophyllotoxins, 11q23 abnormalities, topoisomerase II inhibitors, and infant leukemia (modified from Ross, *et al.* (1994) [11]).

coffee, black tea, green tea, cocoa, and wine) is associated with an elevated risk of infant AML, but not ALL. Thus, the *a priori* hypothesized association exists, and is specific to infant AML.

Table 1.3: Maternal exposure during pregnancy to potential dietary topoisomerase II inhibitors and leukemia in offspring (modified from Ross, *et al.* (1996) [12]).

Category of exposure	Leukemia		ALL only		AML only	
	OR[b]	95 % CI[c]	OR	95 % CI	OR	95 % CI
Low	1.0		1.0		1.0	
Medium	2.1	0.9–5.0	1.3	0.4–4.2	9.8	1.1–84.8
High	1.1	0.5–2.3	0.5	0.2–1.4	10.2	1.1–96.4
p for trend	ns		ns		p = 0.04	

ALL = acute lymphoblastic leukemia; AML = acute myeloid leukemia.
[a] A combined exposure variable was created *a priori* from the following foods: beans, fresh vegetables, canned vegetables, fruit, soy, regular coffee, black tea, green tea, cocoa, and wine. (Responses to each category were summed based on the following categories: 0 = never, 1 = <1 m, 2 = 1–3 m, 3 = 1–3 wk, 4 = 4–6 wk, and 5 = daily). Low exposure was defined as a total aggregate score of less than 15 (27.7 % of respondents), medium exposure 15–19 (40.1 % of respondents), and high exposure 20+ (32.2 % of respondents). Subjects with missing data were excluded.
[b] OR = odds ratio – adjusted for maternal education.
[c] 95 % CI = 95 % confidence interval.

1.3.3 Implications

The apparent paradox that diets high in plant foods might be associated with a reduced risk of most adult cancers, but an elevated risk of infant leukemia, has several implications. First, many of the compounds present in plant foods are biochemically potent, and obviously selected for the plant's benefit, not ours. Therefore, a combination of beneficial and deleterious consequences is not unexpected. Second, topoisomerase II inhibitory activity has not, to this point, been a focus of investigation among the bioactive compounds thought to be important in adult cancer; it may be worth attention. One important consideration in comparing the way in which topoisomerase II inhibitors might influence the risk of cancers in infants *vs* adults is to consider that topoisomerase II is key to the control of DNA structure, in both transcription and replication. As is clear from their use as effective chemotherapeutic agents, topoisomerase inhibitors are useful in controlling rapid cell replication. In adults, rapid replication is a feature largely of cancer cells; thus, inhibitors are more likely to be beneficial. In infants, however, rapid replication is the norm for a wide variety of tissues, and there is a generally pro-replication signalling environment. Preventing replication through inhibiting an essential enzyme such as topoisomerase II may select for cells whose replication is outside normal controls, i. e., may select for malignant cells.

1.4 Obesity and cancer

1.4.1 Breast cancer

That obesity has a paradoxical relationship with risk of breast cancer has been known for about two decades [13–15]. Although obesity is associated with an increased risk of postmenopausal breast cancer, it is inversely associated with the premenopausal disease. More recently, it has been clear that the distinction may not be determined primarily by the difference between pre- and postmenopausal status, but more by the differences in the hormone receptor status (both ER and PR) of the breast cancer itself [16].

1.4.2 Colon cancer

Colon cancer, too, has a paradoxical relation with obesity and other aspects of energy balance. As shown in Table 1.4, whereas men show a strong relationship between excess energy intake, high BMI, and low physical activity on the one hand and risk of colon cancer on the other, women do not [17]. We have additional unpublished data that show a similar relationship between these measures of energy balance and colorectal adenomatous polyps.

1.4.3 Implications

It is clear from these findings that aspects of hormonal status are important in resolving this paradox. This probably involves receptors in the case of breast cancer [16], and perhaps the different metabolic roles that adipose tissue plays in men and women for colon cancer. A plausible influence on the etiology of colon cancer may be insulin or insulin-like growth factors as a stimulus to neoplastic growth. Indeed, McKeown-Eyssen has proposed that Syndrome X (obesity, elevated triglycerides, insulin resistance, hypertension) may be a general metabolic stimulus to neoplastic growth that is particularly important in colon cancer [18]. In postmenopausal women, however, peripheral adipose tissue is also a source of estrogens. It is now clear that estrogens play a significant role in reducing the risk of both adenomatous polyps and colon cancer [19–22]. It seems plausible that, whereas energy imbalance in men is associated with excess risk, obesity in women provides additional estrogen that may offset the elevated risk, resulting in essentially a null association.

Table 1.4: Long-term vigorous leisure-time physical activity, energy intake, Body Mass Index (BMI),[a] and risk of colon cancer (modified from Slattery, *et al.* (1997) [17]).

Long-term Vigorous Physical Activity[c]	Energy Intake[d]	BMI[b] Low OR (95 % CI)[e]	Intermediate OR (95 % CI)	High OR (95 % CI)
All subjects				
High	Low	1.0	1.1 (0.7–1.6)	1.1 (0.7–1.6)
	High	1.2 (0.8–2.0)	1.3 (0.8–2.2)	1.3 (0.8–2.0)
Intermediate	Low	1.1 (0.8–1.5)	1.3 (1.0–1.8)	1.9 (1.4–2.6)
	High	1.6 (1.1–2.4)	1.5 (1.0–2.2)	2.2 (1.5–3.3)
Low	Low	1.4 (0.9–2.0)	1.5 (1.1–2.2)	2.1 (1.5–3.0)
	High	2.0 (1.1–3.5)	2.2 (1.2–3.9)	3.4 (2.1–5.4)
Men				
High	Low	1.0	1.2 (0.7–2.1)	1.6 (1.0–2.7)
	High	1.7 (0.9–3.2)	1.7 (0.9–3.4)	1.9 (1.0–3.4)
Intermediate	Low	1.4 (0.9–2.2)	2.0 (1.3–3.0)	2.9 (1.9–4.4)
	High	2.0 (1.1–3.5)	1.9 (1.1–3.4)	3.0 (1.8–5.0)
Low	Low	1.6 (0.9–2.7)	2.0 (1.2–3.4)	2.4 (1.4–3.9)
	High	2.9 (1.3–6.9)	1.5 (0.5–4.1)	7.2 (3.4–15.2)
Women				
High	Low	1.0	0.9 (0.5–1.7)	0.6 (0.3–1.2)
	High	0.8 (0.4–1.6)	0.9 (0.4–1.9)	0.7 (0.3–1.5)
Intermediate	Low	0.7 (0.5–1.2)	0.8 (0.5–1.3)	1.0 (0.6–1.6)
	High	1.1 (0.6–2.1)	1.0 (0.5–1.8)	1.4 (0.8–2.6)
Low	Low	1.0 (0.6–1.7)	1.1 (0.6–1.8)	1.6 (0.9–2.6)
	High	1.2 (0.6–2.6)	1.9 (0.9–3.9)	1.6 (0.8–3.1)

[a] Risk estimates adjusted for age, sex, family history of colorectal cancer, use of aspirin and/or nonsteroidal anti-inflammatory drugs, dietary intake of fiber, calcium, and dietary cholesterol.
[b] BMI stratified by tertiles.
[c] Physical activity stratified by tertiles.
[d] Energy intake stratified as lower 2/3 *vs* upper 1/3.
[e] OR = odds ratio; 95 % CI = 95 % confidence interval.

1.5 Summary

Increased and decreased risks of cancer are heavily influenced by environmental, by lifestyle, and by metabolic factors. However, not all risk factors are consistent across age, sex, and cancer site. Paradoxes can cast light on mechanisms providing data that may be valuable in understanding etiology and instituting prevention.

References

[1] World Cancer Research Fund Panel (Potter, J. D., Chair) (1997) *Food, nutrition and the prevention of cancer: A global perspective*. American Institute for Cancer Research, Washington, DC.

[2] Potter, J. D. (1997) Hazards and benefits of alcohol. *N. Engl. J. Med.* **33**, 1763–1764.

[3] Steinmetz, K.; Potter, J. D. (1991) A review of vegetables, fruit, and cancer I: Epidemiology. *Cancer Causes and Control* **2**, 325–357.

[4] Steinmetz, K. A.; Potter, J. D. (1996) Vegetables, fruit and cancer prevention: A review. *J. Am. Diet Assoc.* **96**, 1027–1039.

[5] Steinmetz, K.; Potter, J. D. (1991) A review of vegetables, fruit and cancer II: Mechanisms. *Cancer Causes and Control* **2**, 427–442.

[6] Potter, J. D.; Steinmetz, K. A. (1996) Vegetables, fruit and phytoestrogens as preventive agents. In: Stewart, B.; McGregor, D.; Kleihues, P. (Eds.) *Principles of chemoprevention*, Lyon, International Agency for Research on Cancer, p. 61–90.

[7] Pui, C. H.; Behm, F. G.; Raimondi, S. C.; *et al.* (1989) Secondary acute myeloid leukemia in children treated for acute lymphoid leukemia. *N. Engl. J. Med.* **321**, 136–142.

[8] Kumar, L. (1993) Epipodophyllotoxins and secondary leukemia. *Lancet* **342**, 819–820.

[9] Djabali, M.; Selleri, L.; Parry, P.; *et al.* (1992) A trithorax-like gene is interrupted by chromosome 11q23 translocations in acute leukemias. *Nat. Genet.* **2**, 13–18.

[10] Felix, C. A.; Winick, N. J.; Negrini, M.; *et al.* (1993) Common region of ALL-1 gene disrupted in epipodophyllotoxin-related secondary acute myeloid leukemia. *Cancer Res.* **53**, 2954–2956.

[11] Ross, J. A.; Potter, J. D.; Robison, L. R. (1994) Infant leukemia, topoisomerase II inhibitors, and the MLL gene. *J. Natl. Cancer Inst.* **86**, 1678–1680.

[12] Ross, J. A.; Potter, J. D.; Reaman, G. H.; Pendergrass, T. W.; Robison, L. L. (1996) Maternal exposure to potential inhibitors of DNA topoisomerase II and infant leukemia (United States): A report from the Children's Cancer Group. *Cancer Causes Control* **7**, 581–590.

[13] Choi, N. W.; Howe, G. R.; Miller, A. B.; Matthews, V.; Morgan, R.; Munan, L.; Burch, J.; Feather, J.; Jain, M.; Kelly, A. (1978) An epidemiologic study of breast cancer. *Am. J. Epidemiol.* **107**, 510–521.

[14] Paffenbarger, R. S. Jr.; Kampert, J. B.; Chang, H. (1980) Characteristics that predict risk of breast cancer before and after the menopause. *Am. J. Epidemiol.* **112**, 258–268.

[15] Potter, J. D. (1987) Reproduction, sex steroid hormones and cancer. In: Maskens, A. P.; *et al.* (Eds.) *Concepts and theories in carcinogenesis*. Elsevier Science Publishers, Amsterdam, p. 243–255.

[16] Potter, J. D.; Cerhan, J. R.; Sellers, T. A.; McGovern, P. G.; Drinkard, C.; Kushi, L. R.; Folsom, A. R. (1995) Progesterone and estrogen receptors and mammary neoplasia in the Iowa Women's Health Study: How many kinds of breast cancer are there? *Cancer Epi. Biomarkers Prev.* **4**, 319–326.

[17] Slattery, M.; Potter, J.; Caan, B.; Edwards, S.; Coates, A.; Ma, K.-N.; Berry, T. D. (1997) Energy balance and colon cancer – Beyond physical activity. *Cancer Res.* **57**, 75–80.

[18] McKeown-Eyssen, G. (1994) Epidemiology of colorectal cancer revisited: are serum triglycerides and/or plasma glucose associated with risk? *Cancer Epidemiol. Biomarkers Prev.* **3**, 687–695.

[19] Potter, J.; Bostick, R.; Grandits, G.; Fosdick, L.; Elmer, P.; Wood, J.; Grambsch, P.; Louis, T. (1996) Hormone replacement therapy is associated with lower risk of adeno-

matous polyps of the large bowel: the Minnesota CPRU Case-Control Study. *Cancer Epidemiol. Biomarkers Prev.* **5**, 779–784.
[20] McMichael, A. J.; Potter, J. D. (1980) Reproduction, endogenous and exogenous sex hormones and colon cancer: a review and hypothesis. *J. Natl. Cancer Inst.* **65**, 1201–1207.
[21] Kampman, E.; Potter, J.; Slattery, M.; Caan, B.; Edwards, S. (1997) Hormone replacement therapy, reproductive history, and colon cancer: a US multi-center case-control study. *Cancer Causes Control* **8**, 146–157.
[22] Potter, J. D. (1995) Editorial: Hormones and colon cancer. *J. Natl. Cancer Inst.* **87**, 1039–1040.

2 Evidence from Migrant Studies for Dietary Effects on Cancer Incidence and Mortality

Margaret McCredie

2.1 Abstract

Migrant studies generally use routinely collected cancer incidence or mortality data to evaluate changes in the patterns of disease. They do not provide direct evidence of dietary effects on cancer risk but can assess whether the differences are more likely to reflect environmental or genetic factors by taking into account the degree and rate of change. If an environmental influence seems probable then this, together with national dietary consumption patterns or dietary surveys of individuals, may be used as evidence supporting a dietary effect. Studies in migrant populations yield good evidence supporting the role of dietary factors in colorectal and stomach cancers. In the case of the former, the detrimental effect of a presumed change in diet can be seen relatively soon after migration and therefore acts at a late stage in carcinogenesis. However, for stomach cancer the residual excess risk persists longer indicating the importance of exposures early in life or of retained cultural factors, likely to be dietary, which carry a higher risk. The late amelioration of the risk of liver cancer indicates an environmental cofactor for a cancer that is primarily related to persistent viral infection. Environmental factors which appear to be culturally determined and may be dietary are important in breast cancer, but seem relatively less important for prostate cancer where a role for genetic factors is indicated.

2.2 Introduction

Migrant studies have taken advantage of the large variation in cancer risk between countries and cultures, racial groups, socio-economic classes, and climatic regions [1] to assess the relative importance of environmental and genetic influences in cancer etiology and to gain insight into the time relationships between postulated exposures and appearance of cancer. Differences in environmental exposures and host susceptibility are alternative possible explanations for true heterogeneity in cancer rates while availability and accessibility of diagnostic and treatment services will account for some of the apparent variations in incidence and mortality. Cancer risks may be compared between populations of similar genetic background but living under different physical, social and cultural circumstances and between populations of different genetic backgrounds living in the same geographical environment [2, 3].

2.3 Characteristics of migrant studies

Australia, Israel, the USA and Canada, each with large immigrant populations, have been fruitful sources of investigation. Cancer rates in migrants, obtained from routinely collected incidence or mortality statistics, are compared with those in the country of origin and those of the native-born in the host country. The change with time since migration or age at migration, and in successive generations of migrants [4] is assessed. The results are interpreted in the light of possible differences in socio-economic status and the degree of cultural assimilation.

Such studies must be interpreted with caution as migrants are self-selected and not representative of the population in their country of birth. They may come from particular regions within that country or from social or religious groups with distinctive cancer patterns. They are likely to be healthy in order to face the rather daunting prospect of translocation to a new environment, and to pass the medical examination that may be required by the host country. The majority of migrants travel from a poorer to a more industrialized country.

The chief contribution of the earliest formal migrant studies, which were of Japanese [5] and Chinese [6] in the USA, was to demonstrate that for most cancer sites the risk changed towards that of the host country, over time and from one generation to the next. This occurred irrespective of whether the cancer was common or rare in the country of origin. Migrant studies do not provide direct evidence of dietary effects on cancer rates but can indicate whether observed differences are more likely to reflect environmental or genetic factors.

Rapid changes in cancer risk following migration indicate that lifestyle or environmental factors are of over-riding importance in etiology, and that either the relevant agents act late in carcinogenesis or preventive strategies may be introduced effectively at this stage. However, if the change was slow so that the rate in the migrants remained closer to that in the country of birth, either the migrants retained lifestyle factors which modify cancer progression or else exposures early in life (i. e. at an early stage in carcinogenesis) were more relevant. Classical epidemiology alone cannot distinguish between the two. If an environmental influence seems probable then this, together with national dietary consumption patterns or dietary surveys of individuals, may be used as evidence supporting a dietary effect. Additional information obtained from individuals in case-control studies can enhance knowledge gained from data at the population level.

2.4 Examples from incidence and mortality rates of cancer at selected sites

Convergence or lack of convergence with rates in the US-born white population will be illustrated using data from migrants to the USA and their US-born offspring [3]. As duration of residence is available for migrants to Australia and Israel, rapidly or slowly changing risks with time since migration will be portrayed using published data for Israel (incidence for the period 1961–81) [7] and Australia (mortality for 1962–71) [8]. In addition, findings from a new analysis of mortality data from New South Wales (NSW), Australia, for a more recent period (1975–95) will be described for migrants from the British Isles, Southern (S) Europe (chiefly Italy, Greece, Yugoslavia and Malta), Eastern (E) Europe (chiefly Poland, former USSR, Hungary and Czechoslovakia) and East and South-East (E/SE) Asia (chiefly China, Vietnam and the Philippines) for cancers of the colon/rectum, stomach, liver, breast and prostate [9; McCredie *et al.* unpublished]. By international standards [1], the Australia-born have high rates of colorectal, breast and prostate cancers and low rates of stomach and liver cancer [10].

2.4.1 Colorectal cancer

Although most European countries had rates of colon cancer which were lower than those for US whites, rates for male European migrants to the USA were slightly greater than that of US white men while those for female European migrants generally approached that of US white women [3]. Rates of colon cancer were lower in non-European than in European migrants but the US-born descendants of migrants from Latin America, China or Japan had rates close to those of the US whites [3]. Migrants from S Europe and E/SE Asia brought low rates that rose with time spent in Australia until, after 20 years, they were not significantly different from those in the Australia-born [8; McCredie *et al.* unpublished]. Among S European migrants rates tended to increase more slowly and to remain somewhat lower in women than in men [8, 11; McCredie *et al.* unpublished]. By contrast, rates in migrants from E Europe and the British Isles showed no trend over time, remaining slightly less than in the Australia-born. In Israel there was no change in risk of colon cancer, relative to that in the Israel-born, with increasing duration of residence in Israel for migrants from Europe or America, West Asia or North Africa, the risk remaining significantly low in the latter [7].

2.4.2 Stomach cancer

Consistent with the higher rates in their countries of origin, migrants to NSW from the British Isles, S Europe, E Europe or E/SE Asia initially had risks of stomach cancer some 2- to 3-fold higher than in the Australia-born. With the passage of time in Australia their rates fell but even after 30 years some residual excess persisted ranging from 20 % (British Isles) to 80–100 % (S Europe; E Europe; E/SE Asia). Similarly an excess incidence of stomach cancer persisted among migrants to Israel 30 years after migration [7]. Italians who migrated to Australia when aged less than 15 years had a risk of stomach cancer similar to that of the Australia-born, in contrast to those who migrated as adults [12]. In the offspring of Asian migrants to the USA mortality rates were closer to rates in the US-born than those of their foreign-born parents but, except in the Chinese, some residual excess risk persisted [3]. The residual excess in rates was not related to the level of risk in the country of origin.

2.4.3 Liver cancer

The higher rates of liver cancer in E/SE Asia, S Europe and E Europe were reflected in the initial rates in their migrants relative to the Australia-born. After 30 years in Australia, an excess risk remained in each of these migrant groups, but a significant falling trend was seen only in those born in E/SE Asia. By contrast, migrants from the British Isles, where mortality rates are similar to those in Australia, had no excess of deaths from liver cancer at any time since migration. An excess in rates was seen in the offspring of Asian migrants to the USA [3] and persisted in West Asian migrants after 30 years in Israel [7].

2.4.4 Breast cancer

In European but not Asian migrants breast cancer rates generally approached that of US-born white women; moreover, they were still only half that of the US-born white population in the daughters of Asian migrants [3]. However, daughters of Italian migrants to Canada had rates no different from Canada-born women [13]. In Australia in 1962–71, breast cancer mortality in S European migrants increased with the number of years spent in Australia, until after 17 years it was generally at least as high as in the Australia-born [8]. Although this pattern was also seen in E/SE Asian migrants in 1975–95 [9], it was no longer true of S European migrants in whom mortality from breast cancer was similar to that in the Australia-born from the time of migration, and did not change appreciably with time spent in Australia. This change in migrant breast cancer rates reflects the convergence of rates in Australia and, for example, Italy between the mid 1960s, when Australian rates were higher, and the mid 1980s when there was little difference [14]. Migrants from E Europe and the British Isles showed similar rates to Australia-born women and no trend with duration of residence in either period [8; McCredie *et al.* unpublished]. Rates increased with time spent in Israel for North African migrants until they were identical with those in Israel-born women after 30 years but remained significantly low in West Asian migrants to Israel [7].

2.4.5 Prostate cancer

European but not Asian migrants had rates of prostate cancer close to that in US-born white men whereas sons of Asian migrants still had lower rates than in US-born whites [3]. Compared with men born in Australia, the risk of dying from prostate cancer in migrants from S Europe, E Europe and the British Isles was low and remained significantly reduced. However, rates in E/SE Asian migrants, initially low, increased with duration of residence until, after 30 years, they were close to that in the Australia-born [9]. Migrants from Europe/America showed a pattern of rising incidence with respect to the Israel-born, while rates in migrants from West Asia or North Africa remained close to that in the Israel-born throughout the 30-year period [7].

2.5 The effect of migration on diet

Data concerning diets in the countries of origin and among the migrants themselves have been brought together by McMichael and colleagues for southern

European migrants to Australia [11, 15]. During the 1950s sources of dietary fibre (vegetables, fruit, pulses, nuts, cereals), vegetable oils and wine were more commonly consumed in Italy and Greece than in Australia, while animal (saturated) fats were eaten less. An Australian national dietary survey in 1983 showed that migrants from southern Europe continued the dietary habits of their home countries (green leafy vegetables, tomatoes, pasta, citrus fruit, and red wine) but added in more meat, especially beef and veal [16 cited in 15]. Smaller dietary surveys within Australia supported these data. Italian migrants drank more red wine, ate more pasta, bread, salami and fish, and obtained more of their protein and fat intake from vegetables than did the Australian-born [17]. Greek migrants increased meat intake soon after arrival [18] but otherwise tended to maintain their native cuisine, for example, by eating wild green leafy vegetables and olive oil and by making wine at home [19].

The lower mortality from all causes in East and South-East Asian migrants relative to the Australian-born [20] has been correlated with their higher expenditure on vegetables, fruit, and fish and lower expenditure on dairy products, tobacco and alcohol [21]. Vietnamese women who have migrated to Australia by and large retain their traditional diet, but with less fish, rice and vegetables and more meat, cereals, fruit and dairy products than in their homeland [22].

Because of its Anglo-Saxon heritage it is not surprising that food eaten in Australia in the 1950s was similar to that in the British Isles as indicated by national food consumption patterns, an exception being that the British and Irish have eaten more potatoes [11, 15]. In Poland (a representative source of E European migrants to Australia) the diet comprised more potatoes, vegetables, cereals and spirits than in Australia, less meat, fruit and pulses, and a similar amount of animal fats [11, 15].

A population-based case-control study of colorectal cancer among Chinese migrants in western North America collected information that included time since migration, diet, and physical activity [23]. Colorectal cancer risk increased with number of years in North America adjusted for diet and physical activity. The findings suggested that colorectal cancer risk increased with longer duration of exposure to a sedentary lifestyle and to a diet rich in saturated fat, and that the higher incidence among Chinese-American men compared with women was due to a longer duration of these habits among men.

2.6 Are dietary factors implicated?

The S European and E/SE Asian migrants to Australia share an initially low rate of colorectal cancer which is lost after 20–30 years in the Australian environment. Many believe that diet is responsible [11, 15, 24–26]. Both communities have a high intake of fresh fruit and vegetables, and a low intake of saturated fats. They differ in that S Europeans consume more pasta, red wine and olive oil, and the E/SE Asians more rice and fish.

The fall in rates of stomach cancer in first generation migrants indicates that exposures in adult life are etiologically important, but the persisting excess mortality after many years suggests that retained cultural factors, or genetic or early environmental exposures play some part. From soon after the Second World War almost every Australian household had a refrigerator, reducing the dependence upon chemical preservatives [27] and increasing the availability of fresh fruit and vegetables, and probably accounting for the observed improvements in mortality from stomach cancer [28]. That some migrants retained a liking for salted and cured foods [22] may explain the residual excess mortality in these populations.

As most liver cancer in migrants to Australia is associated with hepatitis B or C infection acquired early in life [29] before migration, the fall in mortality with duration of residence in Australia (evident for E/SE Asians; S and E Europeans had too few cases among short-term migrants to show a significant trend) suggests that the Australian environment may confer a degree of protection against this cancer. This may merely be from factors related to a generally higher standard of living, or it may be due to elimination of a specific risk such as the contamination of food by aflatoxins [30].

For migrants moving from a low- to a high-risk country, breast cancer rates increase (more slowly for non-European than European migrants) and may converge with that of the host country indicating that exposures in adult life can affect the risk. Findings of a case-control study among Asian American women in the USA also pointed to cultural influences having a much stronger effect than genetic factors in the etiology of breast cancer [31]. Amongst Asian American women born in the USA, there was a steadily increasing risk according to whether the mother and one or both grandmothers also had been born in the USA, but the place of birth of fathers or grandfathers did not influence the risk. Whether the migrants came from an urban or rural environment in Asia also was relevant with those migrating from urban centres in East Asia having a 30% higher risk of breast cancer than those from rural regions. Whether these cultural influences are dietary is not known.

Of all the cancers considered here, migrant studies have least to offer in explaining the etiology of prostate cancer. That rates are low in most migrant groups, even those from relatively high risk countries, may increase to some degree [7] but tend to remain low long-term [7, 8, McCredie *et al.* unpublished], suggests that genetic factors may be relatively important.

2.7 Genetic susceptibility

Could the persisting excess or deficit of some cancers in migrants many years after migration be explained by genetic factors, such as racial differences in metabolism or DNA repair mechanisms, or altered expression of oncogenes or tumor suppressor genes? Indeed, variation between races has been demonstrated within the prostate where the conversion of testosterone to dihydroxytestosterone by the enzyme 5-α-reductase controls cell division. Since cell proliferation is an important influence in human carcinogenesis, racial variation in the secretion or metabolism of testosterone may be responsible for the worldwide heterogeneity in prostate cancer risk. A series of studies in African American, US-born white and Asian American men has measured: two serum markers of 5-α-reductase activity ($3\alpha,17\beta$ androstanediol glucuronide and androsterone glucuronide) [32]; a polymorphic marker in the *SRD5A2* gene [33] which encodes the type II steroid 5-α-reductase; and the highly polymorphic androgen receptor (AR) [34] which is required to translate the androgen response. The pattern of bio-activity parallels the variation of prostate cancer risk in these three populations.

Asians and Caucasians differ in their patterns of mutations for *N*-acetyltransferase [35] which plays a role in the etiology of colon cancer. Fast acetylators appear to have a greater ability to activate heterocyclic aromatic amines (occurring in meat) to carcinogenic derivatives within the mucosa of the colon [24, 25, 36]. This is an example of one type of gene-environment interaction in which a genetic risk factor has an effect on the disease only in the presence of the exposure [37]. Aspects of diet may be involved in other such gene-environment interactions.

2.8 Conclusion

Studies in migrant populations yield good evidence supporting the role of dietary factors in colorectal and stomach cancers. In the case of the former, the detrimental effect of a presumed change in diet can be seen relatively soon after migration and therefore acts at a late stage in carcinogenesis. However, for stomach cancer the residual excess risk persists longer indicating the importance of exposures early in life or of retained cultural factors, likely to be dietary, which carry a higher risk. The late amelioration of the risk of liver cancer indicates an environmental cofactor for a cancer that is primarily related to persistent viral infection. Environmental factors which appear to be culturally determined and may be dietary are important in breast cancer, but seem relatively less important for prostate cancer where a role for genetic factors is indicated.

References

[1] Parkin, D. M.; Whelan, S. L.; Ferlay, J.; Raymond, L.; Young, J. (Eds.) (1997) *Cancer incidence in five continents. Volume VII.* IARC Sci. Publ. No 143. International Agency for Research on Cancer, Lyon.

[2] Parkin, D. M.; Khlat, M. (1996) Studies of cancer in migrants: Rationale and methodology. *Eur. J. Cancer* **32A**, 761–771.

[3] Thomas, D. B.; Karangas, M. R. (1996) Migrant studies. In: Schottenfeld, D.; Fraumeni, J. F. Jr. (Eds.) *Cancer epidemiology and prevention. 2nd edition.* Oxford University Press, New York, p. 236–254.

[4] Parkin, D. M.; Iscovich, J. (1997) Risk of cancer in migrants and their descendants in Israel II. Carcinomas and germ-cell tumors. *Int. J. Cancer* **70**, 654–660.

[5] Locke, F. B.; King, H. (1980) Cancer mortality among Japanese in the United States. *J. Natl. Cancer Inst.* **65**, 1149–1156.

[6] King, H.; Haenszel, W. (1973) Cancer mortality among foreign- and native-born Chinese in the United States. *J. Chronic. Dis.* **26**, 623–646.

[7] Parkin, D. M.; Steinitz, R.; Khlat, M.; Kaldor, J.; Katz, L.; Young, J. (1990) Cancer in Jewish migrants to Israel. *Int. J. Cancer* **45**, 614–621.

[8] Armstrong, B. K.; Woodings, T. L.; Stenhouse, N. S.; McCall, M. G. (1983) *Mortality from Cancer in Migrants to Australia – 1962 to 1971.* University of Western Australia, Nedlands, Western Australia.

[9] McCredie, M.; Williams, S.; Coates, M. (1998) Cancer mortality in East and Southeast Asian migrants to New South Wales, Australia, 1975–1995. *Br. J. Cancer*, in press.

[10] McCredie, M.; Coates, M. S.; Ford, J. M. (1990) Cancer incidence in migrants to New South Wales. *Int. J. Cancer* **46**, 228–232.

[11] McMichael, A. J; McCall, M. G.; Hartshorne, J. M.; Woodings, T. L. (1980) Patterns of gastrointestinal cancer in European migrants to Australia: the role of dietary change. *Int. J. Cancer* **25**, 431–437.

[12] Balzi, D.; Khlat, M.; Matos, E. (1993) Australia: mortality study. In: Geddes, M.; Parkin, D. M.; Khlat, M.; Balzi, D.; Buiatti, E. (Eds.) *Cancer in Italian migrant populations.* IARC Sci. Pub. No 123, International Agency for Research on Cancer, Lyon, p. 125–137.

[13] Balzi, D.; Geddes, M.; Brancker, A.; Parkin, D. M. (1995) Cancer mortality in Italian migrants and their offspring in Canada. *Cancer Causes Control* **6**, 68–74.

[14] Aoki, K.; Kurihara, M.; Hayakawa, N.; Suzuki, S. (1992) *Death rates for malignant neoplasms for selected sites by sex and five-year age group in 33 countries. 1953–57 to 1983–87.* University of Nagoya Coop Press, Japan.

[15] McMichael, A. J.; Giles, G. G. (1988) Cancer in migrants to Australia: extending the descriptive epidemiological data. *Cancer Res.* **48**, 751–756.

[16] Cashel, K.; English, R.; Bennett, S. *et al.* (1986) *National dietary survey of adults: 1983. 1. Foods consumed.* Commonwealth Department of Health, Canberra.

[17] Hopkins, S.; Margetts, B. M.; Armstrong, B. K. (1980) Dietary change among Italians and Australian in Perth. *Comm. Health Stud.* **4**, 67–75.

[18] Rutishauser, I.; Wahlquist, M. L. (1983) Food intake patterns of Greek migrants to Melbourne in relation to duration of stay. *Proc. Nutr. Soc. Aust.* **8**, 49–55.

[19] Powles, J.; Ktenas, D.; Sutherland, C. *et al.* (1986) *Food habits in southern european migrants: a case-study of migrants from the Greek island of Levkada.* Department of Social and Preventive Medicine, Monash Medical School, Prahran, Victoria, Australia.

[20] Young, C. (1986) *Selection and survival: Immigrant mortality in Australia.* Department of Immigration and Ethnic Affairs, Canberra.

[21] Powles, J.; Hage, B.; Cosgrove, M. (1990) Health-related expenditure patterns in selected migrant groups: data from the Australian household expenditure survey, 1984. *Comm. Health Stud.* **14**, 1–7.

[22] Baghurst, K. I.; Syrette, J. A.; Tran, M. M. (1991) Dietary profile of Vietnamese migrant women in South Australia. *Nutr. Res.* **11**, 715–725.

[23] Whittemore, A. S.; Wu-Williams, A. H.; Lee, M.; Shu, Z.; Gallagher, R. P.; Dengao, J.; Lun, Z.; Xianghui, W.; Kun, C.; Jung, D.; The, C.-Z.; Chengde, L.; Yao, X. J.; Paffenbarger, R. S.; Henderson, B. E. (1990) Diet, physical activity, and colorectal cancer among Chinese in North America and China. *J. Natl. Cancer Inst.* **82**, 915–926.

[24] McMichael, A. J. (1997) Colon cancer: the evolution of causal concepts. *Cancer Causes Control* **8**, 541–543.

[25] Bingham, S. (1998) Diet and cancer causation. In: Mann, J.; Truswell, A. S. (Eds.) *Essentials of human nutrition.* Oxford University Press, Oxford, p. 309–326.

[26] Howe, G.; Aronson, K. J.; Benito, E. and 23 others. (1997) The relationship between dietary fat intake and risk of colorectal cancer: evidence from the combined analysis of 13 case-control studies. *Cancer Causes Control* **8**, 215–228.

[27] Nomura, A. (1996) Stomach cancer. In: Schottenfeld, D.; Fraumeni, J. F. Jr. (Eds.) *Cancer epidemiology and prevention. 2nd edition.* Oxford University Press, New York, p. 707–724.

[28] Howson, C. P.; Hiyama, T.; Wynder, E. L. (1986) The decline of gastric cancer: epidemiology of an unplanned triumph. *Epidemiol. Rev.* **8**, 1–27.

[29] Pisani, P.; Parkin, D. M.; Muñoz, N.; Ferlay, J. (1997) Cancer and infection estimates of the attributable fraction in 1990. *Cancer Epidemiol. Biomarkers Prev.* **6**, 387–400.

[30] Higginson, J.; Muir, C. S.; Muñoz, N. (1992) *Human cancer: Epidemiology and environmental causes.* Cambridge University Press, Cambridge, 296–310.

[31] Ziegler, R. G.; Hoover, R. N.; Pike, M. C.; Hildeheim, A.; Nomura, A. M. Y.; West, D. W.; Wu-Williams, A. H.; Kolonel, L. N.; Horn-Ross, P. L.; Rosenthal, J. F.; Hyer, M. B. (1993) Migration patterns and breast cancer risk in Asian-American women. *J. Natl. Cancer Inst.* **85**, 1819–1827.

[32] Ross, R. K.; Bernstein, L.; Lobo, R. A.; Shimizu, H.; Stanczyk, F. Z.; Pike, M. C.; Henderson, B. E. (1992) 5-α-reductase activity and risk of prostate cancer among Japanese and US white and black males. *Lancet* **339**, 887–889.

[33] Reichardt, J. K. V.; Makridakis, N.; Henderson, B. E.; Yu, M. C.; Pike, M. C.; Ross, R. K. (1995) Genetic variability of the human SRD5A2 gene: implications for prostate cancer risk. *Cancer Res.* **55**, 3973–3975.

[34] Coetzee, G. A.; Ross, R. K. (1994) Re: Prostate cancer and the androgen receptor. *J. Natl. Cancer Inst.* **86**, 872–873.

[35] Zahm, S. H.; Fraumeni, J. F. Jr. (1995) Racial, ethnic, and gender variations in cancer risk: considerations for future epidemiologic research. *Environ. Health Perspect.* **103**, 283–286.

[36] Vineis, P.; McMichael, A. (1996) Interplay between heterocyclic amines in cooked meat and metabolic phenotype in the etiology of colon cancer. *Cancer Causes Control* **7**, 479–486.

[37] Kouhry, M. J.; Wagener, D. K. (1993) Population and familial relative risks of disease associated with environmental factors in the presence of gene-environment interaction. *Am. J. Epidemiol.* **137**, 1241–1250.

3 Caloric Intake as a Modulator of Carcinogenicity and Anticarcinogenicity

Ronald W. Hart, T. Bucci, J. Seng, A. Turturro, J. E. A. Leakey, R. Feuers, P. Duffy, J. James, B. Lyn-Cook, J. Pipkin and S. Y. Li

3.1 Abstract

Carcinogenicity is characterized by a set of complex endpoints, which appear as a series of molecular events. Many of these events can be modified by caloric intake. Since most of these processes determine an organisms ability to cope with various environmental stressors it is not surprising that a relationship (in the presence of a constant nutrient density) exists between caloric intake and time to tumor. Our studies have clearly shown that the greater the body weight (generally in rodents directly related to extent of caloric intake) the higher the incidence of spontaneous tumor occurrence, the greater the susceptibility to chemical carcinogens and the shorter the life span. We have focused our attention on the questions of how and to what extent does caloric intake modify those homeostatic processes believed to be critical in determining the ability of an organism to cope with endogenous and exogenous stress such as chemical, physical and biological carcinogens.

The response of an organism can be classified into four categories – physiological, metabolic, molecular and cellular.

We have found, from a physiological perspective, that body temperature in rodents is decreased by 0.5 to 1.8 °C, water consumption is increased by 40 to 80 % as is running activity, however, metabolic output per gram of lean body mass is not altered with decreasing caloric intake. Reproductive capacity declines whereas the ECG waveform is preserved as caloric intake decreased. Alterations in these and other physiological functions suggest that energy intake serves as a signal to up regulate or down regulate functions related to the flight or fright response observed in placental mammals.

A number of key metabolic pathways are altered as a function of caloric intake despite the observation that food consumption per gram lean body mass remains similar due to decreased body weight with decreased caloric intake. Pharmacological compartmentalization however is altered. As caloric intake declines changes occur in the expression of number of drug metabolizing enzymes; with the most striking effect being seen on sex specific, growth hormones-dependent liver enzymes. Additionally, oxidation stress (free radical production) appears to decrease as a function of caloric intake and antioxidant activity increases concurrent with an up regulation in the activity of a number of key enzymes of intermediary metabolism.

A number of molecular processes also change with changes in energy consumption. Our studies have shown that regardless of the source of DNA damage DNA repair is preserved and/or enhanced as caloric consumption decreases, in addition the fidelity of DNA replication increases and oncogene expression is stabilized, P53 gene expression is increased as a function of the decrease in circulating glucose and apoptosis is elevated by up to 500%.

At the cellular level cellular proliferation is decreased proportionate to energy intake in some but not all tissues. Studies both in our and associate laboratories have also shown an enhancement in immune capacity, changes in IGF_1, and accelerated rates of wound healing proportionate with declines in energy consumption. Our most recent findings however have shown that the benefits associated with decreases in caloric intake only occur in the presence of sufficient nutrient density. In the absence of proper nutrition sensitivity to carcinogens appears to be enhanced. These observations have led us to conclude that response to a decrease in caloric intake involves an up regulation of those processes that modulate the response of organisms to a wide range of environmental stressors.

3.2 Introduction

Reduced caloric intake has been repeatedly shown to increase both mean and maximum life span within all species of placental mammals while concurrently reducing the severity and retarding the onset of both spontaneous and chemically induced pathologies. Reflective of this is the strong correlation that exists between food intake and body weight as well as between body weight and spontaneous tumor occurrence and shorter life span in rodents and non-human primates [1]. The mechanisms by which reduction of caloric intake, in the presence of adequate nutrition, is able to extent longevity and reduce the frequency and lengthen the time to occurrence of degenerative diseases is not known [1]. Since altering caloric intake can concurrently alter so many fundamental physiological processes ranging from free radical formation/inactivation to the induction of DNA damage, its repair and the fidelity of its replication and expression to the functionality of various cellular processes including expression of stress proteins, frequency of cellular replication, apoptosis, immune capacity to the modification of specific physiological functions such as thermo regulation and water consumption [1] it is not unexpected that several hypothesis have been put forth to explain the beneficial nature of reduced caloric intake [1].

The observation that so many fundamental physiological, metabolic and biochemical systems and functions are significantly altered by changes in energy intake suggests that a relationship may exist at a fundamental level between the availability of energy and the evolutionary success of a species [2]. Indeed energy related functions are so fundamental to the survival success of the organism that individual and experimental differences in food consumption appear to be responsible for a significant degree of the variability observed among animal bioassays [3, 4]. Contrary to the previous dogma which assumed that animals on study fed "ad libitum" (AL) would be similar to one another [5] it now appears that this factor may be responsible for much if not most of the present variability observed between studies. Thus, despite early attempts at standardization of most environmental variables found in animals studies including feed type, animal maintenance procedures, animal strains, and engineering parameters, it now appears that one of the few factors not controlled for may have been the one factor most important to have controlled for – namely caloric intake.

3.3 Relationship of body weight to cancer

Over twenty years ago Ross [6] reported that body weight (BW) was inversely proportional to survival. Our laboratory showed that even relatively minor changes in BW correlated well not only with survival, but also the incidence of a number of pathologies observed in the B6C3F1, C57B6 and DBA mouse as well as the BN, F344 and BN x F344 rat [7]. In the B6C3F1 male mouse for example we identified that BW at 12 months on test provided a strong correlation, with a coefficient of 0.72, with liver tumors at 24 months on test in male animals [8]. When the database is further refined to take into account the experimental conditions of the control experiments (e.g. inhalation, feed, corn oil gavage etc.), the correlation is improved significantly. For example, when inhalation studies are analyzed separately the correlation coefficient is 0.92 for BW 12 versus liver tumors [8], thus in this case variability in BW12 can account for up to 90 % of the variability in tumor occurrence in male mice at 24 months of age [8]. It is a reasonable assumption, due to the existence of subpopulations and individual cell differences that with the addition of test chemical variability would be even greater than in control populations.

It is our belief that while in the cases above the animals examined were non-treated controls, the observed average weight differences appear to result from AL feeding practices and not from differences in experimental design features such as single *vs* groups housing or gavage *vs* feeding nor within limits diet composition. These findings strongly suggest that differences in body weight are central to and account for the majority of the differences observed in tumor occurrence as a function of body weight (as a surrogate of food consumption) across studies. Similar results with different levels of correlation occur for other tumor types and other classes of pathology not only in mice and rats but other species of placental mammals including non-human primates' [9]. The question then becomes two-fold: 1) What physiological, cellular and molecular processes are altered to induce such a wholestic response on the part of the organism and 2) What might be the evolutionary basis for the existence of such a system?

3.4 Homeostasis

A series of processes exist, the individual failure of which appears to be related to the onset of cancer or certain other degenerative processes. In this concept presented almost thirty years ago [10], there exist a series of processes between exposure of an organism to agents which may induce macromolecular damage and the pathological expression which may modify or modulate each of these steps. Failure of these systems is believed to be in part responsible for an increased risk of cancer; whereas, up regulation of these systems is believed to be protective [10]. Our studies suggest that these systems are multigenic and can be influenced by exogenous factors such as diet.

3.4.1 Physiological effects

A number of physiological and behavioral changes are effected by alterations in caloric intake. Table 3.1 summarizes a few of these changes as observed in Fischer 344 rats and B6C3F1 mice. Total feeding time per day becomes higher as a function of caloric intake; however when adjusted for lean BW, food consumption per gram lean body mass is the same among test groups and water

Table 3.1: Physiological and behavioral biomarkers for male B6C3F1 mice and Fischer 344 rats fed a reduced caloric intake.

Biomarker	Fischer-344 rats	B6C3F1 mice
Average Body Temperature	↓ 0.8 °C	↓ 1.2 °C
Respiratory Quotient (range)	↑ 514 %	↑ 100 %
Activity	↑ 15 %	↑ 73 %
Water Consumption	↑ 43 %	↑ 74 %

Range = Maximum Value – Minimum value
All variable were statistically significant at $p \leq 0.05$.

consumption is increased [11]. As caloric intake decreased there was a corresponding increase in the skin's permeability to water, thereby altering both water input and output [12]. Caloric intake thus can impact transport across the skin, excretion rate and compartmentalization parameters. Caloric intake can result in changes in timing, frequency, and duration of feeding and thus in metabolic output, motor activity and body temperature. As shown in Table 3.1 the average daily body temperature decreases as a function of decreasing caloric intake, suggesting that there is a decrease in the rate of occurrence of temperature-induced apurinic and apyrimidinic DNA damage in the organism's DNA with decreased food intake.

3.4.2 Drug metabolism

In addition to the physiological factors, which might impact on the pharmacological response of an organism to agent exposure a number of the key enzymes of drug metabolism, are also impacted by differences in caloric intake. One way to characterize the effect of caloric intake on the drug metabolizing enzyme system is to classify these changes into three categories.

- Direct effects: those directly resulting from the organism's adaptation to reduced caloric intake;
- Age-related effects: those resulting from the reduction of the physiological age of the organism relative to its chronological age; and
- Circadian effects: those resulting from the organism are altered feeding behavior.

3.4.2.1 Direct effects

The most striking direct effect of reduced caloric intake is on sex-specific, growth hormone-dependent liver enzymes. In almost all cases examined so far, reduced caloric intake decreased the expression of sex-specific enzymes in both sexes. For example male–specific isoforms such as CYP 2C11 are decreased in male rat liver concurrently with increases in certain female–selective activities such as corticosterone sulfotransferase and androgen 5α-reductase. Conversely as caloric intake decreases there is a concurrent decrease in these same female–selective activities in female rat liver [13].

3.4.2.2 Age-related effects

The most striking age related effect of decreased caloric intake involves the sex-specific, growth hormone-dependant rat liver enzymes. For example reduced caloric intake appears to slow the rate of decline in hepatic CYP 2C11-dependent testosterone metabolism [14] as well as the expression of CYP 2E1 and the UDP–glucuronosyltransferase isoform that conjugates bilirubin.

3.4.2.3 Circadian effects

It is also interesting that as caloric intake is decreased there appears to be, in certain drug metabolizing enzymes, an intensification of circadian profiles. For example, hepatic monooxygenase activities that are selectively catalyzed by CYP 1A, CYP 2A1, CYP 2B, and CYP 2E1 isoforms are all increased at certain time points as a function of a decrease in caloric intake in Fisher 344 rats [14]. In other cases however, reduced caloric intake suppresses the intensity of circadian rhythms such as in the expression of testicular CYP 2A1.

It should be noted that while decreased caloric intake results in readily apparent changes when isoforms selective activities are used, only small changes are observed with less specific substrates such as aminopyrine or ethylmorphine [15]. It is a reasonable assumption that the greater the number of isoforms capable of catalyzing a reaction the more it will be resistant to changes in dietary (caloric) intake since changes in one isoform might be compensated for by other isoforms.

3.4.3 Intermediary metabolism

In addition to drug metabolism and physiological functions being altered by caloric intake a number of the enzyme functions of intermediary metabolism are also effected by changes in caloric intake. Of the various enzymatic effects induced by decreased caloric intake is its ability to modulate free radical metabolism both by reducing the formation of free radicals scavengers and by a stimulation of free radical scavenger enzymes. As caloric intake is decreased the ability of catalase to protect against auto-oxidation increases [16]. The same appears to be true of the liver free radical scavenger enzymes and their activity [16]. Such observations are suggestive that caloric restriction may at least in part operate through the prevention of oxidative damage in cellular DNA. However, the answer may not be that simple since other studies in Emory mice have shown that another free radical scavenger plasma ascorbate is reduced by up to 50 % [17]. Thus it would appear that while decreased energy intake may in part produce its beneficial effects by decreasing the level and rate of oxidative damage to cellular DNA this relationship is not a simple one.

3.4.4 Genetic effects

DNA damage, its repair, expression and replication. Another way to explain the beneficial effects of reduced energy intake on degenerative disease processes would be if such changes result in either a decrease in the rate of occurrence of either or both spontaneous/induced DNA damage as discussed above, increased the repair of such damage or enhanced the fidelity of DNA replication. Decrease in caloric intake does indeed appear to reduce the induction of oxidative damage and some but not all forms of chemically induced DNA damage [1]. Relative to DNA repair it appears to be a consistent finding that various forms of DNA repair are elevated by decreased caloric intake. For example, in cells isolated from the kidney and liver of Fischer 344 rats the age related decline in UV-induced DNA repair was postponed by caloric restriction [18, 19]. A concurrent study by Lipman [20] showed a similar response in rats for repair of DNA damage induced by chemical agents. More recent studies by Li and associated [21] have further clarified these early observations. They have shown that rather than simply slowing the rate of lost of DNA repair capacity, decreased energy intake under the proper set of conditions, actually increases the level of DNA repair. This was shown however to be dependent upon the nutritional condition of the patient, in that patients exhibiting malnutrition or lost of certain key nutrients failed to exhibit an increase in DNA repair capacity [21].

Even in the absence of DNA damage, DNA replication can induce errors in the sequential integrity of DNA [22]. The fidelity of DNA replication declines with age presumably as a result of a decrease in DNA polymerase α (pol α) expression. These changes have been used to explain the increase in cancer observed with age [22]. As shown in Table 3.2, the rate of loss of fidelity in pol α as a function of age has been demonstrated to be slowed by decreasing caloric intake [22]. These data are consistent with the concept that transient loss of polymerases in the cellular pool may contribute to impaired base selection and decreased accuracy of DNA synthesis. Also important is the observation [23] that the rate of loss of DNA pol α expression and loss of fidelity of DNA replication are significantly slowed as caloric intake is decreased and that this appears to be associated with the appearance of an α-accessory protein with ATP-dependent helicase activity. The role of the α-accessory protein appears to be the stabilization of the DNA replication process, thereby enhancing fidelity of DNA replication by reducing the probability of misinsertion.

A number of investigators have now demonstrated that expression of various oncogenes is directly related to the extent of caloric intake [24]. Himeno *et al.* [25] showed that in rat C-Ha-*ras*, C-K_1-*ras* and C-Fos, but not c-myc expression was altered by caloric intake. Lyn-Cook and co-workers [24] subsequently confirmed and extended these findings and showed that the expression of a number of other genes known to be associated with cancer occurrence were also altered by caloric intake. Thus it would appear whether ones evaluates the induction of spontaneous or induced DNA damage, DNA repair, fidelity of DNA replication or the expression of certain key genes, the greater the caloric intake

Table 3.2: Classification of the effect of caloric intake on the enzymes of drug metabolism in male and female Fischer 344 rats.

Effect	Male	Female
Direct Effects	CYP 2C11 ↓ CYP 2A2 ↓ CYP 2CB ↓ CYP 3A2 ↔ CYP 2D ↔ Phenol UGT2 ↔ Aryl sulfotransferase IV ↓ Androgen 5α-reductase ↑ Corticosterone sulfotransferase ↑	CYP 2C12 ↓ Androgen 5α-reductase ↓ Corticosterone sulfotransferase ↓ CYP 2D ↔ Phenol UGT ↔
Age-related Effects	CYP 2C11 ↑ CYP 3A2 ↑ CYP 2E ↑ Bilirubin UGT ↑ Androgen 5α-reductase ↓ Corticosterone sulfotransferase ↓	
Circadian Effects	CYP 1A1 ↑ CYP 2A1 ↑ CYP 2B1 ↑ CYP 2E1 ↑ CYP 2A1 ↑ (testis)	CYP 1A1 ↑ CYP 2A1 ↑ CYP 2B1 ↑ CYP 2E1 ↑

Unless stated otherwise all isoforms are hepatic.

the less the level of genetic homeostasis and the higher the probability of a loss in the organism's ability to deal with stress.

3.5 Evolutionary perspective

Consideration of the evolutionary role that reduced caloric intake might play in extending survival by decreasing the incidence or lengthening the time to occurrence of degenerative disease processes while concurrently lengthening the period in which reproduction is possible may help in developing a broader understanding of the role that the environment plays in evolutionary selection. As caloric intake decreases a series of changes occur and these may be classified into at least five categories. First, there is a curtailment of energy-intensive nonfood gathering activities; second, the ability to extract energy from food is increased, with physiological and metabolic systems becoming more efficient; third, food acquisition activity is increased; fourth, the reproductive life span of the organism is significantly increased and fifth, those processes which are protective of genomic function are significantly improved. Overall, these changes appear to be a means by which organisms preserve the integrity of the genome and its host during periods of food scarcity until food becomes available at which time these systems are down regulated and reproductive activity enhanced. Our observed increase in expression of hsp 90 during these periods of food deprivation and decrease once food is again provided suggests that during these periods mutations may accumulate only to be expressed once food again becomes abundant. Thus the aforecited changes would not only enhance survival of the individual but due to the rapid release of stored up mutations potentially increases the successful propagation of a diversity of offspring that might occupy a diversity of ecological niches.

References

[1] Hart, R.; Neumann, D.; Robertson, R. (Eds.) (1995) *Dietary Restriction: Implications for the design and interpretation of toxicity and carcinogenicity studies.* ILSI Press, Washington, DC.

[2] Hart, R.W.; Turturro, A. (1998) Evolution and dietary restriction. *Exp. Gerontol.* **33**, 53–60.

[3] Hart, R.; Turturro, A. (1995) Dietary Restriction: An Update. In: Hart, R.; Neumann, D.; Robertson, R. (Eds.) *Dietary restriction: Implications for the design and interpretation of toxicity and carcinogenicity studies.* ILSI Press, Washington, DC, p. 1–12.

[4] Turturro, A.; Duffy, P.; Hart, R. (1995) The effect of caloric modulation on toxicity studies. In: Hart, R.; Neumann, D.; Robertson, R. (Eds.) *Dietary restriction: Implications for the design and interpretation of toxicity and carcinogenicity studies.* ILSI Press, Washington, DC, p. 143–161.

[5] Hart, R. (Ed.) (1986) Interagency staff group: Chemical carcinogens: A review of the science and its associated principles. *Environ. Health Perspect.* **67**, 201–282.

[6] Ross, M. (1976) Nutrition and longevity in experimental animals. In: Winick, M. (Ed.) *Nutrition and ageing.* J. Wiley and Sons, New York, p. 23–41.

[7] Bucci, T.; Hart, R.; Turturro, A. (1999) Reduction in spontaneous disease in seven genotypes of calorically restricted rodents. *Exp. Toxicol. Pathol.*, accepted.

[8] Hart, R.; Leakey, J.; Duffy, P.; Feuers, R.; Turturro, A. (1996) The effects of dietary Restriction on drug testing and toxicity. *Exp. Toxicol. Pathol.* **48**, 121–127.

[9] Weindruch, R.; Rolsal, R.S. (1997) Caloric intake and ageing. *New Engl. J. Med.* **337**, 986–994.

[10] Brash, D.; Hart, R. (1977) Molecular biology of aging. In: J. Bemke (Ed.) *The Biology of ageing.* Plenum Press Publishing Corp., New York, p. 57–81.

[11] Duffy, P.; Feuers, R.; Pipkin, J.; Berg, T.; Divine, B.; Leakey, J.; Hart, R. (1995) The effects of caloric restriction and ageing on the physiological response of rodents to drug toxicity. In: Hart, R.; Neumann, D.; Robertson, R. (Eds.) *Dietary restriction: Im-*

plications for the design and interpretation of toxicity and carcinogenicity studies, ILSI Press, Washington, DC, p. 127–140.
[12] Lehman, P. Franz, T. (1993) Effect of age and diet on stratum-corneum barrier function in the Fischer 344 female rat. *J. Invest. Dermatol.* **100**, 200–204.
[13] Manjgaladze, M.; Chen, S.; Frame, L, Seng, J.; Duffy, P.; Feuers, R. J.; Hart, R.; Leakey, J. (1993) Effects of caloric restriction on rodent drug and carcinogen metabolizing enzymes: implications for mutagenesis and cancer. *Mutat. Res.* **295**, 201–222.
[14] Leakey, J. E. A.; Seng, J.; Manjgaladze, M.; Kozlovskaya, N.; Xia, S.; Lee, M-Y.; Frame, L. T.; Chen, S.; Rhodes, C. L.; Duffy, P. H.; Hart, R. W. (1995) Influence of caloric intake on drug metabolizing enzyme expression: Relevance to tumorigenesis and toxicology testing. In: Hart, R.; Neumann, D.; Robertson, R. (Eds.) *Dietary Restriction: Implications for the design and interpretation of toxicity and carcinogenicity studies*, ILSI Press, Washington, DC, p. 167–180.
[15] Leakey, J. A.; Cunny, H. C.; Bazare, J., Jr.; Webb, P. J.; Lipscomb, J. C.; Feuers, R. J.; Duffy, P. H.; Hart, R. W. (1989) Effects of ageing and caloric restriction on hepatic drug metabolizing enzymes in the Fischer 344 rat. II: Effects on conjugating enzymes. *Mech. Ageing Dev.* **48**, 157–166.
[16] Feuers, R.; Duffy, P. H.; Chen, F.; Desai, V.; Oriaku, E.; Shaddock, J. G.; Pipkin, J. L.; Weindruch, R.; Hart, R. W. (1995) Intermediary metabolism and antioxidant systems. In: Hart, R.; Neumann, D.; Robertson, R. (Eds.) *Dietary restriction: Implications for the design and interpretation of toxicity and carcinogenicity studies*, ILSI Press, Washington, DC, p. 181–195.
[17] Taylor, A.; Lipman, R. D.; Jahngen-Hodge, J. (1995) Dietary caloric restriction in the emory mouse: Effects on lifespan, eye lens cataract prevalence and progression, levels of ascorbate, glutathione, glucose and glycohemoglobin, tail collagen breaking time, DNA and RNA oxidation, fecundity and cancer. *Mech. Ageing Dev.* **79**, 33–57.
[18] Weraarchakull, N.; Strong, R.; Wood, W. E. (1989) The effect of ageing and dietary restriction on DNA repair. *Exp. Cell Res.* **181**, 197–204.
[19] Licasto F.; Weindruch R.; Davis, L. (1988) Effect of dietary restriction upon the age-associated decline of lymphocyte DNA-repair activity in mice. *Age* **11**, 48–52.
[20] Lipman, J.; Turturro, A.; Hart, R. (1989) The influence of dietary restriction on DNA repair in rodents: a preliminary study. *Mech. Ageing Dev.* **48**, 135–143.
[21] Hart, R.; Li, S. Y.; unpublished observations.
[22] Srivastava, V. K.; Miller, S.; Schroeder, M.; *et al.* (1993) Age-related changes in expression and activity of DNA polymerase some effects of dietary restriction. *Mutat. Res.* **295**, 265–280.
[23] Busbee D.; Miller, S.; Schroede, M.; *et al.* (1995) DNA polymerase function and fidelity: dietary restriction as it affects age-related enzyme changes. In: Hart, R.; Neumann, D.; Robertson, R. (Eds.) *Dietary restriction: Implications for the design and interpretation of toxicity and carcinogenicity studies*, ILSI Press, Washington, DC, p. 118–131.
[24] Lyn-Cook, B. D.; Hass, B. S.; Hart, R. W. (1995) Oncogene expression and cellular transformation: The effects of dietary restriction. In: Hart, R.; Neumann, D.; Robertson, R. (Eds.) *Dietary restriction: Implications for the design and interpretation of toxicity and carcinogenicity studies*, ILSI Press, Washington, DC, p. 183–198.
[25] Himeno, Y.; Engleman, R.; Good, R. (1992) Influences of caloric restriction on oncogene expression and DNA synthesis during liver regeneration. *Proc. Natl. Acad.* Sci. (USA) **89**, 5497–5505.
[26] Pipkin, J.; Hart, R.; unpublished observations.

4 The Role of Nutritional Factors: Colon Cancer

Robert W. Owen

4.1 Abstract

Colorectal cancer is a major disease of western civilisations and diet may account for approximately 35 % of cases. Epidemiologic studies reveal that the major dietary constituents implicated in the disease process are fat/red meat (causative) and calcium/fibre (protective). From this standpoint toxicologists have evaluated a plethora of dietary and intestinal biochemical characteristics and steroids, the bile acids especially have received a great deal of attention. Formerly, bile acids and their bacterial metabolites were implicated, as either mutagens and/or carcinogens but this was not proven. In recent times, based mainly on animal model systems, opinion favours that if bile acids and their metabolites have a role to play in colon carcinogenesis, they act at the promotion stages of the adenoma-carcinoma sequence. The secondary bile acids deoxycholic and lithocholic acids being of major importance here. From this standpoint, studies have been designed either, to lower total bile acid concentration or, at least to reduce dehydroxylation of the primary bile acids via high fibre dietary regimens.

Results of short-term human and animal model intervention studies with either calcium or fibre have proven to be effective in producing the required profiles and in animal model systems also to ameliorate the development of colonic neoplasia. Current data however indicate that bile acids have a negligible influence at any stage of the adenoma-carcinoma sequence of events. This is exemplified by the results of several recently completed case control and calcium intervention studies involving both adenoma and colon cancer patients in which faecal bile acid concentration does not correlate with any of the known factors influencing the development of adenomas and pathological features associated with frank tumor risk. Obviously the link between diet and colon cancer is not as straightforward as previously thought and there is a dire need for a different approach.

Reactive oxygen species are implicated in the causation of a range of human diseases especially cancer and evidence is accumulating that they may also be of importance in the aetiology of colon cancer. A plausible mechanism is via lipid peroxidation processes, yielding etheno-base DNA adducts leading to mutation. To this end we have developed HPLC methods for the analyses of reactive oxygen species by the faecal matrix. The evidence shows that the faecal matrix is capable of supporting free radical generation in abundance. The active principle is a soluble factor as evinced by the ability of filter sterilised and freeze-dried faeces but not faecal bacteria to support Fenton chemistry. Studies are in progress to assess the effect of dietary manipulation on the production of reactive oxygen species and systems are being developed to study their generation in more realistic oxygen-depleted environments.

4.2 Introduction

Colorectal cancer is a disease of affluent societies and epidemiologic studies [1–3] have clearly shown that nutritional factors are a major contributory factor. High consumption of fat and red meat have been regarded important in the promotion of colorectal cancer perhaps through an additive effect on intestinal metabolism via increased hepatic secretion of biliary components and fermentation by the intestinal microflora [4, 5].

An additional supporting hypothesis was forthcoming from Newmark *et al.* [6] when it was suggested that calcium may have considerable chemopreventive potential against colorectal cancer by chelating and precipitating bile acids in the colonic lumen rendering them non-toxic. This again led to an upsurge in interest in the bile acid/colorectal cancer hypothesis and was accompanied by the development of more sophisticated chromatographic methods for lipid analyses in biological samples [7–10]. Although these methods have allowed the detailed evaluation of faecal bile acid profiles in various patient groups [10–17] the importance of these metabolites in the aetiology of colorectal cancer is still largely unresolved.

By contrast high consumption of dietary fibre is regarded ameliorative [18–19]. The protective role of dietary fibre against colorectal cancer has its origins in the classical study of Burkitt [20]. This was based on epidemiologic studies in Africa which showed that a high intake of dietary fibre was highly correlated with faecal weight. It was proposed that dietary fibre is protective against colorectal cancer by the simple mechanisms of stool bulking, acceleration of transit and dilution of potential endogenous carcinogens. This theory has received qualified support in the meta-analyses of Howe *et al.* [21] and Cummings *et al.* [22] and a recent consensus [23] was reached implying that of dietary fiber the more consistent signs of beneficial effects are derived from high intakes of cereal fibres.

A more definitive mechanism has recently been forwarded by Nair *et al.* [24] to explain the relation between diet and cancer. In a dietary intervention study with either sunflower oil (high in polyunsaturated fatty acids, especially linoleic acid) or rapeseed oil (high in monounsaturated fatty acids, especially oleic acid) a dramatic increase in DNA bridged etheno-adducts was detected in whole white blood cells in females intervened with the former. Etheno adducts are produced as the end-point of lipid peroxidation processes, initiated and propagated by reactive oxygen species (ROS) and thereby indicate that diets which are rich in linoleic acid may be non-beneficial to health. These observations show for the first time a plausible and tangible link between an unhealthy diet and the carcinogenesis process.

Alongside these discoveries ROS are implicated in a range of human diseases especially inflammatory bowel disease and prompted us to develop methods which would enable the study of diet and free radical generation in situations relevant to the large intestine.

4.3 Fat and bile acids

The notion that bile acid metabolites may be involved in the aetiology of colorectal cancer dates back to keynote publications early in the 1970s. The principal hypotheses [25–26], based largely on epidemiologic data, were that high dietary intakes of animal fat (common in western countries) elicited a response within the intestinal hepatic circulation to produce greater quantities of bile acids to aid the digestive process. As a result of an increased proportional loss of these bile acids to the large bowel, and metabolism by the indigenous microflora, an increase in cytotoxic secondary bile acids might be expected to occur in the intestinal lumen. These hypotheses were vindicated by the data of Hill *et al.* [4] and Reddy and Wynder [5] who showed that colorectal cancer patients in England and the USA respectively, excreted significantly elevated levels of faecal secondary bile acids.

A plausible mechanism was also presented by Hills' group by which bile acids could be converted by the intestinal microflora, the *Clostridia* especially, to potential carcinogens. In a series of reports [27–30] the sequential dehydrogenation of the bile acid nucleus to a structure resembling a polyunsaturated hydrocarbon was demonstrated the inference being that similar structural transformations could take place in the intestinal lumen resulting in the formation of a benzo[a]pyrene carcinogenic type structure. The series of reactions are depicted in Figure 4.1 but to date only three of these transformations have been verified, namely oxido-reduction of the hydroxyl group at C3, desaturation of the A-ring to give a δ4-unsaturated steroid and dehydroxylation of a 7-hydroxylated δ4-unsaturated steroid to give the δ4,6-dienone structure. While these reactions can be unequivocally demonstrated *in vitro* evidence for the presence of unsaturated bile acids in intestinal digesta and faeces is sadly lacking.

In later years support was gained from both *in vitro* [31, 32] and animal model [33–34] experiments showing both a co-carcinogenic and promotional effect of these substances on colorectal neoplasia. However the hypotheses were nevertheless considered controversial because of a number of contradictory case-control studies [35–37] and this is exemplified by a study conducted by the author in conjunction with St Marks Hospital London.

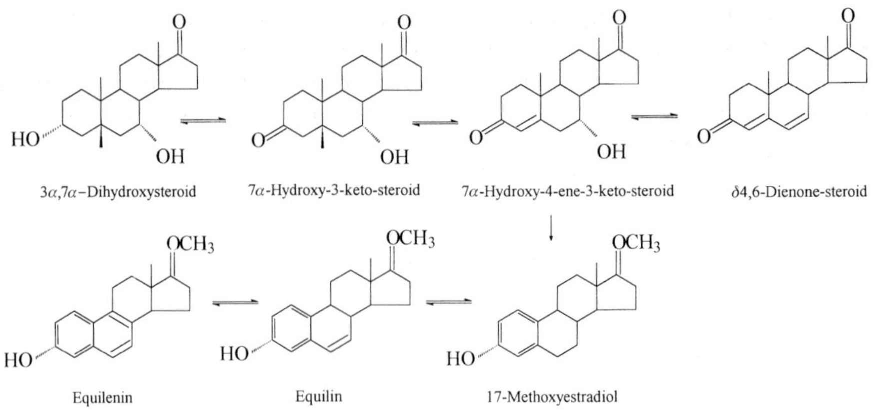

Figure 4.1: Metabolism of steroids by *Clostridia*.

In a case control pilot study [12] in which the faecal individual lipid profiles and total excretion were compared, no significant difference was observed in the concentration of total bile acids between healthy controls and colorectal cancer cases. This was also true for the major secondary bile acids lithocholic and deoxycholic acid but a significant association was detected between the ratio of lithocholic acid and deoxycholic acid and the presence of cancer.

To verify these observations a larger case control study was conducted along the same lines. Seventy eight patients [38] referred to St Marks Hospital for surgery were entered into the study along with 40 healthy controls. Spot stool samples were collected from the patients at least three to five days prior to bowel preparation and surgery. Likewise faecal samples were obtained from 40 healthy volunteers. Details of the study groups are given in Table 4.1 and whilst

Table 4.1: Details of the St Marks colorectal cancer case control study.

Group	Number	Age ± SEM	Sex	
			Male	Female
Colorectal cancer	78	63 ± 1	47	31
Control	40	59 ± 1	20	20
Dukes grade				
A	10	64 ± 2	6	4
B	27	63 ± 2	15	12
C_1	27	62 ± 2	18	9
C_2	12	62 ± 3	7	5
Sub-site				
Caecum	9	63 ± 4	6	3
Rectosigmoid	15	65 ± 3	8	7
Rectum	54	63 ± 1	33	21

they were well matched for age this was not the case for gender. All samples were immediately frozen to –40 °C and transported to the laboratory for analysis. Both wet and dry weights were ascertained and the dried samples were analysed according to the methods of Owen *et al.* [9, 11]. A brief overview of the protocol is given in Figure 4.2.

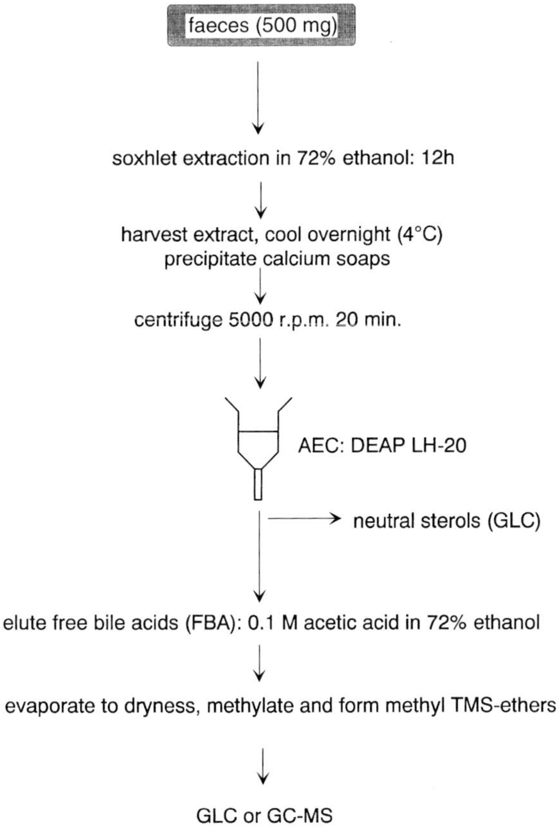

Figure 4.2: Protocol for the analysis of faecal steroids.

The data (Table 4.2) show that there was no significant difference in the faecal concentration of total bile acids but a significant difference between the cases and controls was again observed (Table 4.3) for the ratio of lithocholic to deoxycholic acid ($P = 0.002$). When the patients were sub-grouped on Dukes grading (A–C_2) and site of the cancer (caecum, rectosigmoid and rectum) the only significant association between the cases and controls for patients regarding bile acid concentration was that Dukes grade C_2 cancers excreted significantly lower amounts ($P = 0.014$). This is not surprising because with Dukes C_2 cancer it is known that metastases of the liver are prevalent which decreases

Table 4.2: Composition of the major faecal bile acids in the St Marks colorectal cancer case control study.

Group	Major bile acid				Total bile acids
	LCA	DCA	CDCA	CA	
CRC (78)	3.20 ± 0.33	2.94 ± 0.34	*0.24*	*0.10*	7.04 ± 0.67
Control (40)	3.03 ± 0.28	3.99 ± 0.43	*0*	*0*	6.94 ± 0.68
Dukes Grade					
A (10)	2.67 ± 0.69	2.84 ± 1.21	*0.13*	*0*	6.06 ± 1.92
B (27)	4.30 ± 0.75	3.88 ± 0.70	*0.27*	*0.08*	8.94 ± 1.42
C_1 (27)	2.93 ± 0.39	2.53 ± 0.41	*0.41*	*0.17*	6.48 ± 0.96
C_2 (12)	1.48 ± 0.21	0.99 ± 0.25	*0.01*	*0.02*	3.14 ± 0.43*
Sub-site					
Caecum (9)	3.16 ± 0.78	2.39 ± 0.58	*0*	*0*	5.74 ± 1.33
RS (15)	3.13 ± 0.57	3.78 ± 1.03	*0.04*	*0.11*	7.79 ± 1.63
Rectum (54)	3.23 ± 0.43	2.80 ± 0.40	*0.33*	*0.12*	7.05 ± 0.84

Data expressed in mg/g dry faeces
* Significant compared to control group P = 0.014
Numbers in parentheses represents presence in 15 % of samples or less
CRC – Colorectal cancer

Table 4.3: Ratio of lithocholic to deoxycholic acid in the St Marks colorectal cancer case control study.

Group	LCA:DCA	% Abnormal
CRC	1.81 ± 0.19[a]	72
Control	0.94 ± 0.09	25
Dukes grade		
A	2.33 ± 0.79[b]	50
B	1.47 ± 0.20[c]	74
C_1	1.57 ± 0.16[d]	81
C_2	2.46 ± 0.72[e]	92
Sub-site		
Caecum	1.46 ± 0.24[f]	89
Rectosigmoid	1.41 ± 0.27[g]	47
Rectum	1.98 ± 0.26[h]	76

CRC – Colorectal cancer
Abnormal – Ratio of lithocholic to deoxycholic acid equal to or greater than 1.0
[a] Significantly different from control group $P = 0.002$
[b] Significantly different from control group $P = 0.002$
[c] Significantly different from control group $P = 0.009$
[d] Significantly different from control group $P = 0.0005$
[e] Significantly different from control group $P = 0.0007$
[f] Significantly different from control group $P = 0.03$
[g] Significantly different from control group $P = 0.04$
[h] Significantly different from control group $P = 0.001$

bile acid synthesis and secretion into the intestine. Removal of the Dukes C_2 cancers from the comparison between cases and controls had little effect on the statistics however. The significant difference between the ratio of lithocholic to deoxycholic acid was maintained (Table 4.3) for each Dukes grade (Dukes A, $P = 0.002$; Dukes B, $P = 0.009$; Dukes C_1, $P = 0.0005$; Dukes C_2, $P = 0.007$) and at each sub-site (caecum, $P = 0.03$; rectosigmoid, $P = 0.04$; rectum, $P = 0.001$).

Taking unity for the ratio of lithocholic acid to deoxycholic acid the discriminating power of this marker indicates that it would detect over 70 % of colorectal cancer cases (Figure 4.3) in a screening program but has a fairly high false positive rate for apparently healthy controls. At this point the ratio of lithocholic acid to deoxycholic acid appeared to have some promise as a marker of colorectal cancer but it has not been upheld in further large case control studies involving controls who have been either similarly hospitalised or endoscoped to ratify that they were free of colorectal adenomas. Such studies are currently being reported.

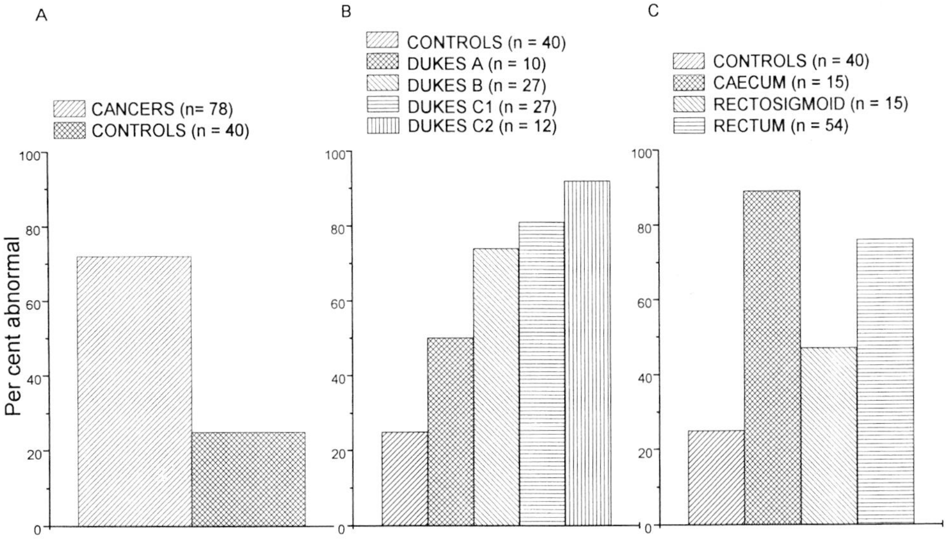

Figure 4.3: Per cent of subjects with an anormal ratio of lithocholic to deoxycholic acid (equal to or greater than 1.0). A: Cancers versus controls; B: Based on Dukes grading; C: Based on sub-site.

4.4 Calcium and bile acids

Over the last decade there has been considerable interest in the study of calcium as a chemopreventive agent against colorectal cancer. The interest stems from epidemiological studies [39, 40] which show that the incidence of colorectal cancer is lower in regions of western industrialized societies where the consumption of calcium containing dietary constitutents such as dairy products is relatively high. A hypothesis was first mooted by Newmark *et al.* [6] based on the considerations that the consumption of calcium leads to a higher concentration of this ion in the intestinal lumen which reacts with endogenous secretions of the liver, namely the bile acids and also dietary long chain fatty acids forming insoluble complexes (Figure 4.4), thereby nullifying the purported cytotoxic effects of these lipids. This hypothesis was also based on several experimental animal studies [41, 42] which showed that calcium can ameliorate the damaging effects of these agents when it is applied intracolonically and was supported by later animal studies investigating the role of calcium in preventing the cytotoxic effect of intestinal lipids and thereby influencing the development of colorectal cancer [43, 44].

These studies were followed by calcium intervention trials in humans but the results overall have been equivocal. Several studies [45–47] have shown that supplementing the human diet with calcium can reduce intestinal cell pro-

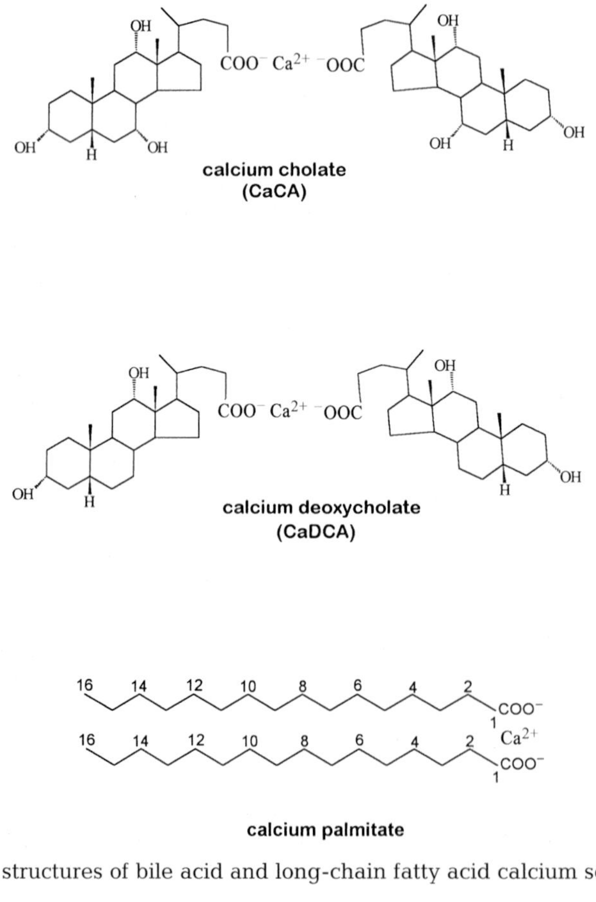

Figure 4.4: The structures of bile acid and long-chain fatty acid calcium soaps.

liferation, an intermediate biomarker of colorectal cancer, others [48, 49] have shown no effect, whilst one [50] has shown the opposite.

In addition to the hypothetical and intervention considerations, Van der Meer and De Vries [51] have provided a "plausible" molecular mechanism *in vitro* to explain how calcium may mediate its effects and the message conveyed from a series of *in vivo* studies [52–54] is that calcium intervention leads to a significant reduction in cell proliferation and the mechanism is clear "supplementary dietary calcium stimulates the formation of insoluble calcium phosphate in the intestine which results in increased binding of cytotoxic luminal bile acids". Furthermore Van der Meer *et al.* [55] have suggested that the amount of bile acid associated with calcium can be determined by the resolubilization of calcium soaps with ethylenediaminetetraacetic acid (EDTA). The rationale indicated, is that because EDTA is a potent chelator of divalent cations especially calcium, incubation of faeces in the presence of EDTA will solubilise those bile acids complexed to calcium.

The theoretical and molecular mechanisms forwarded to explain the protective effect of calcium against colorectal cancer appear to be flawed however. This is exemplified by a recent study by Owen *et al.* [56] which shows, using organically synthesised calcium soaps of the bile acids that the EDTA method described to account for the appreciable levels of calcium bile acid soaps in faeces gives almost identical results for free bile acids (Figure 4.5 a) and calcium bound bile acids (Figure 4.5 b) in that both are solubilised to a similar extent and that the solubilisation of faecal bile acids by this method is significantly dependent on both total and primary bile acid concentration.

In consideration of the equivocal clinical data mentioned above a long-term, double-blind intervention trial was undertaken by Weisgerber *et al.* [57] in polypectomised, sporadic adenoma patients in which the putative role of calcium (2 g/day) as a protective factor in colon carcinogenesis was studied. The results show that despite differences in stool biochemistry elicited by supplementary calcium after 9 months intervention, a similar non-significant decrease of total PI% in sigmoidal mucosa was evident in both the calcium (13.5 down to 11.4) and placebo groups (13.7 down to 10.7).

An increase in the concentration and daily excretion of *total* bile acids, primary bile acids, long chain fatty acids and long chain fatty acid soaps was observed in the calcium group whilst there was no significant reduction in the concentration of the potentially toxic *free* bile acids and long-chain fatty acids. This tends to indicate that even if calcium is beneficial it does not mediate its effects via chelation of intestinal lipid.

There may be several reasons why intervention with dietary calcium gives conflicting results. The capacity of calcium to reduce cell proliferation may be short-term and only a transient effect is observed prior to intestinal adaptation. This would explain why most short-term studies appear effective and long-term studies do not. In the absence of short-term placebo-controlled trials it is probable that the apparent positive results with calcium are a result of inappropriate study design. An appropriate study design is essential in calcium intervention trials especially in adenoma patients who have undergone polypectomy because

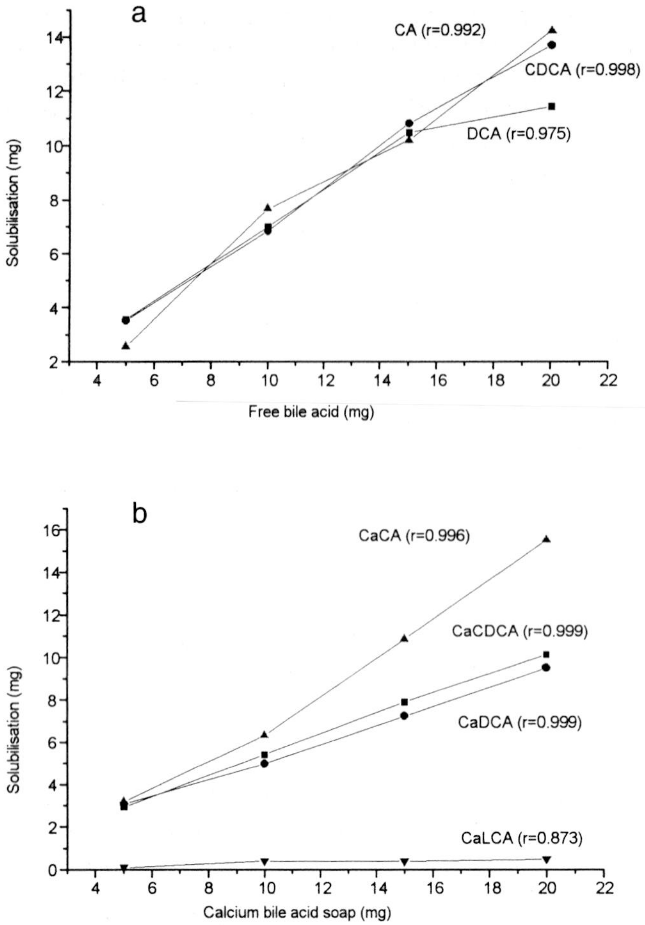

Figure 4.5: The solubilisation of bile acids by EDTA: a. Free bile acids; b. Calcium bound bile acids.

it has been shown that a significant temporal decrease in colonic cell proliferation occurs in adenoma-free colons over a 2-year time period [58]. This observation may explain why calcium intervention in polypectomised patients without a placebo-control group appears effective.

Alternatively calcium may exert its effect only in younger people who are able to maintain or enhance the calcium gradient within the mucosal crypt and colonocytes, thereby facilitating differentiation and apoptosis as described by Whitfield [59]. That the Whitfield model only applies to younger people is supported by the data of Weisgerber *et al.* [57] on relatively elderly people. At entry to the study mean PI% of all the patients was significantly positively associated with soluble calcium in faeces. After intervention PI% was decreased in both the calcium and placebo groups to a similar extent despite a significant increase in faecal soluble calcium in the former group. This is again at odds with the Whitfield model and because increased cell proliferation also positively correlates with patient age it indicates that high luminal calcium concentration as mentioned above may only be effective in reducing cell proliferation in younger people who are able to maintain the necessary calcium gradient that stimulates differentiation and apoptosis.

These conclusions are further supported by the results of calcium balance in the study group. At entry to the study, dietary intake of calcium by the patient group as a whole was over 900 mg/day. Under these conditions obligatory renal loss of calcium (which reflects absorption capacity) was only 86 mg/day compared to an expected 155 mg/day. This represents almost a 100 % decrease in absorption of calcium and dietary supplementation with 2 g/day calcium had little effect on urinary loss. According to the Whitfield model therefore an increase in soluble calcium would be of no benefit if intestinal transmembrane reflux of calcium is diminished in older people. This appears to be the case here and is probably a valid reason why calcium intervention was no more effective in lowering PI% than placebo. This has support in that decreased calcium absorption with increased age is a well recognised phenomenon and may be one general reason why elderly patients exhibit a higher proliferative activity in the colonic epithelium than younger people.

Because calcium absorption in the large intestine is dependent on the supply of vitamin D and is stimulated by the active form of this vitamin it may be prudent in future studies with elderly people to incorporate this vitamin into the intervention protocol. This may re-establish the ability of colonocytes to maintain a positive calcium gradient and effectively reduce cell proliferation.

The results also indicate that calcium does not operate via an indirect antitropic effect in older patients because the calcium balance data show that calcium absorption was severely impaired and therefore an effective gradient in the colonocytes could not be maintained. This is probably an age-related phenomenon and indicates that calcium supplementation is unlikely to be of major therapeutic use in the reduction of cell proliferation and adenoma recurrence in elderly people.

In another slightly larger study by Hofstad *et al.* [60] intervention with calcium and antioxidants had a small but insignificant effect on the repression of growth of adenomas left *in situ* but a significant effect on the formation of new adenomas.

4.5 Fibre and bile acids

It is proposed that dietary fibre exerts its protective effects against colorectal cancer by the simple mechanisms of stool bulking, acceleration of transit and dilution of potential endogenous carcinogens and supplying substrates for bacterial fermentation in the colon which in addition to the above effects also leads to the formation of butyric acid which is regarded to have a antineoplastic effect on the colonic mucosa.

If bacterial metabolites of intestinal lipids are important in the aetiology of colorectal cancer then chemopreventive methods in the form of dietary supplementation may be of some value in this respect. However because it has proven difficult to shift the stance of Western populations towards the generally recommended guidelines of reducing fat intake to 30 % of gross energy and increasing fibre consumption to at least 30 g/day to reduce the incidence of colorectal cancer a more beneficial approach may be to recommend supplementation or intake of prebiotics. That prebiotics influence the biochemistry of the large intestine in humans was first shown by Bown *et al.* [61] where intakes of lactulose depressed caecal pH to 4.5 as measured by a radiotelemetry device. This was of importance because high pH is associated with colorectal cancer [62] and prompted us [63, 64] to study the influence of fibre and prebiotics on bile acid metabolism utilising both an *in vitro* and an *in vivo* approach.

For the *in vitro* studies the effect of wheat bran, pea fibre and lactulose (synthetic fibre) on the dehydroxylation of chenodeoxycholic acid to lithocholic acid, bacterial composition and physical parameters such as pH and Eh in batch and continuous culture respectively were determined [63, 64].

The data showed that both wheat bran and pea fibre effectively reduced bacterial 7α-dehydroxylation of bile acids in batch culture in a dose-dependent manner and lactulose had similar effects in the continuous culture systems. In batch culture a significant increase in the concentration of short-chain fatty acids including butyric acid also occurred while the opposite effect was observed in the continuous fermenters. The reason for this is not clear.

For the *in vivo* approach we studied the effect of fibre intervention on biochemical parameters in both pigs and humans [65]. The data showed that supplementation of the diet with lactitol in mini-pigs had a clear inhibitory effect on the metabolism of cholesterol (Figure 4.6) and bile acids (Figure 4.7). Without supplementation little dehydrogenation of cholesterol was evident at the jejunal and ileal sites, but was increased substantially in the caecum with further small increases aborally to an average of around 60 % in the rectosigmoid. Addition of lactitol to the diet reduced dehydrogenation in the caecum by 40 % and this inhibition was maintained in the mid-colon and to a lesser extent in the rectosigmoid colon. In comparison metabolism of bile acids in pigs without supplementation was fairly extensive at both the jejunal and ileal sites reflecting considerable deconjugation of bile acid amidates and 7α-dehydroxylation of primary bile acids. Metabolism was virtually complete in the caecum and reached 100 % in the rectosigmoid colon. Supplementation with lactitol also had a similar inhibitory effect on metabolism of bile acids. Dehydroxylation was reduced by over 50 % in the caecum and by about 30 % in the mid-colon, but this inhibition unlike with cholesterol did not persist in the rectosigmoid colon.

In the human study only faecal samples were analysed. Lactitol feeding reduced the concentration of faecal sterols in all eight subjects and on average this attained statistical significance (Table 4.4; $P<0.02$, Wilcoxon Signed Rank

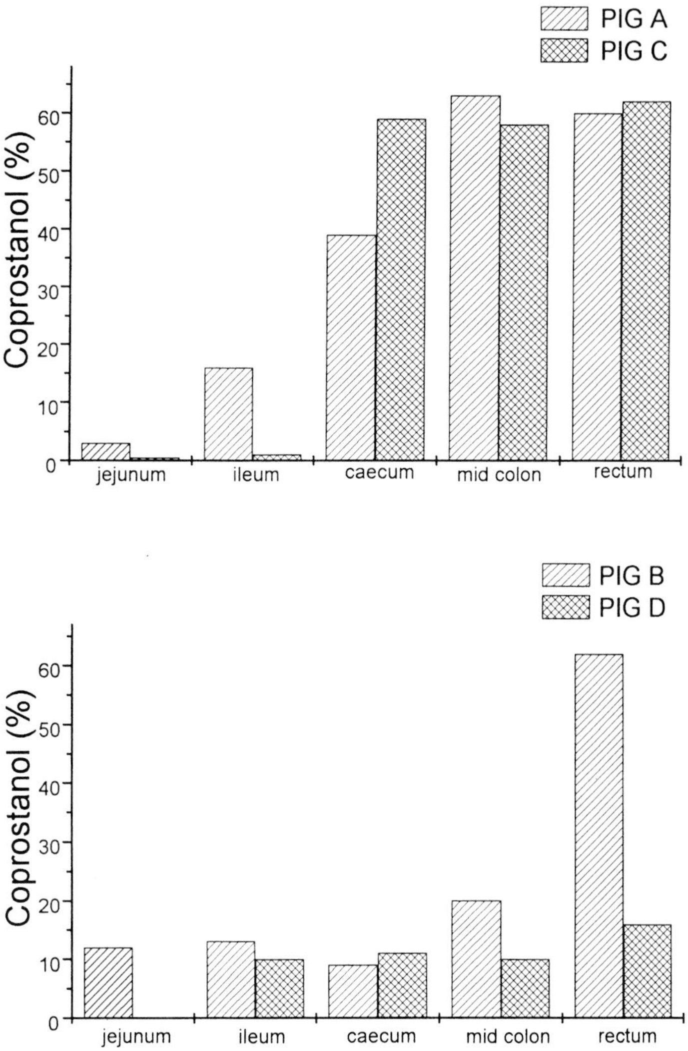

Figure 4.6: Effect of lactitol on the hydrogenation of cholesterol to coprostanol in the miniature pig. Pigs A and C: control diet; Pigs B and D: lactitol supplemented diet.

test) with an even more emphatic decrease in the concentration of bile acids (Table 4.4; lactitol 0.7 ± 0.2 (SEM) mg/g dry faeces versus controls 3.1 ± 0.6: $P < 0.02$, Wilcoxon Signed Rank test). Lactitol supplementation had no effect on the extent of 7α-dehydroxylation, however.

These data show that feeding the prebiotic lactitol to both pigs and man has a significant influence on the metabolism of steroids in the intestinal tract. Overall in the pig model lactitol had a considerable inhibitory effect on the activity of steroid enzymes with little influence on steroid concentrations at any

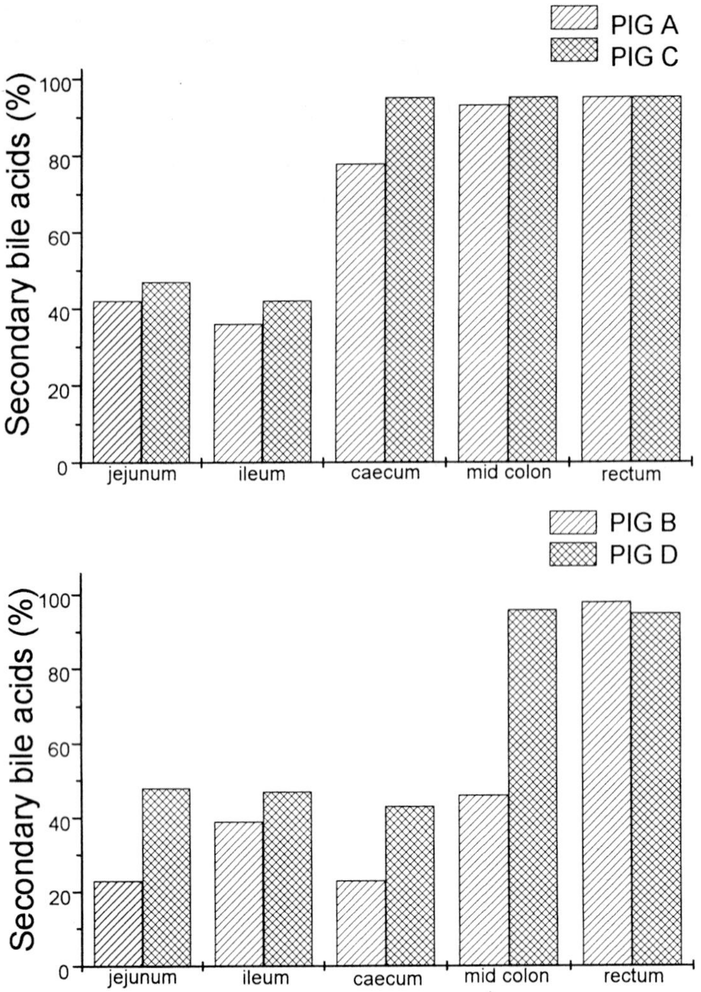

Figure 4.7: Effect of lactitol on 7α-dehydroxylation of primary to secondary bile acids in the miniature pig. Pigs A and C: control diet; Pigs B and D: lactitol supplemented diet.

site. By contrast the activity of the marker enzymes in humans was not significantly effected but a highly significant reduction in steroid concentration was elicited by lactitol administration. An automatic assumption from these data is that dietary fibre may protect against colorectal cancer via the mechanisms observed above and is supported by various animal models of carcinogenesis and metabolic epidemiology studies in vegetarian populations. The proof in humans is not definitive however.

The data from the recently completed pan-European calcium/fibre intervention study in sporadic adenoma patients (n = 665) should help to clarify this situation [66].

Table 4.4: Faecal steroids in human volunteers fed lactitol or control (sucrose) diets.

		Neutral steroids		Bile acids	
Subject	Diet	Total	Metabolites (%)	Total	Secondary (%)
1 (M)	Control	7.2	0	3.7	0
	Lactitol	6.6	0	1.0	4
2 (M)	Control	12.5	63	6.1	100
	Lactitol	9.6	53	0.4	100
3 (F)	Control	13.5	61	1.7	100
	Lactitol	10.7	71	0.1	0
4 (F)	Control	13.7	60	2.1	100
	Lactitol	10.3	25	0.7	100
5 (F)	Control	17.2	64	5.2	100
	Lactitol	9.7	29	1.7	100
6 (F)	Control	16.4	75	2.4	100
	Lactitol	9.8	73	0.2	100
7 (F)	Control	12.3	63	2.1	100
	Lactitol	7.4	7	0.7	100
8 (M)	Control	8.3	75	1.5	100
	Lactitol	7.1	24	0.5	100

Results expressed in mg/g dry weight
M – Male
F – Female

4.6 Phytic acid

Because of certain inconsistencies in the dietary fibre data a new perspective on the dietary fibre colorectal cancer relation was brought to our attention by the hypothesis of Graf and Eaton [67] who suggested that the protective effect of fibre may be due more to the phytate content in cereal fibres especially. In this keynote publication they drew on previously conducted studies of theirs which showed that phytate in less than equimolar amounts in an iron containing solution was capable of almost totally suppressing the generation of $HO^{\bullet}$ and lipid peroxidation [68, 69]. Phytic acid (PA) a hexaphosphorylated sugar is a ubiquitous plant component that may constitute 1–5 % by weight of most cereals, nuts, legumes and oil seeds [70]. Graf and Eaton [71] suggested that a major reason for this is that PA maintains iron in the Fe(III) oxidation state and obstructs generation of reactive oxygen species e. g. the hydroxyl radical ($HO^{\bullet}$)

thereby preventing oxidative damage, especially of unsaturated fatty acids which are a major component of seeds.

The influence of PA in the genesis of colorectal cancer has received substantial support from animal model studies. Nielson *et al.* [72] studied the effect of adding PA at varying concentrations to the normal diet of the rat. Phytic acid at concentrations of 1.2% and 2% significantly lowered colonic cell proliferation, an intermediate biomarker of colorectal cancer in these rats (Table 4.5). Shamsuddin and co-workers [73–75] have shown in a series of studies utilizing the azoxymethane (AOM)-rodent model that addition of PA (1–2%) to the diet significantly reduces tumor burden and volume (Table 4.6). The anticancer effect of PA was found to be dose-dependent and temporal in that it significantly reduced the incidence and colon tumor volume when administered 5 months after the last injection of AOM.

Table 4.5: Effect of phytic acid on labelling indices (LI) in the rat colon (Data of Nielson *et al.* [72]).

Dietary group	Descending colon	
	Labelled cells/crypt	Labelling index
Basal	6.0 ± 0.7^{a}	9.1 ± 1.0^{a}
0.6% phytic acid	4.8 ± 0.5^{a}	7.5 ± 0.8^{a}
1.2% phytic acid	$3.6 \pm 1.9^{a,d}$	$5.5 \pm 1.0^{a,d}$
2.0% phytic acid	$2.4 \pm 0.3^{a,c,e}$	$3.9 \pm 0.4^{a,c,e}$
0.25% cholic acid	10.5 ± 0.6	12.5 ± 0.6

[a] Significantly different from 0.25% cholic acid group (P < 0.001)
[b] Significantly different from 0.25% cholic acid group (P < 0.05)
[c] Significantly different from Basal group (P < 0.001)
[d] Significantly different from Basal group (P < 0.001)
[e] Significantly different from 0.6% PA group (P < 0.001)

Table 4.6: Effect of phytic acid on azoxymethane induced CRC in rats (Data of Shamsuddin and Ullah [73]).

Parameter	AOM only (n = 16)	AOM + PA (n = 28)	Significance level
Number of tumors/rat	7.1 ± 0.6	5.2 ± 0.6	$p < 0.02$
Tumor volume (mm^3)	570 ± 110	200 ± 60	$p < 0.01$
Tumor load/unit area	0.7 ± 0.1	0.4 ± 0.1	$p < 0.001$
Cells/crypt	56.6 ± 0.5	43.2 ± 0.5	$p < 0.001$

Thompson and Zhang [76] have shown that PA can reduce cell proliferation in colonic (Table 4.7) tissue in the mouse and furthermore subverts the promotional effect of added iron and calcium. Nelson *et al.* [77] studied the com-

Table 4.7: Effects of various diets on labelling indices (LI) in the mouse colon (Data of Thompson and Zhang [76]).

Dietary group	Colon labelling index
Low fat	$10.15 \pm 0.28^{b,c}$
High fat	10.98 ± 0.31^{b}
High fat + PA	9.03 ± 0.10^{a}
High fat + Fe	13.92 ± 0.57^{a}
High fat + Fe + PA	9.11 ± 0.39^{c}
High fat + calcium	13.49 ± 0.32^{c}
High fat + calcium + PA	9.22 ± 0.34^{c}

Means with different superscripts significantly different ($p < 0.05$)
PA = phytic acid (1.2 %); Fe = (535 ppm)
Calcium (1.5 %)

Table 4.8: Effect of iron on DMH induced CRC in the rat colon (Data of Nelson *et al.* [77]).

Group	Colon tumors/rat		Colon tumor incidence	
Fe (15 mg/kg)	2.62 ± 1.39		92 %	
Vehicle	1.40 ± 1.06	$p = 0.014$	73 %	$p = 0.21$

Parenteral iron supplementation (17 rats/group)

Group	Colon tumors/rat		Colon tumor incidence	
Basal	0.25 ± 0.55		20 %	
Fe (580 mg/kg)	0.63 ± 0.75		63 %	
Fe (580 mg/kg + PA 2.5 g/kg)	0.25 ± 0.55	$p = 0.09$	20 %	$p = 0.09$

Oral iron supplementation (20 rats/group)

bined effects of iron and PA (Table 4.8) and while the former was shown to promote 1,2-dimethylhydrazine (DMH)-induced colorectal cancer in the rat these effects were nullified by simultaneous administration of the latter.

Furthermore iron-enriched diets caused an increase of tumor rate in two models of DMH-induced colon tumorigenesis in mice [78, 79]. The effect was independent of the time the diet was fed, was dose-dependent and enhanced both tumor initiation and promotion.

To date neither experimental nor intervention trials with PA have been conducted in humans but further indirect support for the hypothesis of Graf and Eaton [67] comes from studies on iron status and risk of cancer in humans. Stevens *et al.* [80] reported, increased body iron stores were significantly associated with colon cancer in a national survey (NHANES 1) and this has been supported by a recent case control study [81] showing that adenoma risk is positively associated with elevated serum ferritin levels (Table 4.9).

Table 4.9: Quartile analysis of effect of serum ferritin on adenoma risk by location of adenoma in colon: subjects with less than 400 ng/ml (Data of Nelson *et al.* [81]).

Ferritin*: quartiles (range)	OR		
	Right colon	Left colon	Rectum
1st (0–43 ng/ml)			
2nd (44–83 ng/ml)	3.0	1.3	5.0
3rd (84–156 ng/ml)	4.7	1.9	7.0
4th (157–400 ng/ml)	6.0	2.9	7.0
Collapsed OR comparing high versus low ferritin (> or < 83 ng/ml)	2.7	2.1	2.3

* For subjects with ferritin levels <400 ng/ml, the median ferritin level = 83 ng/ml
Patients with a serum ferritin level >400 ng/ml were regarded possible genetically predisposed haemochromatosis individuals and so are omitted from the above analysis.

Many HPLC methods have been described for the analysis of PA in various foodstuffs [82–85] with varying success. Apparently the more impressive of these is that described by Sandberg and Adherrine [85] which not only separates PA but also the lower inositol phosphates (IP5, IP4, IP3 and IP2). However in our hands this method could not be reproduced and possible reasons for this are discussed in Owen *et al.* [86]. Recently however a further method has been published [87] which separates PA from IP5 and IP4. Of interest is that considerable quantities of PA and the lower inositol phosphates are detected in rapeseeds. Hopefully this method will be reproducible and of use for a more comprehensive analysis of PA and its metabolites in faeces.

Despite all the circumstantial evidence, until recently, no study had been conducted in the clinical domain to bring together the various factors of the phytic acid/reactive oxygen species hypothesis [71] and its implications for cancer. To test whether or not this hypothesis has relevance in humans a high-performance liquid chromatography (HPLC) system has been developed by Owen *et al.* [86] and utilized for the analysis of phytic acid in selected foods and human faeces.

Therefore cell proliferation rate, faecal lipid, mineral and PA content of patients with sporadic adenomatous polyps [57] have been measured to establish any interrelation which may have a bearing on the aetiology of colorectal cancer.

Analysis of phytic acid in selected foodstuffs showed a good correlation with other methods in that the highest phytic acid content of those compared was found in wheat bran and the lowest in soy beans. Phytic acid levels were also determined in foodstuffs which have not been previously reported and of these coriander, tomato and green pepper seeds were found to contain appreciable amounts (Table 4.10).

The HPLC method was also found to be appropriate for the measurement of phytic acid in human faeces. Phytic acid was detected in faecal extracts of the adenoma patients in the range 0.68–4.00 μmol/g wet faeces and 55–2038 μmol/day. Linear regression analyses of phytic acid versus faecal lipid and

Table 4.10: Phytic acid (%, dry weight) content of a range of foods.

Food sample	External standard[a]			Literature values
	Phytic acid	Sodium phytate	Sodium phytate[b]	
Sesame seeds	2.21 ± 0.02	7.83 ± 0.07	5.56 ± 0.05	5.36 ± 0.09[c]
Wheat bran	2.28 ± 0.08	8.14 ± 0.37	5.78 ± 0.22	5.03 ± 1.85[d]
Peanuts	0.86 ± 0.02	3.05 ± 0.07	2.17 ± 0.04	1.88 ± 0.05[c]
Soy beans	0.89 ± 0.06	3.09 ± 0.22	2.20 ± 0.16	1.84 ± 0.03[c]
Tomato seeds	1.24 ± 0.14	4.36 ± 0.49	3.10 ± 0.35	–
Chilli seeds	0.56 ± 0.06	1.90 ± 0.21	1.34 ± 0.15	–
Coriander seeds	1.11 ± 0.03	3.93 ± 0.10	2.79 ± 0.07	–
Pepper seeds	0.57 ± 0.01	1.96 ± 0.04	1.39 ± 0.03	–
Millett	0.18 ± 0.01	0.62 ± 0.03	0.44 ± 0.02	–

[a] Results expressed as % (mean ± SD) of 2–4 samples: lyophilised, pulverised and defatted with pentane (not corrected for defatting). [b] Sodium phytate corrected for salt content. Literature values from [c] Graf and Dintzis [83] and [d] Camire and Clydesdale [82].

mineral content and intestinal cell proliferation showed that the amount of phytic acid in the stool was strongly correlated (Figure 4.8a) with faecal iron ($r = 0.52$; $P = 0.00004$), unsaturated fatty acids ($r = 0.35$; $P = 0.004$) and total calcium content of the stool ($r = 0.34$; $P = 0.01$). The association between phytic acid and minerals was even stronger when analysed on a daily basis: phytic acid versus iron, $r = 0.76$, $P = 5.5 \times 10^{-12}$ (Figure 4.8b); phytic acid versus total calcium, $r = 0.59$, $P = 1.36 \times 10^{-6}$.

This study shows clearly for the first time that a strong association exists between the presence of phytic acid and iron in the large intestine of humans as evaluated by faecal biochemistry. High concentrations of phytic acid chelate intestinal iron thereby preventing it from partaking in the generation of ROS which may damage DNA leading to mutation and cancer. The sequestration of iron in this way may also lower absorption and prevent excessive accumulation of body iron stores which is also associated with colorectal cancer. Further population, case control and clinical trials are warranted in this area to fully evaluate the chemopreventive potential of phytic acid.

4.7 Nonsteroidal anti-inflammatory drugs

Nonsteroidal anti-inflammatory drugs (NSAIDs) are believed to mediate their effects in colorectal neoplasia by altered metabolism of arachidonic acid via inhibition of cyclooxygenase enzymes, thereby reducing the production of prosta-

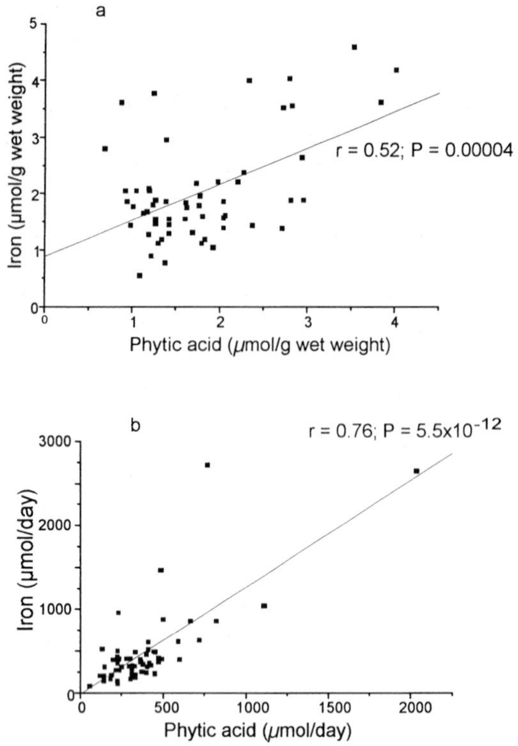

Figure 4.8a: Correlation between faecal phytic acid and iron concentration; b: Correlation between daily faecal phytic acid and iron excretion.

glandins, prostacyclin and thromboxanes. It is now known that two forms of COX enzymes exist namely COX-1 and COX-2 [88]. COX-1 is expressed in most tissues and is thought to be the product of a constitutively expressed gene [89]. It is involved in cellular homeostasis, synthesising prostaglandins in response to physiological stimuli at a rate proportional to the availability of the substrate arachidonic acid. Prostaglandins in the gastrointestinal tract have an important protective role in the maintenance of microvascular integrity, the regulation of cell division and the production of mucous. COX-2 on the other hand has been shown to be inducible [90] in nontransformed rat intestinal epithelial cells *in vitro* by growth factors and tumor promoters at sites of inflammation. Macrophages and other inflammatory cells have abundant COX-2 activity. Levels of COX-2 are also increased several fold in 90 % of human colorectal carcinomas and 40 % in colorectal adenomas and therefore the products of COX-2 may drive inflammatory processes and carcinogenesis [91]. Quite recently it has been shown that the recombinant COX-2 enzyme has a different sensitivity to NSAID inhibition compared to COX-1 [92]. Therefore it is possible that the COX enzymes have different modalities with respect to colorectal neoplasia and the development of NSAIDs which are specific inhibitors of COX-2 may be an important landmark in the already vast potential for NSAIDs in the prevention of colorectal cancer.

Despite this there is a growing literature on two NSAIDs in particular, aspirin and sulindac, which appear to be extremely promising as chemopreventive agents although they both preferentially inhibit the COX-1 enzyme over COX-2.

Epidemiologic case control and cohort studies suggest that the use of aspirin is associated with an approximately 50 % reduction in the development of colorectal cancer. The data for humans has recently been reviewed by Little [93] and in this report 3 of 4 cohort and 3 case control studies showed an inverse relation with the use of aspirin. Similar associations were also described in 3 studies of aspirin use and the incidence of colorectal polyps. In this review it was also noted however that in the one available randomised trial no protective effect of low dose aspirin (325 mg) taken every other day against invasive colorectal cancer or colorectal polyps was found. Since this publication a further prospective cohort study has been described [94]. The setting of this study was male health professionals throughout the USA, with 47,900 respondents to a mailed questionnaire. Regular users of aspirin (≥ twice/week) had a significant lower risk for total colorectal cancer (Relative Risk = 0.68; 95 % CI, 0.52–0.92) and advanced metastatic and total colorectal cancer (RR = 0.51; 95 % CI, 0.32–0.84) after controlling for confounding variables such as age, history of polyp, previous endoscopy, parenteral history of colorectal cancer, smoking, body mass, physical activity and intakes of red meat, vitamin E and alcohol. This well designed study is impressive evidence for the protective effects of aspirin against colorectal cancer.

Sulindac on the other hand unlike aspirin is not available for general prescription and has been utilised predominantly for the treatment of colorectal adenomatous polyps in patients with familial adenomatous polyposis (FAP). This is an autosomal dominant disorder the hallmark of which is the development of hundreds to thousands of colorectal polyps in adults usually below the age of 30. The pioneers of sulindac use for the regression of colorectal polyps were Waddell and Loughry [95] and since then a number of groups have reported the daily administration of 150–400 mg/day of sulindac with varying degrees of success. To date the more impressive data has been demonstrated by Winde *et al.* [96] in which by contrast to previous studies low dose sulindac maintenance therapy was applied intrarectally as opposed to the oral route.

Twenty five patients with histologically confirmed FAP were entered into the study at least 3 years after colectomy with ileorectal anastomosis. Fifteen of these served as the study group and 10 as a non-randomised control group. At entry sulindac suppositories (150 mg) were administered twice daily for 6 weeks and thereafter on visible reduction in the number of polyps by endoscopy the daily dose of sulindac was lowered sequentially. The data from this group show that the number of polyps was reduced significantly during sulindac therapy. In the study group after treatment with sulindac (300 mg/day) for 6 weeks the number of polyps (Σ 208) was reduced by 83 % (Σ 36) and thereafter by 96 % after 18 weeks (Σ 8) such that no polyps were visible after 24 weeks. The number of polyps in the control group remained relatively constant during this period (initial-Σ 99 *vs* final-Σ 92). Furthermore no relapse of polyps was observed after 132 weeks during which the average daily dose of sulindac was reduced significantly from 300 to 67 mg/day.

Clearly the administration of sulindac as a low dose suppository in the rectum of FAP patients is an unmitigated success in the eradication of polyps in these patients and further reports from this group will be eagerly awaited.

It is interesting to note the varying degrees of success between the use of sulindac in FAP patients compared to those with with sporadic polyps. To date there are only two reports for comparison. In one small study by Hixson *et al.* [97] involving 5 patients only 1 patient had partial regression of a polyp after 3 months of sulindac treatment. In the other very recent double blind placebo-controlled study the effect of sulindac (300 mg/day) for 4 months was compared in 44 patients (22 cases, 22 controls) with sporadic adenoma [98]. The data showed using an intention to treat analysis 5 patients (23 %) receiving sulindac and 3 subjects receiving placebo had complete regression of their polyps after 4 months. When this was based on adenomatous polyps 5 of 14 patients (36 %) taking sulindac and 3 of 15 (20 %) taking placebo had adenoma regression. By neither analysis were the differences significant however. The authors conclude it is unlikely that the rate of regression of polyps in sporadic adenoma patients on sulindac therapy is 50 % or more.

Taking these two studies together it is obvious that there is a paucity of data regarding the modality of sulindac for the treatment of sporadic polyps and that rectal suppositories are far more effective in FAP patients than the oral route. This may also apply in the treatment of sporadic adenoma. Therefore further studies are recommended in FAP patients to ascertain if the data of Winde *et al.* [96] can be reproduced and it would be of great interest if such a regime used by this group was also applied to patients with sporadic adenoma.

4.8 Reactive oxygen species (ROS)

Reactive oxygen species (ROS) such as superoxide anion ($O_2^{\bullet -}$), $HO^{\bullet}$ and the non-radical hydrogen peroxide (H_2O_2) are possibly involved in a range of human diseases namely athersclerosis [99], ischaema/reperfusion injury [100], rheumatoid arthritis [101], inflammatory bowel disease [102, Table 4.6, 103], oral cancer [104, 105] and colorectal cancer [106, Table 4.11 a, b]. Of the ROS spectrum $HO^{\bullet}$ appears to be particularly dangerous due to its rapid and non-specific reactivity with most biomolecules.

The HPLC method developed for phytic acid analyses was adapted for the determination of ROS [107] and proved to be ideal for monitoring the dynamics of both the hypoxanthine/xanthine oxidase system and the hydroxylation of salicylic acid (SA) by $HO^{\bullet}$. Utilising an aqueous mobile phase (pH 4.3) containing the ion-pair reagent tetrabutylammonium hydroxide (0.005 M) baseline separa-

Table 4.11 a: ROS production in colon cancer tissues (7 patients). Data of Keshavarzian *et al.* [103].

Tissue	Luminol amplified chemoluminescence (photons · mg^{-1} · min · 10^{-3})	Catalase inhibition		
		2 μg/ml	4 μg/ml	8 μg/ml
non-cancerous	2175 ± 1111	–74 %	–85 %	–71 %
cancerous	4808 ± 2282*	–11 %	–61 %	–53 %

* $p < 0.05$

Table 4.11 b: ROS production in rectal tissue from adenoma and cancer patients. Data of Keshavarzian *et al.* [106].

Controls (n = 20)		Adenoma patients (n = 23)		Cancer (n = 14)	
without	with	without	with	without	with
Indomethacin		Indomethacin		Indomethacin	
883 ± 106	327 ± 34*	2040 ± 339	1023 ± 108*	2711 ± 430	1364 ± 188**

* $p < 0.05$; ** $p < 0.01$

tion of hypoxanthine (9.0 min), and its hydroxylated products xanthine (10.3 min) and uric acid (12.4 min) is effected (Figure 4.9 a). Furthermore the incorporation of methanol (40 %) in this mobile phase facilitates the separation (Figure 4.9 b) of 2,5-DHBA (4.2 min) and 2,3-DHBA (4.6 min) from SA (7.00 min). Under normal conditions (complete buffer system) xanthine oxidase hydroxylated hypoxanthine completely via xanthine to uric acid and in the presence of an excess of SA (2 mM) produced hydroxylated SA (diphenol). The major product of $HO^{\bullet}$ attack was 2,5-DHBA in contrast to the report of Richmond *et al.* [108] who quoted 2,3-DHBA as the major product. The hydroxylated products consisted of 2,5-DHBA to 2,3-DHBA in a ratio of approximately 5 : 1 and is consistent with the reports of Kettle and Winterbourn [109] and Coudray *et al.* [110].

The general behaviour of the system and tests against the various classical inhibitors/scavengers required for validation of ROS generation are given in Table 4.12 and Figures 4.10 a, b respectively.

Because phytic acid has been mooted as an important constituent of dietary fibre with regard to its protective effect against colorectal cancer this hexaphosphorylated sugar was also tested in the system. Graf and Eaton [71] have demonstrated using the detection of formaldehyde as an end-point of $HO^{\bullet}$ attack on DMSO in the hypoxanthine/xanthine oxidase system that phytic acid significantly reduces the generation of ROS. While free-radical scavengers in the mM range were required to inhibit $HO^{\bullet}$ attack on SA in the current system, the concentration of free phytic acid and sodium phytate required was an order of magnitude less in an EDTA-deplete system: phytic acid and sodium phytate in the range 50–2000 μM reduced diphenol production in a dose-dependent

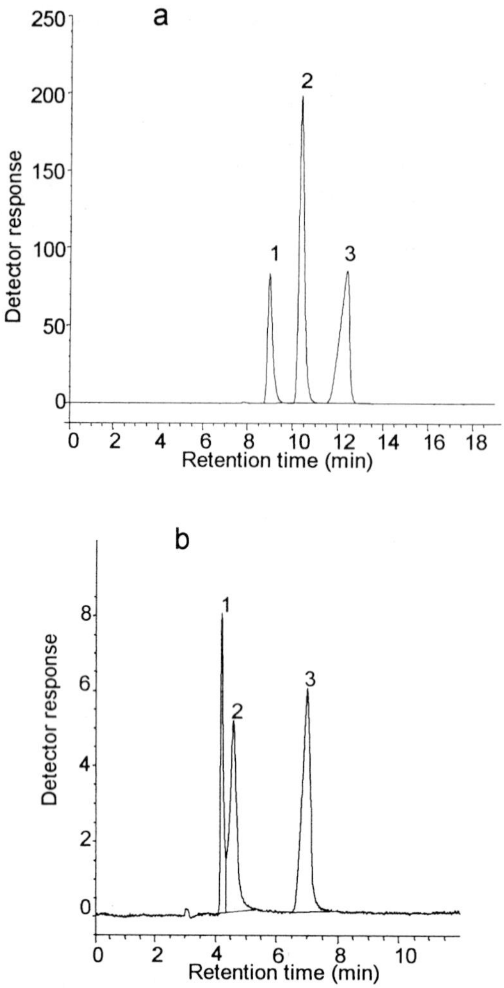

Figure 4.9a: High performance liquid chromatogram of hypoxanthine (1) and its hydroxylated metabolites xanthine (2) and uric acid (3). b: High performance liquid chromatogram of salicylic acid (3) and its hydroxylated products 2,5-dihydroxy benzoic acid (1) and 2,3-dihydroxy benzoic acid (2).

manner (Figure 4.10c). In the complete system containing the strong iron chelator EDTA however, neither phytic acid nor sodium phytate had any discernible effect on diphenol production.

The only description of the faecal matrix being involved in the generation of ROS is that of Babbs [111] who forwarded the hypothesis that this phenomenon may be important in the genesis of colorectal cancer. Babbs showed using DMSO as a probe that 1:100 dilutions of rat faeces generated abundant amounts of ROS (1700 nM/g) and concluded since autoclaved faeces were ineffective that this was the end product of bacterial metabolism. This was a very

Table 4.12: Hydroxylation of salicylic acid by a hypoxanthine/xanthine oxidase system as determined by HPLC.

Buffer system	Diphenol produced (μmol/h)[a]			Inhibition (%)
	2,5-DHBA	2,3-DHBA	Total diphenol	
Complete (Iron III)	8.84 ± 0.01	1.83 ± 0.03	10.66 ± 0.02	0
Complete(Iron II)	7.45 ± 0.37	0	7.45 ± 0.37	30
EDTA omitted	4.81 ± 0.20	0	4.81 ± 0.20	55
Iron omitted	3.28 ± 0.14	0	3.28 ± 0.14	69
Amino acids*	3.76 ± 0.32	0	3.76 ± 0.32	65
Urea (100 mM)	9.03 ± 0.31	1.42 ± 0.01	10.45 ± 0.33	0
DTPA (1 mM)	9.39 ± 0.24	1.32 ± 0.12	10.71 ± 0.36	0
Hypoxanthine omitted	0	0	0	100
Salicylic acid omitted	0	0	0	100
Xanthine oxidase omitted	0	0	0	100

[a] Results expressed as mean ± SEM (μM diphenol/h) of duplicate experiments at 37 °C.
* EDTA substituted by aspartic and glutamic acids (2:7; 500 μM).
DTPA = diethylenetriaminopentaacetic acid.

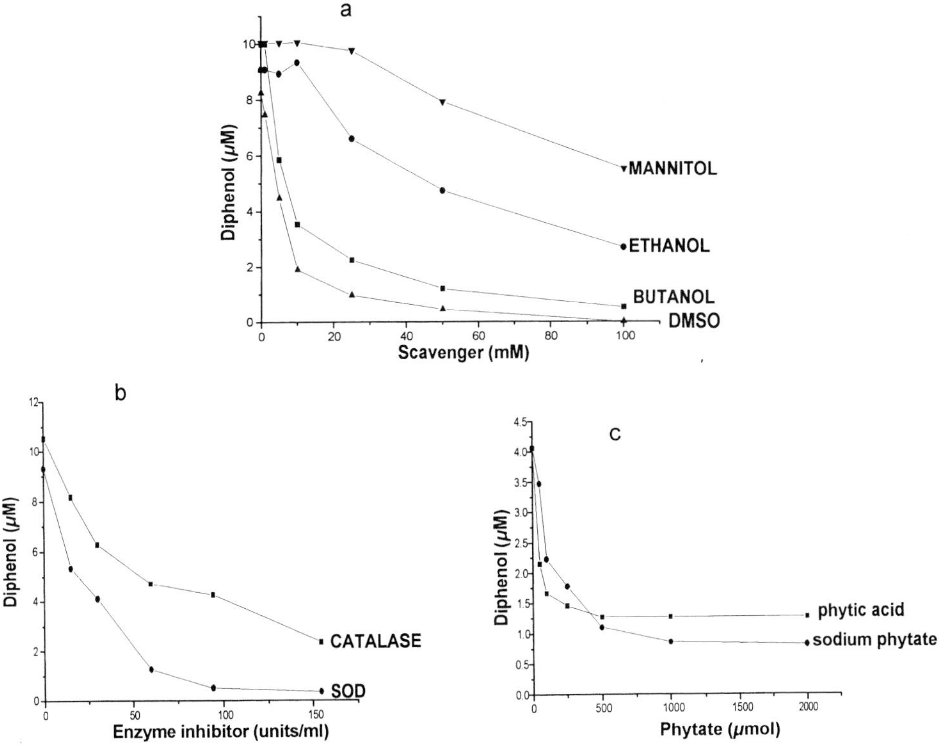

Figure 4.10a: Attenuation of reactive oxygen species detection by scavengers. b: Inhibition of reactive oxygen species generation by enzymes. c: Attenuation of reactive oxygen species detection by phytates.

interesting discovery but the method used to detect ROS i.e. production of methyl sulphinic acid from DMSO [112] and reaction of this with fast blue and subsequent colourimetric determination is rather laborious. Furthermore while the method works reasonably well in an *in vitro* system in our hands this was not the case when working with the faecal matrix.

Therefore we applied the HPLC method described for the detection of $HO^{\bullet}$ production in human faeces. The data show (Table 4.13) that faecal samples (n = 58) from adenoma patients incubated under aerobic conditions are capable of producing considerable quantities of $HO^{\bullet}$ (92 ± 40 μM/g wet faeces/18 h) as evinced by the production of diphenols from SA. This value is substantially higher than that reported by Babbs [111] for faecal dilutions of rat faeces (1700 nM/g wet faeces/16 h) treated in a similar fashion but assayed by a different technique. The production of ROS in these faecal samples however did not show any correlation with the other parameters assayed (Table 4.13).

Table 4.13: Comparison of faecal phytate, iron and cell proliferation in relation to ROS generation in faecal samples.

Parameter	Value	Correlation with ROS
PA (μMol/g)	1.78 ± 0.75	r = 0.02; P = 0.88
Fe (μMol/g)	2.11 ± 1.13	r = 0.03; P = 0.80
PA/Fe	0.97 ± 0.43	r = −0.01; P = 0.92
LI (%)	11.86 ± 3.99	r = 0.17; P = 0.19
ROS (μMol/g)	92.28 ± 40.39	–

[a] Results expressed as mean ± SD (n = 58) of duplicate experiments.

The production of ROS by the faecal matrix was initially attributed to the aerobic bacteria present but assay of the isolated major genera present gave no indication of ROS generation. This was confirmed when faecal suspensions were either autoclaved or filter sterilised. Activity although a little lower in these cases was still abundant. Studies are in progress to identify the nature of this soluble component [113] and to evaluate ROS production in more realistic oxygen-depleted environments.

4.9 Conclusions

The conclusions that can be drawn from the subject content of this chapter which spans almost 30 years of research is that the mechanisms related to nutritional factors which lead to colorectal cancer are still somewhat of a mystery.

The early work which focused on macrocomponents of the diet namely fat and fibre, although lacking support from actual proven mechanisms may still be relevant in terms of general recommendations to the public, in that a significant reduction in colorectal cancer incidence would probably accrue, if the content of the diet was reduced to no more than 30 % fat and fibre (cereals especially) intake increased to at least 30 g/day.

Although steroid metabolites have been implicated in the aetiology of colorectal cancer for a considerable time it must now be regarded that they do not have a major influence on neoplasia of the intestinal mucosa. To date the concepts have been rather too simplified and definitive mechanisms are sadly lacking. Hopefully faster progress in this very important area of cancer research can be made in the near future and the relation between nutritional factors, reactive oxygen species generation and colorectal neoplasia studied in far greater depth.

References

[1] Armstrong, B.; Doll, R. (1975) Environmental factors and cancer incidence and mortality in different countries, with special reference to dietary practices. *Int. J. Cancer* **15**, 617–631.

[2] Willett, W. C.; Stampfer, M. J.; Colditz, G. A.; Rosner, B. A.; Speizer, F. E. (1990) Relation of meat, fat and fiber intake to the risk of colon cancer in a prospective study among women. *New Engl. J. Med.* **323**, 1664–1672.

[3] Giovanucci, E.; Rimm, E. B.; Stampfer, M. J.; Colditz, G. A.; Ascherio, A.; Willett, W. C. (1994) Intake of fat, meat and fiber in relation to risk of colon cancer in men. *Cancer Res.* **54**, 2390–2397.

[4] Hill, M. J.; Drasar, B. S.; Williams, R. E. O.; Meade, T. W.; Cox, A. G.; Simpson, J. E. P.; Morson, B. C. (1975) Faecal bile acids and clostridia in patients with cancer of the large bowel. *Lancet* **1**, 535–539.

[5] Reddy, B. S.; Wynder, E. L. (1977) Metabolic epidemiology of colon cancer: faecal bile acids and neutral sterols in colon cancer patients and patients with adenomatous polyps. *Cancer* **39**, 2533–2539.

[6] Newmark, H. L.; Wargovich, M. J.; Bruce, W. R. (1984) Colon cancer, dietary fat, phosphate and calcium: a hypothesis. *J. Natl. Cancer Inst.* **72**, 1323–1325.

[7] Owen, R. W.; Thompson, M. H.; Hill, M. J. (1984) Analysis of metabolic profiles of steroids in faeces of healthy subjects undergoing chenodeoxycholic acid treatment by liquid-gel chromatography and gas-liquid chromatography-mass spectrometry. *J. Steroid. Biochem.* **21**, 593–600.

[8] Setchell, K. D. R.; Lawson, A. M.; Tanida, N.; Sjövall, J. (1983) General methods for the analysis of metabolic profiles of bile acids and related compounds in feces. *J. Lipid Res.* **24**, 1085–1100.

[9] Costa, N. M.; Low, A. G.; Walker, A. F.; Owen, R. W.; Englyst, H. N. (1994) Effect of baked beans (*Phaseolus vulgaris*) on steroid metabolism and non-starch polysaccharide output of hypercholesterolaemic pigs with or without an ileorectal anastomosis. *Br. J. Nutr.* **7**, 871–886.

[10] Korpela, J. T.; Fotsis, T.; Adlercreutz, H. (1986) Multicomponent analysis of bile acids in faeces by anion exchange and capillary column gas-liquid chromatography: application in oxytetracycline treated subjects. *J. Steroid Biochem.* **25**, 277–284.

[11] Owen, R. W.; Henly, P. J.; Thomson, M. H.; Hill, M. J. (1986) Steroids and cancer: faecal bile acid screening for early detection of cancer risk. *J. Steroid Biochem.* **24**, 391–394.

[12] Owen, R. W.; Dodo, M.; Thompson, M. H.; Hill, M. J. (1987) Faecal steroids and colorectal cancer. *Nutr. Cancer* **9**, 73–80.

[13] Owen, R. W.; Day, D. W.; Thompson, M. H. (1992) Faecal steroids and colorectal cancer: steroid profiles in subjects with adenomatous polyps of the large bowel. *Eur. J. Cancer Prev.* **1**, 105–112.

[14] Tanida, N.; Hikasa, Y.; Shimoyama, T.; Setchell, K. D. R. (1984) Comparison of faecal bile acid profiles between patients with adenomatous polyps of the large bowel and healthy subjects in Japan. *Gut* **25**, 824–832.

[15] Hikasa, Y.; Tanida, M.; Ohno, T.; Shimoyama, T. (1984) Faecal bile acid profiles in patients with large bowel cancer in Japan. *Gut* **25**, 833–838.

[16] Imray, C. H.; Radley, S.; Davis, A.; Barker, G.; Hendrickse, C. W.; Donovan, I. A.; Lawson, A. M.; Baker, P. R.; Neoptolemos, J. P. (1992) Faecal unconjugated bile acids in patients with colorectal cancer or polyps. *Gut* **33**, 1239–1245.

[17] Barker, G. M.; Radley, S.; Davis, A.; Imray, C. H. E.; Setchell, K. D. R.; O'Connell, N.; Donovan, I. A.; Keighley, M. R. B.; Neoptolemus, J. P. (1994) Unconjugated faecal bile acids in familial adenomatous polyposis analysed by gas-liquid chromatography and mass spectrometry. *Br. J. Surg.* **81**, 739–742.

[18] Spiller, G. A. (Ed.) (1986) *CRC handbook of dietary fiber in human nutrition.* FL: colorectal cancer Press, Boca Raton, p. 211–280.

[19] Phillips, S. F.; Pemberton, J. H.; Shorter, R. G. (Eds.) (1991) *The large intestine: physiology, pathophysiology and disease.* Raven, New York, p. 52–92.

[20] Burkitt, D. P. (1971) Epidemiology of cancer of the colon and rectum. *Cancer* **28**, 3–13.

[21] Howe, G. R.; Benito, E.; Castelleto, R.; Cornee, J.; Esteve, J.; Gallagher, R. P. *et al.* (1992) Dietary intake of fiber and decreased risk of cancers of the colon and rectum: evidence from the combined analysis of 13 case-control studies. *J. Natl. Cancer Inst.* **84**, 1887–1896.

[22] Cummings, J. H.; Bingham, S. A.; Heaton, K. W.; Eastwood, M. A. (1992) Faecal weight, colon cancer risk and dietary intake of nonstarch polysaccharides (dietary fiber). *Gastroenterology* **103**, 1783–1789.

[23] Hill, M. J.; Giacosa, A.; Beckly, D.; Caygill, C. P. J.; Chaves, P.; Evans, D.; Faivre, J.; Farinati, F.; Franceschi, S.; Gassull, M.; Gerber, M.; Johnson, I. T.; Kritchevsky, D.; La Vecchia, C.; Mainguet, P.; Maskens, A.; Owen, R. W.; Rafter, J.; Rowland, I. N.; Southgate, D.; Stockbrugger, R. (1997) Consensus meeting on cereals, fibre and colorectal and breast cancers. *Eur. J. Cancer Prev.* **6**, 512–514.

[24] Nair, J.; Vaca, C. E.; Velic, I.; Mutanen, M.; Valsta, L. M.; Bartsch, H. (1997) High dietary ω-6 polyunsaturated fatty acids drastically increase the formation of etheno-DNA base adducts in white blood cells of female subjects. *Cancer Epidemiol. Biomarkers Prev.* **6**, 597–601.

[25] Hill, M. J.; Drasar B. S.; Aries, V. C.; Crowther, J. S.; Hawksworth, G. M.; Williams, R. E. O. (1971) Bacteria and the aetiology of large bowel cancer. *Lancet* **1**, 95–100.

[26] Reddy, B. S.; Wynder, E. L. (1973) Etiology of cancer of the colon. *J. Natl. Cancer Inst.* **50**, 1437–1442.

[27] Aries, V. C.; Hill, M. J. (1970) Degradation of steroids by intestinal bacteria II. Enzymes catalysing the oxido-reduction of the 3α-, 7α- and 12α-hydroxyl groups in cholic acid and dehydroxylation of the 7-hydroxyl group. *Biochim. Biophys. Acta* **202**, 535–543.

[28] Aries, V. C.; Hill, M. J. (1970) The formation of unsaturated bile acids by intestinal bacteria. *Biochem. J.* **119**, 57P.

[29] Goddard, P.; Hill, M. J. (1973) The dehydrogenation of the steroid nucleus by human gut bacteria. *Biochem. Soc. Trans.* **1**, 1113–1115.

[30] Hill, M. J. (1975) The role of the colon anaerobes in the metabolism of bile acids and steroids and its relation to colon cancer. *Cancer* **36**, 2387–2400.

[31] Silverman, S. J.; Andrews, A. W. (1977) Bile acids: co-mutagenic activity in the salmonella-mammalian microsome mutagenicity test. *J. Natl. Cancer Inst.* 59, 1557–1559.

[32] Wilpart, M.; Mainguet, P.; Maskens, A.; Roberfroid, M. (1983) Mutagenicity of 1,2-dimethylhydrazine towards *Salmonella typhimurium*: co-mutagenic effect of secondary bile acids. *Carcinogenesis* **6**, 45–48.

[33] Narisawa, T.; Magadia, N. E.; Weisberger, J. H.; Wynder, E. L. (1974) Promoting effects of bile acids on colon carcinogenesis after intrarectal instillation of *N*-methyl-N^1-nitrosoguanidine in rats. *J. Natl. Cancer Inst.* **53**, 1093–1097.

[34] Galloway, D. J.; Owen, R. W.; Jarrett, F.; Boyle, P.; Hill, M. J.; George, W. D. (1986) Experimental colorectal cancer: the relationship of diet and faecal bile acid concentration to tumour induction. *Br. J. Surg.* **73**, 233–237.

[35] Moskowitz, M.; White, C.; Barnett, R. N.; Stevens, S.; Russell, E.; Vargo, D.; Floch, M. H. (1979) Diet, faecal bile acids and neutral sterols in carcinoma of the colon. *Dig. Dis. Sci.* **24**, 746–751.

[36] Mudd, D. G.; McKelvey, S. T. D.; Norwood, W.; Elmore, D. T.; Roy, A. D. (1980) Faecal bile acid concentrations of patients with carcinoma or increased risk of carcinoma in the large bowel. *Gut* **21**, 587–590.

[37] Murray, W. R.; Blackwood, A.; Trotter, J. M.; Calman, K. C.; Mackay, C. (1980) Faecal bile acids and clostridia in the aetiology of colorectal cancer. *Br. J. Cancer* **41**, 923–928.

[38] Owen, R. W. (1997) Faecal steroids and colorectal carcinogenesis. *Scand. J. Gastroenterol.* **32**, Suppl. 222, 76–82.

[39] Garland, C. F.; Barrett-Connor, E.; Rossof, A. H.; Shekelle, R. B.; Criqui, M. H.; Paul, O. (1985) Dietary vitamin-D and calcium and risk of colorectal cancer: a 19-year prospective study in men. *Lancet* **1**, 307–309.

[40] Sorenson, A. W.; Slattery, M. L.; Ford, M. H. (1988) Calcium and colon cancer: a review. *Nutr Cancer* **11**, 135–145.

[41] Wargovich, M. J.; Eng, V. W. S.; Newmark, H. L.; Bruce, W. R. (1983) Calcium ameliorates the toxic effect of deoxycholic acid on colonic epithelium. *Carcinogenesis* **4**, 1205–1207.

[42] Wargovich, M. J.; Eng, V. W. S.; Newmark, H. L. (1984) Calcium inhibits the damaging and compensatory proliferation effects of fatty acids on mouse colon epithelium. *Cancer Lett.* **23**, 253–258.

[43] Bird, R. P.; Schneider, R.; Stamp, D.; Bruce, W. R. (1986) Effect of dietary calcium and cholic acid on the proliferative indices of murine colonic epithelium. *Carcinogenesis* **7**, 1657–1661.

[44] Rafter, J. J.; Eng, V. W. S.; Furrer, R.; Medline, A.; Bruce, W. R. (1986) Effects of calcium and pH on the mucosal damage produced by deoxycholic acid in the rat colon. *Gut* **27**, 1320–1329.

[45] Lipkin, M.; Friedman, E.; Winawer, S. J.; Newmark, H. (1989) Colonic epithelial cell proliferation in responders and nonresponders to supplementary dietary calcium. *Cancer Res.* **29**, 248–254.

[46] Rozen, P.; Fireman, Z.; Fine, N.; Wax, Y.; Ron, E. (1989) Oral calcium suppresses increased rectal proliferation of persons at risk of colorectal cancer. *Gut* **30**, 650–655.

[47] Steinbach, G.; Lupton, J.; Reddy, B. S.; Kral, J. G.; Holt, P. R. (1994) Effect of calcium supplementation on rectal epithelial hyperproliferation in intestinal bypass subjects. *Gastroenterology* **106**, 1162–1167.

[48] Gregoire, R. C.; Stern, H. S.; Yeung, K. S.; Stadler, J.; Langley, S.; Furrer, R.; Bruce, W. R. (1989) Effect of calcium supplementation on mucosal cell proliferation in high risk patients for colon cancer. *Gut* **30**, 376–382

[49] Stern, H. S.; Gregoire, R. C.; Kashtan, H.; Stadler, J.; Bruce, W. R. (1990) Long-term effects of dietary calcium on risk markers for colon cancer in patients with familial polyposis. *Surgery* **108**, 528–533.

[50] Kleibeuker, J. H.; Welberg, J. W. M.; Mulder, N. H.; Van der Meer, R.; Cats, A.; Limburg, A. J.; Kreumer, W. M. T.; Hardonk, M. J.; De Vries, E. G. (1993) Epithelial cell proliferation in the sigmoid colon of patients with adenomatous polyps increases during oral calcium supplementation. *Br. J. Cancer* **67**, 500–503.

[51] Van der Meer, R.; De Vries, H. T. (1985) Differential binding of glycine- and taurine-conjugated bile acids to insoluble calcium phosphate. *Biochem. J.* **229**, 265–268.

[52] Govers, M. J.; Van der Meer, R. (1993) Effects of dietary calcium and phosphate on the intestinal interactions between calcium, phosphate, fatty acids and bile acids. *Gut* **34**, 365–370.

[53] Lapre, J. A.; De Vries, H. T.; Termont, D. S.; Kleibeuker, J. H.; DeVries, E. G.; Van der Meer, R. (1993) Mechanism of the protective effect of supplementary dietary calcium on cytolytic activity of fecal water. *Cancer Res.* **53**, 248–253.

[54] Welberg, J. W. M.; Kleibeuker, J. H.; Van der Meer, R.; Kuipers, F.; Cats, A.; Van-Rijsbergen, H.; Termont, D. M. S. L.; Boersma Van, E. W.; Vonk, R. J.; Mulder, N. H.; De Vries, E. G. E. (1993) Effects of oral calcium supplementation on intestinal bile acids and cytolytic activity of fecal water in patients with adenomatous polyps of the colon. *Eur. J. Clin. Invest.* **23**, 63–68.

[55] Van der Meer, R.; Welberg, J. W. M.; Kuipers, F.; Kleibeuker, J. H.; Mulder, N. H.; Termont, D. S. M. L.; Vonk, R. J.; De Vries, H. T.; De Vries, E. G. E. (1990) Effects of supplementary dietary calcium on the intestinal association of calcium, phosphate and bile acids. *Gastroenterology* **99**, 1653–1659.

[56] Owen, R. W.; Weisgerber, U. M.; Carr, J.; Harrison, M. H. (1995) The analysis of calcium-lipid complexes in faeces. *Eur. J. Cancer Prev.* **4**, 247–255.

[57] Weisgerber, U. M.; Boeing, H.; Owen, R. W.; Raedsch, R.; Wahrendorf, J. (1996). Effect of long-term placebo-controlled calcium supplementation on sigmoidal cell proliferation in patients with sporadic adenomatous polyps. *Gut* **38**, 396–402.

[58] Risio, M.; Lipkin, M.; Candelaresi, G.; Bertone, A.; Coverlizza, S.; Rossini, F. P. (1991) Correlations between rectal mucosa cell proliferation and the clinical and pathological features of nonfamilial neoplasia of the large intestine. *Cancer Res.* **51**, 1917–1921.

[59] Whitfield, J. F. (1992) Calcium signals and cancer. *Crit. Rev. Oncol.* **3**, 55–90.

[60] Hofstad, B.; Almennigen, K.; Vatn, M.; Norheim-Anderson, S.; Owen, R. W.; Larsen, S.; Osnes, M. (1998) Growth of colorectal polyps: effect of antioxidants and calcium on growth and new polyp formation. *Digestion* **59**, 148–156.

[61] Bown, R. L.; Gibson, J. A.; Sladen, G. E.; Hicks, B.; Dawson, A. M. (1974) Effects of lactulose and other laxatives on ileal and colonic pH as measured by a radiotelemetory device. *Gut* **15**, 999–1004.

[62] Thornton, J. R. (1981) High colonic pH promotes colorectal cancer. *Lancet* **1**, 1081–1083.

[63] Fadden, K.; Hill, M. J.; Owen, R. W. (1997) The effect of fibre on bile acid metabolism by human faecal bacteria in batch and continuous culture. *Eur. J. Cancer Prev.* **6**, 175–194.

[64] Fadden, K.; Owen, R. W. (1992) Faecal steroids and colorectal cancer: the effect of lactulose on faecal bacterial metabolism in a continuous culture model of the large intestine. *Eur. J. Cancer Prev.* **1**, 113–27.

[65] Felix, Y. F.; Hudson, M. J.; Owen, R. W.; Radcliffe, B.; van Es, A. J. H.; van Velthuijsen, J. A.; Hill, M. J. (1993) The effect of dietary lactitol on the composition and me-

tabolic activity of the intestinal microflora in the pig and in humans. *Microbial Ecology in Health and Disease* **3**, 259–267.
[66] Faivre, J.; Couillault, C.; Kronborg, O.; Rath, U.; Giacosa, A.; De-Oliveira, H.; Obrador, T.; O'Morain, C.; Buset, M.; Crespon, B.; Fenger, K.; Justum, A. M.; Kerr, G.; Legoux, J. L.; Marks, C.; Matek, W.; Owen, R. W.; Paillot, B.; Piard, F.; Pienkowski, P.; Pignatelli, M.; Prada, A.; Pujol, J.; Richter, F.; Seitz, J. F.; Sturnolio, G. C.; Zambelli, A.; Andreatta, R. (1997) Chemoprevention of metachronous adenoma of the large bowel: design and interim results of a randomized trial of calcium and fibre. *Eur. J. Cancer Prev.* **6**, 132–138.
[67] Graf, E.; Eaton, J. W. (1985) Dietary suppression of colonic cancer. Fiber or phytate? *Cancer* **56**, 717–718.
[68] Graf, E.; Mahoney, J. R.; Bryant, R. G.; Eaton, J. W. (1984) Iron-catalysed hydroxyl radical formation: stringent requirement for iron coordination site. *J. Biol. Chem.* **259**, 3620–3624.
[69] Graf, E. (1986) Chemistry and applications of phytic acid: an overview. In: Graf, E. (Ed.) *Phytic acid: chemistry and applications.* Pilatus Press, Minneapolis, p. 1–21.
[70] Maga, J. A. (1982) Phytate; its chemistry, occurrence, food interactions, nutritional significance and methods of analysis. *J. Agric. Food Chem.* **30**, 1–9.
[71] Graf, E.; Eaton, J. W. (1990) Antioxidant functions of phytic acid. *Free Radical Biol. Med.* **8**, 61–69.
[72] Nielson, B. K.; Thompson, L. U.; Bird, R. P. (1987) Effect of phytic acid on colonic cell proliferation. *Cancer Lett.* **37**, 317–325.
[73] Shamsuddin, A. M.; Elsayed, A. M.; Ullah, A. (1988) Suppression of large intestinal cancer in F344 rats by inositol hexaphosphate. *Carcinogenesis* **9**, 577–580.
[74] Shamsuddin, A. M.; Ullah, A. (1989) Inositol hexaphosphate inhibits large intestinal cancer in F344 rats 5 months after induction by azoxymethane. *Carcinogenesis* **10**, 625–626.
[75] Ullah, A.; Shamsuddin, A. M. (1990) Dose-dependent inhibition of large intestinal cancer by inositol hexaphosphate in F344 rats. *Carcinogenesis* **11**, 2219–2222.
[76] Thompson, L. U.; Zhang, L. (1991) Phytic acid and minerals: effect on early markers of risk for mammary and colon carcinogenesis. *Carcinogenesis* **11**, 2041–2045.
[77] Nelson, R. L.; Yoo, S. J.; Tanure, J. C.; Andrianopoulos, S.; Misumi, A. (1989) The effect of iron on experimental colorectal carcinogenesis. *Anticancer Res.* **9**, 1477–1482.
[78] Siegers, C.-P.; Bumann, D.; Baretton, G.; Younes, M. (1988). Dietary iron enhances the tumor rate in dimethylhydrazine-induced colon carcinogenesis in mice. *Cancer Lett.* **41**, 251–256.
[79] Siegers, C.-P.; Bumann, D.; Trepkan, H. D.; Schadwinkel, B.; Baretton, G. (1992) Influence of dietary iron overload on cell proliferation and intestinal tumorigenesis in mice *Cancer Lett.* **65**, 245–249.
[80] Stevens, R. G.; Jones, D. Y.; Micozzi, M. S.; Taylor, P. R. (1988) Body iron stores and risk for cancer. *New Engl. J. Med.* **319**, 1047–1052.
[81] Nelson, R. L.; Davis, F. G.; Satter, E.; Sobin, L. H.; Kikendall, W.; Bowen, P. (1994) Body iron stores and risk of colonic neoplasia. *J. Natl. Cancer Inst.* **86**, 455–460.
[82] Camire, A. L.; Clydesdale, F. M. (1982) Analysis of phytic acid in foods by HPLC. *J. Food Sci.* **47**, 575–578.
[83] Graf, E.; Dintzis, F. R. (1982) High performance liquid chromatographic method for the determination of phytate. *Anal. Biochem.* **119**, 413–417.
[84] Lee, K.; Abendroth, J. A. (1983) High performance liquid chromatographic determination of phytic acid in foods. *J. Food Sci.* **48**, 1344–1351.
[85] Sandberg, A.-S.; Ahderrine, R. (1986) HPLC method for determination of inositol tri-, tetra-, penta- and hexaphosphates in foods and intestinal contents. *J. Food Sci.* **51**, 547–550.

[86] Owen, R. W.; Weisgerber, U. M.; Spiegelhalder, B.; Bartsch, H. (1996) Faecal phytic acid and its relation to other putative markers of risk for colorectal cancer. *Gut* **38**, 591–597.
[87] Matthäus, B. (1997) Phytic acid in rapeseeds. *GIT Laborzeitschrift* **5**, 448–450.
[88] Marnett, L. J. (1992) Aspirin and the potential role of prostaglandins in colon cancer. *Cancer Res.* **52**, 5575–5589.
[89] O'Neil, G. P.; Ford Hutchinson, A. W. (1993) Expression of *m*RNA for cyclooxygenase-1 and cyclooxygenase-2 in human tissues. *FEBS Lett.* **330**, 156–160.
[90] Dubois, R. N.; Awad, J.; Morrow, J.; Roberts, L. J.; Bishop, P. R. (1994) Regulation of eicosanoid production and mitogenesis in rat intestinal epithelial cells by transforming growth factor-α and phorbol ester. *J. Clin. Invest.* **93**, 493–498.
[91] Eberhart, C. E.; Coffey, R. J.; Radhika, A.; Giardiello, F. M.; Ferrenbach, S.; Dubois, R. N. (1994) Upregulation of cyclooxygenase-2 gene expression in human colorectal adenomas and adenocarcinomas. *Gastroenterology* **107**, 1183–1188.
[92] Meade, E. A.; Smith, W. L.; DeWitt, D. L. (1993) Differential endoperoxide synthase (cyclooxygenase) isozymes by aspirin and other non-steroidal anti-inflammatory drugs. *J. Biol. Chem.* **268**, 6610–6614.
[93] Little, J. (1994) Aspirin and other nonsteroidal anti-inflammatory drugs and colorectal neoplasia. *Int. J. Oncol.* **5**, 151–1162.
[94] Giovanucci, E.; Rimm, E. B.; Stampfer, M. J.; Colditz, G. A.; Ascherio, A.; Willett, W. C. (1994) Aspirin use and the risk of colorectal cancer and adenoma in male healthy professionals. *Ann. Intern. Med.* **121**, 241–246.
[95] Waddell, W. R.; Loughry, R. W. (1983) Sulindac for polyposis of the colon. *J. Surg. Oncol.* **24**, 83–87.
[96] Winde, G.; Schmid, K. W.; Schlegel, W.; Fischer, R.; Osswald, H.; Bünte, H. (1995) Complete reversion and prevention of rectal adenomas in colectomized patients with familial adenomatous polyposis by rectal low-dose sulindac maintenance therapy. *Dis. Colon Rectum* **38**, 813–830.
[97] Hixson, L. J.; Earnest, D. L.; Fennerty, M. B.; Sampliner, R. E. (1993) NSAID effect on sporadic colon polyps. *Am. J. Gastroenterol.* **10**, 1647–1649.
[98] Ladenheim, J.; Garcia, G.; Titzer, D.; Herzenberg, H.; Lavori, P.; Edson, R.; Omary, M. B. (1995) Effect of sulindac on sporadic colonic polyps. *Gastroenterology* **108**, 1083–1087.
[99] Halliwell, B. (1989) Current status review: free radicals, reactive oxygen species and human disease: a critical evaluation with special reference to athersclerosis. *Br. J. Exp. Path.* **70**, 737–757.
[100] McCord, J. M. (1985) Oxygen-derived radicals in post-ischemic tissue injury. *New Engl. J. Med.* **312**, 159–163.
[101] Grootweld, M.; Halliwell, B. (1986) Aromatic hydroxylation as a potential measure of hydroxyl radical formation *in vivo*. *Biochem. J.* **237**, 499–504.
[102] Simmonds, N. J.; Allen, R. E.; Stevens, T. R. J.; Van Someren, R. N. M.; Blake, D. R.; Rampton, D. S. (1992) Chemiluminescence assay of mucosal reactive oxygen metabolites in inflammatory bowel disease. *Gastroenterology* **103**, 186–196.
[103] Keshavarzian, A.; Sedghi, S.; Kanofsky, J.; List, T.; Robinson, C.; Ibrahim, C.; Winship, D. (1992) Excessive production of reactive oxygen metabolites by inflamed colon: analysis by chemiluminescence probe. *Gastroenterology* **103**, 177–185.
[104] Stich, H. F.; Andrews, F. (1989) The involvement of reactive oxygen species in oral cancers of betel-quid/tobacco chewers. *Mutat. Res.* **214**, 47–61.
[105] Nair, U. J.; Obe, G.; Friesen, M.; Goldberg, M. T.; Bartsch, H. (1992) Role of lime in the generation of reactive oxygen species from betel-quid ingredients. *Environ. Health Perspect.* **98**, 203–205.
[106] Keshavarzian, A.; Olyaee, M.; Sontag, S.; Mobarhan, S. (1993) Increased levels of luminol-enhanced chemiluminescence by rectal mucosa of patients with colonic neoplasia: a possible marker for colonic neoplasisa. *Nutr. Cancer* **19**, 201–206.

[107] Owen, R. W.; Wimonwatwatee, T.; Spiegelhalder, B.; Bartsch, H. (1996) A high performance liquid chromatography system for quantification of hydroxyl radical formation by determination of dihydroxy benzoic acids. *Eur. J. Cancer Prev.* **5**, 233–240.
[108] Richmond, R.; Halliwell, B.; Chauhan, J.; Darbre, A. (1981) Superoxide-dependent formation of hydroxyl radicals: detection of hydroxyl radicals by the hydroxylation of aromatic compounds. *Anal. Biochem.* **118**, 328–335.
[109] Kettle, A. J.; Winterbourn, C. (1994) Superoxide-dependent hydroxylation by myeloperoxidase. *J. Biol. Chem.* **269**, 17146–17151.
[110] Coudray, C.; Talla, M.; Martin, S.; Fatome, M.; Favier, A. (1995) High-performance liquid chromatography electrochemical determination of salicylate hydroxylation products as an *in vivo* marker of oxidative stress. *Anal. Biochem.* **227**, 101–111.
[111] Babbs, C. F. (1990) Free radicals and the etiology of colon cancer. *Free Radical Biol. Med.* **8**, 191–200.
[112] Babbs, C. F.; Gale, M. J. (1987) Colorimetric assay for methanesulfinic acid in biological samples. *Anal. Biochem.* **163**, 67–73.
[113] Owen, R. W.; Spiegelhalder, B.; Bartsch, H. (1997) Production of reactive oxygen species by a soluble factor in the faecal stream. *J. Cancer Clin. Oncol.* **123**, (Suppl. 1), 2 (abstract).

5 Controversies Surrounding Diet and Breast Cancer

Lenore Arab and Michelle Mendez

5.1 Abstract

Breast cancer, which continues to rob the world of approximately half a million women per year, is surprisingly poorly understood. Despite application of all the tools of epidemiology: ecological studies, migrant studies, case-control and cohort studies, and measurement of multiple exposures at various times of life, we still cannot advise women on how to protect themselves or their daughters from breast cancer. The disease has a familial and a hormonal component, but current knowledge of these components does not adequately explain breast-cancer risk. In the search for strategies to prevent this devastating disease, potential food-borne exposures have been studied in both animal models and in women. These include studies of the relationship of the macronutrients (fat, and alcohol, as well as total energy intake), studies of antioxidant micronutrients, studies of

Reprint: This paper was published first by

body fat and estrogen levels, and studies of specific foods (such as cruciferous vegetables and soyabean products). Even substances carried by food, such as the pesticide dichloro-diphenyltrichlorethane (DDT), have received and continue to receive scrutiny. The present paper will attempt to summarize current knowledge and belief in the area of these food-borne exposures.

5.2 Dietary fat, energy intakes and breast cancer

The involvement of dietary fat as a risk factor for breast cancer remains controversial. There seems to be a strong and consistent relationship between per capita average total fat disappearance and breast-cancer incidence, when assessed in international ecological studies [1]. However, results of case-control and cohort studies in which diet was assessed by food-frequency questionnaire are inconsistent. The controversy surrounding this has focused in part on colinearity with energy intakes [2]. Particularly after adjustment for energy intakes, the independent effect of total dietary fat intakes appears to be minimal if existent at all [3].

The importance of fat intake to breast cancer needs to be examined both in its role as a contributor of energy, and in its potential for influencing risk independent of an energy effect. Non-energy-adjusted models ask whether absolute intakes of fat contribute to breast cancer, either through an association with energy intakes or otherwise. Energy-adjusted models ask something much more subtle: to what extent does the amount of fat consumed, beyond that expected of a women at her reported energy intake level, relate to breast-cancer risk, when fat is exchanged isoenergetically for other nutrients? Some animals, such as rats, do appear to adjust their energy intake to the nutrient density of their feed [4]. Hamsters, on the other hand, will substantially increase body weight if fed on a fat-rich diet *ad libitum* [5]. Human subjects are unfortunately more like hamsters: in free-living human populations, as with caged hamsters, dietary fat is rarely exchanged isoenergetically for other macronutrients. Instead, people consuming more fat tend to consume more total energy and become fatter [6]. Therefore, the public health importance of the contribution of fat to total energy intakes remains strong, even if the effect is through total energy intake.

Energy-adjusted models are justified when biological hypotheses suggest that fat, above and beyond its contribution to energy intake, contributes to breast-cancer risk. This might be the case if fat carries carcinogens (for example, pesticides) or if certain types of fat are particularly biologically active. Examples of the difference in hypotheses are expressed in Figure 5.1.

In animal models, total energy intake remains one of the most important predictors of carcinogenesis [7]. In epidemiological studies, which are subject to biased reporting of intakes, the association is less consistent [8, 9]. This may be

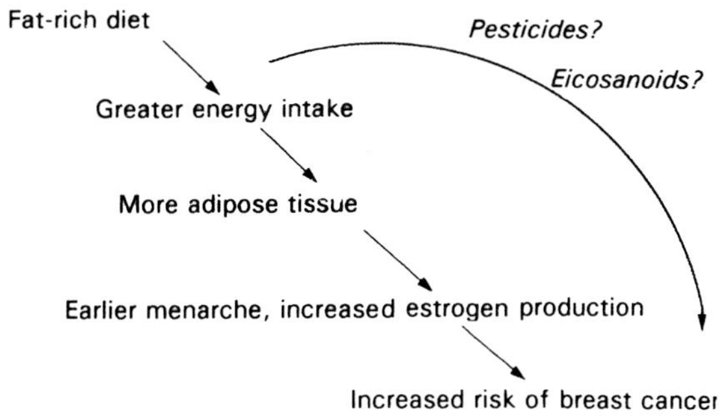

Figure 5.1: Alternative mechanisms underlying a fat-breast cancer relationship.

due to a differential under-reporting of food consumption among more obese women [10], or greater energy intakes among the most-physically-active women.

Pooled analyses of case-control studies from eight countries with approximately 4400 cases of breast cancer show a 25 % increase in breast-cancer risk associated with a 100 g/d increase in fat intake. Since 100 g fat represents 3.8 MJ or approximately 40 % of the total energy requirement of a normal woman, changes of this magnitude are unlikely [8]. A pooled analysis of cohort studies of breast cancer from four countries and including approximately 4980 cases found a relative risk of a similar magnitude, with an increase in risk of 7 % per 25 g fat after deattenuation [9]. This increase in risk, however, was not statistically significant. For the most part, these cohort studies examine post-menopausal diets in relation to post-menopausal breast-cancer risk. None of these studies addresses the question as to whether fat-rich diets in childhood and during puberty are related to breast-cancer risk in adulthood.

5.3 Individual fatty acids and breast cancer

Epidemiological studies that calculate risk ratios for total fat fail to take into account the biological non-equivalence of the various types of fatty acids. Studies of the metabolic relationships of fat intake and human blood lipids have shown how important it is to allow for the differences in individual fatty acids, since different families and specific fatty acids within families have biological effects which differ both in their magnitude and the direction of effect on eicosanoid production as well as on carcinogenesis in animal models. The confusion and conflicting results regarding fat intake may be due to the physiological importance of individual fatty acids which are generally inadequately separated in human studies.

Few studies have examined the components of fat, individual fatty acids, or classes such as short-, medium- and long-chain fatty acids, *trans*-fatty acids, or various of mono- and polyunsaturated fatty acids in relation to breast cancer. This is largely due to the difficulty in estimating intakes, which are a function of sources of oils and fats, types of fish, and other facets of the diet which are poorly known to the consumer. Dietary assessment tends to remain at the level of consumption of types of foods, such as olive oil or fish, and not their fatty acid compositions. These foods can carry other substances, such as phenols, which may account for any association with risk or any protective effects.

It is generally difficult to obtain accurate information on dietary intakes in epidemiological studies [11]. Intake of foods, when assessed by interview or questionnaire, may be recalled differentially between cases and controls. The use of food-composition tables may also introduce errors, because of large variations in nutrient content of some foods. Furthermore, when measuring intake, individual variation in absorption and metabolism is not taken into account. Intakes of fatty acids pose even greater problems than many other nutrients.

Where available, molecular markers of intake often are preferable. Plasma or serum concentrations are used most often [12]. Although adipose-tissue analysis is a reliable alternative to dietary records and probably superior for the assessment of diet in the distant past, few biomarker studies of the profiles of individual fatty acid stores in adipose tissue or other media exist to date [13, 14]. The relative abundance of individual fatty acids in adipose tissue reflects the composition of habitual diet. One of the few large studies of breast cancer to use biomarkers of adipose tissue as a long-term integrated biomarker of dietary intakes is the European Community Multicentre Study on Antioxidants, Myocardial Infarction and Cancer of the Breast (EURAMIC) Study.

5.4 EURAMIC study

Between 1990 and 1992 a multi-national case-control study was conducted in five European countries: Northern Ireland, Germany, The Netherlands, Spain and Switzerland. The incidence of breast cancer has an approximately twofold range between countries (38–72 per 100000) with the lowest risk of disease in Spain. The purpose of this study was to examine potential protective roles of antioxidants in the development of breast cancer. A common protocol for recruitment, sampling and exposure assessment was developed. The participating physicians in this study were trained in the sampling techniques with written and videotaped instructions. Technicians were trained in sample handling, and all samples were transported to the coordinating centre. Central laboratory analyses were the core of this study: the levels of carotenoids, tocopherols and fatty acid composition of the gluteal-adipose-tissue samples were determined in the laboratories of the Toxicology and Nutrition Institute in Zeist, The Netherlands.

The specific purpose of the EURAMIC study was to relate the levels of fat-soluble antioxidant nutrients and Se in tissues with a low turnover, which provide biomarkers of the integrated long-term intake of these nutrients, to risk of breast cancer. The central hypothesis was that concentration of α-tocopherol and β-carotene in adipose tissue, and of Se in toenails is lower in newly-diagnosed breast cancer cases as compared with healthy population controls [15]. In addition, the study developed a component to examine relationships between stores of individual fatty acids and risk of breast cancer. 700 cases and controls from five diverse European countries were enrolled in this multi-centred case-control study. The biomarker-based findings from this study on the relationship between dietary factors and breast-cancer risk are presented below.

5.5 Monounsaturated fats and breast cancer

Most of the studies on monounsaturated fats actually focus on the differences in consumption of olive oil between groups of women. A number of epidemiological studies of olive oil and breast-cancer risk have pointed to a protective effect among women who consume olive oil more frequently. Women consuming olive oil more than once daily have breast-cancer risks which are up to one-third lower than that of the non-consumers in Italy, Greece and Spain [16–18]. Olive oil is, however, a complex product containing various levels of monounsaturates, and vastly differing amounts of vitamin E, depending on the source tree and where it is located. The consumption of olive oil may also be strongly associated with socio-economic status, or other dietary characteristics, such as lower usage of butter or margarine. In epidemiological studies of diet which calculate all sources of monounsaturates, and in biomarker studies, intake of monounsaturates *per se* is not associated with lower breast-cancer risks [14].

5.6 Polyunsaturated fatty acids

Animal models of cancer using chemical carcinogens can generally not produce tumors in the absence of *n*-6 fatty acids. This source of essential fatty acids is known to regulate synthesis of prostaglandins, which in turn can regulate tumor growth, metastases, and can suppress host immune responsiveness [19].

n-3 Fatty acids (largely fish oils) are seen, when consumed in high amounts, to be inhibitory of tumorigenesis. As with olive oil, consumption of the source food (fish), is often taken as a surrogate for *n*-3 consumption. The problem with this is that, depending on the type of fish consumed, and its preparation style, consumption can be poorly, and, in some cases, negatively associated with erythrocyte membrane levels of long-chain *n*-3 fatty acid concentrations [20]. This may be a function of low intake levels, or the contribution of other fats being consumed, for example, with breaded, fried fish products.

It is becoming increasingly clear that the production of active eicosanoids is regulated by the type of precursor fatty acid. Prostaglandins of the 2-series and leukotrienes of the 4-series are produced from *n*-6 fatty acids, whereas the respective 3- and 5-series are derivatives of *n*-3 fatty acids (Figure 5.2). The latter are known to reduce chemically-induced breast-cancer growth in rats [19]. This may be through inhibition of oxidative metabolism of arachidonic acid via the cyclooxygenase pathway of prostaglandin synthesis. However, the actual function of individual fatty acid metabolites in women is not clear, and the mechanisms by which fatty acids stimulate or inhibit tumors remain controversial.

The rate-limiting enzyme, δ-6-desaturase, shows a greater affinity for fatty acids of the *n*-3 family. The result of this is production of eicosanoids which are very similar in structure, but can be antagonistic to those produced from the *n*-6 fatty acid family. Examples of this are the aggregating and vasocontricting effects of thromboxane A2 from the *n*-6 family, and the anti-aggregating effects of thromboxane A3 from the *n*-3 family.

Evidence is accumulating which suggests that more important than the absolute amount of a fatty acid is its concentration relative to that of its competitors in lipooxygenase (*EC* 1.13.11.12) and cyclooxygenase pathways of eicosanoid production. Preliminary results from the EURAMIC study suggest that a higher relative concentration of *n*-3 fatty acids may be protective [21].

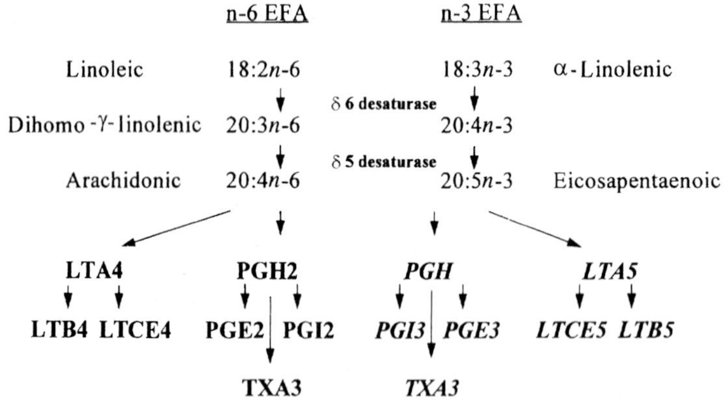

Figure 5.2: Pathways of *n*-6 and *n*-3 essential fatty acid (EFA) metabolism. LT, leukotrienes; PG, prostaglandins; TX, thromboxanes.

5.7 *Trans*-fatty acids and breast cancer

Trans-fatty acids are formed largely through the hydrogenation of vegetable oils to enhance stability. They are produced enzymically to some degree naturally, but the majority of *trans*-fats in Western populations are consumed in processed foods. These fatty acids are known to have the ability to impair microsomal desaturation and elongation of essential fatty acids to their long-chain polyunsaturated metabolites. They compete for desaturase enzyme sites and potentially reduce production of the 3-series eicosanoids [22–26]. Although there has been considerable study of the effects of intakes of *trans*-fatty acids on lipoprotein metabolism and cardiovascular disease, only two studies have examined relationships between *trans*-fatty acid stores and cancer in human subjects. In a study of Americans in whom stores of *trans*-fatty acids are twice as high as those in Europe, increases in risk were found at some levels of greater *trans*-fatty acid intake, although without an overall linear trend [27]. Within the EURAMIC study an increase in risk of breast cancer has been found among those women with greater stores of *trans*-fatty acids [28].

5.8 Antioxidants and breast cancer

Serious adverse effects arise in the presence of excessive concentrations of oxygen beyond the normal need for oxidative processes to support life, or in the absence of adequate antioxidant protection. These effects include peroxidation of membrane lipids, the hydroxylation of nucleic acid bases and the oxidation of protein moieties.

Oxidative stress is enhanced by inflammation, exercise, radiation exposure, surgery, air pollution, reperfusion, cigarette smoking and dietary intakes of polyunsaturated fatty acids. The endogenous sources of oxidants include normal respiration, phagocytosis, cytochrome P450 enzymes, and peroxisomes. The degradation of fatty acids by peroxisomes can cause the release of peroxide into other cell compartments and result in oxidative DNA damage. Fatty acid oxidation can also directly cause oxidative damage through lipid peroxidation, the creation of mutagenic lipid epoxides, lipid hydroperoxides, lipid alkoxyl and peroxyl radicals and enals [29]. As quenchers of free radicals, naturally-occurring antioxidants such as carotenoids and tocopherols can theoretically block potential DNA damage caused by free radicals.

A multitude of epidemiological studies have provided evidence that eating behaviour patterns consistent with a high fruit and vegetable intake carry lower cancer risks [30]. However, the association between intake of fruits and vegetables rich in carotenoids, or plasma carotene levels, and a decreased risk of cancer appears to be most consistent for lung cancer and stomach cancer [31–33] and less consistent for cancers of the breast [34–38]. Since carotenoids serve as markers of other dietary factors in the same source foods, which may actually be responsible for the protective effects seen for certain dietary patterns, and there is little evidence of a specific β-carotene association, it appears unlikely that any of the polyphenols in fruits and vegetables currently under investigation will turn out to be strong predictors of breast-cancer risk.

Attention in epidemiological studies has focused on studying whether β-carotene intakes are protective. At least seven studies have reported on the plasma concentrations of β-carotene as a function of breast-cancer risk [32, 37, 39–43]. Lower mean concentrations were found in cases in three of these studies [31, 37, 42]; however, the difference was only significant in one of these studies [37]. Many of these studies were rather small and so lacked the power to detect a difference. In most studies, the diets of the women enrolled were quite homogeneous, which might mask a true effect. To assess the public health significance of antioxidants, and to provide recommendations on their use as preventives, requires epidemiological research on the modulators and effects of these substances. It is also vital that studies have adequate power, and are conducted in populations with sufficient variance in exposure using objectively-determined exposure levels. The EURAMIC study, which addresses a number of the limitations of geographically-insular epidemiological studies, also found little evidence of an important role for β-carotene in breast-cancer risk [44].

Analyses from the EURAMIC study revealed two surprises: first, the concentrations of β-carotene in the adipose tissue were lowest among women in the southern-most countries, where rates of breast cancer were the lowest. It would have been expected that levels in these women would be the highest from their diets, which are richer in fruits and vegetables than in the northern countries. Second, no protective effect of high adipose stores of either β-carotene or α-tocopherol, or from a combination of antioxidants, was seen. In fact, the direction of the α-tocopherol effect, although not statistically significant, was contrary to expectations.

5.9 Cruciferous vegetables and breast cancer

The modulation of carcinogenesis by the consumption of cruciferous vegetables has been investigated in laboratory animals that were exposed to known carcinogens. Tumor development has often been compared in animals with and without cabbage-supplemented diets [45–49]. A number of these studies provide evidence of an effect on the production of some tumors. Incidence of mammary tumors, particularly in mice and rats, seemed to be reduced on the addition of 5–20 % of cabbage by weight to the diet [50, 51].

The active ingredients in Crucifera have not been definitively determined. Among the more than 2000 naturally-occurring compounds in Brassica [52], phenolic compounds have drawn attention. The action of some flavonoids is mediated by binding to estrogen receptors which may regulate gene transcription and protect against certain cancers [53]. Among the phenols, dithiolthiones and the isothiocyanates formed on cleavage of glucosinolates are among the strongest inducers of quinone reductase (*EC* 1.6.99.2), an important detoxifying phase-II-enzyme [54]. Isothiocyanates stimulate acutely the activity of a wide range of phase-II-enzymes. The induction of glutathione S-transferase (*EC* 2.5.1.18) and glucuronidase may enhance the excretion of carcinogens through the bile and urine. The net effect of this induction may be chemoprotective, but that remains to be proven. 1,2-Dithiol-3-thiones, which occur naturally in Brassica, are under study in clinical trials as chemoprotective agents [55]. It should be noted that both pro- and anti-carcinogenic activities have been ascribed to indole derivatives and isothiocyanates [56].

More closely related to breast-cancer risk in particular is the study of the effect of indole-3-carbinol on estrogen metabolism. Indole-3-carbinol has been shown to be antiproliferative in cell systems [57]. The balance of C-16α-hydroxyestrone to C2 hydroxyestrone, i.e. the ratio, fully-potent: weakly-anti-estrogenic estrogens, is being studied as a potential mechanism. Feeding of pure indole-3-carbinol reduces the ratio 16α2-hydroxylation:2-hydroxylation of estradiol [58, 59]. This is the inverse of the effect expected from estrogenic pesticides, as described in Section 5.10.1.

5.10 Xeno-estrogens

Recently, it has been suggested that exposure to environmental estrogens may be related to risk of breast cancer. These xeno-estrogens have been linked to the hypothesis that cumulative exposure to estrogens plays a role in breast-cancer etiology. Two of the xeno-estrogens under study involve exposure through the diet: phyto-estrogens, which are naturally-occurring plant compounds, and organochlorine pesticides such as DDT, which persist in the environment and may be ingested in foods as well as through the skin and respiratory tract. Both groups of compounds have been shown to affect estrogen metabolism *in vivo* and *in vitro*. While phyto-estrogens are believed to reduce the risk of breast cancer of their low potency relative to endogenous estradiol, organochlorine pesticides have been hypothesized to amplify estrogen activity, increasing risk.

5.10.1 Phyto-estrogens

Interest in phyto-estrogens as potential anti-carcinogens has evolved largely in response to the suggestion that high intakes of soyabean foods may contribute to the lower incidence of breast cancer in Japan, China, and other countries in Asia. These foods are not a traditional part of Western diets. Soyabeans are a rich source of isoflavones, one of two classes of phyto-estrogens which are important in the human diet. Soyabeans have by far the highest concentration of isoflavones, which are found mainly in legumes; chick peas (*Cicer arietinum*) are another dietary source. Genistein, the most extensively studied isoflavone, is the main isoflavone in soyabeans, accounting for two-thirds of soyabean isoflavone content [60]. Lignans, another important type of phyto-estrogen, are derived from the metabolism of lignins (found in plant cell walls), and are produced from almost all cereals and vegetables.

Phyto-estrogens bind to estrogen receptors, but induce less potent estrogen activity than endogenous estrogens. Plant estrogens are believed to reduce risk of breast cancer through their net anti-estrogenic effect. Effects of these compounds on hormone metabolism have been shown in both human subjects and animals. Soyabean consumption increases menstrual cycle length and decreases circulating levels of estradiol in human subjects [61]. A 1994 review of epidemiological studies on phyto-estrogens and breast cancer reported that three of four studies on intakes of various soyabean foods and breast cancer found statistically significant inverse relationships [60]. One of these found protective associations only in premenopausal women. The study which found no relationship used fat from soyabean products as the measure of exposure [62]. Data from animal and *in vitro* experiments with mammary cancers also suggest that soyabean may be protective [60]. However, results of recent studies on genistein reveal an estrogen agonistic effect resulting in proliferation of estrogen-dependent human breast-cancer cells and enhanced tumor growth at low concentrations [63]. Thus, here too, the story remains unclear regarding a net protective effect of consumption of soyabean in breast-cancer incidence or recurrence.

At this time, associations between breast-cancer risk and consumption of soyabean foods are not definitively linked to estrogen bioavailability. Since soyabean foods contain other substances which may be anti-carcinogens, it is possible that non-estrogenic properties of flavonoid-containing foods (e.g. antioxidant or anti-mutagenic properties) may contribute to a reduction in cancer risk. Evidence of a protective effect of soyabean products on non-hormonal cancers lends support to this alternative hypothesis, and argues against an estrogen-related effect. Associations with soyabean product consumption may also be confounded by other dietary patterns or specific food items for which consumption is strongly correlated with soyabean intakes. Further study is needed to confirm the existence of a protective effect of soyabean and other foods rich in phyto-estrogens, and to determine what properties of soyabean foods may be responsible for any effects.

5.10.2 Organochlorine pesticides

Research conducted to date has focused on the pesticide DDT and its metabolite, 1,1-dichloro-2,2-bis(*p*-chlorophenyl)ethylene (DDE). DDT use has been banned for some time in most industrialized countries, but it continues to be widely-used in many developing nations. Although levels are declining in countries where use has been banned, DDT remains ubiquitous in the environment due to its long half-life. Because its use has been discontinued, current levels of DDT in the food chain in industrialized countries are believed to be low. The low level of exposure, and the relatively low potency of organochlorine chemicals compared with other estrogens, has led some researchers to argue that these chemicals are unlikely to be important health hazards [64].

DDT and DDE are fat-soluble and, therefore, accumulate and persist in the environment and in human adipose tissue. These compounds are believed to mimic estrogen activity in the body. Unlike phyto-estrogens, organochlorines are expected to increase net exposure to biologically-active estrogen. In the laboratory, organochlorines increase the hydroxylation of estradiol to 16-α-hydroxyestrone, a potently active form of estrogen, altering the ratio, 16-α-hydroxyestrone : 2-hydroxyestrone, a less-active metabolite [65]. Elevated levels of the more active form of estrogen have been found in breast tumors, suggesting a possible causal link [65].

Epidemiological studies on DDE and risk of breast cancer have provided equivocal results. Positive relationships have been reported in three case-control studies [66–68], but these results have been based on small sample sizes (from 35 to 229 subjects). Odds ratios reported in these studies ranged from 2 to greater than 10. These positive relationships have not been confirmed by three larger, more recent studies (300–606 subjects), including two studies which measured DDE levels an average of 14 years or more before diagnosis [69, 70]. In one of these studies [69], however, there were positive relationships within individual ethnic group strata. Serum DDE levels were elevated in Caucasian and in African-American women with breast cancer relative to controls of the same ethnicity, while higher DDE levels appeared protective in Asians [69]. Preliminary results from the EURAMIC study, which measured levels in gluteal adipose tissue, also fail to support a causal link [71].

To date, no strong and consistent effect of pesticide exposure has been identified. Two of the three studies which found positive associations between breast cancer and pesticide exposure [67, 68] relied on measures of DDE in serum or plasma obtained shortly before (6 months) or after diagnosis. In these studies, the possible influence of cancer-related changes in metabolism or weight cannot be excluded as reasons for the elevated levels in cases. Furthermore, exposures earlier in life may be more relevant for risk of breast cancer [72]. The studies which measured blood levels prospectively did not confirm positive associations. The two studies using adipose tissue levels of DDE also found inconsistent results [66, 71]. However, because of the small size of the pilot study, Falck *et al.* [66] and others, who found elevated DDE in cases relative to controls, were not able to adjust for potential confounding factors.

5.11 Alcohol

The epidemiological literature on alcohol consumption and risk of breast cancer generally supports a weak causal association [73, 74]. Relative risks reported in the literature are variable, but on the whole suggest that mean intakes of as little as one drink daily may be harmful. In a recent meta-analysis, the odds ratio for one drink daily was estimated to be 1.11 (95% CI 1.07–1.16), and 1.24 (95% CI 1.15–1.34) for two drinks daily, with stronger relationships for increasing levels of consumption [73]. Even a small increase in risk is potentially of public health significance, since moderate drinking is a common, modifiable exposure.

A number of pathways through which ethanol, the active constituent of alcoholic beverages, may increase risk of breast cancer have been suggested, although at this time the mechanisms are not well understood. Ethanol itself is not a carcinogen, but is converted to a potentially carcinogenic metabolite (acetaldehyde) in the body. Alcohol has also been associated with oxidative stress in the liver, which may induce the activity of other enzymes, resulting in increased production of carcinogenic metabolites. Ethanol has also been shown to increase the permeability of cells to other potential carcinogens [75]. Animal studies suggest that ethanol consumption can stimulate cell proliferation in mammary tissue, which may increase risk of benign breast disease and cancer. Recently, ethanol has been shown to increase circulating plasma estrone and estradiol levels in premenopausal women [76]. This suggests that drinking alcohol elevates exposure to bioavailable estrogens, providing another possible link with the hormonal pathways hypothesized to be involved in the etiology of breast cancer.

Poor understanding of the mechanisms complicates measurement of the relevant dimension of alcohol drinking for breast-cancer research: the timing, amounts and types of alcoholic beverages which are relevant for risk of breast cancer are not known. Most studies have measured average consumption of ethanol in the recent past, perhaps 5–10 years before diagnosis. However, adolescent drinking, or drinking during early adulthood, may be more important determinants of risk. A few studies [77–81] have explored the effects of early drinking habits; most of these studies suggest that there may be risk associated with drinking before age 30–35 years independent of later drinking. Past drinking was associated with increased risk of breast cancer in the EURAMIC study (odds ratio 1.73, 95% CI 1.07–2.79) [82]. Patterns of drinking, bingeing *vs* regular consumption, may also be important, but have not been explored. Types of beverages with different constituents, wine, beer and spirits, may also have different effects on breast-cancer risk, although the literature to date does not seem to indicate important heterogeneity by type [73].

A few studies have suggested the effect of a threshold at two to three drinks (30 g) daily [83–85], but this hypothesis has not been consistently supported in the literature. Results from the EURAMIC study, which excluded heavy drinkers, may be consistent with a threshold effect. Mean alcohol intakes among current drinkers ranged from 3.0 to 14.0 g/d in the EURAMIC population, and there was no significant increase in risk of breast cancer associated with current drinking [82].

5.12 Summary and conclusion

What we know about prevention of breast cancer is related to lifetime estrogen exposure and exposures to specific estrogens at vulnerable periods of life. This can be influenced by diet. The strongest indicator of a diet-related effect to date is the fairly consistent increase in breast cancer among women who are tall or obese [86]. The other dietary factors, summarized in Table 5.1 are less strongly associated with breast-cancer risk in epidemiological studies.

Table 5.1: Diet-related factors and the risk of breast cancer.

Energy	↑
Height	↑
Fat	↑
Energy-adjusted fat residuals	–
Individual fatty acids	
n-3	(↓)
n-6	↑?
n-9	↓?
trans-	↑?
Carbohydrates	↑?
Antioxidants	–
Cruciferous vegetables	?
Alcohol	↑
Xeno-estrogens	?

↑, Increased risk; ↓, reduced risk; –, no evidence of a relationship.

The relationship between fat and breast-cancer risk has been extensively studied but remains somewhat uncertain. Fat, as a contributor to energy intakes and energy imbalance, is probably a factor in the higher breast-cancer rates in

Western countries. Beyond its role as an energy source, the evidence for an independent effect of dietary fat on breast-cancer risk is weak. More focused analyses of the role of individual fatty acids, and on lipid-related pesticide exposures, may reveal strong effects which are currently masked by the use of inadequate exposure measures, as well as by measurement error.

Currently, there is substantial evidence of a weak relationship with alcohol consumption, even at frequencies of drinking of less than once daily. The evidence of a protective role for antioxidants is weaker for breast cancer than for other cancers. This might by expected in a cancer which is not strongly associated with cigarette smoking. Specific foods are being studied for other potentially-active ingredients which may be involved in hormone metabolism, but conclusive results for soyabean or cruciferous vegetables are not yet available. Studying these relationships will continue to be a challenge for researchers because of the difficulties in measuring dietary exposures, which is complicated by the uncertainty of the relevant time frame for exposure assessment.

While substantial attention has been focused on studying diet in relation to incidence, the potential for diet to reduce recurrence of breast cancer is thoroughly under-studied. There is little reason to believe that the factors which influence the incidence of breast cancer, perhaps during childhood and puberty, are the same as those which affect recurrence in adulthood. In this area, the very limited evidence available suggests that study of biologically-active fatty acids is promising.

References

[1] Prentice, R. L.; Sheppard, L. (1989) Validity of international, time trend, and migrant studies of dietary factors and disease risk. *Prev. Med.* **18**, 167–179.

[2] Palmgren, J. (1993) Controlling for total energy intake in regression models for assessing macronutrient effects on disease. *Eur. J. Clin. Nutr.* **47**, Suppl. 2, S46–S50.

[3] Hulka, B. S. (1989) Dietary fat and breast cancer: Case-control and cohort studies. *Prev. Med.* **18**, 180–193.

[4] Braden, L. M.; Carroll, K. K. (1986) Dietary polyunsaturated fat in relation to mammary carcinogenesis in rats. *Lipids* **21**, 285–288.

[5] Birt, D. F.; Julius, A. D.; White, L. T.; Pour, P. M. (1989) Enhancement of pancreatic carcinogenesis in hamsters fed a high-fat diet *ad libitum* and at a controlled calorie intake. *Cancer Res.* **49**, 5848–5851.

[6] Astrup, A. V.; Buemann, B.; Western, P.; Toubro, S.; Raben, A.; Christensen, M. J. (1995) Objective assessment of the habitual dietary fat content in patients with obesity (In Danish). *Ugeskr. Laeger* **16**, 291–294.

[7] Albanes, D. (1987) Total calories, body weight and tumor incidence in mice. *Cancer Res.* **47**, 1987–1992.

[8] Howe, G. R.; Hirohata, T.; Hislop, T. G.; Iscovish, J. M.; Yuan, J. M.; Katsouyanni, K.; Lubin, F.; Marubini, E.; Modan, B.; Rhohan, T.; Toniolo, P.; Shunzhang, Y. (1990)

Dietary factors and risk of breast cancer: combined analysis of 12 case-control studies. *J. Natl. Cancer Inst.* **82**, 561–569.

[9] Hunter, D. J.; Spiegelman, D.; Adami, H.-O.; Beeson, L.; Van den Brandt, P. A.; Folsom, A. R.; Fraser, G. E.; Goldbohm, A.; Graham, S.; Howe, G. R.; Kushi, L. H.; Marshall, J. R.; McDermott, A.; Miller, A. B.; Speizer, F. E.; Wold, A.; Yuan, S.-S.; Willett, W. (1996) Cohort studies of fat intake and the risk of breast cancer – a pooled analysis. *New Engl. J. Med.* **334**, 356–361.

[10] Lichtman, S. W.; Pisarska, K.; Berman, E. R.; Prestone, M.; Dowling, H.; Offenbacher, E.; Weisel, H.; Heshka, S.; Matthews, D. E.; Heymsfield, S. B. (1992) Discrepancy between self-reported and actual caloric intake and exercise in obese subjects. *New Engl. J. Med.* **327**, 1893–1898.

[11] Barrett-Connor, E. (1991) Nutrition epidemiology: how do we know what they ate? *Am. J. Clin. Nutr.* **54**, 182–187.

[12] Bates, C. J.; Thurnham, D. I. (1991) Biochemical markers of nutrient intake. In: *Design Concepts in Nutritional Epidemiology*, p. 192–265 [B. M. Margetts and M. Nelson, editors]. Oxford: Oxford University Press.

[13] Kohlmeier, L.; Kohlmeier, M. (1995) Adipose tissue as a medium for epidemiologic exposure assessment. *Environ. Health Perspect.* **103**, Suppl. 3, 99–106.

[14] Kohlmeier, L. (1997) Biomarkers of fatty acid exposure and breast cancer risk. *Am. J. Clin. Nutr.*

[15] Kardinaal, A. F. M.; van't Veer, P; Kok, F. J.; Kohlmeier, L.; Martin-Moreno, J. M.; Huttunen, J. K.; Hallen, M.; Aro, A.; Galvex, R.; Gomez-Aracena, J.; Kark, J. D.; Martin, B. C.; Mazaev, V. P.; Riemersma, R. A.; Ringstad, J.; Strain, J. J.; Zatonsky, W. (1993) EURAMIC Study: Antioxidants, myocardial infarction and breast cancer. *Eur. J. Clin. Nutr.* **47**, Suppl. 2, 64–71.

[16] Martin-Moreno, J. M.; Willett, W. C.; Gorgojo, L.; Banegas, J. R.; Rodriguez-Artalejo, F.; Fernandez-Rodriguez, J. C.; Maisonneuve, P.; Boyle, P. (1994) Dietary fat, olive oil intake and breast cancer risk. *Int. J. Cancer* **58**, 774–780.

[17] LaVecchia, C.; Negri, E.; Franceschi, S.; Decarli, A.; Giacosa, A.; Lipworth, L. (1995) Olive oil, other dietary fats, and the risk of breast cancer (Italy). *Cancer Causes Control* **6**, 545–550.

[18] Trichopolou, A.; Katsouyanni, K.; Stuver, S.; Tzala, L.; Gnardellis, C.; Rimm, E.; Trichopoulos, D. (1995) Consumption of olive oil and specific food groups in relation to breast cancer risk in Greece. *J. Natl. Cancer Inst.* **87**, 110–116.

[19] Nogushi, M.; Rose, D. P.; Earashi, M.; Miyazaki, I. (1995) The role of fatty acids and eicosanoid synthesis inhibitors in breast carcinoma. *Oncology* **52**, 265–271.

[20] Olsen, S. F.; Hansen, H. S.; Sandström, B.; Jensen, B. (1995) Erythrocyte levels compared with reported dietary intake of marine *n*-3 fatty acids in pregnant women. *Br. J. Nutr.* **73**, 387–395.

[21] Simonsen, N.; Strain, J. J.; van't Veer, P.; Kardinaal, A.; Fernandez-Crehuet, J.; Huttunen, J.; Martin-Moreno, J. M.; Martin, B.; Thamm, M.; Kok, F.; Kohlmeier, L. (1996) Adipose tissue ω-3 fatty acids and breast cancer in a population of European women. *Am. J. Epidemiol.* **143**, Suppl. 135, Abstr.

[22] Kinsella, J. E.; Hwang, D. H.; Yu, P.; Mai, J.; Shrimp, J. (1979) Prostaglandins and their precursors in tissues from rats fed on *trans,trans*-lineolate. *Biochem. J.* **184**, 701–704.

[23] de Schrijver, R.; Privett, O. S. (1982) Interrelationship between dietary *trans*-fatty acids and the 6- and 9-desaturases in the rat. *Lipids* **17**, 27–34.

[24] Lawson, L. D.; Hill, E. G.; Holman, R. T. (1983) Suppression of arachidonic acid in lipids of rat tissue by dietary mixed isometrics and *trans*-octadecenoates. *J. Nutr.* **113**, 1827–1835.

[25] Mahfouz, M. M.; Smith, T. L.; Kummerow, F. A. (1984) Effect of dietary fats on desaturase activities and the biosynthesis of fatty acids in rat-liver microsomes. *Lipids* **19**, 214–222.

[26] Koletzko, B. (1992) *trans*-Fatty acids may impair biosynthesis of long-chain polyunsaturates and growth in man. *Acta Paediatr.* **81**, 302–306.
[27] London, S. J.; Sacks, F. M.; Stampfer, M. J.; Henderson, I. C.; Madwe, M.; Tomita, A.; Wood, W. C.; Remine, C.; Robert, N. J.; Dmochowski, R.; Willett, W. (1993) Fatty acid composition of subcutaneous adipose tissue and risk of proliferative benign breast disease and breast cancer. *J. Natl. Cancer Inst.* **85**, 785–793.
[28] Kohlmeier, L.; Simonsen, N.; Margolin, B.; Thamm, M. (1995) Stores of *trans*-fatty acids and breast cancer risk. *Am. J. Clin. Nutr.* **61**, 896.
[29] Ames, B. N.; Shigenaga, M. K.; Hagen, T. M. (1993) Oxidants, antioxidants, and the degenerative disease of aging. *Proc. Natl. Acad. Sci. USA* **90**, 7915–7922.
[30] Block, G.; Patterson, B.; Subar, A. (1992) Fruit, vegetables, and cancer prevention: A review of the epidemiological evidence. *Nutr. Cancer* **18**, 1–29.
[31] Gey, K. F.; Brubacher, G. B.; Sthelin, H. B. (1987) Plasma levels of antioxidant vitamins in relation to ischemic heart disease and cancer. *Am. J. Clin. Nutr.* **45**, 1368–1377.
[32] Wald, N. J.; Thompson, S. G.; Densem, J. W.; Boreham, J.; Bailey, A. (1988) Serum β-carotene and subsequent risk of cancer: Results from the BUPA Study. *Br. J. Cancer* **57**, 428–433.
[33] Connett, J. E.; Kuller, L. H.; Kjelsberg, M. O.; Polk, B. F.; Collins, G.; Rider, A.; Hulley, S. B. (1989) Relationship between carotenoids and cancer. The Multiple Risk Factor Intervention Trial (MRFIT) Study. *Cancer* **64**, 126–134.
[34] Wald, N. J.; Boreham, J.; Hayward, J. L.; Bulbrook, R. D. (1984) Plasma retinol, β-carotene and vitamin E levels in relation to future risk of breast cancer. *Br. J. Cancer* **49**, 321–324.
[35] Paganini-Hill, A.; Chao, A.; Ross, R. K.; Henderson, B. E. (1987) Vitamin A, β-carotene, and the risk of cancer: a prospective study. *J. Natl. Cancer Inst.* **79**, 443–448.
[36] Rohan, T. E.; McMichael, A. J.; Baghurst, P. A. (1988) A population-based case-control study of diet and breast cancer in Australia. *Am. J. Epidemiol.* **128**, 478–489.
[37] Potischman, M.; McCulloch, C. E.; Byers, T.; Nemoto, T.; Stubbe, N.; Milch, R.; Parker, R.; Rasmussen, K. M.; Root, M.; Graham, S.; Campbell, T. C. (1990) Breast cancer and dietary and plasma concentration of carotenoids and vitamin A. *Am. J. Clin. Nutr.* **52**, 909–915.
[38] Comstock, G. W.; Helzlsouer, K. J.; Bush, T. L. (1991) Prediagnostic serum levels of carotenoids and vitamin E as related to subsequent cancer in Washington County, Maryland. *Am. J. Clin. Nutr.* **53**, 260S–264S.
[39] Willett, W. C.; Polk, F.; Underwood, B. A.; Stampfer, M. J.; Pressel, S.; Rosner, B.; Taylor, J. O.; Schneider, K.; Hames, C. G. (1984) Relation of serum vitamin A and E and carotenoids to the risk of cancer. *New Engl. J. Med.* **310**, 430–434.
[40] Knekt, P. (1988) Serum vitamin E level and risk of female cancers. *Int. J. Epidemiol.* **17**, 281–288.
[41] Marubini, E.; Decarli, A.; Costa, A.; Mazzoleni, C.; Andreoli, C.; Barbieri, A.; Capitelli, E.; Carlucci, M.; Cavallo, F.; Monferroni, N.; Pastorino, U.; Salvini, S. (1988) The relationship of dietary intake and serum levels of retinol and β-carotene with breast cancer. *Cancer* **61**, 173–180.
[42] Basu, T. K.; Hill, G. B.; Ng, D.; Abdi, E.; Temple, N. (1989) Serum vitamins A and E, β-carotene, and selenium in patients with breast cancer. *J. Am. College Nutr.* **8**, 524–528.
[43] London, S. J.; Sacks, F. M.; Caesar, J.; Stampfer, M. J.; Siguel, E.; Willett, W. C. (1991) Fatty acid composition of subcutaneous adipose tissue and diet in postmenopausal US women. *Am. J. Clin. Nutr.* **54**, 340–345.
[44] van't Veer, P.; Strain, J. J.; Fernandez-Crehuet, J.; Martin, B. C.; Thamm, M.; Kardinaal, A. F. M.; Kohlmeier, L.; Huttunen, H. K.; Martin-Moreno, J. M.; Kok, F. J. (1996) Tissue antioxidants and postmenopausal breast cancer: The European Community multicentre study on antioxidants, myocardial infarction and cancer of the breast (EURAMIC). *Cancer Epidemiol. Biomarkers Prev.* **5**, 441–447.

[45] Wattenberg, L. W. (1983) Inhibition of neoplasia by minor dietary constituents. *Cancer Res.* **43**, 2448S–2453S.
[46] Birt, D. F.; Pelling, J. C.; Pour, P. M.; Tibbels, M. G.; Schweichkert, L.; Bresnick, E. (1987) Enhanced pancreatic and skin tumorigenesis in cabbage-fed hamsters and mice. *Carcinogenesis* **8**, 913–917.
[47] Temple, N. J.; Basu, T. K. (1987) Selenium and cabbage and colon carcinogenesis in mice. *J. Natl. Cancer Inst.* **79**, 1131–1134.
[48] Temple, N. J.; el Khatib, S. M. (1987) Cabbage and vitamin E: their effect on colon tumor formation in mice. *Cancer Lett.* **35**, 71–77.
[49] Scholar, E. M.; Wolterman, K.; Birt, D. F.; Bresnick, E. (1989) The effect of diets enriched in cabbage and collards on urine pulmonary metastasis. *Nutr. Cancer* **12**, 121–126.
[50] Stoewsand, G. S.; Anderson, J. L.; Munson, L. (1988) Protective effect of dietary Brussels sprouts against mammary carcinogenesis in Sprague-Dawley rats. *Cancer Lett.* **39**, 199–207.
[51] Bresnick, E.; Birt, D. F.; Wolterman, K.; Wheeler, M.; Markin, R. S. (1990) Reduction in mammary tumorigenesis in the rat by cabbage and cabbage residue. *Carcinogenesis* **11**, 1159–1163.
[52] Huang, M.-T.; Ferraro, T. (1992) Phenolic compounds in food and cancer prevention. In: *Phenolic Compounds in Food and Their Effects on Health. vol. 2: Antioxidants and Cancer Prevention. ACS Symposium Series* 502, p. 8–33 [M.-T. Huang, C.-T. Ho and C. Y. Lee, Eds.]. Washington, DC: American Chemical Society.
[53] Baker, M. E. (1992) Evolution of regulation of steroid-mediated intercellular communication in vertebrates: insight from flavonoids, signals that mediate plant-rhizobia symbiosis. *J. Steroid Biochem. Mol.* **41**, 301–308.
[54] Prestera, T.; Holtzclaw, W. D.; Zhang, Y.; Talalay, P. (1993) Chemical and molecular regulation of enzymes that detoxify carcinogens. *Proc. Natl. Acad. Sci. USA* **90**, 2965–2969.
[55] Ansher, S. S.; Dolan, P.; Bueding, E. (1986) Biochemical effects of dithiolthiones. *Food Chem. Toxicol.* **24**, 405–415.
[56] McDanell, R.; McLean, A. E.; Hanley, A. B.; Heaney, R. K.; Fenwick, G. R. (1989) The effect of feeding Brassica vegetables and intact glucosinolates on mixed-function-oxidase activity in the livers and intestines of rats. *Food Chem. Toxicol.* **27**, 289–293.
[57] Newfield, L.; Goldsmith, A.; Bradlow, H. L.; Auborn, K. (1993) Estrogen metabolism and human papillomavirus-induced tumors of the larynx: Chemo-prophylaxis with indole-3-carbinol. *Anticancer Res.* **13**, 337–341.
[58] Michnovicz, J. J.; Bradlow, H. L. (1991) Altered estrogen metabolism and excretion in human following consumption of indole-3-carbinol. *Nutr. Cancer* **16**, 59–66.
[59] Telang, N. T.; Bradlow, H. L.; Osborne, M. P. (1992) Molecular and endocrine biomarkers in non-involved breast: Relevance to cancer chemoprevention. *J. Cell. Biochem.* **16**G, Suppl. 161–169.
[60] Messina, M. J.; Persky, V.; Setchell, K. D. R.; Barnes, S. (1994) Soy intake and cancer risk: A review of the *in vitro* and *in vivo* data. *Nutr. Cancer* **21**, 113–131.
[61] Lu, L. W.; Anderson, K. E.; Grady, J. J.; Nagamani, M. (1996) Effects of soya consumption for one month on steroid hormones in premenopausal women: Implications for breast cancer risk reduction. *Cancer Epidemiol. Biomarkers Prev.* **5**, 63–70.
[62] Hirohata, T.; Shigematsu, T.; Nomura, A.; Nomura, Y.; Horie, A.; Hirohata, I. (1985) Occurrence of breast cancer in relation to diet and reproductive history: a case-control study in Fukuoka, Japan. *Natl. Cancer Inst. Monographs* **69**, 187–190.
[63] Helferich, W. G. (1996) Paradoxical effects of the soy phytoestrogen, genistein, on growth of human breast cancer cells *in vitro* and *in vivo*. *Proceedings of a Meeting on Modulation of Chemical Toxicity and Risk Assessment* (in press).
[64] Safe, S. H. (1995) Environmental and dietary estrogens and human health: Is there a problem? *Environ. Health Perspect.* **103**, 346–351.

[65] Davis, D. L.; Bradlow, H. L. (1995) Can environmental estrogens cause breast cancer? *Scientific American* October issue, 166–172.

[66] Falck, F. Y.; Ricci, A. Jr.; Wolff, M. S.; Godbold, J.; Deckers, J. (1992) Pesticides and polychlorinated biphenyl residues in human breast lipids and their relation to breast cancer. *Arch. Environ. Health* **47**, 143–146.

[67] Wolff, M. S.; Toniolo, P.; Lee, E.; Rivera, M.; Dubin, N. (1993) Blood levels of organochlorine residues and risk of breast cancer. *J. Natl. Cancer Inst.* **85**, 648–652.

[68] Dewailly, E.; Dodin, S.; Verreault, R.; Ayotte, P.; Sauve, L.; Morin, J. (1994) High organochlorine body burden in women with estrogen receptor positive breast cancer. *J. Natl. Cancer Inst.* **86**, 232–234.

[69] Krieger, N.; Wolff, M. S.; Hiatt, R. A.; Rivera, M.; Vogelman, J.; Orentreich, N. (1994) Breast cancer and serum organochlorines: a prospective study among white, black and Asian women. *J. Natl. Cancer Inst.* **86**, 589–599.

[70] Sutherland, S. E.; Benard, V. B.; Keil, J. E.; Austin, H.; Hoel, D. G. (1996) Pesticides and twenty-year risk of breast cancer. *Am. J. Epidemiol.* **143**, Suppl. 135, abstract.

[71] van't Veer, P.; Lobbezzoo, I. E.; Martin-Moreno, J. M.; Guallar, E. L.; Gomez-Aracena, J.; Kardinaal, A. F. M.; Kohlmeier, L.; Martin, B. C.; Strain, J. J.; Thamm, M.; van Zoonen, P.; Baumann, B. A.; Kok, F. J. (1997) DDT (dicophane) and post menopausal breast cancer in Europe: case control study. *Br. Med. J.* **315**, 81–85.

[72] Wolff, M. S.; Toniolo, P. (1995) Environmental organochlorine exposure as a potential etiologic factor in breast cancer. *Environ. Health Perspect.* **103**, Suppl. 7, 141–145.

[73] Longnecker, M. P. (1994) Alcoholic beverage consumption in relation to risk of breast cancer: meta-analysis and review. *Cancer Causes Control* **5**, 73–82.

[74] Hunter, D. J.; Willett, W. C. (1996) Nutrition and breast cancer. *Cancer Causes Control* **7**, 56–68.

[75] Thomas, D. B. (1995) Alcohol as a cause of cancer. *Environ. Health Perspect.* **103**, 153–160.

[76] Reichman, M. E.; Judd, J. T.; Longcope, C.; Schatzkin, A. (1993) Effects of alcohol consumption on plasma and urinary hormone concentrations in premenopausal women. *J. Natl. Cancer Inst.* **85**, 722–727.

[77] Harvey, E. B.; Schairer, C.; Brinton, L. A.; Hoover, R. N.; Faumeni, J. F. Jr. (1987) Alcohol consumption and breast cancer. *J. Natl. Cancer Inst.* **78**, 657–661.

[78] van't Veer, P.; Kok, F. J.; Hermus, R. J.; Sturmans, F. (1989) Alcohol dose, frequency and age at first exposure in relation to the risk of breast cancer. *Int. J. Epidemiol.* **18**, 511–517.

[79] Young, T. B. (1989) A case control study of breast cancer and alcohol consumption habits. *Cancer* **64**, 552–558.

[80] Longnecker, M. P.; Newcomb, P. A.; Mittendorf, R.; Greenberg, R. E.; Clapp, R. W.; Bogdan, G. F.; Baron, J.; MacMahon, B.; Willett, W. C. (1995*a*) Risk of breast cancer in relation to lifetime alcohol consumption. *J. Natl. Cancer Inst.* **12**, 923–929.

[81] Longnecker, M. P.; Paganini-Hill, A.; Ross, R. K. (1995*b*) Lifetime alcohol consumption and breast cancer risk among postmenopausal women in Los Angeles. *Cancer Epidemiol. Biomarkers Prev.* **4**, 721–725.

[82] Royo-Bordonana, M. A.; Martin-Moreno, J. M.; Guallar, E.; Gorgoja, L.; van't Veer, P.; Mendez, M.; Huttunen, J. K.; Martin, B. C.; Kardinaal, A. F. M.; Fernandez-Crehuet, J.; Thamm, M.; Strain, J. J.; Kok, F. J.; Kohlmeier, L. (1977) Alcohol Intake and Risk of Breast Cancer: The Euramic Study. *Neoplasma* **44**, 150–156.

[83] Garfinkel, L.; Boffetta, P.; Stellman, S. D. (1988) Alcohol and breast cancer: a cohort study. *Prev. Med.* **17**, 686–693.

[84] Toniolo, P.; Riboli, E.; Protta, F.; Charrel, M.; Cappa, A. P. M. (1989) Breast cancer and alcohol consumption: a case-control study in Northern Italy. *Cancer Res.* **49**, 5203–5206.

[85] Howe, G.; Rohan, T.; DeCarli, A.; Iscovich, J.; Kaldor, J.; Katsouyanni, K.; Marubini, E.; Miller, A.; Riboli, E.; Toniolo, P.; Trichopoulos, D. (1991) The association between alcohol and breast cancer risk: evidence from the combined analysis of six dietary case control studies. *Int. J. Cancer* **47**, 707–710.
[86] Hunter, D. J.; Willett, W. C. (1993) Diet, size and breast cancer. *Epidemiol. Rev.* **15**, 110–132.

6 Molecular Epidemiology: Identification of Susceptible Subgroups

Christine B. Ambrosone

6.1 Abstract

Carcinogenesis is a complex process, with multiple factors interacting to drive the neoplastic cascade. Historically in epidemiologic studies, associations between exposures and disease outcomes have been made with the assumption that all individuals are equally susceptible to the effects of those exposures, although modified by factors such as age, gender, ethnicity and menopausal status. Pharmacogenetic studies have shown, however, that individuals vary in their metabolism of drugs, dietary components and carcinogens. There is also increasing evidence that variability in DNA repair capabilities, cell cycle control, and immune response may impact on cancer risk. Because of the interindividual genetic variability in a number of factors that will predict the carcinogenic potential of an exposure, risk factors may only be identifiable if the association between the exposure and the disease is quite strong, as in the case of smoking and lung cancer. Assessment of risk based on populations that are heterogeneous in response to an exposure may be biased by this variability in susceptibility, resulting in estimates of risk that are diluted or masked. Identification of susceptible subsets of the population, based on polymorphisms in genes involved in the line of defense between exposures and the initiation of carcinogenesis in cells, may more clearly identify factors that may increase cancer risk among some, but not all, individuals. However, the incorporation of molecular markers of susceptibility into epidemiologic studies may present a number of methodologic issues.

While molecular epidemiologic studies should more clearly elucidate carcinogenic mechanisms and our understanding of disease etiology, much of the molecular epidemiologic literature is rife with inconclusive data. Inconsistent study findings may be due to a number of factors, including biases and flaws in study design and analyses. Null findings in studies could be the result of inadequate power, that is, the ability to detect a true effect, resulting from sample sizes that are too small. On the other hand, small study numbers may also result in apparently strong exposure-disease relationships. When stratification by genotype or exposure exceeds the limits of the data, the exposure effect estimates begin to get further and further from the null. Odds ratios become enormous as a result of bias due to applying large-sample methods to excessively sparse data. Disparate results between molecular epidemiologic studies may also be related to characteristics of the control group to whom cases are compared, such as hospital, neighborhood, or population controls, or a convenience sample.

However, differences found between molecular epidemiologic studies may be real differences, particularly if there is ethnic variability between groups. Heterogeneity between study populations in relation to differential exogenous exposures that could affect enzyme induction or activity may also explain inconsistencies between studies. Because of all of the possible pitfalls and sources of bias in molecular epidemiologic studies, it is advisable that molecular epidemiologists seriously critique their own data and question every p value, in an effort to support the null hypothesis, until it is proven otherwise.

6.2 Introduction

Epidemiology is traditionally defined as the study of the distribution and determinants of disease in specified populations, for the good of public health [1]. For a number of years, this work was considered "black box epidemiology". It was not necessary to understand mechanisms or events in the continuum between exposures and disease outcomes to identify disease risk factors or to determine preventive strategies. For example, John Snow, often hailed as the "Father of Epidemiology" traced the source of cholera in a specific neighborhood to use of water from a particular pump [2]. There was little understanding of bacteria and contamination from poor sanitation; removal of the pump handle alleviated the problem. Black box epidemiology was sufficient even with a complex disease such as cancer. When physicians observed a dramatic increase in lung cancer incidence, formerly a rare disease, and noted that it was occurring primarily among cigarette smokers, it became clear that smoking was related to lung cancer etiology, even though mechanisms were unknown [3]. However, there are few exposure/disease relationships that are as straightforward as tobacco smoke and lung cancer. Particularly in the area of dietary carcinogens and anticarcinogens, relationships may be much less clear and associations difficult to tease out in epidemiologic studies.

Furthermore, even though we know that smoking causes lung cancer, the fact remains that only one in ten smokers gets the disease [4]. Clearly, not all smokers are equally susceptible to the carcinogenic effects of tobacco smoke. This variability in susceptibility may be due to a number of exogenous and endogenous factors. Epidemiologic, clinical, biochemical and molecular studies demonstrate that cancer is multifactorial and that the effects of multiple exposures are modulated by variability in metabolism of potentially carcinogenic agents, scavenging of reactive intermediates by "anticarcinogens", DNA repair, cell cycle control, and apoptosis.

Molecular epidemiology seeks to clearly elucidate these interactions by use of biomarkers of exposure, internal dose, biologically effective dose, early molecular changes, and preclinical disease endpoints. The role of inter-individual variability at a molecular level is crucial for our understanding of gene-environment interactions and for identification of subgroups of individuals who may be particularly sensitive to certain environmental exposures.

6.3 Identification of susceptible subgroups

As reviewed by Perera, it is clear that individuals are not uniformly susceptible to exposures that may contribute to carcinogenesis [5]. While it has been known for some time that individuals vary in their response to pharmaceutical agents from ten to several hundred-fold [6], this realization has not been consistently applied to epidemiologic studies. There has been an implied assumption in most epidemiologic studies that, with the exception of factors such as age, gender, ethnicity and menopausal status, which are often entered in regression models as possible confounding or interactive variables, all humans are equally susceptible to putative carcinogenic exposures. However, variability in activity levels of enzymes involved in the activation and detoxification of carcinogenic and anticarcinogenic substances is likely to modify cancer risk associated with exogenous risk factors. This variability in susceptibility could have important implications for public health, in that risk assessment needs to take into account the tremendous heterogeneity in target populations that may be specific for certain classes of chemical carcinogens.

This concept was illustrated in a study of colorectal polyps and cancer by Lang and Kadlubar [7]. In a majority of epidemiologic studies of colorectal cancer, meat consumption has been identified as a risk factor for colon cancer [8]. When meat is cooked, particularly at high temperatures or for a long period of time, a number of heterocyclic amines are formed [9]. Several of these are potent colon carcinogens in rodent models [10]. Metabolism of heterocyclic amines varies among individuals, depending, in part, on polymorphisms in enzymes involved in their metabolism, including cytochrome P450 1A2 (CYP 1A2), and *N*-acetyltransferases NAT1 and NAT2 [11]. Aromatic amines are also metabolized through these pathways. Several polymorphic sites at the *NAT2* locus result in decreased *N*-acetyltransferase activity [12, 13], and slow NAT2 phenotype and genotype has been associated with increased risk for bladder cancer [14, 15] and breast cancer associated with cigarette smoking [16]. Heterocyclic amines appear to be poor substrates for *N*-acetylation at the liver, however, and may, instead, be activated by an hepatic oxidative process by CYP 1A2. These metabolites may then circulate and be further activated in the target tissue by NAT2, among other enzymes [11]. In the above cited study, individuals who had rapid *N*-oxidation by CYP 1A2 *and* rapid *O*-acetylation by NAT2 had almost 3-times the colon cancer risk of persons with the slow phenotypes [7]. Risk was even greater for those with the 'at-risk' genotypes who were also in the highest tertile of consumption of well-done red meat. If these findings are replicated in a number of well-designed studies, the ultimate goal would be for individuals with 'at-risk' genotypes or phenotypes to be strongly advised to reduce red meat consumption.

6.4 Identification of etiologic agents in carcinogenesis

The identification of susceptible subgroups of individuals in relation to certain exposures may have important implications beyond risk assessment. By identification of an increased risk in certain subgroups, disease risk factors may be more clearly elucidated. For example, in a study by Welfare and colleagues [17], associations between colorectal cancer, *N*-acetyltransferase 2 (NAT2) genetic polymorphisms, and putative risk factors for the disease were evaluated. Several studies of the role of risk factors such as cigarette smoking and *N*-acetyltransferase 2 (NAT2) genotypes in risk of colorectal cancer have resulted in inconsistent findings. In the Welfare study, no association was observed between NAT2 genotype alone and cancer risk, but risk was increased among recent smokers with slow NAT2 genotype. Alternately, rapid NAT2 genotype increased risk among frequent consumers of red meat. The concept that risk is only observed among those with susceptible genotypes was also noted in our study of breast cancer, NAT2 and cigarette smoking [16]. As reviewed by Palmer and Rosenberg [18], there have been numerous studies of smoking and breast cancer with inconsistent results, and in this study, neither smoking nor NAT2 genotype increased breast cancer risk. However, when women were stratified by NAT2 genotype, those with slow NAT2 were at dose-dependent risk with smoking, particularly if they began smoking at an early age. These studies illustrate that heterogeneity in response to carcinogenic exposures may dilute or mask the true effects that exist among susceptible populations. They indicate that rather than forsaking pursuit of an effect in subgroups when there is not a main effect in the overall population, one should do precisely that to identify clear exposure-disease associations.

6.5 Caveat emptor!

The emergence of molecular epidemiology as a discipline has enabled researchers to study determinants of disease risk at a new and different level, and for epidemiologists, molecular biologists, toxicologists, clinicians, and others to cross disciplines in the study of carcinogenesis. This merging of fields may present a number of problems and issues, however, resulting not only from lack of cross-training, but also due to the nature of molecular epidemiology itself. While "molecular epidemiologic" studies performed by researchers from each of these perspectives may make important contributions to the literature, each working independently of the other may result in sometimes serious flaws in study design, implementation and interpretation.

For example, as the complexity of carcinogenesis becomes evident, it is clear that epidemiologists need to understand the pathophysiology of disease to evaluate exposure/disease associations. Studies performed to evaluate gene-environment interactions should be based on a solid knowledge of toxicology, biochemical metabolism, and molecular genetics. Too often in molecular epidemiology, study design and interpretation of data may be flawed by a lack of understanding of fundamentals of substrate and organ specificity in relation to susceptibility to carcinogenic exposures. Similarly, hospital or population-based studies performed by clinicians or primarily laboratory scientists may be hampered by the lack of training in methods of epidemiologic and statistical design. Perhaps more importantly, new issues arise within the field that may be particular to studies of this nature and that should be considered in interpretation of study results.

The molecular epidemiologic study of susceptibility inferred by genetic polymorphisms, and the elucidation of gene/environment interactions, should more clearly elucidate carcinogenic mechanisms and our understanding of etiology. But in fact, for many studies of polymorphisms in xenobiotic-metabolizing enzymes and cancer risk, and those of gene/environment interactions, results are in conflict with each other. The molecular epidemiologic literature is rife with inconclusive data. Disparate results between molecular epidemiologic studies may reflect true differences between study populations, or they may be due to flaws in study design and methodological issues.

One reason for inconsistencies in studies may be small sample sizes, resulting in inadequate power or spurious findings. Small sample sizes are common to molecular epidemiologic studies, not only because the molecular assays are expensive and thus, restrict the number of subjects recruited, but also because the means of analysis automatically cuts the population in half or in many instances, well below that, in stratified analyses. Effect modification, also described as interaction, occurs when the association between an exposure and disease outcome varies at different levels of a third variable. When it is hypothesized that an exposure/disease relationship may differ between subgroups of individuals, data are often stratified on the variable that is thought to cause the effect modification. This has been a common practice in epidemiology, with data analyzed separately by variables that may modify associations. For example, in studies of smoking and lung cancer, data are often evaluated separately for males and females, based on the evidence that, dose for dose, women may be more susceptible to the effects of tobacco smoke than men. Molecular epidemiology has extended this model of analysis to stratification by genotype for polymorphisms that may affect carcinogen activation and detoxification, DNA repair, and cell cycle control. Within specific genetic categories, associations may be evaluated between groups who are putatively "at risk" and those who are not. Similarly, the effect of a polymorphism in an enzyme involved in carcinogen metabolism may be evaluated separately within groups who are exposed to that carcinogen and those who are not. While this method of studying gene/environment interaction may allow opportunities to detect effects in subsets when no main effect is observed overall, it also is more likely to result in Type I (false positive) or Type II (false negative) errors. This can occur even in large studies, due to the fact that the numbers of subjects in each cell will be drastically reduced. Statistical power, that is, the ability to detect a true effect if it exists, is dependent upon sample size, the size of the effect to be detected, and the variability within the study population. If the study population is small and with extensive variability within the group, or the effect is weak, it is likely that no exposure/disease relationship will be noted, despite the fact that one does exist. On the other hand, the likelihood of "false positive" relationships is also heightened when stratification results in small numbers. Theoretically, one would stratify by genotype and, in a multiple logistic regression model, calculate odds ratios to estimate risk associated with an exposure, adjusting for multiple possible confounding factors or other known disease risk factors. As pointed out by Greenland and K. Rothman [19], however, when stratification has exceeded the limits of the data, the exposure effect estimates begin to get further and further from the null, and the odds ratio becomes enormous. This occurs as the result of bias due to applying large-sample methods to excessively sparse data. Thus, what may be interpreted as a strong effect because of the specificity of the exposure in an "at risk" population may, in fact, be a large odds ratio due to inappropriate use of complex analyses in small data sets. Particularly for molecular epidemiologic studies, it is very important to reconsider the necessity of application of proper statistical fundamentals to study design, and to understand the implications of performing complex data analysis on unavoidably small sample sizes. Clearly, innovative advances in methods for testing hypotheses in molecular epidemiologic studies are necessary.

Before long, the entire human genome will have been identified. DNA chips will be used to identify polymorphisms in a multitude of genes, resulting in the *ability* to evaluate complex interactions of numerous genetic polymorphisms and environmental exposures. With the tools to more clearly understand cancer etiology and to identify susceptible subgroups of individuals, also comes the responsibility to devise strategies to produce results that are valid. Because of all of the possible pitfalls and sources of bias in molecular epidemiologic studies, it is imperative that we as researchers seriously critique our own data and question every p value, in an effort to support the null hypothesis, until it is proven otherwise.

6.6 Conclusion

Molecular epidemiologic studies within the field of cancer research provide the potential for elucidation of the carcinogenic cascade at a molecular level. The use of multifactorial molecular epidemiologic models can simultaneously evaluate environmental exposures, genetic and hormonal factors that influence susceptibility. Identification of susceptible subsets of the population, based on polymorphisms in genes involved in the line of defense between exposures and the initiation of carcinogenesis in cells, may more clearly delineate factors that may increase cancer risk among some, but not all, individuals. Risk assessment and targeting of susceptible populations may be key to the effectiveness of strategies designed for cancer prevention.

References

[1] Last, J. M. (1988) *A dictionary of epidemiology*, 2nd ed., International Epidemiological Association, Oxford University Press, New York, p. 42.

[2] Snow, J. (1855) *On the mode of communication of cholera.* 2nd ed., London, England, J. Chjurchill. Reprinted as: *Snow on cholera.* (1936) New York, NY, Commonwealth Fund.

[3] Blum, A. (1993). Curtailing the tobacco pandemic. In: DeVita, V. T.; Hellman, S.; Rosenberg, S. A. (Eds.) *Cancer: Principles and practice of oncology*, 4th ed., J. B. Lippincott Co., Philadelphia, p. 480–482.

[4] Ginsberg, R. J.; Kris, M. G.; Armstrong, J. G. (1993) Cancer of the lung. In: DeVita, V. T.; Hellman, S.; Rosenberg, S. A. (Eds.) *Cancer: Principles and practice of oncology*, 4th ed., J. B. Lippincott Co., Philadelphia, p. 673.

[5] Perera, F. P.; Mooney, L. S.; Dickey, C. P.; *et. al.* (1996) Molecular epidemiology: Insights into cancer susceptibility, risk assessment, and prevention. *J. Natl. Cancer Inst.* **88**, 496–509.

[6] Nebert, D. W. (1991) Role of genetics and drug metabolism in human cancer risk. *Mutat. Res.* **247**, 267–281.

[7] Lang, N. P.; Butler, M. A.; Massengill, J.; *et. al.* (1994) Rapid metabolic phenotypes for acetyltransferase and cytochrome P450 1A2 and putative exposure to food-borne heterocyclic amines increase the risk for colorectal cancer or polyps. *Cancer Epidemiol. Biomarkers Prev.* **3**, 675–682.

[8] Potter, J. D. (1996) Nutrition and colorectal cancer. *Cancer Causes and Control* **7**, 127–146.

[9] Felton, J. S.; Knize, M.; Dolbeare, F. A.; Wu, R. (1994). Mutagenic activity of heterocyclic amines in cooked foods. *Environ. Health Perspect.* **106**, 201–204.

[10] Layton, D. W.; Bogen, K. T.; Knize, M. G.; Hatch, F. T.; Johnson, V. M.; Felton, J. S. (1995) Cancer risk of heterocyclic amines in cooked foods: an analysis and implications for research. *Carcinogenesis* **16**, 39–52.

[11] Kadlubar, F. F.; Butler, M. S.; Kaderlik, K.; Chou, H.-C.; Lang, N. P. (1992) Polymorphisms for aromatic amine metabolism in humans: relevance for human carcinogenesis. *Environ. Health Perspect.* **98**, 69–74.
[12] Bell, D. A.; Taylor, J. A.; Butler, M. A.; *et. al.* (1993) Genotype/phenotype discordance for human arylamine *N*-acetyltransferase (NAT2) reveals a new slow-acetylator allele common in African-Americans. *Carcinogenesis* **14**,1689–1692.
[13] Blum, M.; Demierre, A.; Grant, D. M.; Heim, M.; Meyer, U. A. (1991) Molecular mechanism of slow acetylation of drugs and carcinogens in humans. *Proc. Natl. Acad. Sci. USA* **88**, 5237–5241.
[14] Cartwright, R. A.; Omenn, G. S.; Gelboin, H. V. (Eds.) (1984) Epidemiological studies on *N*-acetylation and C-center ring oxidation in neoplasia. In: *Genetic variability in responses to chemical exposure,* Cold Spring Harbor Press, Cold Spring Harbor, NY, p. 359–68.
[15] Hanssen, H. P.; Agarwal, D. P.; Goedde, H. W.; Bucher, H.; Huland, H.; Brachmann, W.; Ovenbeck, R. (1985) Association of *N*-acetyltransferase polymorphism and environmental factors with bladder carcinogenesis. Study in a north German population. *Eur. Urol.* **11**, 263–266.
[16] Ambrosone, C. B.; Freudenheim, J. L.; Graham, S.; *et. al.* (1996). Cigarette smoking, *N*-acetyltransferase 2 genetic polymorphisms, and breast cancer risk. *J. Am. Med. Assoc.* **276**, 1494–501.
[17] Welfare, M. R.; Cooper, J.; Bassendine, M. F.; Daly, A. K. (1997) Relationship between acetylator status, smoking, diet and colorectal cancer risk in the north-east of England. *Carcinogenesis* **18**, 1351–1354.
[18] Palmer, J. R.; Rosenberg, L. (1993) Cigarette smoking and the risk of breast cancer. *Epidemiol. Rev.* **15**, 145–156.
[19] Greenland, S.; Rothman, K. (1998). *Introduction to stratified analysis. Modern epidemiology.* 2nd edition, Lippincott-Raven Publishers, Philadelphia, p. 258.

B Carcinogenic Factors: Exogenous

7 Relative Contributions of Chemical Carcinogens in the Diet *vs* Overnutrition – The Role of Individual Dose and Susceptibility

Werner K. Lutz and Josef Schlatter

7.1 Abstract

Based on average values for dietary exposure levels in Switzerland and carcinogenic potencies derived from animal bioassays, chemical carcinogens could by far not explain the cancer incidence attributed by epidemiologists to dietary factors. The discrepancy was explained by overnutrition to which a carcinogenic potency had been assigned from dietary restriction experiments [1]. Here, additional factors are added to the discussion. They focus on using individual rather than averaged data, both for exposure and susceptibility. First, for carcinogens which exhibit a sublinear (convex) dose response, the cancer incidence obtained by using an average exposure level will be lower than if individual exposure levels associated with particular dietary habits are taken into account. Second, carcinogenic factors, including those unrelated to the diet, can act synergistically, in particular at high dose. For instance, alcohol has a higher carcinogenic potency in the heavy smoker as compared with the non-smoker or the light smoker. Third, if an individual's diet is low in protective factors, such as fruits and vegetables, the potency of a diet-related carcinogen will be higher than for an individual who consumes more. The cancer-chemopreventive effect of a dietary constituent could, however, be limited to a certain dose range, and high dose levels might even result in an increased tumor incidence. This is exemplified with caffeic acid which produced a J-shaped dose response for the rate of cell division in the target organs for tumor induction. Taking all aspects into account, exposure to dietary carcinogens could be responsible for more cancer cases than based on estimates of average exposure and carcinogenic potency.

7.2 Chemical carcinogens in the diet *vs* overnutrition

In their review on "Quantitative Estimates of Avoidable Risks of Cancer", Doll and Peto stated that "it may be possible to reduce U.S. cancer death rates by practicable dietary means by as much as 35%" [2]. With cancer accounting for about one quarter of all deaths, this means that 8% of all deaths (i.e., 80,000 of 1 million lives) should be attributable to avoidable dietary carcinogens. In 1992, we published a quantitative analysis of the known risk factors, by compiling average daily intake estimates for Switzerland [3, 4] and calculating the risk of cancer on the basis of estimates of carcinogenic potencies in rodent bioassays [1]. Table 7.1 gives an updated summary of the published results, in descending order of estimated importance, for those factors which appeared to be responsible for 1 or more cancer cases per 1,000,000 lives.

Alcohol emerged as the most important carcinogenic factor even if its contribution is expected not to be as large as listed, in the absence of the synergism with smoking. For caffeic acid and saccharin, the theoretical values must be considered to overestimate the actual risk, in view of the available information on the mode of action and the respective dose-response relationships. For arsenic and cadmium, which also rank high, the potency at low dose is still an open question. Somewhat surprisingly, the wellknown genotoxic classes of polynuclear aromatic hydrocarbons (PAH), nitroso compounds (NOC), aromatic amines, or fungal and microbial contaminants did not account for more than a total of about 100 cancer cases per one million lives.

When we published this analysis six years ago, our favoured explanation to explain the discrepancy was the idea that overnutrition is likely to be the most important dietary carcinogen in industrialized countries. This view was based on an evaluation of dietary restriction experiments in rats and mice, where a dramatic reducing effect on spontaneous tumor formation had been seen. The 1200-rat BIOSURE study [5, 6] formed the basis to estimate a carcinogenic potency of overfeeding: The rats which were restricted to a food intake of 80% of *ad libitum* fed groups were assumed to be the "untreated" controls, and the increased tumor incidence seen in the animals fed *ad libitum* was assumed to be induced by the additional feed. The difference in tumor incidence, 23% in males and 18% in females, was associated with an additional food intake of 3.2 and 2.9 g per day per animal. A "carcinogenic potency" of the excess feed, expressed as a tumorigenic dose 50% (TD_{50} value) of 16 g/kg/d was deduced as an average value for males and females. Caloric overnutrition in Switzerland

Table 7.1: Theoretical estimate of the tumor incidence attributed to various dietary factors on the basis of average exposure levels in Switzerland and linear dose response for tumor incidence.

Chemical or factor	Daily intake [ng/kg]	Carcinogenic potency (inverse; TD_{50}) [mg/kg/d]	Tumor incidence [Cases per one million lives]
Alcohol		(Epid.)	8000(-)
Caffeic acid	1×10^6	>400	<1000
Arsenic	150	(Epid.)	400
Saccharin	5×10^5	>2000	<100
Cadmium	200	1.3	80
Polynucl. Arom. Hydrocarb.	200 (BaP equiv.)	3	30
2,3,7,8-TCDD	0.002	0.00007	10
Estragole	1000	50	10
Nitroso compounds	14	1	8
Aflatoxin B_1	0.25	0.02	6
Dieldrin	15	2	4
Zearalenone	100	30	2
Aromatic amines	50	15	2
Estradiol	2	1	1
Diethylhexylphthalate	2000	>1000	<1
"Overnutrition"	2×10^9	16,000	60,000

Adapted from W. K. Lutz and J. Schlatter (1992) Chemical carcinogens and overnutrition in diet-related cancer. *Carcinogenesis* **13**, 2211–2216.

was estimated to be about 5.5 kcal/kg per day, which is equivalent to a daily excess of 1.9 g "feed" per kg body weight. Taking into account the TD_{50} value derived above, almost all "diet-related" cancer deaths could be explained (60,000 in 1 million lives; bottom line of Table 7.1).

While there is growing evidence indeed that overnutrition is a prime carcinogenic dietary risk factor [7], a number of aspects have to be discussed which affect our analysis. All are related to the problem that we used average values for both dose and carcinogenic potency instead of estimating cancer risks on an individual basis.

7.3 Individual dietary habits resulting in high dose levels

The use of average exposure levels results in an underestimation of the cancer risk, if the dose-response curve is sublinear (convex). If some individuals in a population have a low intake and some have a high intake, the cancer risk in the population will be larger than with all individuals exposed to the average amount. As an example, the dose response for the induction of mouse bladder cancer by the arylamine 2-acetylaminofluorene is clearly sublinear, with a 1% incidence at 30 ppm in the diet, about 10% at 60 ppm, and 60% at 90 ppm (cumulative over 33 months) [8]. If all individuals in a population were exposed to 30 ppm, the incidence would be 1%. If, however, one third of the population is exposed to 0, 30, and 60 ppm each, the tumor incidence would be about 4% ($[1 \times 0 + 1 \times 1 + 1 \times 10]/3$), with the same total amount of arylamines ingested.

For many food items, large interindividual differences are expected. For instance, while the average daily per capita consumption of mushrooms reported for Switzerland is 5 grams, it is well understood that some people like to eat mushrooms and consume a lot while others will be close to dose zero. Exposure to mushroom-related carcinogens, e.g., hydrazine derivatives, is therefore expected to show very large interindividual variability. The same is expected to hold for the pyrolysis products PAH, nitroso compounds, and arylamines, associated, above all, with meat. While the average daily consumption is close to 200 g, the span between individuals will go from 0 (vegetarians) to 500 g and more.

In addition to the above-mentioned voluntary differences in carcinogen exposure, local habits and contaminations contribute further to widening the dose range of individual exposure. In the case of a pronounced non-linear dose response, the effect on cancer incidence in the population can be considerable.

7.4 Synergistic effects of exposure to other carcinogens

Exposure to other dietary or nondietary carcinogens can result in an increased sensitivity of individuals. Synergistic combination effects (in the meaning of supra-additive effects) are best known for smokers exposed to additional carcinogens. Smokers are much more susceptible to asbestos fibers in connection with the induction of lung cancer or to alcohol with respect to esophageal cancer. The data which demonstrate the latter point [9] can be used to show that the slope of the dose response for the induction of esophageal cancer by alcohol is much steeper for heavy smokers than for light smokers or nonsmokers, particularly in the lower dose range. Therefore, the carcinogenic potency of alcohol is individually different and a value taken on the basis of average levels will result in an underestimation.

7.5 Anticarcinogenicity of dietary factors

There is ample epidemiological evidence that a diet rich in fruits and vegetables is associated with a decreased cancer incidence. There can therefore be no doubt that malnutrition in the sense of a lack of protective factors can result in an increased susceptibility against carcinogenic factors of both endogenous and exogenous origin. With respect to the evaluation summarized in Table 7.1 this means that the potency of a carcinogen is dependent on the intake of protective factors. The use of an average potency estimate resulted in an underestimation of the importance of chemical carcinogens if the potency of the carcinogen is nonlinearly dependent on the intake of protective factors.

7.6 J-shaped dose response for tumor induction: Caffeic acid

While nonlinear dose-response relationships could be considered to be the rule rather than the exception for chemical carcinogenesis [10], J-shaped shapes receive increasing attention [11]. A J-shaped dose response describes the situation of a chemical being carcinogenic at high dose but anticarcinogenic at low dose. Caffeic acid could be discussed as an example. According to Table 7.1, caffeic acid represents the second most important dietary carcinogen, under the assumption of a linear dose response. Caffeic acid induced tumors in the forestomach and the kidney of both rats and mice, when administered at a 2 % level in the diet. Tumor induction was associated with toxicity and hyperplasia in the forestomach. Most tests for genotoxicity and mutagenicity were negative so that a higher rate of cell proliferation in connection with regenerative hyperplasia could be the driving force in tumor formation. This endpoint was investigated in rats as a function of dose [12], in order to establish a biologically based dose extrapolation from the tumorigenic dose in rats to a human exposure level.

After 4 weeks of feeding, the unit length labeling index for incorporation of bromodeoxyuridine into DNA of forestomach lining cells was increased at the higher dose levels of 0.4 and 1.64 %, but, astonishingly, was reduced by 46 percent at 0.14 % caffeic acid in the diet (Figure 7.1). The decrease was paralleled by a thickening of the epithelium of the forestomach at high dose and a thinning at low dose. In the proximal tubular cells of the kidney, a similar J-shape of the dose response was observed, while no effect was seen in the glandular stomach or the liver, both non-target organs for the tumor induction (data not shown).

If hyperplasia is indeed the main mode of carcinogenic action of caffeic acid, a cancer risk at low dose cannot be deduced from linear extrapolation of

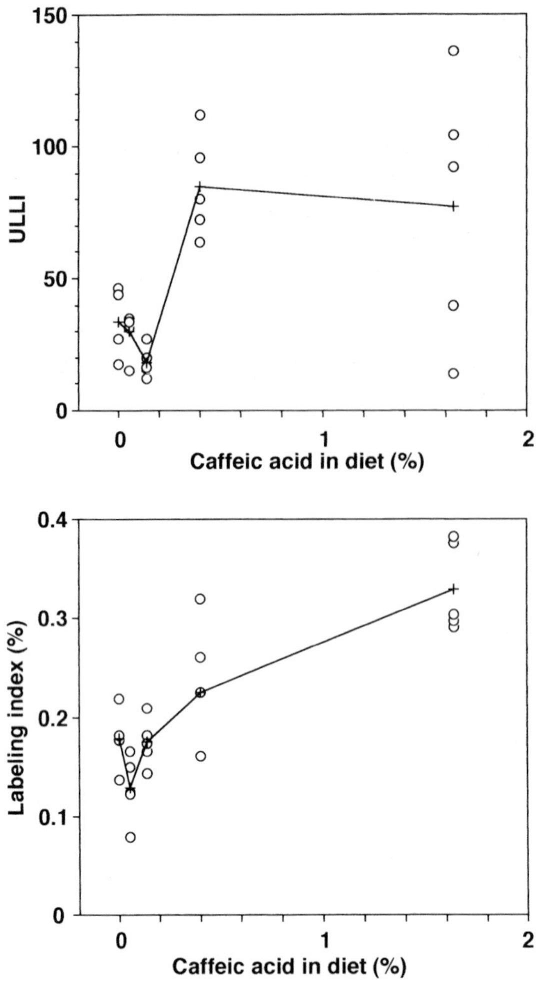

Figure 7.1: J-shaped dose response for the inhibition/stimulation of DNA replication in the forestomach (upper chart) and the kidney (lower chart) of male F344 rats, after 4 weeks of feeding with caffeic acid (CA) at different dietary concentrations. 5-Bromo-2'-deoxyuridine (BrdU) was administered *via* i.p. injection 2 hours before killing. o: individual animals; +: mean value (n=5 [4 in control group]). Statistical analyses based on Student's t-test (two-tailed). Data adapted from U. Lutz *et al.* (1997) *Fundam. Appl. Toxicol.* **39**, 131–137. *Upper chart:* Unit length labeling index (ULLI, number of BrdU-positive cells per mm section length) in the forestomach. Control *vs* 0.14 % CA: p=0.06 (decrease); control *vs* 0.4 % CA: p=0.003 (increase). *Lower chart:* Labeling index (number of BrdU-stained nuclei divided by the total number of nuclei) in proximal tubular cells of the kidney. Control *vs* 0.05 % CA p=0.085 (decrease); control *vs* 1.64 % CA p=0.0004 (increase).

the high dose data. The figure shown in Table 7.1 for caffeic acid, 1,000 cancer cases per one million lives, is unlikely to be correct, if a biologically founded dose response is used.

7.7 Conclusions

A diet-related carcinogen does not have a single potency for all individuals in a population. Therefore, the contribution of a particular carcinogen to the observed tumor incidence in the population cannot be estimated on the basis of an average exposure level and an average carcinogenic potency. Individual dose and individual susceptibility have to form the basis to estimate an individual cancer risk first. The cancer risk attributable to a single dietary factor in a given population then is the sum of the individual risks. In view of the non-linearities of dose-response relationships and synergistic effects, exposure to dietary carcinogens could be responsible for more cancer cases than based on averaged estimates of exposure and carcinogenic potency.

References

[1] Lutz, W. K.; Schlatter, J. (1992) Chemical carcinogens and overnutrition in diet-related cancer. *Carcinogenesis* **13**, 2211–2216.

[2] Doll, R.; Peto, R. (1981) The causes of cancer: Quantitative estimates of avoidable risks of cancer in the United States today. *J. Natl. Cancer Inst.* **66**, 1191–1308.

[3] Staehelin, H. B.; Luethy, J.; Casabianca, A.; Nonnier, N.; Mueller, H. R.; Schutz, Y.; Sieber, R. (1991) *Dritter Schweizerischer Ernährungsbericht*, Bundesamt für Gesundheitswesen, Bern, Switzerland.

[4] Aeschbacher, H. U. (Ed.) (1991) Potential carcinogens in the diet. *Mutat. Res., Spec. Issue* **259**, issue 3/4, p. 203–410.

[5] Roe, F. J. C. (1991) 1200-rat BIOSURE study: design and overview of results. In: Fishbein, L. (Ed.) *Biological effects of dietary restriction*. Springer, Berlin, p. 287–304.

[6] Roe, F. J. C.; Lee, P. N.; Conybeare, G.; Kelly, D.; Matter, B.; Prentice, D.; Tobin, G. (1995) The BIOSURE study: influence of composition of diet and food consumption on longevity, degenerative diseases and neoplasia in Wistar rats studied up to 30 months post weaning. *Food Chem. Toxic.* **33**, Suppl. 1, 1S–100S.

[7] Hart, R. W.; Neumann, D. A.; Robertson, R. T. (Eds.) (1995) *Dietary restriction: Implications for the design and interpretation of toxicity and carcinogenicity studies*. ILSI Press, Washington, DC, p. 396.

[8] Littlefield, N. A.; Farmer, J. H.; Gaylor, D. W. (1980) Effects of dose and time in a long-term, low-dose carcinogenic study. *J. Environ. Pathol. Toxicol.* **3**, 17–34.

[9] Tuyns, A. J.; Pequignot, G.; Jensen, O. M. (1977) Le cancer de l'oesophage en Ille-et-Vilaine en fonction des niveaux de consommation d'alcool et de tabac. Des risques qui se multiplient. *Bull. Cancer (Paris)* **64**, 45–60.

[10] Lutz, W. K. (1990) Dose response relationship and low dose extrapolation in chemical carcinogenesis. *Carcinogenesis* **11**, 1243–1247.

[11] Lutz, W. K. (1998) Dose-response relationships in chemical carcinogenesis: superposition of different mechanisms of action, resulting in linear-sublinear curves, practical thresholds, J-shapes. *Mutat. Res.* **405**, 117–124.
[12] Lutz, U.; Lugli, S.; Bitsch, A.; Schlatter, J.; Lutz, W. K. (1997) Dose response for the stimulation of cell division by caffeic acid in forestomach and kidney of the male F344 rat. *Fundam. Appl. Toxicol.* **39**, 131–137.

8 Alcohol and Cancer

Helmut K. Seitz and Nils Homann

8.1 Abstract

A great number of epidemiological data have identified chronic alcohol consumption as a significant risk factor for upper alimentary tract cancer including cancer of the oropharynx, larynx, esophagus, and for the liver. In contrast to those organs, the risk by which alcohol consumption increases cancer in the large intestine and in the breast is much lower. Nevertheless, carcinogenesis can already be enhanced with relatively low daily doses of ethanol. Considering the high prevalence of these tumors, even a small increase in cancer risk is of great importance, especially in those individuals who exhibit already a higher risk for other reasons. The epidemiological data on alcohol and other organ cancers are controversial and there is at present not enough evidence for a significant association. Although the exact mechanisms by which chronic alcohol ingestion stimulates carcinogenesis are not known, experimental studies in animals support the concept that ethanol is not a carcinogen, but under certain experimental conditions a cocarcinogen and/or a tumor promotor. The metabolism of ethanol leads to the generation of acetaldehyde and free radicals. These highly active compounds bind rapidly to cell constituents and to DNA. Acetaldehyde decreases the DNA repair mechanisms and the methylation of cytosine in DNA. It also traps glutathion, an important peptide in detoxification. Furthermore, it leads to chromosomal aberration and seems to be associated with tissue damage and secondary compensatory hyperregeneration. The production of acetaldehyde is, among others, genetically determined due to polymorphysm of alcohol dehydrogenase and acetaldehyde dehydrogenase. Other mechanisms by which alcohol stimulates carcinogenesis include induction of cytochrome P450 2E1 associated with an enhance activation of various procarcinogens, a change in the metabolism and distribution of carcinogens, alterations in cell cycle behavior such as cell cycle duration, leading to hyperregeneration, nutritional deficiencies such as methyl, vitamin A, folate, pyridoxale phosphate, zinc and selene deficiency, and an alteration of the immune system possibly resulting in increased susceptibility to certain viral infections such as hepatitis B and hepatitis C virus.

8.2 Introduction

The concept that chronic alcohol consumption enhances cancer risk in certain organs is not new. Almost a century ago, French pathologists discovered the association between heavy chronic alcohol consumption and the development of esophageal cancer [1]. This early observation was followed by a great number of epidemiological studies, which showed a striking positive correlation between chronic alcohol ingestion and the occurrence of cancer in the oropharynx, larynx and esophagus. Alcohol intake also favours the development of liver cancer in the cirrhotic liver. In addition, during the last decade countless numbers of case-control and prospective studies have identified the large intestine, especially the rectum and the female breast as additional target organs, in which alcohol even at lower doses stimulates cancer growth. In 1978 the first workshop on Alcohol and Cancer was held at the NIH, and at this time the mechanisms by which alcohol affects carcinogenesis were completely unclear. Meanwhile, intensive research has focused on such mechanisms and has elucidated some co-carcinogenic and promoter effects of ethanol. This review summarizes the epidemiology on alcohol and cancer and describes possible mechanisms by which chronic alcohol consumption stimulates carcinogenesis.

8.3 Epidemiology and experimental carcinogenesis

8.3.1 Upper alimentary tract

In France, Lamu already reported at the beginning of this century that absinth drinkers have an increased risk of developing esophageal cancer [1]. Meanwhile a great number of epidemiological studies have demonstrated a significant correlation between alcoholism and the development of oropharyngeal, laryngeal, and esophageal cancer [2]. It was demonstrated that heavy drinkers of highly concentrated alcoholic beverages have a 10- to 12-fold increased risk to develop tumors in the mouth, pharynx and larynx, while this risk was significantly lower when beer and wine were consumed [3]. In addition, alcohol abuse is often associated with heavy smoking. These factors have a synergistic effect on carcinogenesis in the upper alimentary tract. In a carefully designed French study Tuyns was able to demonstrate that alcohol consumption of more than 80 g per day (approximately 1 bottle of wine) increases the relative risk (RR) of esophageal cancer by a factor of 18, while smoking alone of more than 20 cigarettes has an increased RR by a factor of 5. Both together stimulate the risk synergistically by a factor of 44 [4]. It was calculated that 76% of all cancers could be prevented by avoiding smoking and alcohol consumption [2]. More recently, an epidemiological study by Maier and coworkers showed that 90% of all patients with head and neck cancer consumed alcohol regularly in amounts almost double as high as in a control group [5]. They found a significant dose-response relationship. If the RR for a person with a daily alcohol consumption of 25 g was assumed to be 1, the controlled RR increases steadily with increasing alcohol dosage and reaches a value of 32.4 when 100 g alcohol per day were consumed. These RR values are comparable with the dose reported by others. Tuyns and coworkers found an RR of 12.5 for hypopharynx carcinoma, of 10.6 for epipharynx carcinoma, of 2.0 for supraglottic larynx carcinoma, and of 3.4 for glottic and subglottic larynx carcinoma when 121 g alcohol were consumed daily [6]. Furthermore, Bruguere and coworkers [7] found a significantly higher RR for oral cancer which was 13.5, when 100–159 g alcohol were consumed daily. They found an RR of 15.2 for oropharynx carcinoma and of 28.6 for hypopharynx carcinoma. It is noteworthy that even with those high daily alcohol doses the alcohol-associated cancer risk is not saturable. If alcohol is consumed excessively with more than 160 g per day there is a further increase in cancer risk (oral cancer RR=70, oropharyngeal cancer RR=70, hypopharyngeal cancer RR=143). Chronic alcohol consumption and smoking have an independent risk on cancer development in the head and neck area. Tuyns *et al.* emphasized that 68% of the risk of those tumors can be attributed to alcohol alone [6]. As expected, smoking has a higher risk compared to alcohol abuse for the oral cavity and the pharynx, while this relationship is vice versa for the esophagus.

Table 8.1 summarizes the effect of chronic ethanol consumption on chemically induced upper aerodigestive tract carcinogenesis. It shows that local application of alcohol on the oral and esophageal mucosa increases the occurrence of tumors probably due to an irritational effect of alcohol [8–12]. When ethanol was given systemically [13–22], in most of the studies with some exceptions [17, 18, 20, 21, 23] a stimulatory effect on carcinogenesis was noted. Surprisingly, both, an enhancement of tumor initiation [15] and promotion [20] has been reported.

Table 8.1: Influence of ethanol on chemically induced upper aerodigestive tract carcinogenesis.

Species	Carcinogen	Ethanol application	Target organ	Ethanol effect	References
local effect on skin and mucosa					
mouse	BP, oral	as a solvent	esophagus	stimulation	21
hamster	DMBA, local	as a solvent	oral mucosa	stimulation	22
hamster	DMBA, local	as a solvent	oral mucosa	stimulation	23
mouse	DMBA, local	as a solvent	skin	stimulation	24
hamster	DMBA, local	before ethanol diet	oral mucosa	stimulation	25
aerodigestive tract					
rat	DENA, i.g.	30 % i.g., with carcinogen	esophagus	stimulation	26
rat	NMBA, i.g.	4 % DW, continously, Zn-deficiency	esophagus	stimulation	27
rat	MPNA, s.c.	25 % DW., continously	esophagus	no effect	35
rat	NNP, in diet	50 % intrapharyngeal and 10 % d. w.	esophagus	no effect	36
rat	NMBA	5 % LD, before and with carcinogen	esophagus	inhibition	33
rat	NMBA	5 % LD, after carcinogen	esophagus	stimulation	33
rat	DMNA, p.o.	10 % DW with carcinogen	esophagus	stimulation	28
mouse	NMBA, i.g.	6 % LD, after carcinogen	esophagus	stimulation	29
hamster	NPYR, i.p.	5 % LD, before and with carcinogen	nasal cavity	stimulation	30
rat	NNN, p.o.	6 % LD, before and with carcinogen	nasal cavity	stimulation	31
rat	NNN, s.c.	6 % LD, before and with carcinogen	nasal cavity	no effect	31
hamster	NNN, i.p.	5 % LD, before and with carcinogen	trachea	no effect	30
mouse	DENA, p.o.	10 % DW with carcinogen	forestomach	stimulation	32
rat	MNNG, p.o.	20 % i.p., after carcinogen	stomach	stimulation	34

BP = Benzo[α]pyrene, DMBA = Dimethylbenzanthracene, DENA = Diethylnitrosamine, MPNA = Methylphenylnitrosamine, NNP = *N*-Nitrosopiperidine, NPYR = *N*-Nitrosopyrrolidine, NNN = *N*-Nitrosonornicotin, DMNA = Dimethylnitrosoamine, NMBA = Nitrosomethylbenzylamine, LD = liquid diet, DW = Drinking Water, i.g. = intragastrically, i.p. = intraperitoneally, p.o. = orally, s.c. = subcutaneously

8.3.2 Liver

Cirrhosis of the liver is the major prerequisite for the development of hepatocellular cancer (HCC). Since infection with hepatitis B (HBV) and C virus (HCV) also leads to cirrhosis of the liver followed by an increased occurrence of HCC and since alcoholics are often infected by those viruses, the exact risk of alcohol as compared to HBV and HCV etiology in the development of HCC is still not exactly defined. Almost all prospective and retrospective case-control studies in western countries indicate that the incidence of HCC among alcoholics is above the expected level [24]. However, variable prevalences of HCC in alcoholic cirrhosis have been reported. With some exceptions generally lower incidence rates have been reported in western countries (<15 %) which showed some increased trends within the last two decades, while in Japan the prevalence of HCC in alcoholic cirrhosis increased at the rate of 1.0 % per annum during 1976 to 1985, reaching a 25 % incidence rate [25]. The higher prevalence in Asia may be linked to the increased concomitant viral infection. The observed increase of the incidence of HCC worldwide may be partially related to the prolongation of survival time for patients with alcoholic cirrhosis due to improved treatment.

Although the effect of abstinence on the development of HCC was variable in various studies, it has been reported that after cessation of drinking the risk to develop HCC increased and therefore it was speculated, that this may be due to changes in cell regeneration after alcohol withdrawal which will be discussed below. However, abstinence improves hepatocellular damage due to alcohol and prolongs survival time significantly which by itself increases the chance to develop HCC.

Most animal experiments with respect to hepatocarcinogenesis have been performed with nitrosamines as inducing agents. Almost all these studies showed an inhibition of carcinogenesis with alcohol, but on the other hand an enhancement in the incidence of extrahepatic tumors such as those in the nasal cavity, trachea and esophagus [13, 17, 18, 26–28]. Only if additional manipulations were added such as the administration of a methyl-deficient [29] or low carbohydrate [30] diet or partial hepatectomy [31], hepatic carcinogenesis was stimulated by alcohol. A striking enhancement of hepatic carcinogenesis was also observed when alcohol and the procarcinogen were given strictly alternatively to avoid an interaction between alcohol and carcinogen metabolism [32, 33].

8.3.3 Large intestine

In 1974 Breslow and Enstrom were the first who considered the possibility of an association between beer consumption and the occurrence of rectal cancer [34]. Up to date 7 correlational studies, 34 case-control studies and 17 prospective cohort studies have been performed to elucidate the role of alcohol in the development of colorectal cancer [35]. An association was found in five of the seven correlational studies and in half of the 34 case-control studies. In the majority of the case-control studies (10 out of 12) using community controls such a correlation was found, suggesting that the absence of an association when hospital controls were used is due to a high prevalence of alcohol consumption/alcohol related diseases in the hospital controls. Eleven of the 17 cohort studies also demonstrate a positive association with alcohol. A positive trend with respect to dose-response was found in 5 of the 10 case-control studies and in all prospective cohort studies in which this factor had been taken into consideration.

Six studies have investigated the effect of chronic alcohol consumption on the occurrence of adenomatous polyps of the large bowel [35]. In 5 of them such a correlation was observed. In addition, an RR increase of hyperplastic polyps of the distal colon and rectum was also observed with increasing amounts of alcohol [36]. When more than 30 g alcohol per day were consumed, the RR for men was 1.8 and for women 2.5.

Finally, alcohol may influence the adenoma-carcinoma sequence at different early steps as very recently reported by Boutron *et al.* [37]. In conclusion, epidemiological data are still somewhat controversial, but it seems that chronic ethanol ingestion even at low daily intake (10–40 g), especially consumed as beer, results in a 1.5 to 3.5-fold risk of rectal and to a lesser extent of colonic cancer in both sexes, but predominantly in males. Most recently, these data have been reviewed in detail by a panel of European experts at the WHO Consensus Conference on Nutrition and Colorectal Cancer in Stuttgart, Germany and it was stated that more than 20 g alcohol per day increases the risk of colorectal cancer.

Table 8.2 summarizes the effect of chronic alcohol consumption on colorectal cancer. In two of the eight studies ethanol was given in the drinking water [38, 39], and the results of these experiments have therefore to be questioned. When the two procarcinogens dimethylhydrazine (DMH) and azoxymethane (AOM) were used to induce colorectal tumors, different results were reported depending on the experimental conditions [40–43]. In these studies it is important to note that both compounds need metabolic activation by cytochrome P450 dependent microsomal enzymes to become carcinogenic. The results of these studies depend on the ethanol dose used and on the timing of ethanol administration. The conclusions derived from those experiments are as follows:

- The modulation of experimental colonic tumorigenesis by chronic dietary beer and ethanol consumption is due to alcohol rather than to other beverage constituents.
- The tumorigenesis in the right and left colorectum is affected differently by alcohol and may depend on the levels of alcohol consumption. Thus, high alcohol intake (18–33 % of total calories) inhibits carcinogenesis in the right colon and has no effect on the left colon, while lower ethanol consumption (9–12 % of total calories) enhances tumor development in the left colon without effect on the right colon.
- Ethanol affects carcinogenesis during the preinduction and/or induction phase, including carcinogen metabolism, but not in the postinduction phase (promotion).
- An interaction between ethanol and procarcinogen metabolism does occur and this may influence tumor incidence.

Table 8.2: Effect of ethanol on chemically induced colorectal carcinogenesis in rats.

Carcinogen	Ethanol administration	Ethanol effect	References
DMH, s.c.	6 % l.d. (36 % total calories), preinduction	Increased rectal but not colonic tumors	39
DMH, s.c.	5 % DW, induction	No effect	37
DMH, s.c.	5 % DW, preinduction/induction	No effect	38
DMH, s.c.	6 % LD (36 % total calories), preinduction	No effect	40
AMMN, i.r.	6 % LD (36 % total calories), preinduction/induction	Increased rectal tumors.	43
AOM, s.c.	LD (11 %, 22 %, 33 % total calories), preinduction/induction, postinduction	Inhibition of tumor development in the left but less in the right colon. Higher ethanol intake has a stronger inhibitory effect. No effect, when ethanol is given in the postinduction phase.	41
AOM, s.c.	LD (9 %, 18 % total calories ethanol), (12 %, 23 % total calories beer), preinduction/induction	High ethanol inhibits tumors in the right, but not in the left colon, while low ethanol enhances tumors in the left colon, but not in the right colon. No effect of beer.	42
AMMN, i.r	i.g. (4.8 g/kg body weight per day), preinduction/induction	Increased rectal tumors. Carcinogenesis was further stimulated when cyanamide, an acetaldehyde dehydrogenase inhibitor, was administered additionally.	44

DMH = 1,2-dimethylhydrazine; AMMN = Azetoxymethyl-methylnitrosamine; AOM = Azoxymethane; s.c. = subcutaneously; i.r. = intrarectally; i.g. = intragastrically, LD = Liquid Diet; DW = Drinking Water.

It must be emphasized that in one experiment with DMH, ethanol ingestion only enhanced tumor development in the rectum, but not in the remaining large intestine [40]. In this study ethanol was given during acclimatization and initiation, but at the time of procarcinogen application ethanol was not present in the body. In a similar study by McGarrity and coworkers, these results could not be confirmed [41]. In addition, in two other animal experiments the primary carcinogen acetoxymethylmethylnitrosamine (AMMN) was used to induce rectal tumors [44, 45]. This carcinogen does not need metabolic activation to exert its carcinogenic effect. It is applied locally to the rectal mucosa of rats and the animals were endoscoped regularly. Since chronic ethanol administration either as liquid diet or intragastrically accelerates the appearance of rectal tumors induced by AMMN, it seems most likely that alcohol enhances carcinogenesis, at least in part, by local mechanisms in the rectal mucosa and not only by increasing the activation of procarcinogens.

8.3.4 Breast

During the last decade a great number of epidemiological studies have identified alcohol as one of several risk factors for breast cancer, usually but not always showing a 1.2 to 2.0-fold RR as reported in 1994 in a large metaanalysis of the data available at this time [46]. Strong evidence was found supporting a dose-response relation in both case-control and follow-up epidemiological data. The relative risk of breast cancer at an alcohol intake of 24 g of absolute alcohol per day relative to non-drinkers was found to be 1.4 in the case-control and 1.7 in the prospective studies. In a recent review article on breast cancer various risk factors for breast cancer were compared and it was interesting to note that the RR for breast cancer was about 2 when three drinks or more per day were consumed as compared to a relative risk of 3 associated with the exposure to radiation due to the atomic bomb explosion in Hiroshima [47]. At the 8th Congress of the International Society for Biomedical Research on Alcoholism (ISBRA) a symposium was held on the interaction between alcohol and other risk factors for breast cancer and it was concluded that there is enough epidemiological evidence to support a modest association between alcohol ingestion and breast cancer risk [48]. An international panel of experts at a recent WHO Consensus Conference on Nutrition and Cancer in Stuttgart, Germany, came to a similar conclusion.

Animal experiments with respect to breast cancer are limited. There is no consistent evidence that alcohol enhances the formation of neither spontaneous or of dimethylbenzanthracene (DMBA) induced mammary tumors when ethanol is consumed continuously in both stages of tumor development [48]. When ethanol consumption is limited to only the initiation or promotion stage there is evidence that ethanol may be a weak cocarcinogen and/or a promoter in the methylnitrosourea-(MNO) and DMBA-induced mammary tumor models [49, 50]. However, there is no consistent dose-response relationship between alcohol intake and carcinogenesis. Ethanol also seems to augment mammary tumor progression [51].

8.3.5 Other organs

Epidemiological research has also found other organs in which chronic alcohol consumption increases cancer risk including stomach, pancreas, lungs, bladder, prostate and skin (melanoma). However, these data are controversial and up to now there is not enough evidence to link chronic alcohol consumption as a risk factor to those cancer sites.

8.4 General mechanisms by which alcohol modulates carcinogenesis

Experiments in which alcohol was given chronically to rodents have shown that alcohol per se is not a carcinogen since animals with lifelong chronic exposure to alcohol do not develop cancer more frequently than controls [52]. Since ethanol modulates chemically induced carcinogenesis, it has to be defined as a tumor promoter and/or cocarcinogen. Multiple mechanisms increase alcohol-associated cancer development (Figure 8.1). Some of these, including the consequences of alcohol metabolism, cytochrome P450 2E1 (CYP 2E1) induction, modulation of cell regeneration, and nutritional deficiencies, may be relevant in a variety of tissues with local quantitative and qualitative differences.

8.4.1 Sources of carcinogen intake

Alcoholics who smoke have an increased carcinogen intake with two sources of carcinogens: firstly, certain alcoholic beverages like some types of whisky, vermouth, sherry, beer and wine may contain carcinogenic substances including polycyclic hydrocarbons, nitrosamines and asbestos fibers [53]; secondly, an increased carcinogen load through smoking since tobacco smoke contains a great number of various carcinogens including polycyclic hydrocarbons and nitrosamines. In addition, dietary carcinogens have also been considered. An epidemiological study on aflatoxin exposure demonstrated that the daily consumption of 24 g of ethanol or more increases the risk of developing HCC induced by 4 μg of dietary aflatoxin B1 (AFB1) by a factor of 35 [54]. Finally, simultanous exposure to vinylchloride (VC) at the working place and to alcohol has an enhancing effect on the occurrence of HCC in men [55].

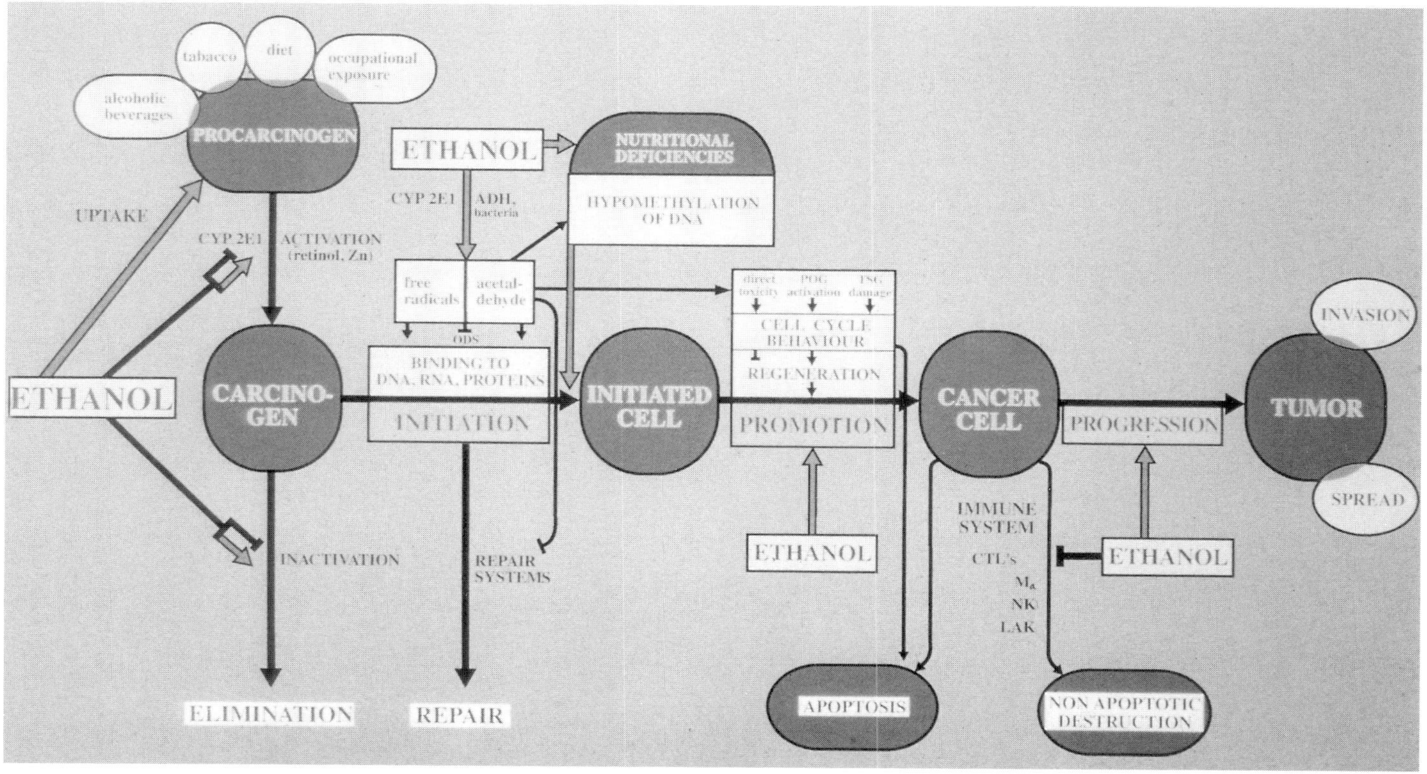

Figure 8.1: Effect of ethanol on carcinogenesis. Ethanol affects procarcinogen activation and possibly carcinogen inactivation. It also leads to an increased uptake of carcinogens into the cells. Ethanol is metabolized by CYP 2E2 and ADH to free radicals and AA which both can bind to cell macromolecules including DNA, especially since detoxification systems are depressed. Alcohol also leads to nutritional deficiencies, some of which (including vitamin A, E and zinc deficiency) stimulate carcinogenesis. Methyl and folate deficiency together with the action of AA result in hypomethylation of DNA and, in addition, AA also hinhibits the nuclear repair system. Direct toxicity (alcohol/AA) modulates cell cycle behavior resulting in regenerativity changes. Alcohol also acts as a promoter, in particular in the liver, leads to an inhibition of the immune system and may stimulate cancer progression (for more details see text).
AA = Acetaldehyde, CYP 2E1 = Cytochrome P450 2E1, ADH = Alcohol dehydrogenase, ODS = Oxidative defense system, POG = Proto-oncogen, TSG = Tumor suppressor gene, CTL = Cytotoxic thymus-dependent lymphocates, Ma = Activated macrophages, NK = Natural killer cells, LAK cells = Lymphokine activated killer cells.

8.4.2 Ethanol metabolism and its link to carcinogenesis

Ethanol is predominantly metabolized in the liver by alcoholdehydrogenase (ADH) and CYP 2E1. Although quantitatively much lower, ethanol metabolism also occurs in a variety of other tissues and in gastrointestinal bacteria. Acetaldehyde (AA), the extremely toxic first intermediate of alcohol metabolism, binds rapidly to cellular proteins and possibly to DNA [56] which results in morphological and functional impairment of the cell. In addition, AA-adducts represent neoantigens leading to the production of specific antibodies and to the stimulation of the immune system. AA has well known mutagenic and carcinogenic effects which include inflammation and metaplasia of tracheal epithelium, induction of laryngeal carcinoma in animals, inhibition of DNA repair, delay in cell cycle progression, stimulation of apoptosis, and enhanced cell injury associated with hyperregeneration [57–60]. According to the International Agency for Research on Cancer there is sufficient evidence to identify AA as a carcinogen in animals [61].

8.4.2.1 Generation of acetaldehyde via ADH and by gastrointestinal bacteria

ADH is not only present in the liver, but also in the gastrointestinal mucosa. In contrast to the liver gastrointestinal mucosa contains not only class I ADH, but also class IV ADH (σ-ADH). Its highest activity exists in the epithelium of the esophagus and the oropharynx. Although its Km is high, because of the high ethanol concentration in the upper alimentary tract, the enzyme is completely saturated after ethanol intake. σ-ADH also metabolizes longer alcohols such as butanol or propanol to their corresponding aldehydes which have been shown to affect mucosal integrity. σ-ADH also detoxifies the dietary carcinogen nitrobenzaldehyde and it is of considerable interest that a large percentage of Japanese population lack this enzyme since nitrobenzaldehyde occurs in food and gastric cancer is the most common cancer in Japan [62]. Thus, the production of AA via σ-ADH is especially high in the upper alimentary tract resulting in elevated AA levels. AA is also produced via class I ADH, especially ADH3 for which a polymorphism exists. Most recently, it has been shown that individuals with ADH3*1 which has a significantly higher V_{max} compared to ADH3*2 are at an increased risk for esophageal cancer when consuming more than 60 g ethanol per day [63]. AA is further metabolized by various AA-dehydrogenases (ALDH). Recently, a deficiency of ALDH2 has been found in alcoholics who are susceptible to esophageal cancer [64]. Those individuals cannot metabolize AA adequately and thus accumulate AA.

Class I and class IV ADH also occur in the colorectal mucosa and it seems at least from one study that its activity is higher in the rectum than in the remaining colon which could lead to higher AA concentration in the rectal mucosa [65]. Most recently a change in ADH pattern during colorectal carcinogenesis has been observed demonstrating an increased expression of σ-ADH in adenomatous polyps as compared to the adjacent normal mucosa [66].

AA can also be produced from ethanol by gastrointestinal bacteria including *Helicobacter pylori* [67] and such present in the oropharynx and in the faeces [68–70]. AA production by oropharyngeal bacteria may be especially important in heavy alcoholics with poor dental status and bacterial overgrowth. More recent data have shown that AA is also produced by colorectal bacteria [45, 70]. This production is extremely high leading to mucosal AA concentrations which are per gram of tissue much higher than those observed in the liver [45]. Since AA concentrations measured in the colorectal mucosa correlated significantly with tissue regenerativity, it was speculated that AA results in mucosal injury leading to secondary compensatory hyperregeneration [60].

8.4.2.2 Induction of cytochrome P450 2E1 and production of free radicals

It is important to note that chronic alcohol consumption leads to an induction of microsomal CYP 2E1 which is capable to metabolize ethanol to AA. This cytochrome is also involved in the metabolism of various xenobiotics including procarcinogens [71]. More recently, it has been shown in the liver that the concentration of CYP 2E1 can be correlated with the generation of hydroxyethyl radicals and thus with lipid peroxidation [72]. These biochemical observations are associated with changes in liver morphology. The induction of CYP 2E1 resulted in an enhanced hepatocellular injury and inhibition of CYP 2E1 was associated with an improvement of the liver lesions. It was concluded that this is mainly due to stimulation and inhibition of free radical formation.

Besides the hydroxyethyl radical, other types of free oxygen radicals and also radicals of unsaturated fatty acids do occur during intermediary metabolism. Other sources for radical generation are the mitochondrial electron transport system of the respiratory chain leading to superoxide radicals, the NADH-dependent cytochrome C reductase, the aldehyde and xanthine oxidase, and the NADPH oxidase in neutrophils [73]. Free radicals initiate predominantly lipid peroxidation, but also react rapidly with cell constituents including DNA and may lead to cancer initiation. Oxygen radicals can indeed lead to copper dependent formation of etheno-DNA adducts in the liver [74]. Under normal conditions such toxic free radicals are detoxified by several defense mechanisms including the action of glutathion, α-tocopherol, superoxide dismutase, catalase, and glutathion peroxidase. Since chronic alcohol consumption leads to a decrease of all these factors [75], an adequate detoxification of free radicals does not occur in the chronic alcoholic.

The role of free radicals in upper alimentary tract cancer has been demonstrated most recently in an animal study. Eskelson and coworker reported that chronic alcohol consumption increases the carcinogenesis induced by *N*-nitrosomethylbenzylamine (NMBA) in the esophagus which was associated with an increased free radical production and which was inhibited by administration of the scavenger α-tocopherol [16].

8.4.3 Enhanced CYP 2E1-mediated procarcinogen activation

The induction of CYP 2E1 also increases the conversion of various xenobiotics including procarcinogens to potentially toxic metabolites including dimethylnitrosamine (DMN), AFB1, VC and possibly nitrosopyrrolidone (NPY) and DMH [71]. The induction of CYP 2E1 not only takes place in the liver, but also in the gastrointestinal tract. Increased CYP 2E1 concentrations after chronic ethanol ingestion up to 3-fold have been reported in the oropharynx [76] and in the mucosa of the small [77] and large intestine [78] of rodents and more recently in the oral mucosa of man [79]. Enhanced activation of many structurally diverse carcinogens by microsomes from various tissues including those from liver, lungs, intestine and esophagus [71, 80–83] has been observed after inductive pretreatment with ethanol. The carcinogens used in these studies have included compounds and mixtures found in tobacco smoke and diets such as amino acid pyrolisates, polycyclic hydrocarbons and nitrosamines. Induction of CYP 2E1 in the esophagus may be particularly relevant to carcinogenesis at this site because of the low concentrations of other detoxifying enzyme systems in this tissue [80]. The lack of such enzyme systems may further enhance DNA alkylations and carcinogenesis.

At the same time ethanol is capable of inhibiting the metabolism of these compounds when present in sufficiently high concentrations. Whether the microsomal metabolism of procarcinogens is enhanced or inhibited depends on the presence or absence of ethanol in the organism. In some instances the inductive effect exhibits tissue, substrate, gender and species specifities [71].

The interaction of ethanol and nitrosamine metabolism has been investigated intensively [28]. It has been shown that chronic alcohol consumption induces low Km-DMN-demethylase activity [81] leading to an increased activation of the carcinogen both in rats [82, 83] and in man [84]. On the other hand, ethanol is an effective competitive inhibitor of DMN-demethylase activity [85–87]. This capacity to act as both an inducer and inhibitor may explain the conflicting results of ethanol influence on DMN mediated carcinogenicity in the liver when the route of exposure and the presence or absence of ethanol at the time of exposure are taken into account. In most of the studies published the coadministration of ethanol with nitrosamines has resulted in a large increase of tumors in extrahepatic target organs [28]. The results are very similar to those observed when the CYP 2E1 inhibitor disulfiram was administered. The tumors which occured were cancer of the nasal cavity and trachea in hamsters, olfactory mucosal, lung, kidney and forestomach tumors in mice and esophageal and nasal cavity tumors in rats. It has been a consistent, reproducible and general finding.

When DMN is given orally, it undergoes a first pass metabolism in the liver up to a dose of 30 μg per kilogram bodyweight [87]. At higher doses the hepatic enzymes are saturated and the methylation of other organs such as the kidney or the esophagus occurs. When ethanol is given to rats at low levels, it inhibits the first pass metabolism of DMN by competing with the hepatic microsomal enzymes. As a result more nitrosamine can bypass the liver and extrahepatic organs are exposed to higher concentrations of the procarcinogen. When given to rats in relatively low amounts equivalent to a person drinking 0.5 liters of beer ethanol prevents the clearance of DMN and can produce a 5-fold increase in the methylation of kidney DNA. Measurements of DNA metabolism in liver slices and esophageal epithelium suggest that the changes in alkylation of esophageal DNA can be the result of a selective inhibition of DMN metabolism in the liver [87]. This goes along with the fact that no increased methylation of hepatic DNA was detected when radioactively labelled DMN was given to ethanol fed and control rats. However, labelling of the esophagus DNA was enhanced after alcohol [88]. Furthermore, ethanol administration also increased DMN-derived O^6-methylguanine (O^6MG) in gastrointestinal mucosal DNA of monkeys [89]. Following the administration of the esophageal carcinogen NMBA the formation of O^6-methyldeoxyguanosine in the esophagus was increased 3-fold by 20 % ethanol. Various alcoholic beverages such as brandy, scotch whiskey, white wine or beer had the same effect. However, red burgundy and calvados exhibited the most striking increase in DNA alkylation [90].

These biochemical data on the interaction between ethanol and nitrosamine metabolism in the liver and extrahepatic tissues may at least in part explain why alcohol does not stimulate the nitrosamine induced hepatocarcinogenesis, but stimulates the development of extrahepatic tumors. This may possibly be related to the fact that the hepatic inactivating enzyme activities are also increased after prolonged ingestion of ethanol or more likely that the presence of ethanol in the liver during procarcinogen application inhibited hepatic activation of nitrosamines.

Some experiments with DMN as tumor inducing agent underline the importance of the induction of CYP 2E1 and the promoting effect of ethanol in the liver. Most recently Tsutsumi *et al.* [33] reported on the occurrence of preneoplastic hepatic changes in the liver of rats which were treated with low amounts of DMN and alcohol chronically. These changes were neither observed in the alcohol nor in the carcinogen treated group alone. It was emphazised that to avoid interaction between ethanol and DMN metabolism both compounds were administered in the diets alternatively. A similar enhancement of DMN-induced hepatocarcinogenesis by ethanol when given during promotion was reported by Driver and McLean [32]. Other experiments show that hepatic carcinogenesis is enhanced by chronic alcohol administration when carcinogenesis was further modulated by other stimulators, namely by the administration of a methyl deficient [29], a low carbohydrate [30] diet or by partial hepatectomy to stimulate hepatic regeneration [31].

Nitrosamine metabolism is also influenced by dietary factors such as zinc and vitamin A [53]. Zinc deficiency leads to a strikingly enhanced rate of NMBA metabolism in mucosal microsomes from the esophagus [91]. Zinc seems to inhibit the CYP 2E1-dependent activation of the nitrosamine. Alcohol consumption also results in a severe depression of hepatic vitamin A levels. Since ethanol, DMN and retinol share the same CYP 2E1 species, chronic ethanol consumption results not only in an enhanced retinol metabolism leading to decreasing hepatic vitamin A levels but also in an increased activation of DMN which is further enhanced by diminishing competitive inhibition of DMN activation due to low vitamin A [92].

An enhanced metabolism of AOM by CYP 2E1 may be an important mechanism of the cocarcinogenic effect of ethanol on the rectum [93]. It has been shown that ethanol inhibits the hepatic microsomal activation of AOM, while the activation of the procarcinogen was strikingly enhanced following chronic ethanol consumption when ethanol was withdrawn. The conversion of AOM to MAM is catalyzed by a microsomal cytochrome P450-dependent *N*-hydroxylase in the liver and in the colon. Pretreatment of animals with microsomal enzyme inducers such as phenobarbital, chrysene or ethanol leads to an increased metabolism of AOM to carbon monoxide, probably through an induction of the microsomal enzyme. On the other hand, agents which inhibit DMH metabolism also inhibit DMH-induced colorectal carcinogenesis *in vivo* [2]. It therefore seems possible that the effect of ethanol observed in the animal experiment with DMH and AOM can be attributed, at least in part, to the alcohol-related changes in the metabolism of the procarcinogen. In the light of these facts, it is understandable why high ethanol intake results in an inhibition of colorectal carcinogenesis whereas low alcohol intake does not [42, 43]. The presence of ethanol during tumor initiation also inhibits tumor development, while its absence at a stage of enzyme induction enhances the carcinogenic process.

Other compounds relevant in carcinogenesis and metabolized by CYP 2E1 are AFB1 and VC. With respect to AFB1 induced hepatocarcinogenesis controversial results have been obtained. While one study reported an enhancement of AFB1 induced carcinogenesis by ethanol [94] which is in agreement with the reported increased activation of AFB1 [95], another study reported no effect [96]. It was suggested that the metabolism of aflatoxin B_1 to aflatoxin M1 is favored while the metabolism to aflatoxin Q1 is depressed by alcohol administration.

Metabolism of VC is also mediated by CYP 2E1 and it is inhibited by ethanol. The animals who received alcohol and VC had a doubling in incidence of angiosarcoma in the liver compared to VC alone [97]. This is consistent with liver blood vessel endothelium representing an intraorgan downstream target receiving increased doses of the carcinogen as a result of inhibition of metabolism in the hepatocyte. On the other hand, in man it has been shown that large quantities of ethanol in addition to exposure to VC led to the development of both angiosarcoma and HCC possibly as a result of an increased activation of VC to its toxic metabolite at the working place where no alcohol was present in the blood [55].

In summary, chronic ethanol consumption increases the capacity of microsomes to activate many classes of chemical carcinogens in different tissues. This effect is gender and species dependent. The significance of this effect of ethanol vis-a-vis actual cancer risk will be influenced by other factors operating *in vivo* including the carcinogen detoxifying capacity of various tissues, the route of carcinogen exposure and, in the alcohol abuser, in particular the presence or absence of ethanol in the circulation at the time of carcinogen exposure.

8.4.4 Effect of alcohol on DNA

As already pointed out AA inhibits DNA methylation [98] and may directly bind to DNA [99]. It has been reported that it induces sisterchromatide exchanges (SCE) in tissue cultures [100]. Also an elevation of chromosomal aberrations in lymphocytes of alcoholics has been found [101]. The potential significance of these observations with respect to tumor promotion is related to the hypothesis that compounds with SCE activity may act as promoters. By increasing the frequency of SCE's, such compounds could theoretically enhance recessive mutations being converted from heterocygous to a homocygous state and thereby lead to tumor development.

More recently, it has been reported that AA induces c-phos and c-jun-proto-oncogen expression in fat storing cells associated with enhanced fibrogenesis [102].

Besides an increased activation of DMN, an enhanced O^6MG DNA adduct formation is additionally due to an inhibition of the capacity of cells to repair carcinogen-induced DNA damage [58]. Indeed, it was reported that DMN induced hepatic DNA alkylation persisted for longer periods in ethanol fed animals than in controls. This effect appeared to be specific for O^6MG repair. The enzyme responsible for the repair of O^6MG adducts is O^6MG transferase which transfers methyl- or ethylgroups from the O^6 position of guanine to a cystein residue located in the enzyme which in turn inactivates the transferase. Chronic ethanol consumption was found to reduce this enzyme activity significantly [58]. Since alkylation at the O^6 position of guanine is associated with both mutagenesis and carcinogenesis the apparently decreased O^6MG transferase activity in alcohol fed rats could be an important mechanism in alcohol associated cancer risk.

8.4.5 The effect of ethanol on cell regeneration and its link to carcinogenesis

Cell proliferation is an important measure and characteristic of tissues. In general there are two different entities of changes in mucosal cell regeneration. One is related to tissue damage and indicates reparative growth and the other is concerned with the acute and chronic progressive changes during cancer development. Abnormal cellular regeneration is a hallmark of neoplasia. Actively proliferating cells are more susceptible to initiators of carcinogenesis and genetic alterations. A substantial body of literature indicates that in the colon a sequence of events after crypt cell production, during migration and differentiation is disordered in carcinogenesis [103].

Morphometric analysis in rats who were fed with alcohol over six months has shown an enlargement of the size of the nuclei of the basal cells of the oral mucosa from the floor of the mouth and from the edge and the base of the tongue [104]. The size of the basal cell layer in these rats was also increased and the stratification of the cells was altered. The percentage of cells in the S-phase of the cell cycle was significantly higher in ethanol fed rats as compared to controls. Mean epithelium thickness of the mucosa from the floor of the mouth was significantly reduced after chronic alcohol ingestion. This indicates an atrophy of the mucosa and it is remarkable that this finding was most pronounced for a location within the oral cavity which is believed to have the most intensive contact with alcoholic beverages. A reduction of epithelial thickness increases the vulnerability of the epithelium towards chemical and physical noxae.

Acute [105] and chronic ethanol consumption [106, 107] also enhances cell replication in the rat esophagus. Eight weeks of feeding an ethanol containing liquid diet doubled the labelling index in the esophagus. The thickness of the epithelium was increased, but no overt changes in morphology were detected. Simanowski *et al.* found a significant increase of cell proliferation in the middle part of the esophagus of male F344 rats. This enhancement of cell proliferation was particularly obvious in young and middle-aged animals. Age alone did not significantly effect cell regeneration [107]. When the effect of ethanol was investigated in Wistar rats with and without sialoadenectomy using PCNA immunohistology, proliferative index values were significantly increased in the intact alcohol consuming animals, whereas this effect of alcohol was completely abolished after sialoadenectomy. No detectable mucosal damage was observed by light microscopy.

In an attempt to mimic more closely the situation as it occurs in humans, alcoholic solutions were administered to rats using an intubation tube long enough to prevent the solution from passing into the lungs, but as short as possible to allow maximum flow through the esophagus. One day after treatment the animals were given bromodesoxyuridine and it was found that the intubation of 64 % ethanol had no detectable effect on basal cell replication, but when 2-methylbutanol was dissolved in ethanol the mixture produced a dramatic increase in replication [108]. This effect could well explain the dependence of the risk on the type of beverage consumed.

A similar hyperregeneration was observed in the rectum after alcohol consumption [45, 60, 109]. This was associated with a marked extension of the proliferative compartment and with a reduced life span of functional epithelium cells. In addition, aging further increased this ethanol-associated proliferation. Furthermore, using PCNA labelling technique cell regeneration was also found to be increased in rectal biopsies from chronic alcoholics as compared to age and sex matched controls [110]. Mucosal hyperregeneration observed after chronic ethanol ingestion is paralleled by a significant increase in rectal mucosal ornithin decarboxylase (ODC) activity, a marker for high risk with respect to colorectal cancer [45].

8.4.6 Alcohol-associated nutritional deficiencies and carcinogenesis

The ethanol induced malnutrition is of clinical significance. Various deficiencies of vitamins and trace elements that occur in chronic alcoholics lead to certain diseases including alcohol-associated cancer [75]. The principle mechanisms involved are:

- *Decrease of the oxidative defense system*

The increased oxidative stress observed during ethanol metabolism leads to an increased requirement for glutathion and α-tocopherol. In addition, glutathion is trapped by AA and its regeneration is restricted by a limited availability of cystein and methionine. Catabolism of methionine is stimulated to generate cystein and replenish glutathion, but this is compensated by an attempt to conserve methionine through a futile cycle of enhanced choline oxidation. As a result, a striking wastage of methyl groups occurs [111]. It has been shown that glutathion inhibits oral carcinogenesis [112].

Ethanol also produces an increased breakdown of other lipid-soluble vitamins such as hepatic α-tocopherol [113], possibly secondary to a marked increase in the formation of α-tocopherol-quinone, a metabolite of α-tocopherol by free radical reaction. This may lead additionally to an increased hepatic lipid peroxidation seen in alcoholics.

- *Methyl deficiency including an alteration of methyl transfer*

Chronic alcoholism increases the requirement for methyl groups [111] and dietary methyl deficiency enhances hepatic carcinogenesis [95]. Methionine obtained from the diet and synthesized by several reactions is the sole precursor of S-adenosylmethionine (SAM), the primary methyl donor in the body. Disruption in methionine metabolism and methylation reaction may be involved in cancer process. SAM is involved in the methylation of a small percentage of cytosine bases of the DNA. Findings suggest that enzymatic DNA methylation is an important component of gene control and may serve as a silencing mechanism for gene function. Some carcinogens interfere with enzymatic DNA methylation and thus may allow oncogene activation. DNA hypomethylation has been observed in many cancer cells and tumors and has been shown recently in colonic mucosal cells following chronic ethanol consumption in rats [114]. Chronic ethanol consumption decreases dietary intake of methionine and its conversion to SAM [115, 116].

Folate deficiency which is common in the alcoholic may additionally contribute to an inhibition of transmethylation, since it is an important factor in One-carbon transport [117]. Decreased folate levels after alcohol intake are due among others to a decreased ability to retain folate in the liver or to an increased breakdown of folate [75]. Most recently it has been demonstrated that vitamine B_6 deficiency which also occurs following chronic ethanol ingestion also leads to a decrease of SAM [118].

Another factor for DNA hypomethylation is the AA-mediated inhibition of methyltransferase activity [98].

- *Vitamin A deficiency*

Vitamin A regulates among others epithelial cell function via metabolism to retinoic acid, a pleiotrophic regulator of gene expression. Vitamin A decreases in the liver after chronic ethanol consumption predominantly due to an enhanced metabolism via CYP 2E1 [73, 92]. It interfers with the metabolism of ethanol via ADH and with that of nitrosamines via CYP 2E1. Extrahepatic tissue exhibits rather an increased vitamin A level after chronic alcohol consumption due to increased mobilization of vitamin A from the liver to peripheral tissues [73, 92, 119]. Since ethanol competes with retinol at the ADH binding site, retinol metabolism via ADH is inhibited resulting in extremely low levels of retinoic acid in the liver, leading to the activation of the Ap-1 gene (c-fos, c-jun) [120]. This seems of special importance with respect to age since cellular vitamin A binding protein decreases with age which may lead to an increased toxicity of vitamin A in those tissues [119]. It has been reported that even β-carotene has an increased toxicity when alcohol is consumed additionally [121]. Epidemiological studies have shown that the supplementation of β-carotene in smokers does not prevent lung cancer. In contrast, those who received β-carotene had an increased occurrence of lung cancer and this has been attributed to the concomitant consumption of alcohol [122]. It has been reported experimentally that chronic alcohol consumption leads to the generation of toxic intermediates of vitamin A and β-carotene due to their increased metabolism via CYP 2E1 [73, 92, 121].

- *Zinc and selenium deficiency*

A deficiency of zinc and selenium may also contribute to cancer development [123]. Besides the effect of zinc on nitrosamine activation by CYP 2E1 [91], it is important to note that zinc deficiency also leads to disturbancies in vitamin A metabolism, since zinc is an important factor in the conversion of retinol to retinal, as well as in the synthesis and secretion of retinol binding protein in the liver. Many other enzyme systems are impaired by zinc deficiency in alcoholic subjects. One classical example is the altered activity of zinc/copper superoxide dismutase which plays an important role in the protection of oxidative tissue damage. Due to the altered zinc status with altered superoxide dismutase activity the hepatocytes become more vulnerable to oxidative stress [75].

Zinc deficiency also reduces glutathion transferase, an enzyme important in the detoxification of carcinogens *in vivo* and increases cell proliferation in the esophageal mucosa [123].

Selenium deficiency leads to a decreased activity of the selenium containing enzyme glutathion peroxidase which guarantees an adequate availability of glutathion [75].

8.5 Specific pathogenesis of alcohol-associated organ cancer

8.5.1 Upper alimentary tract

The effect of alcohol on upper alimentary tract carcinogenesis includes a local production of AA from ethanol either via mucosal σ-ADH or via bacterial metabolism and the production of free radicals through CYP 2E1 induction. Both AA and free radicals may damage mucosal cells which may lead to tumor initiation and could also play a role in tumor promotion by the stimulation of secondary compensatory cell hyperregeneration. In addition, the induction of CYP 2E1 may lead to an increased activation of procarcinogen which are mainly enhaled by smoking. When alcohol is present nitrosamine metabolism is inhibited in the liver and extrahepatic organs as shown in animal experiments are enhanced exposed to these carcinogens. Vitamin deficiencies such as those of riboflavin and zinc may be of additional importance. However, no significant difference in vitamin A and E concentrations could be detected in oral biopsies from alcoholics with oropharyngeal cancer and controls [124].

Ethanol may facilitate the uptake of environmental carcinogens, especially from tobacco smoke, through cell membranes which are damaged and changed in their molecular composition by the direct effect of alcohol. Furthermore, it is postulated that alcohol acts as a solvent which enhances the penetration of carcinogenic compounds into the mucosa [53]. Both factors may be of relevance in the upper gastrointestinal tract, particularly since chronic alcohol abuse leads to atrophy and lipomatic metamorphosis of the parenchyma of the parotic and submandibulary gland and this morphologic alteration results in functional impairment including reduction of saliva flow and increased viscosity of saliva [125]. Thus, the mucosa surface will be insufficently rinsed. Therefore, higher concentrations of locally effective carcinogens in addition to a prolongation of the contact time of those substances with the mucosa can be observed.

Other local mechanisms include the direct toxic effect of highly concentrated alcoholic beverages on the squamous epithelium, the impaired motility of the esophagus due to alcohol and the enhanced gastroesophageal reflux leading to esophagitis and Barrett esophagus.

8.5.2 Liver

The exact role of alcohol itself, of the alcohol-induced cirrhosis of the liver, or of the concomitant HBV or HCV infection, or of a combination of all three has not been determined. In addition, many other variables including geographic exposure to carcinogens, ethnicity and diet all influence the results in studies correlating chronic alcohol ingestion with hepatocarcinogenesis. A major prerequisite for HCC in the alcoholic is the presence of cirrhosis of the liver. Mechanisms by which chronic alcohol consumption leads to a liver cirrhosis are discussed in detail elsewhere in this book. In addition the infection of patients with alcoholic cirrhosis with HBV and HCV is high. Studies from Japan show that HCV infections are more closely related to HCC in heavy drinkers than is HBV infection [24, 25, 126]. Nevertheless, the incidence of HCC was significantly higher among chronic HBV carriers who were drinkers than among HBV carriers who were abstinent [24, 25]. Among drinkers HCC developed at a younger age [127]. It was concluded that hepatic cell injury caused by alcohol may enhance the development of HCC caused by HBV.

It has also been shown that HCV infection is more important for the pathogenesis of HCC and that a high prevalence of HCV antibodies has been detected in alcoholic liver disease especially in cirrhosis and HCC [128–130] that alcohol abuse enhances the development of HCC related to hepatitis virus infection through the interaction with the replication and oncogenicity of HCV and through the promoter action superimposed on HBV oncogenicity. Results from Japan have identified chronic alcohol consumption as a promoter in viral associated hepatic carcinogenesis [24, 25].

In addition to cirrhosis of the liver and concomitant viral infection, alcohol per se may influence hepatic carcinogenesis. The mechanism by which alcohol acts includes possibly an increased activation of procarcinogen through microsomal CYP 2E1, an inhibition of the nuclear DNA repair system by small amounts of AA, nutritional deficiencies such as methyl- and vitamin A deficiency, the promoting action of ethanol and changes in cell proliferation. Some of these factors have been discussed already in detail. Some controversy exist whether alcohol is a promoter in hepatocarcinogenesis. Recent studies have found that numbers and areas of enzyme altered foci were significantly increased in chronically alcohol treated rats. These changes were similar to those in the phenobarbital treated animals. In alcohol and phenobarbital treated groups the numbers of visible nodules were also significantly increased. The visible nodules showed preneoplastic histological changes. These results indicated that ethanol may act as a promoting agent in hepatocarcinogenesis [25, 33].

It is interesting that mallory body (MB) formation is significantly high in HCC and the incidence of HCC is significantly elevated in cirrhosis with MBs than in those without [131]. Therefore, it was hypothesized that MBs may be a phenotypical expression of carcinogenesis of hepatocytes, especially since gammaglutamyltranspeptase activity was observed in those hepatocytes with MB from the early stage of development of HCC [132]. Another histological abnormality is the occurrence of oval cells in the liver after long term alcohol exposure resulting in an alteration of the cellular composition of the liver similar as observed after the administration of a cholin deficient, ethionine supplemented diet which is known to stimulate hepatocarcinogenesis [133].

Finally, it was observed that alcohol inhibits hepatocellular regeneration. The fact that in some studies HCC was observed after absence of alcohol led to the theory that the inhibition of regeneration was omitted and that therefore, the increased cell regenerativity contributed to the carcinogenesis. However, chronic alcohol consumption leads to hepatocellular injury accompanied by cell death and fibrogenesis, a process which is associated with an increased renewal. Such hyperregenerativity during the development of hepatic cirrhosis may indeed be a pathogenetic factor. Rats with hepatic hyperregenerativity after partial hepatectomy developed more liver tumors after nitrosamine application and alcohol [31].

The significance of ethanol mediated changes in immune function including its depressing effect on B- and T-lymphocytes, macrophages, neutrophils, and natural killer cells associated with increased cytokine concentrations in hepatocarcinogenesis needs to be further determined [134].

8.5.3 Large intestine

Although the enhanced activation of procarcinogens due to chronic ethanol consumption could be important in ethanol associated colorectal cocarcinogenesis, local mechanisms may predominate. As already pointed out chronic ethanol consumption leads to hyperproliferation and expansion of the proliferative compartment of the rectal crypt toward the intestinal lumen in the rat model and also in man [45, 60, 109, 110]. ODC activity was found to be significantly enhanced after chronic alcohol consumption [45]. It was also shown that the hyperregenerative effect of chronic ethanol consumption on the rectal mucosa was further increased in old age which by itself is a risk factor in colorectal cancer [60]. Mucosal AA concentrations correlated significantly with cell regeneration. Thus, it was suspected that AA injures the rectal mucosa. The rectal hyperproliferation observed after alcohol ingestion may be of secondary compensatory nature, since light microscopy of rectal mucosa from alcoholics reveals superficial cell damage, which returns to normal following alcohol abstinence for two weeks [135], and since the life span of functional epithelial cells in the rectal crypt is reduced [109].

Significantly high concentrations of AA were found in the distal colon after alcohol application. These AA concentrations were markedly elevated compared to the proximal colon and to the liver when calculated per gram of tissue [45]. Data on the effect of ethanol on AMMN-induced rectal cancer further support the concept that AA is involved in the ethanol-associated rectal carcinogenesis. Animals who received ethanol and cyanamide, a potent AA-dehydrogenase inhibitor, exhibited an earlier occurrence of rectal tumors compared to animals who received ethanol alone [45]. In these experiments AA concentrations were significantly elevated in the serum and in the colonic mucosa following the application of cyanamide. All these experiments underline the role of AA in ethanol rectal cocarcinogenicity.

Since the ethanol concentration in the rectal lumen is similar to that in the blood, it seems unclear how AA is generated and why more AA associated with increased cell regeneration is found in the rectum compared to other large intestinal segments. Increased ADH activity was found in the mucosa derived from the distal colon when compared to the proximal large intestine which may favor AA accumulation through ethanol oxidation [40]. Meanwhile, the presence of increased mucosal ADH activity in the rectum has also been found in man [65]. However, it seems impossible that the rectal ADH with its low activity is capable of producing the striking accumulation of AA seen in the rectum. It is therefore more likely that bacterial production of AA especially in the distal colon (where the highest bacteria count occurs) may be responsible for the AA formation [45]. Indeed, Jokelainen *et al.* reported increasing AA production when human colonic content was incubated with increasing ethanol concentrations starting already at 2.75 mM ethanol [70]. The question whether different types of alcoholic beverages such as beer and wine may affect fecal bacteria quantitatively and qualitatively, affecting AA generation, still remains.

In addition, folate deficiency may also enhance colorectal carcinogenesis [136]. The mechanisms have already been discussed.

8.5.4 Breast

Factors others than those already discussed for gastrointestinal and liver cancer may be involved in alcohol-associated mammary carcinogenesis. Since many breast cancers are estrogen-dependent, the observation of increased serum sex hormone levels in women with breast cancer and elevated estrogen and androgen serum concentrations after alcohol ingestion may have pathogenic importance [48]. Only limited information exists on the effect of chronic ethanol consumption on estrogen receptor positive and negative breast cancer as well as for women who reported ever using estrogen replacement therapy. Another possibility by which alcohol may affect mammary carcinogenesis may be by increasing cell proliferation of the mammary gland. In cell cultures ethanol selectively stimulated cell proliferation of estrogen receptor-positive, but not estrogen receptor-negativ human breast cancer cells [137]. Also, a positive association of ethanol intake with circulating prolactin levels has been reported in animals [48].

The evidence of an effect of ethanol on carcinogen metabolism in mammary tissue is rather small.

Finally, an increased mammary tumor metastasis has been observed following alcohol administration which may be linked to an impairment of natural killer cell activity suggesting that impaired immune function leads to increased tumor progression [51].

References

[1] Lamu, L: (1910) Etude de statistique clinique de 131 cas de cancer de l'oesophage et du cardia. *Arch. Mal. Appar. Dig. Mal. Nutr.* **4**, 451–456.

[2] Seitz, H. K.; Pöschl, G. (1996) Alcohol and cancer: pathogenetic mechanisms. *Addiction Biol.* **2**, 19–33.

[3] Wynder, E. L.; Mabushi, K: (1973) Etiological and environmental factors in esophageal cancer. *J. Am. Med. Assoc.* **226,** 1546–1548.

[4] Tuyns, A: (1978) Alcohol and cancer. *Alcohol Health Res. World* **2**, 20–31.

[5] Maier, H.; Dietz, A.; Zielinski, D; Jünemann, K. H.; *et al.* (1990) Risikofaktoren bei Patienten mit Plattenepithelkarzinomen der Mundhöhle, des Oropharynx, des Hypopharynx und des Larynx. *Dtsch. Med. Wochenschr.* **115**, 843–850.

[6] Tuyns, A. J.; Esteve, J; Raymond, L.; *et al.* (1988) Cancer of the larynx/hypopharynx, tobacco and alcohol: IARC International case-control study in Turin and Varese (Italy), Zaragoza and Navarra (Spain), Geneva (Switzerland), and Calvados (France). *Int. J. Cancer* **41**, 483.

[7] Bruguere, J.; Guenel, P.; Leclerc, A.; *et al.* (1986) Differential effects of tobacco and alcohol in cancer of the larynx, pharynx and mouth. *Cancer* **57**, 391–397.

[8] Horie, A.; Kohchi, S.; Karatsune, M. (1965) Carcinogenesis in the esophagus II. Experimental production of esophageal cancer by administration of ethanolic solution of carcinogens. *Gann* **56**, 429–441.

[9] Henefer, E. P. (1966) Ethanol 30% and hamster pouch carcinogenesis. *J. Dent. Res.* **45**, 838–844.

[10] Elzay, R. R. (1966) Local effect of alcohol in combination with DMBA on hamster cheek pouch. *J. Dent. Res.* **45**, 1788–1795.

[11] Stenback, F. (1969) The tumorigenic effect of alcohol. *Acta Pathol. Microbiol. Scand.* **77**, 325–326.

[12] Nachiappan, V.; Mufti, S. I.; Eskelson, C. D. (1993) Ethanol mediated promotion of oral carcinogenesis in hamsters: Association with lipid peroxidation. *Nutr. Cancer* **20**, 293–302.

[13] Gibel, V. W. (1967) Experimentelle Untersuchungen zur Synkarzinogenese beim Ösophaguskarzinom. *Arch. Geschwulstforsch.* **30**, 181–189.

[14] Gabrial, G.; Schrager, T. F.; Newberne, P. M. (1982) Zinc deficiency, alcohol and retinoid: association with esophageal cancer in rats. *J. Natl. Cancer Inst.* **68**, 785–789.

[15] Aze, Y.; Toyoda, K.; Furukawa, F.; *et al.* (1993) Enhancing effect of ethanol on esophageal tumor development in rats by initiation of diethylnitrosamine. *Carcinogenesis* **4**, 1437–1440.

[16] Eskelson, C. D.; Odeleye, O. E.; Watson, R. R.; *et al.* (1993) Modulation of cancer growth by vitamin E and alcohol. *Alcohol Alcohol.* **28**, 117–126.

[17] McCoy, G. D.; Hecht, S. S.; Katayama, H.; *et al.* (1981) Differential effects of chronic ethanol consumption on the carcinogenicity of *N*-nitrosopyrrolidine and *N*-nitrosonornicotine in male Syrian hamsters. *Cancer Res.* **41**, 2849–2854.

[18] Castonguay, A.; Rivenson, A.; Trushin, N; *et al.* (1984) Effect of chronic ethanol consumption on the metabolism and carcinogenicity of *N*-nitrosonornicotine in F344 rats. *Cancer Res.* **44**, 2285–2290.

[19] Anderson, L. M.; Carter, J. P.; Driver, C. L.; *et al.* (1993) Enhancement of tumorigenesis by *N*-nitrosodiethylamine, *N*-nitrosopyrrolidine and N^6-(methylnitroso)-adenosine by ethanol. *Cancer Lett.* **68**, 61–66.

[20] Mufti, S. I.; Becker, G.; Sipes, I. G. (1989) Effect of chronic dietary ethanol consumption on the initiation and promotion of chemically-induced esophageal carcinogenesis in experimental rats. *Carcinogenesis* **10**, 303–309.

[21] Ishii, H.; Tatsuta, M.; Baba, M.; *et al.* (1989) Promotion by ethanol of gastric carcinogenesis induced by *N*-methyl-*N'*-nitro-*N*-nitroso-guanidine in Wistar rats. *Br. J. Cancer* **59**, 719.
[22] Schmähl, D. (1976) Investigations of esophageal carcinogenicity by methylphenyl nitrosamine and ethyl-alcohol in the rat. *Cancer Lett.* **1**, 215–218.
[23] Konishi, N.; Kitahori, Y.; Shimoyama, T.; *et al.* (1986) Effects of sodium chloride and alcohol on experimental esophageal carcinogenesis induced by *N*-nitrosopiperidine in rats. *Gann* **77**, 446–451.
[24] Ohnishi, K: (1992) Alcohol and hepatocellular carcinoma. In: Watson, R. R. (Ed.) *Alcohol and cancer.* CRC Press, Boca Raton, p. 179–202.
[25] Takada, A.; Takase, S.; Tsutsumi, M. (1993) Alcohol and hepatic carcinogenesis. In: Yirmiya, R.; Taylor, A. N. (Eds.) *Alcohol, immunity and cancer.* CRC Press, Boca Raton, p. 187–210.
[26] Anderson, L. M. (1988) Increased numbers of *N*-nitrosodimethylamine-initiated lung tumors in mice by chronic coadministration of ethanol. *Carcinogenesis* **9**, 1717–1721.
[27] Griciute, L.; Castegnaro, M.; Bereziat, J. C. (1981) Influence of ethyl alcohol on carcinogenesis with *N*-nitrosodimethylamine. *Cancer Lett.* **13**, 345–352.
[28] Anderson, L. M. (1992) Modulation of nitrosamine metabolism by ethanol: Implications of cancer risk. In: Watson, R. R. (Ed.) *Alcohol and cancer*, CRC Press, Boca Raton, p. 17–54.
[29] Porta, E. A.; Markell, N.; Dorado, R. D. (1985) Chronic alcoholism enhances hepatocarcinogenesis of dimethylnitrosamine in rats fed a marginally methyl-deficient diet. *Hepatology* **5**, 1120–1125.
[30] Yonekura, I.; Matsumoto, Y.; Miura, K.; *et al.* (1992) Ethanol ingestion combined with lowered carbohydrate intake enhances the initiation of diethylnitrosamine liver carcinogenesis in rats. *Nutr. Cancer* **17**, 171–178.
[31] Takada, A.; Nei, J.; Takase, S.; *et al.* (1986) Effect of ethanol on experimental carcinogenesis. *Hepatology* **6**, 65–72.
[32] Driver, H. E.; McLean, A. E. M. (1986) Dose-response relationship for initiation of rat liver tumors by dimethylnitrosamine and promotion by phenobarbital and alcohol. *Food Chem. Toxicol.* **24,** 241–245.
[33] Tsutsumi, M.; Matsuda,Y.; Takada, A. (1993) Role of cytochrome P450 2E1 in the development of hepatocellular carcinoma by the chemical carcinogen *N*-nitrosomethylamine. *Hepatology* **18**, 1483–1489.
[34] Breslow, N. E.; Enstrom, J. E: (1974) Geographic correlations between mortality rates and alcohol, tobacco consumption in the United States. *J. Natl. Cancer Inst.* **53**, 631–639.
[35] Kune, G. A.; Vitetta, L. (1992) Alcohol consumption and the etiology of colorectal cancer: a review of the scientific evidence from 1957 to 1991. *Nutr. Cancer* **18**, 97–111.
[36] Kearney, J.; Giavamnucci, E.; Rimm, E. B.; *et al.* (1995) Diet, alcohol and smoking and the occurrence of hyperplastic polyps of the colon and rectum. *Cancer Causes Control* **6,** 45–56.
[37] Boutron, M. C.; Faivre, J.; Dop, M. C.; *et al.* (1995) Tobacco, alcohol and colorectal tumors: a multistep process. *Am. J. Epidemiol.* **141**, 1038–1046.
[38] Howarth, A. E.; Phil, E. (1985) High fat diet promotes and causes distal shift of experimental rat colonic cancer – beer and alcohol do not. *Nutr. Cancer* **6**, 229–235.
[39] Nelson, R. L.; Samelson, S. L. (1985) Neither dietary ethanol nor beer augments experimental colon carcinogenesis in rats. *Dis. Colon Rectum* **28**, 460–462.
[40] Seitz, H. K.; Czygan, P.; Waldherr, R.; *et al.* (1984) Enhancement of 1,2-dimethylhydrazine induced rectal carcinogenesis following chronic ethanol consumption in the rat. *Gastroenterology* **86**, 886–891.

[41] McGarrity, T. J.; Via, E. A.; Colony, P. C. (1986) Changes in tissue sialic acid content and staining in dimethylhydrazine(DMH)-induced colorectal cancer: effects of ethanol (abstract). *Gastroenterology* **90**, 1543.

[42] Hamilton, S. R.; Sohn, O. S.; Fiala, E. S. (1987 b) Effects of timing and quantity of chronic dietary ethanol consumption on azoxymethane induced colonic carcinogenesis and azoxymethane metabolism in Fischer 344 rats. *Cancer Res.* **47**, 4305–4311.

[43] Hamilton, S. R.; Hyland, J.; McAvinchey, D.; *et al.* (1987 a) Effects of chronic dietary beer and ethanol consumption on experimental colonic carcinogenesis by azoxymethane in rats. *Cancer Res.* **47**, 1551–1559.

[44] Garzon, F. T.; Simanowski, U. A.; Berger, M. R.; *et al.* (1987) Acetoxymethyl-methylnitrosamine(AMMN)-induced colorectal carcinogenesis is stimulated by chronic alcohol consumption. *Alcohol Alcohol.* (Suppl) **1**, 501–502.

[45] Seitz, H. K.; Simanowski, U. A.; Garzon, F. Z.; *et al.* (1990) Possible role of acetaldehyde in ethanol related rectal carcinogenesis in the rat. *Gastroenterology* **98**,1–8.

[46] Longnecker, N. (1994) Alcohol beverage consumption in relation to risk of breast cancer: Metaanalysis and review. *Cancer Causes Control* **5**, 73–82.

[47] Harris, J. R.; Lippman, M. E.; Veronesi, U.; *et al.* (1992) Breast cancer. *New Engl. J. Med.* **327**, 319–328.

[48] Singletary, K. W.; Meadows, G. G. (1996) Alcohol and breast cancer: Interaction between alcohol and other risk factors. *Alcohol. Clin. Exp. Res.* **20** (Suppl.), 57A–61A.

[49] Singletary, K.; Nelshoppen, J.; Wallig, N. (1995) Enhancement by chronic ethanol intake of *N*-methyl-nitrosourea-induced rat mammary tumorigenesis. *Carcinogenesis* **15**, 959–964.

[50] Singletary, K.; McNary, M.; Odoms, A.; *et al.* (1991) Ethanol consumption and DMBA induced carcinogenesis in rats. *Nutr. Cancer* **16**, 13–21.

[51] Yimaya, R.; Ben-Eliyahu, S.; Gale, R.; *et al.* (1992) Ethanol increases tumor progression in rats: Possible involvement of natural killer cells. *Brain, Behav. Immun.* **6**, 74–86.

[52] Ketcham, A. S.; Wexler, H.; Mantel, N. (1963) Effects of alcohol in mouse neoplasia. *Cancer Res.* **23**, 667–670.

[53] Seitz, H. K.; Simanowski, U. A. (1988) Alcohol and carcinogenesis. *Annu. Rev. Nutr.* **8**, 99–119.

[54] Bulatao-Jayme, J.; Almero, E. M.; Castro, C. A.; *et al.* (1981) A case control dietary study of primary liver cancer risk from aflatoxin exposure. *Int. J. Epidemiol.* **11**, 112–119.

[55] Tamburro, C. H.; Lee, H. M. (1981) Primary hepatic cancer in alcoholics. *Clin. Gastroenterol.* **10**, 457–477.

[56] Fang, J. L.; Vaca, C. E. (1995) Development of a ^{32}P-postlabelling method for the analysis of adducts arising through the reaction of acetaldehyde with 2′-deoxyguanosine-3′-monophosphate and DNA. *Carcinogenesis* **16**, 2177–2185.

[57] Appelman, L. M.; Wouterson, R. A.; Feron, V. J. (1982) Toxicity of acetaldehyde in rats. Acute and subacute studies. *Toxicology* **23**; 293–307.

[58] Garro, A. J.; Espina, N.; Farinati, F.; *et al.* (1986) The effect of chronic ethanol consumption on carcinogen metabolism and on O^6-methylguanine transferase-mediated repair of alkylated DNA. *Alcohol. Clin. Exp. Res.* **10**, 73S–77S.

[59] Zimmerman, B. T.; Crawford, G. D.; Dahl, R.; *et al.* (1987) Mechanism of acetaldehyde-mediated growth inhibition: delayed cell cycle progression and induction of apoptosis. *Alcohol. Clin. Exp. Res.* **19**, 434–440.

[60] Simanowski, U. A.; Suter, P.; Russell, R. M.; *et al.* (1994) Enhancement of ethanol induced rectal mucosal hyperregeneration with the age in F344 rats. *Gut* **35**, 1102–1106.

[61] International Agency for Research on Cancer (1985) Working group on the evaluation of the carcinogenic risk of chemicals to humans. Acetaldehyde. *IARC Monogr.* **36**, 101–132.

[62] Baraona, E.; Yokoyama, A.; Ishii, H.; *et al.* (1991) Lack of alcohol dehydrogenase isoenzyme activities in the stomach of Japanese subjects. *Life Sci.* **49**, 1929–1934.
[63] Harty, L. C.; Caparaso, N. E.; Hayes, R. B.; *et al.* (1997) Alcohol dehydrogenase-3 genotype and risk of oral cavity and pharyngeal cancer. *J. Natl. Cancer Inst.* **89**, 1698–1705.
[64] Jokoyama, A.; Muramatsu, T.; Ohmori, T.; *et al.* (1996) Multiple primary esophageal and concurrent upper aerodigestive tract cancer and the aldehyde dehydrogenase-2 genotype of Japanese alcoholics. *Cancer* **77**, 1986–1990.
[65] Seitz, H. K.; Egerer, G.; Oneta, C.; *et al.* (1996) Alcohol dehydrogenase in the human colon and rectum. *Digestion* **57**, 105–108.
[66] Egerer, G.; Schulitz, R.; Gebhardt, A.; *et al.* (1997) Change of alcohol dehydrogenase phenotypes during colorectal carcinogenesis in men (abstract). *Gastroenterology* **112**, A1260.
[67] Salmela, K. S.; Roine, R. P.; Höök-Nikanne, J.; *et al.* (1994) Acetaldehyde and ethanol production by *helicobacter pylori. Scand. J. Gastroenterol.* **29**, 309–312.
[68] Pikkarainen, P. H.; Baraona, E.; Jauhonen, P.; *et al.* (1979) Contribution of oropharynx microflora and of lung microsomes to acetaldehyde in exspired air after alcohol ingestion. *J. Lab. Clin. Med.* **97**, 617–621.
[69] Homann, N.; Jousimies-Somer, H.; Jokelainen, K.; Heine, R.; Salaspuro M. (1997) High acetaldehyde levels in saliva after ethanol consumption: Methological aspects and pathogenetic implications. *Carcinogenesis* **18**, 1739–1743.
[70] Jokelainen, K.; Roine, R. P.; Väänänen, H.; *et al.* (1994) *In vitro* acetaldehyde formation by human colonic bacteria. *Gut* **35**, 1271–1274.
[71] Seitz, H. K.; Osswald, B. R. (1992) Effect of ethanol on procarcinogen activation. In: Watson, R. R. (Ed.) *Alcohol and cancer.* CRC Press, Boca Raton, p. 55–72.
[72] Albano, E.; Clot, P. (1996) Free radicals and ethanol toxicity. In: Preedy, V. R.;Watson, R. R. (Eds.) *Alcohol and the gastrointestinal tract.* CRC Press, Boca Raton, New York, London, Tokyo, p. 57–68.
[73] Lieber, C. S. (1994) Alcohol and the liver: 1994 update. *Gastroenterology* **106**, 1085–1105.
[74] Nair, J.; Sone, H.; Nagao, M.; *et al.* (1996) Copper dependent formation of miscoding etheno-DNA adducts in the liver of Long Evans Cinnamon (LEC) rats developing hereditary hepatitis and hepatocellular carcinoma. *Cancer Res.* **56**, 1267–1271.
[75] Seitz, H. K.; Suter, P. M. (1994) Ethanol toxicity and nutritional status. In: Kotsonis, F. M.; McKey, M.; Hjelle, J. (Eds.) *Nutritional toxicology.* Raven Press, New York, p. 95.
[76] Shimizu, M.; Lasker, M.; Tsutsumi, M.; *et al.* (1990) Immunohistochemical localization of ethanol inducible cytochrome P450 2E1 in the rat alimentary tract. *Gastroenterology* **93**, 1044–1050.
[77] Seitz, H. K.; Korsten, M.; Lieber, C. S. (1978) Effect of chronic ethanol ingestion on intestinal metabolism and mutagenicity of benzo-α-pyrine. *Biochem. Biophys. Res. Commun.* **85**, 1061–1066.
[78] Hakkak, R.; Korourian, S.; Ronis, M. J.; *et al.* (1996) The effects of diet and ethanol on the expression and localisation of cytochromes P450 2E1 and P450 2C7 in the colon of male rats. *J.* Chem. Pharmacol. **51**, 61–69.
[79] Baumgarten, G.; Waldherr, R.; Stickel, F.; *et al.* (1996) Enhanced expression of cytochrome P450 2E1 in the oropharyngeal mucosa in alcoholics with cancer (abstract); *Annu. Meeting Int. Soc. Biomed. Res. Alcohol.*, Washington DC, June 22–27.
[80] Farinati, F.; Lieber, C. S.; Garro, A. J. (1989) Effects of chronic ethanol consumption on carcinogen activating and detoxifying systems in rat upper alimentary tract tissue. *Alcohol. Clin. Exp. Res.* **13**, 357–360.
[81] Garro, A. J.; Seitz, H. K.; Lieber, C. S. (1981) Enhancement of dimethylnitrosamine metabolism and activation to a mutagen following chronic ethanol consumption in the rat. *Cancer Res.* **41**, 120–124.

[82] Farinati, F.; Zhou, Z.; Bella, H. C.; *et al.* (1985) Effect of chronic ethanol consumption on activation of nitrosopyrolidine to a mutagen by rat upper alimentary tract, lung and hepatic tissue. *Drug Metab. Dispos.* **13**, 210–216.
[83] Seitz, H. K.; Garro, A. J.; Lieber, C. S. (1981) Enhanced pulmonary and intestinal activation of procarcinogens and mutagens after chronic ethanol consumption. *Eur. J. Clin. Invest.* **11**, 33–38.
[84] Amelizad, S.; Appel, K. E.; Schoepke, M.; *et al.* (1989) Enhanced dimethylase and dinitrosation of *N*-nitrosodimethylamine by human liver microsomes from alcoholics. *Cancer Lett.* **46**, 43–48.
[85] Peng, R.; Yong-Tu, J.; Yang, C. S. (1982) Induction and competitive inhibition of a high affinity microsomal nitrosodimethylamine dimethylase by ethanol. *Carcinogenesis* **3**, 1457–1461.
[86] Hauber, G.; Frommberger, R.; Remmer, G.; *et al.* (1984) Metabolism of low concentrations of *N*-nitrosodimethylamine in isolated liver cells of guinea pig. *Cancer Res.* **44**, 1343–1348.
[87] Swann, P. F.; Koe, A. M.; Mace, R. (1984) Ethanol and dimethylnitrosamine metabolism and disposition in the rat. Possible relevance in the influence of ethanol on human cancer incidence. *Carcinogenesis* **5**, 1337–1343.
[88] Kouros, M.; Mönch, W.; Reifer, F. J. (1983) The influence of various factors on the methylation of DNA by the esophageal carcinogen *N*-nitrosomethylbenzylamine: 1. The importance of alcohol. *Carcinogenesis* **4**, 1081–1084.
[89] Anderson, L. M.; Souliotis, V. L.; Chhabra, S. K.; *et al.* (1996) *N*-nitrosodimethylamine-derived O^6-methylguanine in DNA of monkey gastrointestinal and urogenital organs and enhancement by ethanol. *Int. J. Cancer* **66**, 130–134.
[90] Yamada, Y.; Weller, R. O.; Kleihues, P.; *et al.* (1992) Effects of ethanol and various alcoholic beverages on the formation of O^6-methyldeoxyguanosine from concurrently administered *N*-nitrosomethylbenzylamine in rats: a dose-response study. *Carcinogenesis* **13**, 1171–1175.
[91] Barch, D. H.; Kuemmerle, S. C.; Holenberg, P. F.; *et al.* (1984) Esophageal microsomal metabolism of *N*-nitrosomethylbenzylamine in the zinc deficient rat. *Cancer Res.* **44**, 5629–5633.
[92] Lieber, C. S.; Garro, A. J.; Leo, M. A.; *et al.* (1986) Alcohol and cancer. *Hepatology* **6**, 1005–1019.
[93] Sohn, O. S.; Fiala, E. S.; Puz, C.; *et al.* (1987) Enhancement of rat liver microsomal metabolism of azoxymethane to methylazoxymethanol by chronic ethanol administration: similarity to the microsomal metabolism of *N*-nitrosomethylamine. *Cancer Res.* **47**, 3123–3129.
[94] Tanaka, T.; Nishikara, A.; Iwata, H. (1989) Enhanced effect of ethanol of aflatoxin B_1 induced hepatocarcinogenesis in male ACI/N rats. *Jpn. J. Cancer Res.* **80**, 526–530.
[95] Seitz, H. K.; Simanowski, U. A.; Hörner, M.; *et al.* (1989) Alcohol and liver carcinoma. In: Bannasch, P.; Keppler, D.; Weber, G. (Eds.) *Liver cell carcinoma.* Kluwe Acad. Publ., p. 227–242.
[96] Mendenhall, C. L.; Chedid, L. A. (1980) Peliosis hepatis. Its relationship to chronic alcoholism, aflatoxin B_1 and carcinogenesis in male Holtzman rats. *Dig. Dis. Sci.* **25**, 587–594.
[97] Radike, M. J.; Stemmer, K. L.; Brown, P. B.; *et al.* (1977) Effect of ethanol and vinylchloride on the induction of liver tumors. *Environ. Health Perspect.* **21**, 153–155.
[98] Garro, A. J.; McBeth, D. L.; Lima, V.; *et al.* (1991) Ethanol consumption inhibits fetal DNA methylation in mice: Implications for the fetal alcohol syndrome. *Alcohol. Clin. Exp. Res.* **15**, 395–398.
[99] Fang, J.-L.; Vaca, C. E. (1997) Detection of DNA adducts of acetaldehyde in peripheral white blood cells of alcohol abusers. *Carcinogenesis* **18**, 627–632.

[100] Obe, G.; Ristow, H. (1977) Acetaldehyde but not alcohol induces cystochromatide exchanges in Chinese hamster cells *in vitro*. *Mutat. Res.* **56**, 211–213.

[101] Obe, G.; Ristow, H. (1979) Mutagenic, carcinogenic and teratogenic effects of alcohol. *Mutat. Res.* **65**, 229–259.

[102] Casini, A.; Galli, G.; Salcano, R.; *et al.* (1994) Acetaldehyde induces C-phos and C-jun proto-oncogenes in fat storing cell cultures through protein kinase C activation. *Alcohol Alcohol.* **29**, 303–314.

[103] Seitz, H. K.; Simanowski, U. A. (1996) Cell turnover in the gastrointestinal tract and the effect of ethanol. In: Preedy. V. R.; Watson, R. R. (Eds.) *Alcohol and the gastrointestinal tract.* CRC Press, Boca Raton, New York, London, Tokyo, p. 273–288.

[104] Maier, H.; Weidauer, H.; Zöller, J.; *et al.* (1994) Effect of chronic alcohol consumption on the morphology of the oral mucosa. *Alcohol. Clin. Exp. Res.* **18**, 387–391.

[105] Haentjens, P.; DeBacker, A.; Willems, G. (1987) Effect of an apple brandy from Normandy and of ethanol on epithelial cell proliferation in the esophagus of rats. *Digestion* **37**, 184–192.

[106] Mak, K. M.; Leo, M. A.; Lieber, C. S. (1987) Effect of ethanol and vitamin A deficiency on epithelial cell proliferation and structure in the rat esophagus. *Gastroenterology* **93**, 362–370.

[107] Simanowski, U. A.; Suter, P.; Stickel, F.; *et al.* (1993) Esophageal epithelial hyperregeneration following chronic ethanol consumption: effect of age and salivary gland function. *J. Natl. Cancer Inst.* **85**, 2030–2033.

[108] Craddock, V. M. (1992) Ethiology of esophageal cancer: some operative factors. *Eur. J. Cancer Prev.* **1**, 89–92.

[109] Simanowski, U. A.; Seitz, H. K.; Baier, B.; *et al.* (1986) Chronic ethanol consumption selectively stimulates rectal cell proliferation in the rat. *Gut* **27**, 278–82.

[110] Homann, N.; Seitz, H. K.; Schuhmann, H.; *et al.* (1995) Immunhistochemical studies on cell regeneration, differentiation and regulatory genes in the rectal mucosa of alcoholics (abstract). *Alcohol Alcohol.* **30**, 520.

[111] Trimble, K. C.; Molloy, A. M.; Scott, J. M.; *et al.* (1993) The effect of ethanol on one-carbon metabolism: Increased methionine catabolism and lipotrope methyl group wastage. *Hepatology* **18**, 984–989.

[112] Trickler, D.; Shklar, G.; Schwartz, J. (1993) Inhibition of oral carcinogenesis by glutathione. *Nutr. Cancer* **20**, 139–144.

[113] Meydani, M.; Seitz, H. K.; Blumberg, J.; *et al.* (1991) Effect of chronic alcohol feeding on hepatic and extrahepatic distribution of vitamin E in rats. *Alcohol. Clin. Exp. Res.* **15**, 771–774.

[114] Choi, S. W.; Stickel, F.; Baik, H. W.; *et al.* (1997) Chronic alcohol consumption induces colonic DNA hypomethylation in the rat. *FASEB J.* **18**, A1255.

[115] Lieber, C. S.; Casini, A.; DeCarli, L. M.; *et al.* (1990) *S*-adenosyl-L-methionine attenuates alcohol induced liver injury in the baboon. *Hepatology* **11**, 165–172.

[116] Halsted, C. H.; Villanuewa, J.; Chandler, C. J.; *et al.* (1996) Ethanol feeding of micropigs alters methionine metabolism and increases hepatocellular apoptosis and proliferation. *Hepatology* **23**, 497–505.

[117] Glynn, S. A.; Albanez, D. (1994) Folate and cancer: a review of literature. *Nutr. Cancer* **22**, 101.

[118] Stickel, F.; Kim, Y.; Selhub, J.; *et al.* (1997) Einfluß von hochdosierter Folsäuresubstitution auf die alkoholbedingte Reduktion der Methylierungskapazität bei Ratten. *Ztschr. Gastroenterol.* **35**, A802.

[119] Mobarhan. S.; Seitz, H. K.; Russell, R. M.; *et al.* (1991) Age related effects of chronic ethanol intake and vitamin A status in rats. *J. Nutr.* **121**, 510–517.

[120] Wang, X. D.; Liu, C.; Chung, J.; *et al.* (1998) Chronic alcohol intake reduces plasma retinoic acid concentration and enhances liver AP-1 (c-jun and c-fos) expression in rats. *Hepatology*, in press.

[121] Leo, M. A.; Kim, C.; Lowe, N.; *et al.* (1992) Interaction of ethanol with β-carotene: delayed blood clearance and enhanced hepatotoxicity. *Hepatology* **15**, 883–891.
[122] Albanes, D.; Heinonen, O. P.; Taylor, P. R.; *et al.* (1996) α-tocopherol and β-carotene supplements and lung cancer incidence in the α-tocopherol, β-carotene cancer prevention study: effects of baseline characteristics and study compliance. *J. Natl. Cancer Inst.* **88**, 1560–1570.
[123] Cho, C. H. (1991) Zinc: absorption and role in gastrointestinal metabolism and disorders. *Dig. Dis.* **9**, 49–60.
[124] Leo, M. A.; Seitz, H. K.; Maier, H.; *et al.* (1995) Carotinoid, retinoid and vitamin E status of the oropharyngeal mucosa in the alcoholic. *Alcohol Alcohol.* **30**, 163–70.
[125] Maier, H.; Born, I. A.; Veith, S.; *et al.* (1986) The effect of chronic ethanol consumption on salivary gland morphology and function in the rat. *Alcohol. Clin. Exp. Res.* **10**, 425–427.
[126] Tsutsumi, M.; Ishizaki, M.; Takada, A. (1996) Relative risk for the development of hepatocellular carcinoma in alcoholic patients with cirrhosis: a multiple logistic-regression coefficient analysis. *Alcohol. Clin. Exp. Res.* **20**, 758–762.
[127] Ohnishi, K.; Iida, S.; Iwama, S.; *et al.* (1982) The effect of chronic habituel alcohol intake on the development of liver cirrhosis and hepatocellular carcinoma: relation to hepatitis B surface antigen carriers. *Cancer* **49**, 672–677.
[128] Oshita, M.; Hayashi, N.; Kashara, A.; *et al.* (1994) Increased serum hepatitis C virus RNA levels among alcoholic patients with chronic hepatitis C. *Hepatology* **20**, 1115–1120.
[129] Zignego, A. L.; Foschi, M.; Laffi, G.; *et al.* (1994) Inapparent hepatitis B virus infection and hepatitis C virus replication in alcoholic subjects with and without liver disease. *Hepatology* **19**, 577–582.
[130] Fong, T. L.; Kanel, G. C.; Conrad, A.; *et al.* (1993) Clinical significance of concomitant hepatitis C infection in patients with alcoholic liver disease. *Hepatology* **19**, 554–557.
[131] Nakanuma, Y.; Ohta, G. (1985) Is mallory body formation a preneoplastic change? A study of 181 cases of liver bearing hepatocellular carcinoma and 82 cases of cirrhosis. *Cancer* **55**, 2400–2405.
[132] Tazawa, J.; Irie, T.; French, S. W. (1983) Mallory body formation runs parallel to gammaglutamyl transferase induction in hepatocytes of griseofulvin fed mice. *Hepatology* **3**, 989–996.
[133] Smith, P. G.; Tee, L. B.; Yeoh, G. C. (1996) Appearance of oval cells in the liver of rats after longterm exposure of ethanol. *Hepatology* **23**, 145–154.
[134] Roselle, G.; Mendenhall, C. L.; Grossman, C. J. (1993) Effects of alcohol on immunity and cancer. In: Yirmiya, R.; Taylor, A. N. (Eds.) *Alcohol, immunity and cancer.* CRC Press, Boca Raton, p. 3–22.
[135] Brozinski, S.; Fami, K.; Grosberg, J. J. (1979) Alcohol ingestion-induced changes in the human rectal mucosa: light and electronmicroscopic studies. *Dis. Colon Rectum* **21**, 329–335.
[136] Giovannucci, E.; Rimm, E. B.; Ascherio, A.; *et al.* (1995) Alcohol, low methionine-low folate diets and risk of colon cancer in men. *J. Natl. Cancer Inst.* **87**, 265–273.
[137] Singletary, K.; Yan, W. (1996) Ethanol and proliferation of human breast cancer cells. *FASEB J.* **10**, 712.

9 Heterocyclic Aromatic Amines: Genotoxicity and DNA Adduct Formation

Wolfgang Pfau

9.1 Abstract

Mutagenic heterocyclic aromatic amines (HAA) are formed when meat, fish or poultry is fried, grilled or cooked. In the household levels in the ppb range have been detected but HAA have also been observed in process flavours and meat extracts used in the food industry. HAA have been shown to induce tumors in rodent bioassays and have thus been considered to possibly contribute to human cancer risk. Here we present recent results on the genotoxicity of HAA from our group and give an update of the recent literature.

Research has focused mainly on the HAA with an amino imidazo moiety including 2-amino-3-methylimidazo[4,5-*f*]quinoline (IQ), 2-amino-3,4-dimethylimidazo[4,5-*f*]quinoline (MeIQ), 2-amino-3,8-dimethylimidazo[4,5-*f*]quinoxaline (MeIQx), 2-amino-3,4,8-trimethylimidazo[4,5-*f*]quinoxaline (4,8-DiMeIQx) which exhibit extreme high responses in the Ames *Salmonella*/microsome assay or 2-amino-1-methyl-6-phenylimidazo[4,5-*b*]pyridine (PhIP) the most prevalent HAA in grilled and fried meat and poultry accounting for up to 60 % of the total HAA content. Similar to PhIP, the amino carbolines have a somewhat less pronounced mutagenic effect in the Ames test, these include the amino-α-carbolines 2-amino-9H-pyrido[2,3-*b*]indole (AαC) and 2-amino-3-methyl-9H-pyrido [2,3-*b*]indole (MeAαC), the amino-γ-carbolines 3-amino-1,4-dimethyl-5*H*-pyrido [4,3-*b*]indole (Trp-P-1), 3-amino-1-methyl-5*H*-pyrido[4,3-*b*]indole (Trp-P-2) and the glutamic acid pyrolysis products 2-amino-6-methyl-dipyrido[1,2-*a*:3′,2′-*d*] imidazole (Glu-P-1), 2-amino-dipyrido[1,2-*a*:3′,2′-*d*]imidazole (Glu-P-2).

Recent advancements in the genetic engineering of *Salmonella* strains have resulted in new strains that exhibit particular sensitivity towards HAA. As an example, mutagenicity data of the above mentioned HAA in the strain YG 1019 that over expresses the bacterial *N*-acetyltransferase are shown.

The genotoxic and clastogenic properties of HAA are reviewed and data presented on the cell transforming potential of HAA in the mouse fibroblast cell line M2. Furthermore, we have shown that HAA are potent inducers of DNA strand breaks employing the alkaline single cell-gel (COMET) assay. It was demonstrated that in both assays, cell transformation and COMET formation, that PhIP is the most potent HAA, with amino-α-carbolines being of intermediate activity and the amino imidazoquinoline and -quinoxaline derivatives being the least potent HAA.

HAA have been shown to form covalent DNA adducts in a number of organs in HAA treated animals and *in vitro* upon metabolic activation. Up to date eleven different DNA adducts formed by nine different HAA have been characterised. All of these are guanine adducts and in every case the major adduct is the C8-guanine derivative. This was also true for a RNA adduct formed by IQ *in vitro*.

The mutagenic consequences of HAA adducts have been investigated in a number of studies, including mutational spectra *in vitro* and in transgenic animals, mutations detected in oncogenes or tumor suppressor genes in HAA induced experimental tumors; the most prevalent lesions observed were GC→TA transversions. The major target sites of tumorigenic activity in experimental animals are the intestine and the liver, with the exception of PhIP and MeIQ that induced tumors of the mammary gland. Recent reports of the carcinogenic activity include wider spectra of target organs, e.g. tumors of the prostate have been detected in PhIP treated rats, Trp-P1 has been shown to induce tumors in the urinary bladder of rats and tumors of the liver and pancreas induced by MeAαC in the rat. Furthermore, the application of specific feeding protocols or specific strains of experimental animals resulted in reduced induction periods for malignant neoplasms.

Since HAA are ingested as complex mixtures the evaluation of possible mixture effects is of interest. Mixtures of HAA have been employed in a long term carcinogenicity experiment and also in short term *in vivo* studies measuring liver foci and DNA adduct formation. We have investigated the comutagenic effect of the β-carbolines and amino-α-carbolines. The ability of β-carbolines to enhance the mutagenic activity of aromatic amines has been reported earlier and most recently, the formation of DNA adducts was reported.

A point of concern seems to be the possible exposure of breast-feeding infants since excretion of HAA with breast-milk that has been reported for experimental animals.

Epidemiological evidence supports the hypothesis that dietary intake of HAA might contribute to human cancers at several organ sites including mammary gland, pancreas, prostate and colon.

9.2 Introduction

Heating as a means of food preparation is unique to mankind. The browning of the meat surface and the formation of typical flavours are considered appetising, and heating also makes food digestible. By Maillard reactions not only flavours and colours but also hazardous substances are formed in the crust of cooked meat.

Sixty years ago the tumorigenic activity of roasted meat was reported [1]. It was at that time considered to be due to the content of polycyclic aromatic hydrocarbons. Some forty years later, a strong mutagenic activity in extracts of cooked meat was detected, not explicable by the content of PAHs [2, 3]. It was at the Japanese National Cancer Center Research Institute that the group of Sugimura identified the pyrolysates of amino acids and later the aminoimidazo quinoline and -quinoxaline derivatives as mutagenic principles in cooked meat [4]. Later, Felton *et al.* reported on the identification of the aminoimidazo pyridine-derivative PhIP [5]. See Figure 9.1 for the structures of the most common heterocyclic aromatic amines (HAA).

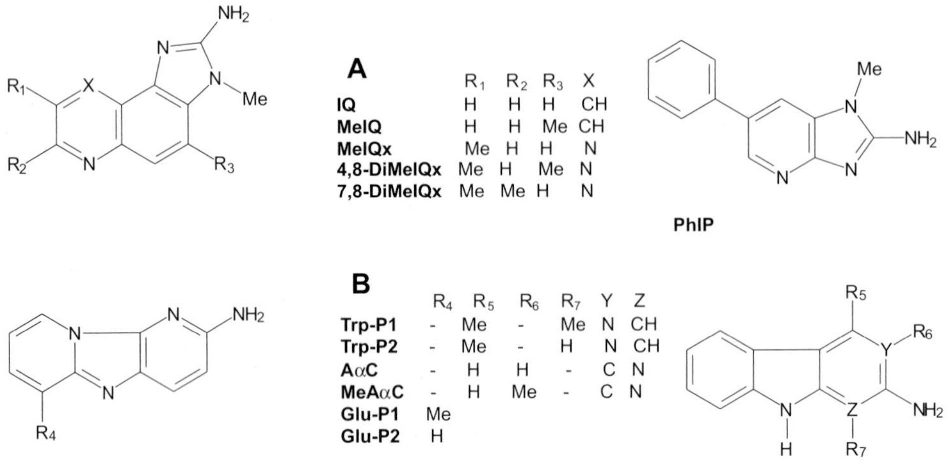
A
R1 R2 R3 X
IQ H H H CH
MeIQ H H Me CH
MeIQx Me H H N
4,8-DiMeIQx Me H Me N
7,8-DiMeIQx Me Me H N
PhIP
B
R4 R5 R6 R7 Y Z
Trp-P1 - Me - Me N CH
Trp-P2 - Me - H N CH
AαC - H H - C N
MeAαC - H Me - C N
Glu-P1 Me
Glu-P2 H

Figure 9.1: Structural formulae of HAA: **A** amino imidazoazarenes and **B** amino carbolines.

A number of commentaries and review articles have summarised the literature on the occurrence, the detection and the genotoxic properties of this group of carcinogenic compounds in the human diet [4, 6–12]. Here it is attempted to update the literature on HAA with an emphasis on the last five years.

9.3 Levels

Since the levels of HAA formed are low analysis of these from the complex matrix of cooked meat is not trivial. Various methods have been described employing different enrichment procedures, chromatographic separation and high sensitivity detection techniques. The enrichment step can be accomplished by extraction with blue cotton [13] or combinations of different solid phase extraction steps combining ion exchange and reversed phase cartridges [14]. Individual HAA have also been enriched by affinity chromatography using monoclonal antibodies to PhIP or MeIQx [15–17]. A tandem cartridge extraction procedure was developed by Gross employing subsequently an extraction from diatomaceous earth, propyl sulfonic acid ion exchange and octadecyl silane reversed phase cartridges, followed by an extraction from TSK gel [18]. The latter procedure is most widely applied and modifications of this method have been described [19, 20]. Separation of the HAA can be accomplished by GC [19, 21, 22] but most often HPLC analysis is used on reversed phase columns [18, 23]. HAA have high coefficient of extinction and, thus, may be detected at nanogram levels by UV absorbance, the carbolines and PhIP are highly fluorescent and can be analysed with a fluorescence detector down to the picomole level [24]. The application of electrochemical detection has been reported [25, 26]. Both, high sensitivity and additional structural conformation is ideally achieved with a mass spectrometric detector [27, 28].

A considerable data base has accumulated on HAA levels detected in heat prepared food and process flavours. A compilation of the data published up 1994 has been presented by Layton *et al.* [29]. During the last five years a number of studies added to the data base including studies from Finland and Sweden on typical local food [30–32], detailed studies on HAA in grilled chicken [33], pork [34] and beef [35, 36], the influence of pre-treatment (marinating [37]) and data from Germany [28, 38]. HAA have also been detected in process flavours [26, 39, 40]. While these analyses focus mostly on the amino imidazoazarenes, we employed the Gross-method [18] and detected in pan-fried turkey meat AαC and MeAαC and, similar to the study on chicken meat, considerable levels of PhIP [38]. These findings on poultry are of interest since this white meat is generally considered as a more healthy alternative as compared to red meat. Although lower in fat and cholesterol there are still comparable amounts of HAA formed when this poultry meat is fried or grilled to the same extent as pork or beef [33, 38].

In order to assess exposure it is necessary to determine the levels that are formed in typical food stuffs. However, an alternative approach is the biomonitoring of HAA levels in human urine. Indeed, HAA (Trp-P-1, Trp-P-2, PhIP and MeIQx) were detected in the urine of healthy volunteers on a regular diet [41, 42] but not in the urine of hospital patients receiving parenteral alimentation. Murray and Lynch reported on the interindividual and intraindividual variation of the renal excretion of MeIQx [43]. In a feeding study Sinha *et al.* [44] observed an excretion of both MeIQx and its metabolites. These results are presented in more detail in a paper by Dr. Sinha in this volume. Again these studies focus on amino imidazoazarenes while little is known about the exposure- and excretion levels of the carbolines.

9.4 Adducts

In common with other genotoxic aromatic amines, the activating metabolism of HAA is considered to proceed via oxidation of the exocyclic amino group, a reaction mediated mainly by the hepatic cytochrome P450 isoenzyme CYP 1A2 [45] and less so by CYP 1A1 [46]. The cytochrome P450 isoenzyme CYP 1B1 was shown to exhibit considerable activity to metabolically activate HAA [36, 47]. This is of importance since this enzyme was detected in extrahepatic tissues including the target organs of HAA carcinogenicity the mammary gland and the pancreas [48]. Indeed, it was shown that human mammary cells are capable of metabolically activating HAA [49–51].

Conjugating reactions such as acetylation or sulfation of the N-OH group [52, 45] may also be required to further activate the HAA. Recent studies suggest that human hepatic microsomes possess higher activity as compared to microsomes isolated from the livers of rats or mice [53–55].

HAA have been shown to form covalent adducts with DNA upon metabolic activation in both *in vivo* and *in vitro* systems [56]. Up to date the structures of twelve nucleotide adducts have been spectroscopically elucidated (these are summarised in Figures 9.2 and 9.3). Employing ^{3}H- or ^{14}C-labelled IQ, MeIQx, 4,8-DiMeIQx or PhIP and their hydroxylamine derivatives revealed that the major DNA adducts (**1**, **3**, **6**, **8**) are formed by the binding of a HAA moiety via the exocyclic amino group to the C-8-position of deoxyguanosine [57–62]. Minor adducts **2**, **4** have been identified for the *N*-hydroxyl-derivatives of IQ and MeIQx as N^2-deoxyguanosine derivatives [58, 60]. Similarly, the major adducts formed by MeIQ with DNA (**5**) and by IQ with RNA (**7**) have been identified as C8-derivatives of guanine [63, 64]. Fluorescence studies lead to the identification of the major adducts of Trp-P2 and Glu-P1 [65, 66] both C8-derivatives of guanine **9** and **10**.

We were able to identify the major DNA adducts formed by AαC and MeAαC as *N*-(deoxyguanosin-8-yl)-2-amino-9H-pyrido[2,3-*b*]indole **11** and as *N*-(deoxyguanosin-8-yl)-2-amino-3-methyl-9H-pyrido[2,3-*b*]indole **12**, respectively, shown in Figure 9.3 [67, 68]. We synthetically prepared the respective guanine derivatives by the method developed by Hashimoto [65] and were able to identify these guanine derivatives as major adducts formed *in vitro* and *in vivo* of the liver and other organs of AαC or MeAαC treated rats. This was accomplished by chromatographic comparison of the standard and products of hydrolysis of DNA adducts (Figure 9.4).

The ^{32}P-postlabelling assay [69, 70] has been successfully applied in numerous studies [71–85]. However, in contrast to the above mentioned studies with ^{3}H-labelled substrates where only one or two major adducts were observed, ^{32}P-postlabelling analysis of amino imidazoazarene-modified DNA has commonly been found to result in a pattern of up to six adduct spots with considerable interlaboratory variation with regard to the number and relative intensities of the spots. Comparable patterns were observed *in vitro* and *in vivo* in

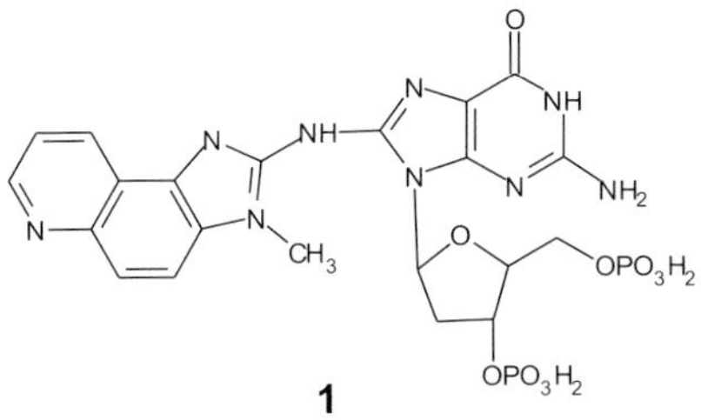

1

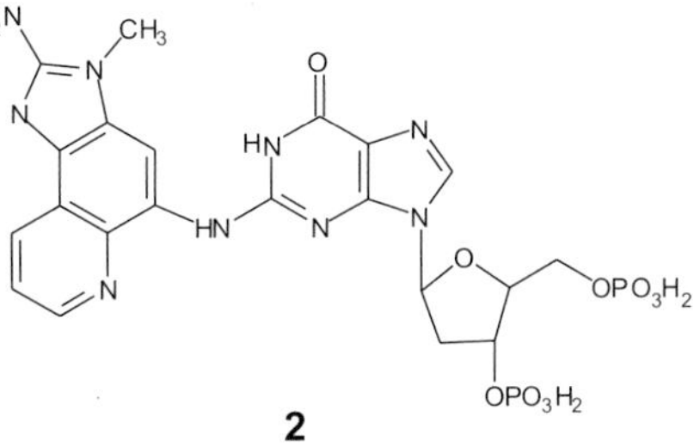

2

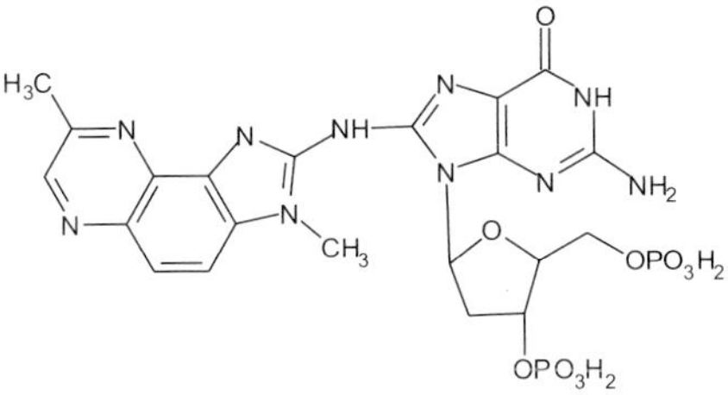

3

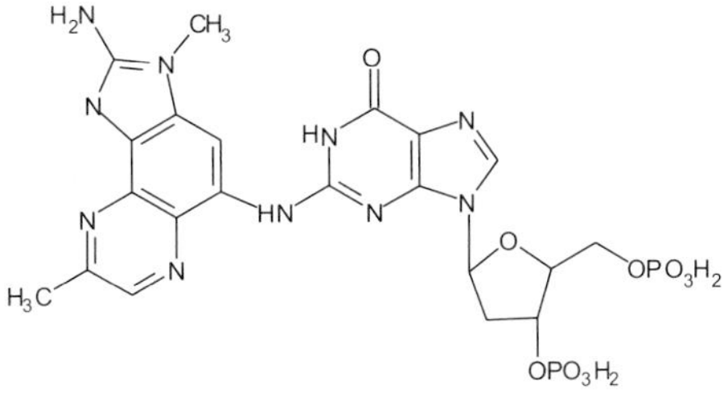

4

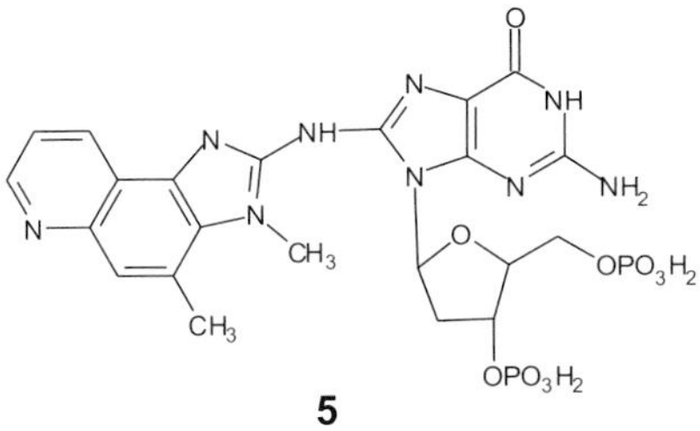

5

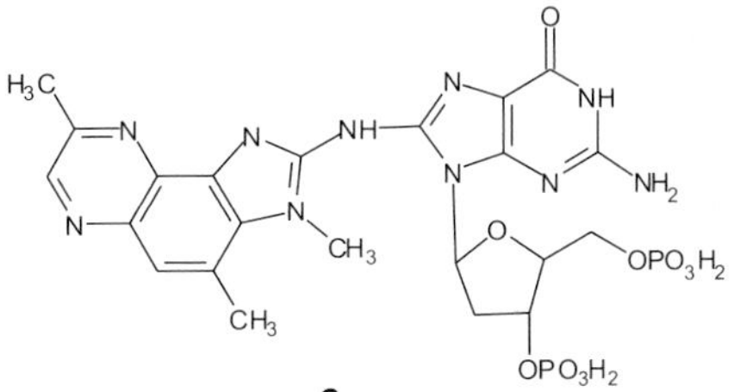

6

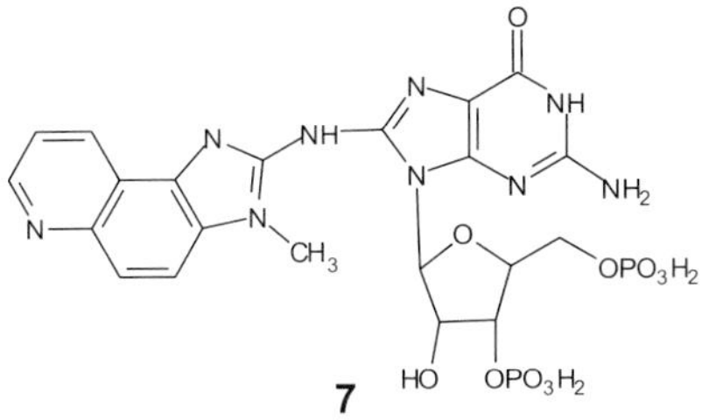

7

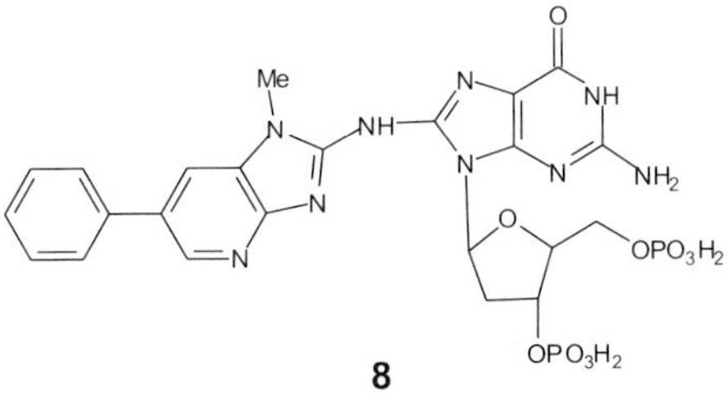

8

Figure 9.2: DNA adducts of amino imidazoazarenes that have been detected *in vivo* and/or *in vitro*. **1**: *N*-(deoxyguanosin-8-yl)-2-amino-3-methylimidazo[4,5-*f*]quinoline [57, 58], **2**: 5-(deoxyguanosin-N^2-yl)-2-amino-3-methylimidazo[4,5-*f*]quinoline [58], **3**: *N*-(deoxyguanosin-8-yl)-2-amino-3,8-dimethylimidazo[4,5-*f*]quinoxaline [58], **4**: 5-(deoxyguanosin-N^2-yl)-2-amino-3,8-dimethylimidazo[4,5-*f*]quinoxaline [58], **5**: *N*-(deoxyguanosin-8-yl)-2-amino-3,4-dimethylimidazo[4,5-*f*]quinoline [63], **6**: *N*-(deoxyguanosin-8-yl)-2-amino-3,4,8-trimethylimidazo[4,5-*f*]quinoxaline [59], **7**: *N*-(guanosin-8-yl)-2-amino-3-methylimidazo[4,5-*f*] quinoline [64], **8**: *N*-(deoxyguanosin-8-yl)-2-amino-1-methyl-6-phenylimidazo[4,5-*b*] pyridine [60–62].

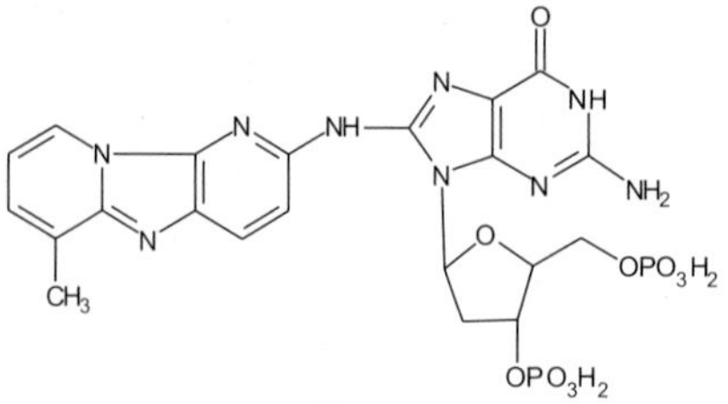

9

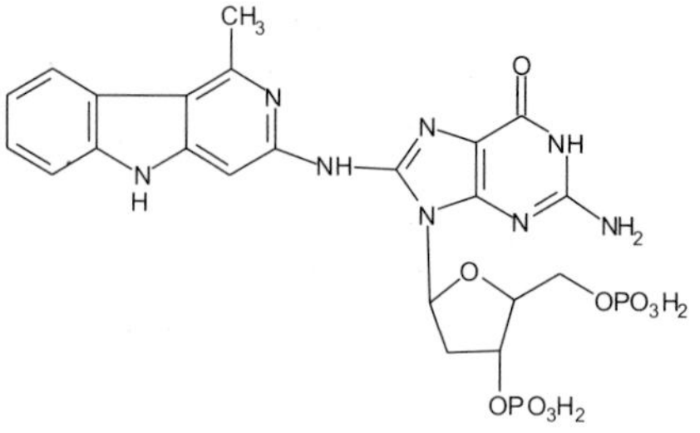

10

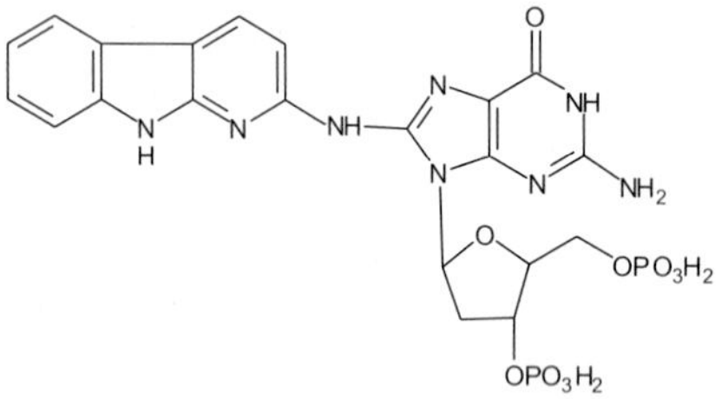

11

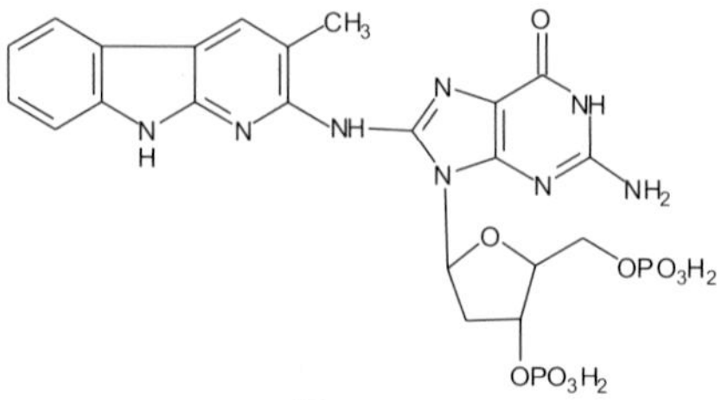

12

Figure 9.3: DNA adducts of aminocarbolines that have been characterised. **9**: *N*-(deoxyguanosin-8-yl)-2-amino-6-methyldipyrido[1,2-*a*:3′,2′-*d*]imidazole [66], **10**: *N*-(deoxyguanosin-8-yl)-3-amino-1-methyl-5H-pyrido[4,3-*b*]indole [65], **11**: *N*-(deoxyguanosin-8-yl)-2-amino-9H-pyrido [2,3-*b*]indole [67], **12**: *N*-(deoxyguanosin-8-yl)-2-amino-3-methyl-9H-pyrido-[2,3-*b*]indole [68].

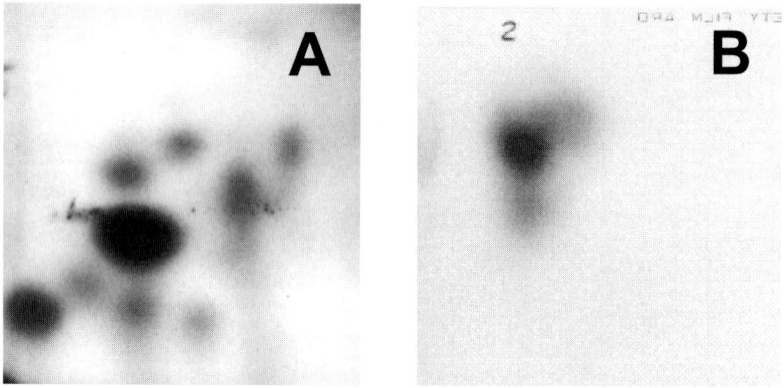

Figure 9.4: Autoradiographic representations of ^{32}P-postlabelling/ion-exchange TLC analyses of DNA isolated from the liver of F344 rats exposed to a diet containing 800 ppm AαC (**A**) or MeAαC (**B**). The origin was in the bottom left hand corner. Major adducts that have been identified as deoxyguanosine-C8-adducts [67, 193].

the DNA of several different tissues and different species [49, 57, 83, 84] treated with an HAA. We have recently shown that by introducing an additional enzymatic hydrolysis step following the labelling reaction, the adduct patterns obtained upon ^{32}P-postlabelling analysis of amino imidazoazarene-modified DNA can be simplified [71]. Thus the additional spots seen in the conventional procedure are probably undigested adducted oligonucleotides. These findings have been confirmed by Fukutome *et al.* [72] and Tada *et al.* [63].

The ^{32}P-postlabelling assay has been employed extensively in studying the organ distribution of DNA adducts induced by amino imidazoazarenes in experimental animals [73, 74] and most recently DNA adducts derived from HAA have been observed in human tissue [75, 76]. PhIP derived adducts were observed in DNA isolated from biopsies of colon, confirmed by GC/MS analysis [75] and low levels of MeIQx adducts were detected in the colon, rectum or kidney of cancer patients [76]. With considerable variation between laboratories, butanol extraction [77, 78] or nuclease P1 [79, 80] enrichment procedures have been applied to the DNA modifications induced by amino imidazoazarenes. In most studies a variation of the assay, using carrier free [γ-^{32}P]-ATP not in excess of nucleotides is employed; this leads to a preferential labelling of adducts and modified oligonucleotides as has been shown by Randerath *et al.* [81]. Recently, Turesky *et al.* [74] presented a reversed-phase solid phase extraction procedure allowing the enrichment of IQ-modified 3′-nucleotides. Employing this procedure it was shown that the IQ-dG-N^2-adduct is underestimated by ^{32}P-postlabelling under intensification conditions and may be the more persistent adduct compared to the C8 adduct [48].

These data are useful in establishing dose-response relationships and comparing organ distribution of adducts [67, 74], modulation of DNA adduct formation [82, 83] and species [84] or gender [85] specific susceptibilities to a single HAA.

When mixtures of adducts induced by mixtures of HAA are analysed the resolution of the conventional TLC separation technique is not sufficient [86, Figure 9.4]. HPLC methods for the analysis of ^{32}P-labelled HAA adducts have been described [49, 76, 78, 87]. In these studies the separation of adducts from different HAA was not described. In order to characterise adducts formed by mixtures of HAA we developed a method that allowed separation of the eight major adducts formed by IQ, MeIQ, MeIQx, 4,8-DiMeIQx, 7,8-DiMeIQx and PhIP [88]. This was achieved on a phenyl modified silica gel column with a gradient of acetonitrile in 0.5 M sodium phosphate buffer at low pH (Figure 9.5).

A higher degree of certainty in the identification of adducts from mixtures or unknown exposures is achieved with the use of mass spectrometric methods. Alkaline hydrolysis of modified DNA resulting in the release of PhIP followed by derivatisation allowed high sensitivity GC/MS analysis [22]. Capillary HPLC in combination with electrospray tandem-MS has been successfully applied to analyse PhIP-derived adducts [89].

The most sensitive method that has been applied to the study of adduct formation, metabolism and toxicokinetics of HAA is accelerator mass spectrometry [90]. While this method requires the use of ^{14}C-labelled test compounds it has been possible to investigate adduct formed by MeIQx and PhIP at the low dose levels that correspond to human dietary exposure [91, 92]. Indeed, adducts were formed at

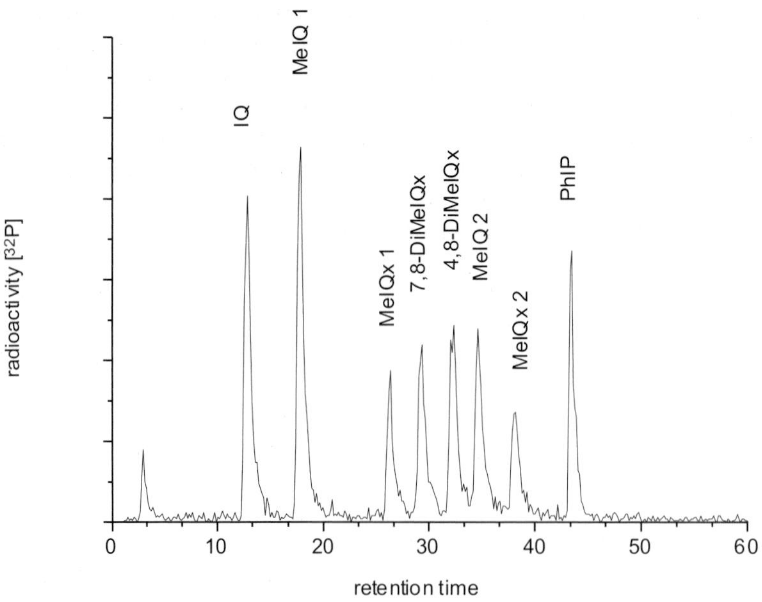

Figure 9.5: HPLC analysis of ^{32}P-labelled amino-imidazoazarene-deoxyguanosine-5′-monophosphate adducts: adduct standards prepared by photoreaction of DNA with azido derivatives of HAA as described [71]; separation was achieved on phenyl-modified reversed phase column with a gradient of acetonitrile in 0.5 M sodium phosphate buffer, pH 2.0 [67, 88].

these low levels and a linear dose response relationship was observed. Furthermore, application of ^{14}C-labelled PhIP and MeIQx to human volunteers lead to the identification of MeIQx adducts in colon biopsies. Interestingly, the adduct levels were about 10-fold higher in human colon as compared to rodents [91–93].

9.5 Mutagenicity of adducts

DNA adducts are generally considered responsible for mutagenic and carcinogenic initiation activities. The mutagenic events induced by HAA have been investigated in some depth:

In IQ-induced tumors of the Zymbal gland in rats 4 out of 15 tumors contained mutations of the p53 gene at guanine sites, two were GC→TA transversions [94]. Similarly, in 4 of 20 liver tumors induced by IQ in cynomolgus monkeys, p53-mutations were detected with three GC→TA transversions [95]. Kudo *et al.* [96] analysed tumors of the Zymbal gland in rats treated with IQ, MeIQ or

MeIQx for mutations in the H-*ras* gene and observed in 10 out of 11 cases GC→TA transversions. These type of mutations were also detected in *H-ras*- and p53 genes of forestomach tumors induced by MeIQ in the rat [94, 97]. In tumors of the mammary gland induced with PhIP in Sprague-Dawley rats no mutations were detected in the oncogenes H-*ras*, Ki-*ras*, *myc* and *neu* (*erB-2*) or the tumor suppressor gene p53 [98], but an increased expression of *neu* (*erB*-2)-, TGF-α and EGFR-*m*RNA was observed.

In colon tumors in F344 rats, 2 of 13 tumors induced by IQ and 4 of 8 PhIP-induced tumors contained mutations of the APC gene, mostly deletions of a guanine base at GGGA-sites [99] but no mutations of *p53* were detected [100]. While mutations in the Ki-*ras* proto-oncogene were detected in 4 out of 37 IQ-induced aberrant crypt foci [101] the frequency of *ras*-mutations in colonic tumors induced by IQ or PhIP was even lower [100].

In an *in vitro* study [102] N-OH-PhIP failed to induce c-*fos*, c-*myc* or *p53* *m*RNA expression in a human bladder epithelial cell line, thus, it was concluded to lack tumor promoting activity in this system.

Employing an array of *E.coli/lacZ* strains Watanabe and Ohta [103] observed mostly frameshift mutations upon treatment with HAA; for MeIQ, IQ and Trp-P2 they detected also GC→TA transversions. Kosarkarn [104] investigated mutations in the *lacZ* gene in the plasmid pKM101 in *E.coli* where also GC→TA transversions predominated. Different to that, Broschard *et al.* [105] reported on a high level of base substitutions (70 %) upon treatment of the plasmid pTU-AC with N-OH-IQ and subsequent transformation into *E.coli*. Mutations occurred almost exclusively at G:C base pairs and were GC→AT transitions (52 %), GC→CG (26 %) or GC→TA transversions (19 %) of 63 mutants analysed.

The mutational spectra induced by PhIP have been investigated in several studies (Table 9.1): Carothers *et al.* [106] found in the *dfhr* gene of N-OH-PhIP-treated CHO cells (20 mutants) predominantly GC→TA transversions (65 %). Morgenthaler and Holzhäuser [107] detected in the *hprt* gene of 54 mutants of the human lymphoblastoid cell line TK6 following incubation with PhIP in the presence of rat liver homogenate GC→TA transversions with an incidence of 54 % but also GC→AT transitions (27 %). Yadollah-Farsani *et al.* analysed the *hprt* gene of 40 mutants of the cell line XEMh1a2-MZ [108]. These V79 cells transfected with

Table 9.1: *In vitro*-studies on the induction of mutations by PhIP-DNA adducts in mammalian cells: mainly GC→TA transversions were observed.

Metabolic activation	Cell line	GC→TA (%)	GC→AT (%)	Gene	Ref.
NOH-PhIP	CHO	65	5	*dhfr*	[106]
rat liver S9	human lymphoblastoid	54.5	27	*hprt*	[107]
murine-CYP 1A2	CHO	80	0	*aprt*	[109]
human-CYP 1A2	CHO	62.5	15	*hprt*	[108]
NOAc-PhIP	human fibroblasts	59	27	*supF*	[110]
	Xeroderma pigmentosum	64	21	*supF*	

c-DNA of the human CYP 1A2 were treated with PhIP and predominantly GC→TA transversions were detected. Wu *et al.* [109] investigated the PhIP-induced mutations in the *aprt* gene of CHO-UV5P3 cells. This cell line has a defect in the excision repair system and is transfected with the mouse CYP 1A2 gene. Here also GC→TA transversions (80 %) were detected in 30 mutants analysed.

Mutations induced by NOH-PhIP and NOH-IQ were compared in the sup*F* shuttle vector in DNA repair proficient cell line GMO637 (SV40) and repair deficient human fibroblasts (Xeroderma pigmentosum); here also GC→TA transversions were predominant [110]. In a corresponding study the distribution of DNA adducts was also analysed in this vector with a polymerase arrest assay [111]. Only adducts at guanine sites were detected for IQ and PhIP. Although there was some correlation between adduct-sites and sites of mutation the hot spots of mutation were not identical with the hot spots of adduct formation. While adduct spectra were similar for both HAA, mutational spectra were distinct. In all studies strand specific mutagenicity was observed, a phenomenon also reported for other chemical carcinogens [112] considered to be due to a preferentially repair of the coding strand. Indeed, this strand specificity was not observed in the study of Wu *et al.* [109] where a repair deficient cell line was employed.

Koch *et al.* [113] analysed the mutations in *Salmonella* tester strains. While only PhIP was mutagenic in TA 1535 and mutations were exclusively GC→AT transitions, both, PhIP and IQ induced predominantly GC→TA transversions in strain TA 100. The authors concluded that the SOS system was required for the induction of point mutations by IQ. On the other hand Maenhaut *et al.* [114] reported that IQ-induced frameshift mutations were independent of the SOS system in *E. coli*.

Site-specifically modified oligodeoxynucleotides containing a single N^2-(deoxyguanosin-8-yl)-PhIP were inserted into single-stranded shuttle vectors and transfected into COS-7 cells. It was shown that mutation frequency and spectra depend on the neighbouring sequence context, and the adduct was found to induce G→T transversions and G→A transitions [115].

Okonogi *et al.* [116] reported on the mutational spectra induced by PhIP, MeIQ and AαC in the *lacI* gene in the colon of Big Blue® mice. In accordance with the major adduct being a deoxyguanosine derivative the majority of mutations (>90 %) was detected at GC base pairs. Similar to PhIP and MeIQ base substitutions were predominant with GC→TA transversions being the most common mutation.

9.6 Genotoxic effects

Originally HAA were detected because of the extremely strong mutagenic activity in the Salmonella/microsome assay (Ames test) [2]. The strongest response was observed for MeIQ, IQ, MeIQx and 4,8-DiMeIQx, especially in tester strains sensitive towards frameshift mutations. Recent advances in the genetic engineering of Salmonella tester strains have led to the development of new variants of *Salmonella typhimurium* that exhibit even higher sensitivity, due to the overexpression of bacterial *N,O*-acetyltransferase (YG1019, YG1024) [117, 118], expression of human *N*-acetyltransferase 1 or *N*-acetyltransferase 2 [119] or expression of cytochrome P450 1A2 [120]. Most recently, a new strain derived from TA1538 with two additional plasmids, one plasmid carrying human cytochrome P450 1A2 and NADPH cytochrome reductase and an expression plasmid containing *N,O*-acetyltransferase was introduced (TA1538/ARO) by Suzuki *et al.* [121]. Specific mutagenic activities reported are MeIQ: 22,000 revertants/ng, IQ: 4500 rev/ng, MeIQx: 455 rev/ng, PhIP: 2.9 rev/ng, MeAαC: 3.8 rev/ng in the tester strain TA 1538/ARO not requiring external metabolic activation. We observed mutagenic responses with similar sensitivity in YG 1019 (preincubation assay in the presence of rat liver homogenate), the dose response curves are shown in Figure 9.6.

In contrast to microbial test systems, the genotoxic effects of amino imidazoquinoline and -quinoxaline-type HAA in eukaryotic cells have been shown to be less pronounced [8–10].

PhIP was more active as compared to IQ or MeIQx in Chinese hamster ovary cells inducing gene mutations [122]. While PhIP was mutagenic *in vivo* in the mouse small intestine MeIQx or AαC were inactive [123–125].

High incidences of mutations were observed in the *lacI* gene isolated from the colon of Big Blue® mice treated with PhIP or AαC [116, 125]. Upon chronic treatment of Big Blue® mice with MeIQ a high frequency of mutations was ob-

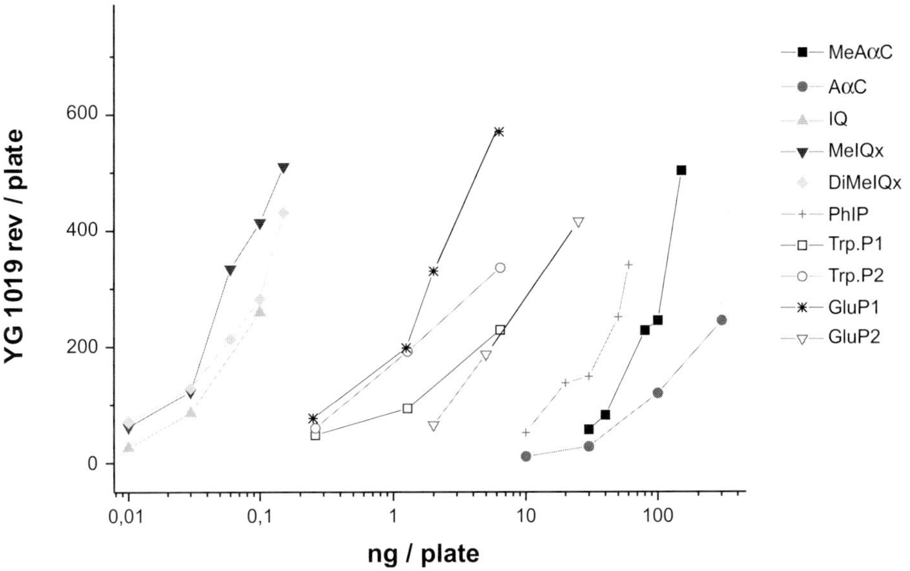

Figure 9.6: Dose response of mutagenic activity of ten of the most prevalent HAA in the Ames assay using *Salmonella* strain YG 1019, that over expresses bacterial *N,O*-acetyltransferase and Aroclor-induced rat liver homogenate in the preincubation assay.

served in the colon and liver, lower frequencies in the bone marrow and forestomach, no mutations were observed in the heart [126]. Mutation frequencies induced by PhIP in the *lacI* gene of the colon of Big Blue® mice and Big Blue® rats were similar but there was a higher incidence of GC→TA transversions in the rats [127]. In this study, similar mutation frequencies were observed in the colon of male and female rats, thus, contradicting the possibility that differences in the number of mutations are responsible for the gender specificity of colon carcinogenicity. While PhIP-induced adduct level in the colonic mucosa was also similar for both sexes, Ochiai *et al.* observed a weak increase in cell proliferation only in the male colonic mucosa and a higher number of aberrant crypt foci (ACF) in male large intestine [128]. Further, this difference in ACF between male and female rats was nullified by castration of the male rats two weeks before the beginning of PhIP administration, while ovarectomy of females had no effect on ACF incidence [129]. Both IQ and PhIP-induced tumors of the colon in F344 rats were shown to contain with high frequency mutations of the β-catenin (CTNNB1) gene, while mutations of this gene were not detected in PhIP induced mammary tumors [130].

An increased frequency of mutations was observed in the *lacZ* gene of the Mutamouse® in the pancreas but not in the liver following a single dose of PhIP (1–25 mg/kg BW) [131].

Sister chromatide exchanges (SCE) and chromosomal aberrations (CA) were detected in Chinese hamster ovary (CHO) cells transfected with cytochrome P450 1A2 and *N*-acetyltransferase activity [132]. No CA were observed upon treatment of CHO cells or human lymphocytes with IQ [133]. But chromosomal aberrations were observed in Chinese hamster lung (CHL) cells at 20 µg IQ/ml [134]. IQ induced CA and SCE in a human fibroblast cell line and human lymphocytes in the presence of rat liver S9 [8]. CA were induced *in vivo* in rat hepatocytes by IQ and MeIQx [135].

PhIP induced micronuclei in Chinese hamster lung cells, this activity was enhanced when these cells were transfected with a *N*-acetyltransferase gene [136]. In the human hepatoma cell line HepG2 micronucleus formation was observed with IQ, MeIQx and PhIP only at high concentrations [137]. The lack of micronucleus formation was reported in peripheral lymphocytes or bone marrow of mice treated with MeIQx or PhIP [138, 139]. We analysed the clastogenic activity of HAA *in vitro* by the micronucleus assay in MCL-5 cells [140, 141]. Dose dependent induction of micronuclei was observed for all HAA tested with 1%–5.4% of cells containing micronuclei at 10 ng/ml. Clastogenic activity increased in the order AαC, PhIP, MeAαC, IQ, MeIQx, 4,8-DiMeIQx.

The induction of unscheduled DNA synthesis (UDS) was investigated in human, rat and mouse liver slices. In human liver PhIP was the most potent compound examined, followed by MeIQx, IQ and then MeIQ, Glu-P1 and Trp-P1 at low concentrations. In rat liver slices only MeIQ significantly induced UDS, while in mouse liver slices apart from MeIQx, all HAA produced significant increases in UDS [142].

Using the alkaline elution assay PhIP was shown to induce DNA strand breaks in V79 cells upon external metabolic activation [143]. When the COMET assay was applied neither PhIP nor IQ induced strand breaks in human colon cells whereas only PhIP was active in rat colonic cells [144]. Following a single dose of Trp-P1, IQ, MeIQ, MeIQx or PhIP the induction of DNA strand breaks was observed in colonic epithel but not in the small intestine or bladder of CF-1 mice [145]. We analysed the induction of DNA strand breaks by six of the most prevalent HAA in metabolically-competent mammalian (MCL-5) cells by the alkaline single cell-gel (COMET) assay [141]. The HAA tested induced significant, dose-dependent increases in COMET tail length at 45.5 µg/ml (PhIP), 94.3 µg/ml (MeAαC, AαC) or 454.5 µg/ml (IQ, MeIQx, 4,8-DiMeIQx), respectively (Figure 9.7).

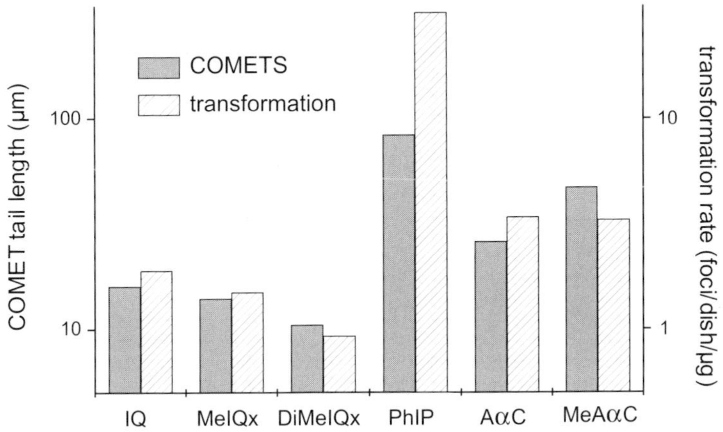

Figure 9.7: Cell transforming activity of six of the most prevalent HAA, induction of transformed foci of M2 fibroblasts following incubation with HAA in the presence of rat liver homogenate and strand breaking activity in MCL-5 cells in the COMET assay [141].

In vitro cell transformation assays are short term predictive tests of tumorigenic activity. IQ induces cell transformation in BALB 3T3 mouse embryo fibroblasts [146]. We analysed the transforming potency of HAA in an *in vitro* assay using a M2/C3H fibroblast cell line derived from murine prostate. This cell line has been employed with a wide range of test compounds and morphologically transformed foci have been shown repeatedly to induce tumors *in vivo* [147]. In order to observe any cytotoxic or genotoxic effects an external metabolic activation system (rat liver homogenate) was required. For IQ and 4,8-DiMeIQx a dose-dependent increase of the transformation rate was observed in the range of 0.2–50 µg/ml. A maximum of activity was detected at 10 µg/ml for AαC, MeAαC and MeIQx with 11.6, 13.1 and 5.5 foci per 10^4 surviving cells, respectively. The most potent compound was PhIP with a transformation rate of 9.2 foci per 10^4 survivors at 3.4 µg/ml and the order of activity being PhIP > AαC = MeAαC > IQ > 4,8-DiMeIQx > MeIQx (Figure 9.6) [141].

Hasegawa *et al.* [148] reported on the induction of preneoplastic foci in the liver of F344 rats following subchronic treatment with HAA. While most HAA induce high levels of foci in the liver, PhIP produced only marginal responses. This corresponds well with the lack of tumorigenic activity of PhIP in this organ [149].

9.7 Carcinogenic activity

Tumorigenic activity of the most prevalent HAA has been investigated in mice and rats and these long term experiments have been reviewed [10, 150].

The amino carbolines have been shown to induce tumors in the liver and the blood vessels of brown adipose tissue in mice and in rats tumors have generally been observed in the liver and for Glu-P1, Glu-P2 and Trp-P2 tumors were also observed in the small and large intestine. Glu-P1 was shown to induce tumors also in the brain, the Zymbal and the clitoral gland of rats. MeAαC was reported to induce tumors in the liver and pancreas with high incidence, with a lower frequency in the urinary bladder, skin and salivary gland of F344 rats [151]. For Trp-P2 tumors were also observed in the urinary bladder, the mammary and clitoral glands [152]. AC failed to induce tumors in the mammary gland of Sprague-Dawley rats (E. Snyderwine *et al.*, submitted).

When IQ, MeIQ or MeIQx were added at doses of 200–600 ppm mice developed tumors in the liver, forestomach and lung. In rats the spectrum of tumorigenic activity was more diverse: IQ induced tumors in the liver, colon, small intestine, Zymbal and clitoral glands and skin of F344 rats and tumors in the liver and mammary gland of female Sprague-Dawley rats. MeIQ (300 ppm) induced tumors of the Zymbal gland, skin and colon, and also the mammary gland (female) of F344 rats. Low incidences of liver and forestomach tumors were detected. MeIQx induced tumors of the liver, Zymbal and clitoral gland and the skin in F344 rats [10, 150].

IQ has been shown to induce liver tumors in cynomolgous monkeys. No tumors were observed when nonhuman primates of this species were treated with MeIQx [153].

Additionally, the tumor initiating activities of HAA in the mouse skin using TPA as promoting agent was investigated: Trp-P1, Trp-P2, MeAαC and Phe-P1 showed significant initiating activity while AαC, IQ, Glu-P1 and Glu-P2 failed to do so [150]. IQ and PhIP were strong inducers of liver tumors in the neonatal mouse assay while MeIQx was less active [154].

The tumorigenic activity of PhIP has been reviewed in detail [155]. Administration of PhIP to F344 rats in the diet at a dose of 400 ppm for 52 weeks resulted in an incidence of colon carcinomas in males of 55 % and of mammary tumors in females of 47 % [150]. Subsequently dose-dependent induction of mammary carcinoma was demonstrated in female Sprague-Dawley rats [155]. Hasegawa *et al.* [156] reported on an high incidence of mammary tumors in Sprague-Dawley rats following transplacental exposure to PhIP. An increase in incidence and a reduced latent period of carcinogenesis was observed when Sprague-Dawley rats were fed a high-fat diet [157]. High incidences of mammary tumors were observed in CD rats receiving eight weekly doses of 50 μmol PhIP [158]. In this study it was shown that the carcinogenic potency in the rat mammary gland of PhIP was equal to benzo[*a*]pyrene [158]. In a study with female Sprague-Dawley rats receiving a diet supplemented with 100 ppm PhIP for fifteen weeks it was shown that a higher incidence of mammary tumors occured in nulliparous rats while gestation and lactation had a protective effect [155]. PhIP also induced preneoplastic lesions, preneoplastic aberrant crypt foci, in male F344 rat colons [128]. In Nagase analbuminemic male rats PhIP induced tumors in the small intestine, the large intestine and the caecum. In CDF1 mice PhIP did not induce intestinal tumors, but lymphomas in both sexes [150]. However low numbers of abberant crypt foci have been reported in PhIP-treated CF1 mice [159]. Moreover, the activity of PhIP to induce tumors in the colon was investigated in three different strains of transgenic mice carrying mutations or deletions in the Apc (adenomous polyposis coli) gene that has been implicated in (human) colon carcinogenesis [160]. PhIP was shown to increase the size of intestinal polyps in $Apc^{\Delta 716}$ knockout mice [161]. In heterozygous multiple intestinal neoplasia mice (*min*/+) PhIP treatment resulted in increased numbers of tumors in the small intestine and abberant crypt foci in the large intestine [162]. In heterozygous transgenic *Apc*1638N mice the numbers of tumors in the small intestine and of abberant crypt foci in the large intestine were reported in males but no effects on intestine or mammary tumor development were detectable in females [163].

9.8 Mixture effects

HAA formed in fried or grilled meat always occur as mixtures. The combinational effect of mixtures of HAA has been investigated in long term feeding studies with rodents [164], the effect of mixtures of five or even ten [165] HAA on the formation of preneoplastic foci in rat liver has been investigated [148, 166, 168]. The effects were mostly additive or slightly elevated. We investigated the transforming activity of mixtures of HAA in M2/C3H cells *in vitro* and the effect of mixtures of HAA in the Ames test; in both assays we observed inhibitory effects [167].

The biologically available dose may be influenced by absorption of HAA to other components of the diet such as chlorophyll [169, 170] and HAA may also bind to proteins in the diet such as casein [171]. It has indeed been demonstrated in animal experiments where simultaneous feeding of PhIP and chlorophyll resulted in a reduction of the bioavailable dose and consequently in a reduction of cancer incidence [83, 172]. Remarkably, it was observed in F344 rats that, while renal excretion was reduced the excretion of PhIP via breast milk was increased when chlorophyllin was coadministered [173].

When fried meat containing 70 ppm PhIP was fed to Sprague-Dawley rats no PhIP derived DNA adducts were observable in the colon [174]; an increase in the incidence of colon tumors in animals fed a diet high in PhIP was only observed when 1,2-dimethylhydrazine was also administered as initiating agent [175].

HAA have been used as model compounds in an immense number of studies on antimutagenic effects including dietary components [176], herbal or synthetical drugs [137] or tea [177]. These have been reviewed elsewhere [178]; the significance of these studies with regard to HAA as human dietary carcinogens is ambiguous.

Comutagenic effects have been reported for the β-carbolines harman and norharman which lack an exocyclic amino group but are observed at levels similar to HAA in cooked meat. These β-carbolines are not mutagenic on their own but enhance the mutagenic activity of HAA when coincubated in the Ames test [179]. We have investigated the influence of norharman on the mutagenic activity of MeAαC (Figure 9.8). Furthermore, mutagenic responses were observed when non-mutagenic monocyclic amines like aniline or ortho-toluidine were applied together with norharman. It has been shown that coupling products of norharman and aniline are responsible for the mutagenicity and the formation of covalent DNA adducts [180]. Norharman has also been detected as normal body

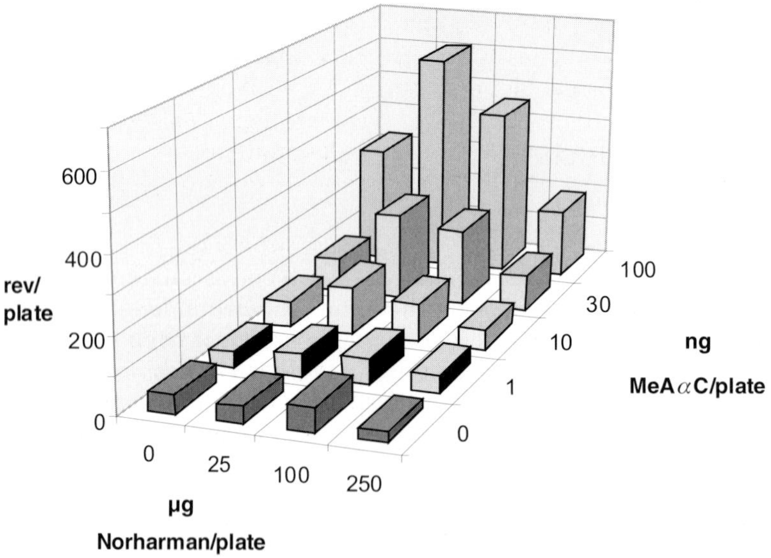

Figure 9.8: Comutagenic activity of MeAαC and norharman in *Salmonella YG1019*, preincubation assay with rat liver homogenate (Aroclor-induced).

constituent in laboratory animals and in humans [181]. Norharman has also been implicated in neurotoxic, psychotropic and physiological effects [182], activates G-proteins [183] and has been considered as a marker of alcohol abuse [184]. It has also been shown to be a selective dopaminergic toxin precursor and may be an underlying factor in idiopathic Parkinson's disease [185].

9.9 Conclusions

The contribution of HAA to human cancer has been a matter of discussion [29, 186–189]. While it is difficult to extrapolate from (high dose) animal experiments to human (low dose) exposure [190] recent experiments using AMS and low dose ^{14}C-labelled HAA indicate that even at dietary doses genotoxic damage occurs [90–93].

The knowledge of HAA levels in the diet is still inadequate, this is especially true for local or ethnic food stuff and home-cooked food.

However, while the high potency of HAA observed in bacterial mutagenicity assays initially raised a great deal of concern about this class of dietary compounds subsequent carcinogenesis experiments with rodents [150] and primates [153] indicated that the carcinogenic potency is not as pronounced. Early experiments suggested that the genotoxic potency in mammalian test systems is moderate [8], but recently, genetically engineered model systems equipped with appropriate enzymatic capacities have been developed that exert extremely high sensitivity towards HAA [132, 136].

Similarly, it has been brought forward that individuals with specific combinations of phenotypes of the xenobiotic metabolism may be at a high risk especially when exposed to a diet with a high content of HAA [188, 189, Sinha *et al.* this volume].

Since excretion of HAA with breast-milk has been reported for experimental animals [156, 173, 192], the possible exposure of breast-feeding infants to HAA via the mother's milk is a possibility that has not yet been investigated thoroughly. Animal experiments [154] suggest that there might be an increased susceptibility of infants towards genotoxic damage. Similarily, HAA fed to pregnant rats have been shown to be transferred via the placenta to unborne experimental animals and to induce DNA damage and increase in tumor incidence in the offspring [155, 156].

The experimental evidence, at least for the most prominent HAA, certainly suggest these compounds to be chemical carcinogens; the low levels of HAA ingested with a conventional diet may still pose a threat to human health [10, 194]. Furthermore, there are groups high at risk and, therefore, measures, such as information of the public, are necessary.

Abbreviations

4,8-DiMeIQx	2-amino-3,4,8-trimethylimidazo[4,5-*f*]quinoxaline
7,8-DiMeIQx	2-amino-3,7,8-trimethylimidazo[4,5-*f*]quinoxaline
AαC	2-amino-9H-pyrido[2,3-*b*]indole (amino-α-carboline)
ACF	aberrant crypt foci
CA	chromosomal aberrations
Glu-P1	2-amino-6-methyldipyrido[1,2-*a*:3′,2′-*d*]imidazole
Glu-P2	2-aminodipyrido[1,2-*a*:3′,2′-*d*]imidazole
HAA	heterocyclic aromatic amine(s)
Harman	1-methyl-9H-pyrido[3,4-*b*]indole (methyl-β-carboline)
IQ	2-amino-3-methylimidazo[4,5-*f*]quinoline
MeAαC	2-amino-3-methyl-9*H*-pyrido[2,3-*b*]indole (methylamino-α-carboline)
MeIQ	2-amino-3,4-dimethylimidazo[4,5-*f*]quinoline
MeIQx	2-amino-3,8-dimethylimidazo[4,5-*f*]quinoxaline
Norharman	9H-pyrido[3,4-*b*]indole (β-carboline)
PhIP	2-amino-1-methyl-6-phenylimidazo[4,5-*b*]pyridine
SCE	sister chromatid exchange
Trp-P1	3-amino-1,4-dimethyl-5H-pyrido[4,3-*b*]indole
Trp-P2	3-amino-1-methyl-5H-pyrido[4,3-*b*]indole
UDS	unscheduled DNA synthesis

References

[1] Widmark, E. M. P. (1939) Presence of cancer producing substances in roasted food. *Nature* **143**, 984.

[2] Nagao, M.; Honda, M.; Seino, Y.; Yahagi, T.; Sugimura, T. (1977) Mutagenicities of smoke condensates and the charred surface of fish and meat. *Cancer Lett.* **2**, 221–226.

[3] Commoner, B.; Vithayathil, A. J.; Dolara, P.; Nair, S.; Madyastha, P.; Cuca, G. C. (1978) Formation of mutagens in beef and beef extract during cooking. *Science* **201**, 913–916.

[4] Sugimura, T.; Sato, S. (1983) Mutagens-carcinogens in foods. *Cancer Res. (Suppl.)* **43**, 2415a–2421s.

[5] Knize, M. G.; Felton, J. S. (1986) The synthesis of the cooked-beef mutagen 2-amino-1-methyl-6-phenylimidazo[4,5-*b*]pyridine (PhIP) and its 3-methyl isomer. *Heterocycles* **24**, 1815–1819.

[6] Tricker, A. R.; Preussmann, R. (1990) Chemical food contaminants in the initiation of cancer. *Proc. Nutr. Soc.* **49**, 133–144.

[7] Snyderwine, E. G.; Schut, H. A. J.; Adamson, R. H.; Thorgeirsson, U. P. (1992) Metabolic activation and genotoxicity of heterocyclic arylamines. *Cancer Res. (Suppl.)* **52**, 2099s–2102s.

[8] Aeschbacher, H-U.; Turesky, R. J. (1991) Mammalian cell mutagenicity and metabolism of heterocyclic aromatic amines. *Mutat. Res.* **259**, 235–250.

[9] Eisenbrand, G.; Tang, W. (1993) Food-borne heterocyclic amines, chemistry, formation, occurrence and biological activities. A literature review. *Toxicology* **84**, 1–82.

[10] IARC (1993) Some naturally occurring substances: Food items and constituents, heterocyclic aromatic amines and mycotoxins. *Monographs on the evaluation of the carcinogenic risk of chemicals to humans* **56**, p. 165–227.

[11] Skog, K. (1993) Cooking procedures and food mutagens: a literature review. *Food Chem. Toxicol.* **31**, 655–675.

[12] Stavric, B. (1994) Biological significance of trace levels of mutagenic heterocyclic aromatic amines in human diet: a critical review. *Food Chem. Toxicol.* **32**, 977–994.

[13] Hayatsu, H.; Hayatsu, T.; Wataya, Y. (1986) Use of the blue cotton for detection of mutagenicity in human faeces excreted after ingestion of cooked meat. *Environ. Health Perspect.* **67**, 31–34.

[14] Knize; G. M.; Felton; S. J.; Gross; A. G. (1992) Chromatographic methods for the analysis of heterocyclic amine food mutagens/carcinogens. *J. Chromat.* **624**, 253–265.

[15] Vanderlaan, M.; Hwang, M.; Djanegara, T. (1993) Immunoaffinity purification of dietary heterocyclic amine carcinogens. *Environ. Health Perspect.* **99**, 285–287.

[16] Stillwell, W. G.; Turesky, R. J.; Gross, G. A.; Skipper, P. L.; Tannenbaum, S. R. (1994) Human urinary excretion of sulfamate and glucuronide conjugates of MeIQx. *Cancer Epidemiol. Biomarker Prev.* **3**, 399–405.

[17] Dragsted, L. O.; Frandsen, H.; Reistad, R.; Alexander, J.; Larsen, J. C. (1995) DNA-binding and disposition of 2-amino-1-methyl-6-phenylimidazo[4,5-*b*]pyridine (PhIP) in the rat. *Carcinogenesis* **16**, 2785–2793.

[18] Gross, G. A. (1990) Simple methods for quantifying mutagenic heterocyclic aromatic amines in food products. *Carcinogenesis* **11**, 1597–1603.

[19] Skog, K; Solyakov, A.; Arvidsson, P.; Jagerstad, M. (1998) Analysis of nonpolar heterocyclic amines in cooked foods and meat extracts using gas chromatography-mass spectrometry. *J. Chromat.* **803**, 227–233.

[20] Jägerstad, M.; Skog, K.; Arvidsson, P.; Solyakov, A. (1998) Chemistry, formation and occurence of genotoxic heterocyclic amines identified in model systems and cooked foods. *Z. Lebensm. Unters. Forsch. A* **207**, 419–427.

[21] Vainiotalo, S.; Matveinen, K.; Reunanen, A. (1993) GC/MS determination of the mutagenic heterocyclic amines MeIQx and DiMeIQx in cooking fumes. *Fresenius J. Anal. Chem.* **345**, 462–466.

[22] Friesen, M. D.; Cummings, D. A.; Garren, L.; Butler, R.; Bartsch, H.; Schut, H. A. J. (1996) Validation in rats of two biomarkers of exposure to the food-borne carcinogen 2-amino-1-methyl-6-phenylimidazo[4,5-*b*]pyridine (PhIP): PhIP-DNA adducts and urinary PhIP. *Carcinogenesis* **17**, 67–72.

[23] Gross, G. A.; Grüter, A.; Heyland, S. (1992) Optimization of the sensitivity of high-performance liquid chromatography in the detection of heterocyclic aromatic amine mutagens. *Food Chem. Toxicol.* **30**, 491–498.

[24] Gross, G. A.; Fay, L. (1992) Quantitative determination of heterocyclic aromatic amines in food products. In: Adamson, R. H.; Gustafson, J. A.; Ito, N.; Nagao, M.; Sugimura, T.; Wakabayashi, K.; Yamazoe, Y. (Eds.) *Heterocyclic amines in cooked foods, Proceedings of the 23rd symposium of the Princess Takamatsu cancer research fund.* Princeton Scientific Publishing Co., Princeton, NJ.

[25] Galceran, M. T.; Pais, P.; Puignou, L. (1993) HPLC determination of ten heterocyclic aromatic amines with electrochemical detection. *J. Chromatogr.* **655**, 101–110.

[26] Vollenbröcker, M.; Eichner, K. Determination of IQ-compounds and HAAs in process flavours heated at high temperatures. *Z. Lebensm. Unters. Forsch.* (in press).

[27] Gross; G. A.; Turesky; J. R.; Fay; B. L.; Stillwell; G. W.; Skipper; L. P.; Tannenbaum, R. S. (1993) Heterocyclic aromatic amine formation in grilled bacon, beef and in grill scrapings. *Carcinogenesis* **14**, 2313–2318.

[28] Richling, E.; Herderich, M.; Schreier, P. (1996) High performance liquid chromatography-electrospray tandem mass spectrometry (HPLC-ESI-MS-MS) for the analysis of heterocyclic aromatic amines (HAA). *Chromatographia* **42**, 7–11.

[29] Layton, D. W.; Bogen, K. T.; Knize, M. G.; Hatch, F. T.; Johnson, V. M.; Felton, J. S. (1995) Cancer risk of heterocyclic amines in cooked foods: an analysis and implications for research. *Carcinogenesis* **16**, 39–52.

[30] Tikkanen, M. L.; Sauri, M. T.; Latva-Kala, J. K. (1993) Screening of heat-processed finnished foods for the mutagens 2-amino-3,8-dimethylimidazo[4,5-*f*]-quinoxaline, 2-amino-3,4,8-trimethylimidazo[4,5-*f*]-quinoxaline and 2-amino-1-methyl-6-phenylimidazo[4,5-*b*]pyridine. *Food Chem. Toxicol.* **31**, 717–721.

[31] Skog, K.; Steineck, G.; Augustsson, K.; Jägerstad, M. (1995) Effect of cooking temperature on the formation of heterocyclic amines in fried meat products and pan residues. *Carcinogenesis* **16**, 861–867.

[32] Skog, K.; Augustsson, K.; Steineck, G.; Stenberg, M.; Jägerstad, M. (1997) Polar and non-polar heterocyclic amines in cooked fish and meat products and their corresponding pan residues. *Food Chem. Toxicol.* **35**, 555–565.

[33] Sinha, R.; Rothman, N.; Brown, E. D.; Salmon, C. P.; Knize, M. G.; Swanson, C. A.; Rosi, S. C.; Mark, S. D.; Levander, O. A.; Felton, J. S. (1995) High concentrations of the carcinogen 2-amino-1-methyl-6-phenylimidazo[4,5-*b*]pyridine (PhIP) occur in chicken but are dependent on the cooking method. *Cancer Res.* **55**, 4516–4519.

[34] Sinha, R.; Knize, M.G.; Salmon, C. P.; Brown, E. D.; Rhodes, D.; Felton, J. S.; Levander, O. A.; Rothman, N. (1998) Heterocyclic amine content in pork products cooked by different methods to varying degrees of doneness. *Food Chem. Toxicol.* **36**, 289–297.

[35] Stavric, B.; Matula, T. I.; Klassen, R.; Downie, R. H. (1995) Evaluation of hamburgers and hot dogs for the presence of mutagens. *Food Chem. Toxicol.* **33**, 815–820.

[36] Crofts, F. G.; Strickland, P. T.; Hayes, C. L.; Sutter, T. R. (1997) Metabolism of 2-amino-1-methyl-6-phenylimidazo[4,5-*b*]pyridine (PhIP) by human cytochrome P450 1B1. *Carcinogenesis* **18**, 1793–1798.

[37] Salmon, C. P.; Knize, M. G.; Felton, J. S. (1997) Effects of marinating on heterocyclic amine carcinogen formation in grilled chicken. *Food Chem. Toxicol.* **35**, 433–441.

[38] Brockstedt, U.; Pfau, W. (1998) Formation of 2-amino-α-carbolines in pan-fried poultry and ^{32}P-postlabelling analysis of DNA adducts. *Z. Lebensm. Unters. Forsch. A* **207**, 472–475.

[39] Jackson, L. S.; Hargraves, W. A.; Stroup, W. H.; Diachenko, G. W. (1994) Heterocyclic amine content of selected beef flavours. *Mutat. Res.* **320**, 113–124.

[40] Schwarzenbach, R.; Gubler, D. (1992) Detection of heterocyclic aromatic amines in food flavours. *J. Chromatogr.* **624**, 491–495.

[41] Ushiyama, H.; Wakabayashi, K.; Hirose, M.; Itoh, H.; Sugimura, T.; Nagao, M. (1991) Presense of carcinogenic heterocyclic amines in urine of healthy volunteers eating normal diet, but not of inpatients receiving parenteral alimentation. *Carcinogenesis* **12**, 1417–1422.

[42] Ji, H.; Yu, M. C.; Stillwell, W. G.; Skipper, P. L.; Ross, R. K.; Henderson, B. E. Tannenbaum, S. R. (1994) Urinary excretion of MeIQx in white, black, and Asian men in Los Angeles county. *Cancer Epidemiol. Biomarkers Prev.* **3**, 407–411.

[43] Lynch; A. M.; Knize; M. G.; Boobis; A. R.; Gooderham; N. J.; Davies; D. S.; Murray; S. (1992) Intra- and interindividual variability in systemic exposure in humans to 2-amino-3,8-dimethylimidazo[4,5-*f*]-quinoxaline and 2-amino-1-methyl-6-phenylimidazo[4,5-*b*]pyridine, carcinogens present in cooked beef. *Cancer Res.* **52**, 6216–6223.

[44] Sinha, R.; Rothman, N.; Brown, E. D.; Mark, S. D.; Hoover, R. N.; Caporaso, N. E.; Levander, O. A.; Knize, M. G.; Lang, N. P.; Kadlubar, F. F. (1994) Pan-fried meat containing high levels of heterocyclic aromatic amines but low levels of polycyclic aromatic hydrocarbons induces cytochrome P450 1A2 activity in humans. *Cancer Res.* **54**, 6154–6159.

[45] Snyderwine, E. G. (1994) Some perspectives on the nutritional aspects of breast cancer research. Food derived heterocyclic amines as etiologic agents in human mammary cancer. *Cancer* **74**, 1070.

[46] Hammons, G. J.; Milton, D.; Stepps, K.; Guengerich, F. P.; Tukey, R. H.; Kadlubar, F. F. (1997) Metabolism of cacinogenic heterocyclic and aromatic amines by recombinant human cytochrome P450 enzymes. *Carcinogenesis* **18**, 851–854.

[47] Shimada, T.; Hayes, C. L.; Yamazaki, H.; Amin, S.; Hecht, S. S.; Guengerich, F. P.; Sutter, T. R. (1996) Activation of chemically diverse procarcinogens by human cytochrome P450 1B1. *Cancer Res.* **56**, 2979–2984.

[48] Alexander, D. A; Eltom, S. E.; Jefcoate, C. R. (1998) Ah receptor regulation of CYP 1B1 expression in primary mouse embryo-derived cells. *Cancer Res.* **58**, 4498–4506.

[49] Pfau, W.; O'Hare, M. J.; Grover, P. L.; Phillips, D. H. (1992) Metabolic activation of the food mutagens 2-amino-3-methylimidazo[4,5-*f*]quinoline (IQ) and 2-amino-3,4-dimethylimidazo[4,5-*f*]quinoline (MeIQ) to DNA binding species in human mammary epithelial cells. *Carcinogenesis* **13**, 907–909.

[50] Carmichael, P. L.; Stone, E. M.; Grover, P. L.; Gusterson, B. A.; Phillips, D. H. (1996) Metabolic activation and DNA binding of food mutagens and other environmental carcinogens in human mammary epithelial cells. *Carcinogenesis* **17**, 1769–1772.

[51] Stone, E. M.; Williams, J. A.; Grover, P. L.; Gusterson, B. A.; Phillips, D. H. (1998) Interindividual variation in the metabolic activation of heterocyclic amines and their *N*-hydroxy derivatives in primary cultures of human mammary epithelial cells. *Carcinogenesis* **19**, 873–879.

[52] Felton, J. S.; Knize, M. G. (1991) Occurrence, identification, and bacterial mutagenicity of heterocyclic amines in cooked food. *Mutat. Res.* **259**, 205–217.

[53] Lin, D. X.; Lang, N. P.; Kadlubar, F. F. (1995) Species differences in the biotransformation of the food borne carcinogen PhIP by hepatic microsomes from humans, rats and mice. *Drug Metab. Dispos.* **23**, 518–522.

[54] Raza, H.; King, R. S.; Squires, R. B.; Guengerich, F. P.; Miller, D. W.; Freeman, J. P.; Lang, N. P.; Kadlubar, F. F. (1996) Metabolism of 2-amino-α-carboline. A food-borne heterocyclic amine mutagen and carcinogen by human and rodent liver microsomes and by human cytochrome P450 1A2. *Drug Metab. Dispos.* **24**, 395–400.

[55] Turesky, R. J.; Constable, A.; Richoz, J.; Varga, N.; Markovic, J.; Martin, M. V.; Guengerich, F. P. (1998) Activation of heterocyclic aromatic amines by rat and human liver microsomes and by purified rat and human cytochrome P450 1A2. *Chem. Res. Toxicol.* **11**, 925–936.

[56] Turesky, R. J. (1994) In: Hemminki, K.; Dipple, A.; Shuker, D. E. G.; Kadlubar, F. F. (Eds.) *DNA adducts – identification and biological significance*, **125**. IARC, Lyon, p. 217.

[57] Snyderwine E. G.; Yamashita, K.; Adamson, R. H.; Sato, S.; Nagao, M.; Sugimura, T.; Thorgeirsson, S. S. (1988) Use of the ^{32}P-postlabelling method to detect DNA adducts of 2-amino-3-methylimidazolo[4,5-*f*]quinoline (IQ) in monkeys fed IQ: identification of the *N*-(deoxyguanosin-8-yl)-IQ adduct. *Carcinogenesis* **9**, 1739–1743.

[58] Turesky, R. J.; Rossi, S. C.; Welti, D. H.; Lay, J. O.; Kadlubar, F. F. (1992) Characterization of DNA adducts formed *in vitro* by reaction of *N*-hydroxy-2-amino-3-methylimidazo[4,5-*f*]quinoline and *N*-hydroxy-2-amino-3,8-dimethylimidazo[4,5-*f*]quinoxaline at the C-8 and N^2 atoms of guanine. *Chem. Res. Toxicol.* **5**, 479–490.

[59] Frandsen, H.; Grivas, S.; Turesky, R. J.; Andersson, R.; Dragstedt, L. O.; Larsen, J. C. (1994) Formation of DNA adducts by the food mutagen 2-amino-3,4,8-trimethyl-

3H-imidazo[4,5-*f*]quinoxaline (4,8-DiMeIQx) *in vitro* and *in vivo*. Identification of a N^2-(2′-deoxyguanosin-8-yl)-4,8-DiMeIQx adduct. *Carcinogenesis* **15**, 2553–2558.

[60] Lin, D.; Kaderlik, K. R.; Turesky, R. J.; Miller, D. W.; Lay, J. O.; Kadlubar, F. F. (1992) Identification of *N*-(deoxyguanosin-8-yl)-2-amino-1-methyl-6-phenylimidazo[4,5-*b*] pyridine as the major adduct formed by the food borne carcinogen, 2-amino-1-methyl-6-phenylimidazo[4,5-*b*]pyridine (PhIP), with DNA. *Chem. Res. Toxicol.* **5**, 692–697.

[61] Frandsen, H.; Grivas, S.; Andersson, R.; Dragsted, L.; Larsen, J. C. (1992) Reaction of the N^2-acetoxy derivative of 2-amino-1-methyl-6-phenylimidazo[4,5-*b*]pyridine (PhIP) with 2′-deoxyguanosine and DNA. Synthesis and identification of N^2-(2′-deoxyguanosin-8-yl)-PhIP. *Carcinogenesis* **13**, 629–635.

[62] Nagaoka, H.; Wakabayashi, K.; Kim, S.-B.; Kim, L.-S.; Tanaka, Y.; Ochiai, M.; Tada, A.; Nukaya, H.; Sugimura, T.; Nagao, M. (1992) Adduct formation at C-8 of guanine on *in vitro* reaction of the ultimate form of 2-amino-1-methyl-6-phenylimidazo[4,5-*b*]pyridine with 2′-deoxyguanosine and its phosphate esters. *Jpn. J. Cancer Res.* **83**, 1025–1029.

[63] Tada, A.; Ochiai, M.; Wakabayashi, K.; Nukaya, H.; Sugimura, T.; Nagao, M. (1994) Identification of *N*-(deoxyguanosin-8-yl)-2-amino-3,4-dimethylimidazo[4,5-*f*]quinoline (dG-C8-MeIQ) as a major adduct formed by MeIQ with nucleotides *in vitro* and with DNA *in vivo*. *Carcinogenesis* **15**, 1275–1278.

[64] Degen, G. H.; Wolz, E.; Gerber, M.; Pfau, W. (1998) Bioactivation of 2-amino-3-methylimidazo[4,5-*f*]quinoline (IQ) by prostaglandin H synthase. *Arch. Toxicol.* **72**, 183–186.

[65] Hashimoto, Y.; Shudo, K.; Okamoto, T. (1982) Modification of DNA with potent mutacarcinogenic Glu-P1 isolated from a glutamic acid pyrolysate: Structure of the modified nucleic acid base and initial chemical event caused by the mutagen. *J. Am. Chem. Soc.* **104**, 7636–7640.

[66] Hashimoto, Y.; Shudo, K. (1985) Chemical modification of DNA with potent mutacarcinogens 3-amino-1-methyl-5H-pyrido[4,3-*b*]indole (Trp-P2) and 2-amino-6-methyldipyrido[1,2-a:3′,2′-*d*]imidazole (Glu-P1): metabolic activation and structure of the DNA adducts. *Environ. Health Perspect.* **62**, 209–214.

[67] Pfau, W.; Brockstedt, U.; Schulze, C.; Neurath, G.; Marquardt, H. (1996) Characterization of the major DNA adduct formed by the food mutagen 2-amino-3-methyl-9H-pyrido[2,3-*b*]indole (MeAαC) in primary rat hepatocytes. *Carcinogenesis* **17**, 2727–2732.

[68] Pfau, W.; Schulze, C.; Shirai, T.; Hasegawa, R.; Brockstedt, U. (1997) Identification of the major hepatic DNA adduct formed by the food mutagen 2-amino-9H-pyrido[2,3-*b*] indole (AαC). *Chem. Res. Toxicol.* **10**, 1192–1197.

[69] Randerath, K.; Reddy, M. V.; Gupta, R. C. (1981) ^{32}P-Postlabelling test for DNA damage. *Proc. Natl. Acad. Sci. USA* **78**, 6126–6129.

[70] Phillips, D. H. (1997) Detection of DNA modifications by the ^{32}P-postlabelling assay. *Mutat. Res.* **378**, 1–12.

[71] Pfau, W.; Brockstedt, U.; Söhren, K. D.; Marquardt, H. (1994) ^{32}P-Postlabelling analysis of DNA adducts formed by food-derived heterocyclic amines: evidence for incomplete hydrolysis and a procedure for adduct simplification. *Carcinogenesis* **15**, 877–882.

[72] Fukutome, K.; Ochiai, M.; Wakabayashi, K.; Watanabe, S.; Sugimura, T.; Nagao, M. (1994) Detection of Guanine-C8–2-amino-1-methyl-6-phenylimidazo[4,5-*b*]pyridine adduct as a single spot on thin-layer chromatography by modification of the ^{32}P-postlabelling method. *Jpn. J. Cancer Res.* **85**, 113–117.

[73] Turesky, R. J.; Gremaud, E.; Markovic, J.; Snyderwine, E. G. (1996) DNA adduct formation of the food-derived mutagen 2-amino-3-methylimidazo-[4,5-*f*]quinoline (IQ) in nonhuman primates undergoing carcinogen bioassay. *Chem. Res. Toxicol.* **9**, 403–408.

[74] Turesky, R. J.; Markovic, J.; Aeschlimann, J.-M. (1996) Formation and differential removal of C-8 and N^2-guanine adducts of the food carcinogen 2-amino-3-methylimidazo[4,5-*f*]quinoline (IQ) in the liver, kidney, and colorectum of the rat. *Chem. Res. Toxicol.* **9**, 397–402.

[75] Friesen, M. D.; Kaderlik, K.; Lin, D.; Garren, L.; Bartsch, H.; Lang, N. P.; Kadlubar, F. F. (1994) Analysis of DNA adducts of 2-amino-1-methyl-6-phenylimidazo[4,5-*b*] pyridine (PhIP) in rat and human tissues by alkaline hydrolysis and gas chromatography/electron capture mass spectrometry: validation by comparison with ^{32}P-postlabelling. *Chem. Res. Toxicol.* **7**, 733–739.

[76] Totsuka, Y.; Fukutome, K.; Takahashi, M.; Takahashi, S.; Tada, A.; Sugimura, T.; Wakabayashi, K. (1996) Presence of 2-(deoxyguanosin-8-)-2-amino-3,8-dimethylimidazo-[4,5-*f*]quinoxaline (dG-C8-MeIQx) in human tissues. *Carcinogenesis* **17**, 1029–1034.

[77] Gupta, R. C. (1985) Enhanced sensitivity of ^{32}P-postlabelling analysis of aromatic carcinogen: DNA adducts. *Cancer Res.* **45**, 5656–5662.

[78] Wohlin, P.; Zeisig, M.; Gustafsson, J. A.; Möller, L. (1996) ^{32}P-HPLC analysis of DNA adducts formed *in vitro* and *in vivo* by 2-amino-1-methyl-6-phenylimidazo[4,5-*b*]pyridine and 2-amino-3,4,8-trimethyl-3H-imidazo[4,5-*f*]quinoxaline, utilizing an improved adduct enrichment procedure. *Chem. Res. Toxicol.* **9**, 1050–1056.

[79] Reddy, M.V.; Randerath, K. (1986) Nuclease P1-mediated enhancement of sensitivity of ^{32}P-postlabelling test for structurally diverse DNA adducts. *Carcinogenesis* **7**, 1543–1551.

[80] Ochiai, M.; Nagaoka, H.; Wakabayashi, K.; Tanaka, Y.; Kim, S.B.; Tada, A.; Nukaya, H.; Sugimura, T.; Nagao, M. (1993) Identification of N^2-(deoxyguanosin-8-yl)-2-amino-3,8-dimethylimidazo[4,5-*f*]quinoxaline 3′,5′-diphosphate, a major DNA adduct, detected by nuclease P1 modification of the ^{32}P-postlabelling method, in the liver of rats fed MeIQx. *Carcinogenesis* **14**, 2165–2170.

[81] Randerath, E.; Agrawal, H.P.; Weaver, J.A.; Bordelon, C.B.; Randerath, K. (1985) ^{32}P-postlabelling analysis of DNA adducts persisting for up to 42 weeks in the skin, epidermis and dermis of mice treated topically with 7,12-dimethylbenz[*a*]anthracene. *Carcinogenesis* **6**, 1117–1126.

[82] Nerurkar, P. V.; Schut, H. A. J.; Anderson, L. M.; Riggs, C. W.; Fornwald, L. W.; Davis, C. D.; Snyderwine, E. G.; Thorgeirsson, S. S.; Weber, W. W.; Rice, J. M.; Levy, G. N. (1996) Ahr locus phenotype in congenic mice influences hepatic and pulmonary DNA adduct levels of 2-amino-3-methylimidazo[4,5-*f*]quinoline in the absence of cytochrome P450 induction. *Mol. Pharmacol.* **49**, 874–881.

[83] Guo, D.; Schut, H. A. J.; Davis, C. D.; Snyderwine, E. G.; Bailey, G. S.; Dashwood, R. H. (1995) Protection by chlorophyllin and 3-carbinol against 2-amino-1-methyl-6-phenylimidazo[4,5-*b*]pyridine (PhIP)-induced DNA adducts and colonic aberrant crypts in the F344 rat. *Carcinogenesis* **16**, 2931–2937.

[84] Davis, C. D.; Schut, H. A. J.; Adamson, R. H.; Thorgeirsson, U. P.; Thorgeirsson, S. S.; Snyderwine, E. G. (1993) Mutagenic activation of IQ, PhIP and MeIQx by hepatic microsomes from rat, monkey and man: low mutagenic activation of MeIQx in cynomolgus monkeys *in vitro* reflects low DNA adduct levels *in vivo*. *Carcinogenesis* **14**, 61–65.

[85] Ochia, M.; Watanabe, M.; Kushida, H.; Wakabayashi, K.; Sugimura, T.; Nagao, M. (1996) DNA adduct formation, cell proliferation and aberrant crypt focus formation induced by PhIP in male and female rat colon with relevance to carcinogenesis. *Carcinogenesis* **17**, 95–98.

[86] Brockstedt, U.; Pfau, W. Separation of ^{32}P-labelled nucleotide adducts formed by food derived carcinogenic aminoimidazoazarenes using reversed-phase high performance liquid chromatography. *Chromatographia*, in press.

[87] Mauthe, R. J.; Marsch, G. A.; Turteltaub, K. W. (1996) Improved high-performance liquid chromatography analysis of ^{32}P-postlabelled 2-amino-1-methyl-6-phenylimi-

dazo[4,5-*b*]pyridine-DNA adducts using in-line precolumn purification. *J. Chromatogr., Biomed. Appl.* **679**, 91–101.

[88] Pfau, W.; Wild, D.; Söhren, K.-D.; Marquardt, H. (1993) HPLC ^{32}P-postlabelling analysis of DNA adducts formed by heterocyclic amines. *Naunyn-Schm. Arch. Pharmacol.* **347S**, R23.

[89] Rindgen, D.; Turesky, R. J.; Vouros, P. (1995) Determination of *in vitro* Formed DNA adducts of 2-amino-1-methyl-6-phenylimidazo[4,5-*b*]pyridine using capillary liquid chromatography/electrospray ionization/tandem mass spectrometry. *Chem. Res. Toxicol.* **8**, 1005–1013.

[90] Turteltaub, K. W.; Felton, J. S.; Gledhill, B. L.; Vogel, J. S.; Southon, J. R.; Caffee, M. W.; Finkel, R. C.; Nelson, D. E.; Proctor, I. D.; Davis, J. C. (1990) Accelerator mass spectrometry in biomedical dosimetry: Relationship between low-level exposure and covalent binding of heterocyclic amine carcinogens to DNA. *Proc. Natl. Acad. Sci. USA* **87**, 5288–5292.

[91] Turteltaub, K. W.; Vogel, J. S.; Frantz, C. E.; Shen, N. (1992) Fate and distribution of 2-amino-1-methyl-6-phenylimidazo[4,5-*b*]pyridine (PhIP) in mice at a human dietary equivalent dose. *Cancer Res.* **52**, 4682–4687.

[92] Turteltaub, K. W.; Frantz, C. E.; Creek, M. R.; Vogel, J. S.; Shen, N.; Fultz, E. (1993) DNA adducts in model systems and humans. *J. Cell Biochem.* **17F**, 138–148.

[93] Turteltaub, K. W.; Mauthe, R. J.; Dingley, K. H.; Vogel, J. S.; Frantz, C. E.; Garner, R. C.; Shen, N. (1997) MeIQx-DNA adduct formation in rodent and human tissues at low doses. *Mutat. Res.* **376**, 243–252.

[94] Makino, H.; Ishizaka, Y.; Tsujimoto, A.; Nakamura, T.; Onda, M.; Sugimura, T.; Nagao, M. (1992) Rat p53 gene mutations in primary Zymbal gland tumors induced by 2-amino-3-methylimidazo[4,5-*f*]quinoline, a food mutagen. *Proc. Natl. Acad. Sci. USA* **89**, 4850–4854.

[95] Sadrieh, N.; Snyderwine, E. G. (1995) Cytochromes P450 in cynomolgus monkeys mutagenically activate 2-amino-3-methyl-imidazo[4,5-*f*]quinoline (IQ) but not 2-amino-3,8-dimethylimidazo[4,5-*f*]quinoxaline (MeIQx). *Carcinogenesis* **16**, 1549–1555.

[96] Kudo, M.; Ogura, T.; Esumi, H.; Sugimura, T. (1991) Mutational activation of c-Ha-*ras* gene in squamous cell carcinomas of rat zymbal gland induced by carcinogenic heterocyclic amines. *Mol. Carcinog.* **4**, 36–42.

[97] Knasmüller, S.; Kienzl, H.; Huber, W.; Schulte Hermann, R. (1992) Organ-specific distribution of genotoxic effects in mice exposed to cooked food mutagens. *Mutagenesis* **7**, 235–241.

[98] Davis, C. D.; Snyderwine, E. G. (1995) Analysis of EGFR, TGFa, neu and c-myc in 2-amino-1-methyl-6-phenylimidazo[4,5-*b*]pyridine-induced mammary tumors using RT-PCR. *Carcinogenesis* **12**, 3087–3092.

[99] Kakiuchi, H.; Watanabe, M.; Ushijima, T,; Toyota, M.; Imai, K.; Weisburger, J. H.; Sugimura, T.; Nagao, M. (1995) Specific 5′-GGGA-3′→5′-GGA-3′ mutation of the Apc gene in rat colon tumors induced by 2-amino-1-methyl-6-phenylimidazo[4,5-*b*] pyridine. *Proc. Natl. Acad. Sci. USA* **92**, 910–914.

[100] Kakiuchi, H.; Ushijima, T.; Ochiai, M.; Imai, K.; Ito, N.; Yachi, A.; Sugimura, T.; Nagao, M. (1993) Rare frequency of activation of the Ki-*ras* gene in rat colon tumors induced by heterocyclic amines: possible alternative mechanisms of human colon carcinogenesis. *Mol. Carcinog.* **8**, 44–48.

[101] Tachino, N.; Hayashi, R.; Liew, C.; Bailey, G.; Dashwood, R. (1995) Evidence for *ras* gene mutation in 2-amino-3-methylimidazo[4,5-*f*]quinoline (IQ)-induced colonic aberrant crypts in the rat. *Mol. Carcinog.* **12**, 187–192.

[102] Scouv, J.; Kryspin-Sörensen, I.; Frandsen, H.; Selzer-Rasmussen, E.; Forchhammer, J. (1995) The reducing agent dithiothreitol (DTT) increases expression of c-myc and c-fos proto-oncogenes in human cells. *ATLA* **23**, 497–503.

[103] Watanabe, M.; Ohta, T. (1993) Analysis of mutational specificity induced by heterocyclic amines in the lacZ gene of *Escherichia coli*. *Carcinogenesis* **14**, 1149–1153.

[104] Kosakarn, P.; Halliday, J. A.; Glickman, B. W.; Josephy, P. D. (1993) Mutational specificity of 2-nitro-3,4-dimethylimidazo[4,5-*f*]quinoline in the lacI gene of *Escherichia coli*. *Carcinogenesis* **14**, 511–517.

[105] Broschard, T. H.; Lebrun-Garcia, A.; Fuchs, R. P. (1998) Mutagenic specificity of the food mutagen 2-amino-3-methylimidazo[4,5-*f*]quinoline in *Escherichia coli* using the yeast URA3 gene as a target. *Carcinogenesis* **19**, 305–310.

[106] Carothers, A. M.; Yuan, W; Hingerty, B. E.; Broyde, S.; Grunberger, D.; Snyderwine, E. G. (1994) Mutation and repair induced by the carcinogen 2-(hydroxyamino)-1-methyl-6-phenylimidazo[4,5-*b*]pyridine (N-OH-PhIP) in the dihydrofolate reductase gene of Chinese hamster ovary cells and conformational modeling of the dG-C8-PhIP adduct in DNA. *Chem. Res. Toxicol.* **7**, 209–218.

[107] Morgenthaler, P. M. L.; Holzhäuser, D. (1995) Analysis of mutations induced by 2-amino-1-methyl-6-phenylimidazo[4,5-*b*]pyridine (PhIP) in human lymphoblastoid cells. *Carcinogenesis* **16**, 713–718.

[108] Yadollahi-Farsani, M.; Gooderham, N. J.; Davies, D. S.; Boobis, A. R. (1996) Mutational spectra of the dietary carcinogen PhIP at the Chinese hamster hprt locus. *Carcinogenesis* **17**, 617–624.

[109] Wu, R. W.; Wu, E. M.; Thompson, L. H.; Felton, J. S. (1995) Identification of aprt gene mutations induced in repair-deficient and P450-expressing CHO cells by the food-related mutagen/carcinogen PhIP. *Carcinogenesis* **16**, 1207–1213.

[110] Endo, H.; Schut, H. A. J.; Snyderwine E. G. (1994) Mutagenic specificity of 2-amino-3-methyl-imidazo[4,5-*f*]quinoline (IQ) and 2-amino-1-methyl-6-phenylimidazo[4,5-*b*]pyridine (PhIP) in the supF shuttle vector system. *Cancer Res.* **54**, 3745–3751.

[111] Endo, H.; Schut, H. A. J.; Snyderwine, E. G. (1995) Distribution of the DNA adducts of 2-amino-3-methylimidazo[5,4-*f*]quinoline and 2-amino-1-methyl-6-phenylimidazo[4,5-*b*]pyridine in the supF gene as determined by polymerase arrest assay. *Mol. Carcinog.* **14**, 198–204.

[112] Chen, R.-H.; Maher, V. M.; Brouwer, J.; van de Putte, P.; McCormick, J. J. (1992) Preferential repair and strand-specific repair of benzo[*a*]pyrene diol epoxide adducts in the HPRT gene of diploid human fibroblasts. *Proc. Natl. Acad. Sci. USA* **89**, 5413–5417.

[113] Koch, W. H.; Wu, R. W.; Cebula, T. A.; Felton, J. S. (1998) Specificity of base substitution mutations induced by the dietary carcinogens 2-amino-1-methyl-6-phenylimidazo[4,5-*b*]pyridine (PhIP) and 2-amino-3-methylimidazo[4,5-*f*]quinoline (IQ) in Salmonella. *Environ. Mol. Mutagen.* **31**, 327–333.

[114] Maenhaut-Michel, G.; Janel-Bintz, R.; Samuel, N.; Fuchs, R. P. (1997) Adducts formed by the food mutagen 2-amino-3-methylimidazo[4,5-*f*]quinoline induce frameshift mutations at hot spots through an SOS-independent pathway. *Mol. Gen. Genet.* **253**, 634–641.

[115] Shibutani, S.; Fernandes, A.; Suzuki, N.; Johnson, F.; Grollman, A. P. (1998) Site-specific mutagenesis studies of PhIP-derived DNA adduct in mammalian cells. *Z. Lebensm. Unters. Forsch. A* **207**, 459–463.

[116] Okonogi, H.; Ushijima, T.; Zhang, X. B.; Heddle, J. A.; Suzuki, T; Sofuni, T.; Felton, J. S.; Tucker, J. D.; Sugimura, T.; Nagao, M. (1997) Agreement of mutational characteristics of heterocyclic amines in lacI of the Big Blue mouse with those in tumor related genes in rodents. *Carcinogenesis* **18**, 745–748.

[117] Watanabe, M.; Ishidate Jr., M.; Nohmi, T. (1990) Sensitive method for the detection of mutagenic nitriarenes and aromatic amines: new derivatives of *Salmonella typhimurium* tester strains posessing elevated *O*-acetyltransferase levels. *Mutat. Res.* **234**, 337–348.

[118] Wild, D. (1995) Verbesserte mikrobiologische Bestimmung heterozyklischer aromatischer Amine in erhitzten Nahrungsmitteln. *Z. Ernährungswiss.* **34**, 22–26.

[119] Wild, D.; Feser, W.; Michel, S.; Lord, H. L.; Josephy, P. D. (1995) Metabolic activation of heterocyclic aromatic amines catalysed by human arylamine *N*-acetyltransferase isozymes (NAT 1 and NAT 2) expressed in *Salmonella typhimurium. Carcinogenesis* **16**, 643–648.

[120] Josephy, P. D.; DeBruin, L. S.; Lord, H. L.; Oak, J. N.; Evans, D. H.; Guo, Z.; Dong, M.-S.; Guengerich, F. P. (1995) Bioactivation of aromatic amines by recombinant human cytochrome P450 1A2 expressed in Ames tester strain bacteria: A substitute for activation by mammalian tissue preparations. *Cancer Res.* **55**, 799–802.

[121] Suzuki, A.; Kushida, H.; Iwata, H.; Watanabe, M.; Nohmi, T.; Fujita, K.; Gonzalez, F. J.; Kamataki, T. (1998) Establishment of a salmonella tester strain highly sensitive to mutagenic heterocyclic amines. *Cancer Res.* **58**, 1833–1838.

[122] Thompson, L. H.; Wu R. W.; Felton, J. S. (1995) Genetically modified Chinese hamster ovary (CHO) cells for studying the genotoxicity of heterocyclic amines from cooked foods. *Toxicol. Lett.* **82–83**, 883–889.

[123] Winton, D. J.; Gooderham, N. J.; Boobis, A. R.; Davies, D. S.; Ponder, B. A. J. (1990) Mutagenesis of mouse intestine *in vivo* using the Dbl-1 specific locus test. *Cancer Res.* **50**, 7992–7996.

[124] Brooks, R. A.; Gooderham, N. J.; Zhao, K.; Edwards, R. J.; Howard, L. A.; Boobis, A. R.; Winton, D. J. (1994) 2-Amino-1-methyl-6-phenylimidazo[4,5-*b*]pyridine (PhIP) is a potent mutagen in the mouse small intestine. *Cancer Res.* **54**, 1665–1671.

[125] Zhang, X. B.; Felton, J. S.; Tucker, J. D.; Urlando, C.; Heddle, J. A. (1996) Intestinal mutagenicity of two carcinogenic food mutagens in transgenic mice: 2-amino-1-methyl-6-phenylimidazo[4,5-*b*]pyridine and amino(α)carboline. *Carcinogenesis* **17**, 2259–2265.

[126] Ochiai, M.; Ishida, K.; Ushijima, T.; Suzuki, T.; Sofuni, T.; Sugimura, T.; Nagao, M. (1998) DNA adduct level induced by 2-amino-3,4-dimethylimidazo[4,5-*f*]quinoline in Big Blue mice does not correlate with mutagenicity. *Mutagenesis* **13**, 381–384.

[127] Okonogi, H.; Stuart, G. R.; Okochi, E.; Ushijima, T.; Sugimura, T.; Glickman, B. W.; Nagao, M. (1997) Effects of gender and species on spectra of mutation induced by 2-amino-1-methyl-6-phenylimidazo[4,5-*b*]pyridine in the lacI transgene. *Mutat. Res.* **395**, 93–99.

[128] Ochiai, M.; Watanabe, M.; Kushida, H.; Wakabayashi, K.; Sugimura, T.; Nagao, M. (1996) DNA adduct formation, cell proliferation and aberrant crypt focus formation induced by PhIP in male and female rat colon with relevance to carcinogenesis. *Carcinogenesis* **17**, 95–98.

[129] Nagao, M.; Ochiai, M.; Ushijima, T.; Watanabe, M.; Sugimura, T.; Nakagama, H. (1998) Genetic determinant and environmental carcinogens. *Mutat. Res.* **402**, 85–91.

[130] Dashwood, R. H.; Suzui, M.; Nakagama, H.; Sugimura, T.; Nagao, M. (1998) High frequency of β-catenin (ctnnb1) mutations in the colon tumors induced by two heterocyclic amines in the F344 rat. *Cancer Res.* **58**, 1127–1129.

[131] Felton, J. S.; Wu, R.; Knize, M. G.; Thompson, L. H.; Hatch, F. T. (1995) Heterocyclic amine mutagenicity/carcinogenicity: influence of repair, metabolism, and structure. *Proc. Princess Takamatsu Symp.* **23**, 50–58.

[132] Wu, R. W.; Tucker, J. D.; Sorensen, K. J.; Thompson, L. H.; Felton, J. S. (1997) Differential effect of acetyltransferase expression on the genotoxicity of heterocyclic amines in CHO cells. *Mutat. Res.* **390**, 93–103.

[133] Aeschbacher, H. U.; Ruch, E. (1989) Effect of heterocyclic amines and beef extract on chromosome abberations and sister chromatid exchanges in cultured human lymphocytes. *Carcinogenesis* **10**, 429–433.

[134] Miura, K. F.; Hatanaka, M.; Otsuka, C.; Satoh, T.; Takahashi, H.; Wakabayashi, K.; Nagao, M.; Ishidate Jr, M. (1993) 2-Amino-3-methyl-imidazo[4,5-*f*]quinoline (IQ), a

carcinogenic pyrolysate, induces chromosomal aberrations in Chinese hamster lung fibroblasts *in vitro*. *Mutagenesis* **8**, 349–354.

[135] Sawada, S.; Yamanaka, T.; Yamatsu, K.; Furihata, C.; Matsushima, T. (1991) Chromosome aberrations, micronuclei and SCEs in rat liver induced *in vivo* by hepatocarcinogens including heterocyclic amines. *Mutat. Res.* **251**, 59–69.

[136] Otsuka, C.; Miura, K. F.; Satoh, T.; Hatanaka, M.; Wakabayashi, K.; Ishidate, M. Jr. (1996) Cytogenetic effects of a food mutagen, 2-amino-1-methyl-6-phenylimidazo[4,5-*b*]pyridine (PhIP), and its metabolite, 2-hydroxyamino-1-methy-6-phenylimidazo[4,5-*b*]pyridine (N-OH-PhIP), on human and Chinese hamster cells *in vitro*. *Mutat. Res.* **367**, 115–121.

[137] Sanyal, R.; Darroudi, F.; Parzefall, W.; Nagao, M.; Knasmüller, S. (1997) Inhibition of the genotoxic effects of heterocyclic amines in human derived hepatoma cells by dietary bioantimutagens. *Mutagenesis* **12**, 297–303.

[138] Breneman, J. W.; Briner, J. F.; Ramsey, M. J.; Director, A.; Tucker, J. D. (1996) Cytogenetic results from a chronic feeding study of MeIQx in mice. *Food Chem. Toxicol.* **34**, 717–724.

[139] Director, A.; Nath, J.; Ramsey, M. J.; Swiger, R. R.; Tucker, J. D. (1996) Cytogenetic analysis of mice chronically fed the food mutagen PhIP. *Mutat. Res.* **359**, 53–61.

[140] Crofton-Sleigh, C.; Doherty, A.; Ellard, S.; Parry, E. M.; Venitt, S. (1993) Micronucleus assays using cytochalasin-blocked MCL-5 cells, a proprietary human cell line expressing five human cytochromes P450 and microsomal epoxide hydrolase. *Mutagenesis* **8**, 363–372.

[141] Pfau, W.; Martin, F. L.; Cole, K. J.; Venitt, S.; Phillips, D. H.; Grover, P. L.; Marquardt, H. (1999) Heterocyclic amines induce DNA strand breaks and cell transformation. *Carcinogenesis* **20**, 545–551.

[142] Beamand, J. A.; Barton, P. T.; Tredger, J. M.; Price, R. J.; Lake, B. G. (1998) Effect of some cooked food mutagens on unscheduled DNA synthesis in cultured precision-cut rat, mouse and human liver slices. *Food Chem. Toxicol.* **36**, 455–466.

[143] Holme, J. A.; Wallin, H.; Brunborg, G.; Soderlund, E. J.; Hongslo, J. K.; Alexander, J. (1989) Genotoxicity of the food mutagen 2-amino-1-methyl-6-phenylimidazo[4,5-*b*] pyridine (PhIP): formation of 2-hydroxamino-PhIP, a directly acting genotoxic metabolite. *Carcinogenesis* **10**, 1389–1396.

[144] Pool-Zobel, B. L.; Leucht, U. (1997) Induction of DNA damage by risk factors of colon cancer in human colon cells derived from biopsies. *Mutat. Res.* **375**, 105–115.

[145] Sasaki, Y. F.; Saga, A.; Yoshida, K.; Su, Y. Q.; Ohta, T.; Matsusaka, N.; Tsuda, S. (1998) Colon-specific genotoxicity of heterocyclic amines detected by the modified alkaline single cell gel electrophoresis assay of multiple mouse organs. *Mutat. Res.* **414**, 9–14.

[146] Cortesi, E.; Dolara, P. (1983) Neoplastic transformation of BALB 3T3 mouse embryo dibroblasts by the beef extract mutagen IQ. *Cancer Lett.* **20**, 43–47.

[147] Marquardt, H.; Kuroki, T.; Huberman, E.; Selkirk, J. K.; Heidelberger, C.; Grover, P. L.; Sims, P. (1972) Malignant Transformation of Cells derived from Mouse Prostate by Epoxides and other Derivatives of Polycyclic Hydrocarbons. *Cancer Res.* **32**, 716–720.

[148] Hasegawa, R.; Kato, T.; Hirose, M.; Takahashi, S.; Shirai, T.; Ito, N. (1996) Enhancement of Hepatocarcinogenesis by Combined Administration of Food-derived Heterocyclic Amines at Low Doses in the Rat. *Food Chem. Toxicol.* **34**, 1097–1101.

[149] Hasegawa, R.; Takahashi, S.; Shirai, T.; Iwasaki, S.; Kim, D. J.; Ochiai, M.; Nagao, M.; Sugimura, T.; Ito, N. (1992) Dose-dependent formation of preneoplastic foci and DNA adducts in rat liver with 2-amino-3-methyl-9H-pyrido[2,3-*b*]indole (MeAαC) and 2-amino-1-methyl-6-phenylimidazo[4,5-*b*]pyridine (PhIP). *Carcinogenesis* **13**, 1427–1431.

[150] Ohgaki, H.; Takayama, S.; Sugimura, T. (1991) Carcinogenicities of heterocyclic amines in cooked food. *Mutat. Res.* **259**, 399–410.

[151] Tamano, S.; Hasegawa, R.; Hagiwara, A.; Sugimura, T.; Ito, N. (1994) Carcinogenicity of a mutagenic compound from food-derived, 2-amino-3-methyl-9H-pyrido[2,3-*b*]indole (MeAαC), in male F344 rats. *Carcinogenesis* **15**, 2009–2015.
[152] Takahashi, M.; Toyoda, K.; Aze, Y.; Furuta, K.; Mitsumori, K.; Hayashi, Y. (1993) The rat urinary bladder as a new target of heterocyclic amine carcinogenicity: Tumor induction by Trp-P2. *Jpn. J. Cancer Res.* **84**, 852–858.
[153] Adamson, R. H.; Thorgeirsson, U. P.; Snyderwine, E. G.; Thorgeirsson, S. S.; Takayama, S.; Sugimura, T. (1992) Interspecies studies on heterocyclic amines and the need for mechanistic understanding. *bga Schriften* **3**, 43–47.
[154] Dooley K. L.; Von Tungeln, L. S.; Bucci, T.; Fu, P. P.; Kadlubar, F. F. (1992) Comparative carcinogenicity of 4-aminobiphenyl and the food pyrolysates, Glu-P-1, IQ, PhIP, and MeIQx in the neonatal B6C3F1 male mouse. *Cancer Lett.* **62**, 205–209.
[155] Ito, N.; Hasegawa, R.; Imaida, K.; Tamano, S.; Hagiwara, A.; Hirose, M.; Shirai, T. (1997) Carcinogenicity of PhIP in the rat. *Mutat. Res.* **376**, 107–114.
[156] Hasegawa, R.; Kimura, J.; Yaono, M.; Takahashi, S.; Kato, T.; Futakuchi, M.; Fukutake, M.; Fukutome, K.; Wakabayashi, K.; Sugimura, T.; Ito, N.; Shirai, T. (1995) Increased risk of mammary carcinoma development following transplacental and trans-breast milk exposure to a food-derived carcinogen, 2-amino-1-methyl-6-phenylimidazo[4,5-*b*]pyridine (PhIP), in Sprague-Dawley rats. *Cancer Res.* **55**, 4333–4338.
[157] Ghoshal, A.; Preisegger, K.-H.; Takayama, S.; Thorgeirsson, S. S.; Snyderwine, G. (1994) Induction of mammary tumors in female Sprague-Dawley rats by the food mutagen PhIP. *Carcinogenesis* **15**, 2429–2433.
[158] El-Bayoumy, K; Chae, Y.; Upadhyaya, P.; Rivenson, A.; Kurtzke, C.; Reddy, B.; Hecht S. S. (1995) Comparative tumorigenicity of BaP, 1-nitropyrene and PhIP administered by gavage to female CD rats. *Carcinogenesis* **16**, 431–434.
[159] Tudek, B.; Bird, R. P.; Bruce, W. R. (1989) Foci of aberrant crypts in the colons of mice and rats exposed to carcinogens associated with foods. *Cancer Res.* **49**, 1236–1240.
[160] Fearon, E. R.; Vogelstein, B. (1990) A genetic model for colorectal tumorigenesis. *Cell* **61**, 759–767.
[161] Oshima, M.; Oshima, H.; Tsutsumi, M.; Nishimura, S.; Sugimura, T.; Nagao, M.; Taketo, M. M. (1996) Effects of 2-amino-1-methyl-6-phenylimidazo[4,5-*b*]pyridine on intestinal polyp development in Apc delta 716 knockout mice. *Mol. Carcinog.* **15**, 11–17.
[162] Steffensen, I. L.; Paulsen, J. E.; Eide, T. J.; Alexander, J. (1997) 2-Amino-1-methyl-6-phenylimidazo[4,5-*b*]pyridine increases the numbers of tumors, cystic crypts and aberrant crypt foci in multiple intestinal neoplasia mice. *Carcinogenesis* **18**, 1049–1054.
[163] Sorensen, I. K.; Kristiansen, E.; Mortensen, A.; van Kranen, H.; van Kreijl, C.; Fodde, R.; Thorgeirsson, S. S. (1997) Short-term carcinogenicity testing of a potent murine intestinal mutagen, 2-amino-1-methyl-6-phenylimidazo[4,5-*b*]pyridine (PhIP), in Apc1638 transgenic mice. *Carcinogenesis* **18**, 777–781.
[164] Hasegawa, R.; Tanaka, H.; Tamano, S.; Shirai, T.; Nagao, M.; Sugimura, T.; Ito, N. (1994) Synergistic enhancement of small and large intestinal carcinogenesis by combined treatment of rats with five heterocyclic amines in a medium-term multiorgan bioassay. *Carcinogenesis* **15**, 2567–2573.
[165] Hasegawa, R.; Miyata, E.; Futakuchi, M.; Hagiwara, A.; Nagao, M.; Sugimura, T.; Ito, N. (1994) Synergistic enhancement of hepatic foci development by combined treatment of rats with 10 heterocyclic amines at low doses. *Carcinogenesis* **15**, 1037–1041.
[166] Ito, N.; Hasegawa, R.; Shirai, T.; Fukushima, S.; Hakoi, K.; Takaba, K.; Iwasaki, S.; Wakabayashi, K.; Nagao, M.; Sugimura, T. (1991) Enhancement of GST-P positive

liver cell foci development by combined treatment of rats with five heterocyclic amines at low doses. *Carcinogenesis* **12**, 767–772.
[167] Brockstedt, U. (1997) *Mutagene und Kanzerogene in Lebensmitteln: DNA Addukte durch heterocyclische aromatische Amine.* Dissertation, Hamburg University.
[168] Hasegawa, R.; Shirai, T.; Hakoy, K.; Takaba, K.; Iwasaki, S.; Hoshiya, T.; Ito, N.; Nagao, M.; Sugimura, T. (1991) Synergistic enhancement of glutathione S-transferase placental form-positive hepatic foci development in diethylnitrosamine-treated rats by combined administration of five heterocyclic amines at low doses. *Jpn. J. Cancer Res.* **82**, 1378–1384.
[169] Dashwood, R.; Guo, D. (1992) Inhibition of 2-amino-3-methylimidazo[4,5-*f*]quinoline (IQ)-DNA binding by chlorophyllin: studies of enzyme inhibition and molecular complex formation. *Carcinogenesis* **13**, 1121–1126.
[170] Dashwood, R. H. (1992) Protection by chlorophyllin against the covalent binding of 2-amino-3-methylimidazo[4,5-*f*]quinoline (IQ) to rat liver DNA. *Carcinogenesis* **13**, 113–118.
[171] Yoshida, S.; Xiuyun, Y. (1992) The binding ability of bovine milk caseins to mutagenic heterocyclic amines. *J. Dairy Sci.* **75**, 958–961.
[172] Hasegawa, R.; Hirose, M.; Kato, T.; Hagiwara, A.; Boonyaphiphat, P.; Nagao, M.; Ito, N.; Shirai, T. (1995) Inhibitory effect of chlorophyllin on PhIP-induced mammary carcinogenesis in female F344 rats. *Carcinogenesis* **16**, 2243–2246.
[173] Mauthe, R. J.; Snyderwine, E. G.; Ghoshal, A.; Freeman, S. P.; Turteltaub, K. W. (1998) Distribution and metabolism of 2-amino-1-methyl-6-phenylimidazo[4,5-*b*]pyridine (PhIP) in female rats and their pups at dietary doses. *Carcinogenesis* **19**, 919–924.
[174] Shen, C. L.; Purewal, M.; San Francisco, S.; Pence, B. C. (1998) Absence of PhIP adducts, p53 and Apc mutations, in rats fed a cooked beef diet containing a high level of heterocyclic amines. *Nutr. Cancer* **30**, 227–231.
[175] Pence, B. C.; Landers, M.; Dunn, D. M.; Shen, C. L.; Miller, M. F. (1998) Feeding of a well-cooked beef diet containing a high heterocyclic amine content enhances colon and stomach carcinogenesis in 1,2-dimethylhydrazine-treated rats. *Nutr. Cancer* **30**, 220–226.
[176] Weisburger, J. H.; Dolan, L.; Pittman, B. (1998) Inhibition of PhIP mutagenicity by caffeine, lycopene, daidzein, and genistein. *Mutat. Res.* **416**, 125–128.
[177] Apostolides, Z.; Balentine, D. A.; Harbowy, M. E.; Weisburger, J. H. (1996) Inhibition of 2-amino-1-methyl-6-phenylimidazo[4,5-*b*]pyridine (PhIP) mutagenicity by black and green tea extracts and polyphenols. *Mutat. Res.* **359**, 159–163.
[178] Knasmüller, S. (1999) Dietary factors influencing the mutagenic/carcinogenic effect of HAs. *Z. Lebensm. Unters. Forsch.* (in press).
[179] Nagao, M.; Yahagi, T.; Sugimura, T. (1978) Differences in effects of norharman with various classes of chemical mutagens and amounts of S-9. *Biochem. Biophys. Res. Commun.* **83**, 373–378.
[180] Mori, M.; Totsuka, Y.; Fukutome, K.; Yoshida, T.; Sugimura, T.; Wakabayashi, K. (1996) Formation of DNA adducts by the co-mutagen norharman with aromatic amines. *Carcinogenesis* **17**, 1499–1503.
[181] Fekkes, D.; Schouten, M. J.; Pepplinhuizen, L.; Bruinvels, J.; Lauwers, W.; Brinkman, U. A. (1992) Norharman, a normal body constituent. *Lancet* **339**, 506.
[182] Verheij, R.; Timmerman, L.; Passchier, J.; Fekkes, D.; Pepplinkhuizen, L. (1997) Trait anxiety, coping with stress, and norharman. *Psychol. Rep.* **80**, 51–59.
[183] Lichtenberg-Kraag, B.; Klinker, J. F.; Muhlbauer, E.; Rommelspacher, H. (1997) The natural *β*-carbolines facilitate inositol phosphate accumulation by activating small G-proteins in human neuroblastoma cells (SH-SY5Y). *Neuropharmacology* **36**, 1771–1778.
[184] Wodarz, N.; Wiesbeck, G. A.; Rommelspacher, H.; Riederer, P.; Boning, J. (1996) Excretion of *β*-carbolines harman and norharman in 24-hour urine of chronic alco-

holics during withdrawal and controlled abstinence. *Alcohol. Clin. Exp. Res.* **20**, 706–710.
[185] Matsubara, K.; Gonda, T.; Sawada, H.; Uezono, T.; Kobayashi, Y.; Kawamura, T.; Ohtaki, K.; Kimura, K.; Akaike, A. (1998) Endogenously occurring β-carboline induces parkinsonism in nonprimate animals: a possible causative protoxin in idiopathic Parkinson's disease. *J. Neurochem.* **70**, 727–735.
[186] Sugimura, T. (1995) Heterocyclic amines: Food mutagens to be or not to be considered human carcinogens. *Proc. Am. Assoc. Cancer Res.* **36**, 702.
[187] Steineck, G.; Gerhardsson de Verdier, M.; Övervik, E. (1993) The epidemiological evidence concerning intake of mutagenic activity from the fried surface and the risk of cancer cannot justify preventive measures. *Eur. J. Cancer Prev.* **2**, 293–300.
[188] Felton, J. S.; Malfatti, M. A.; Knize, M. S.; Salmon, C. P.; Hopmans, E. C.; Wu, R. W. (1997) Health risk of heterocyclic aromatic amines. *Mutat. Res.* **376**, 37–42.
[189] Gooderham, N. J.; Murray, S.; Lynch, A. M.; Yadollahi-Farsani, M.; Zhao, K.; Rich, K.; Boobis, A. R.; Davies, D. S. (1997) Assessing human risk to heterocyclic amines. *Mutat. Res.* **376**, 53–60.
[190] Ames, B. N.; Gold, L. S. (1994) Chemical carcinogenesis: Too many rodent carcinogens. *Proc. Natl. Acad. Sci. USA* **87**, 7772–7776.
[192] Goshal, A.; Snyderwine, E. G. (1993) Excretion of food-derived heterocyclic amine carcinogens into breast milk of lactating rats and formation of DNA adducts in the newborn. *Carcinogenesis* **14**, 2199–2203.
[193] Pfau, W.; Brockstedt, U.; Shirai, T.; Ito, N.; Marquardt, H. (1997) Pancreatic DNA adducts formed *in vitro* and *in vivo* by the food mutagens 2-amino-1-methyl-6-phenylimidazo[4,5-*b*]pyridine (PhIP) and 2-amino-3-methyl-9H-pyrido[2,3-*b*]indole (MeAαC). *Mutat. Res.* **378**, 13–22.
[194] World Cancer Research Fund (1997) *Food, nutrition and the prevention of cancer: A global perspective.* American Institute for Cancer Research, Washington.

10 Heterocyclic Aromatic Amines: Genetic Susceptibility

Montserrat García-Closas and Rashmi Sinha

10.1 Abstract

There is increasing evidence for an association between red meat intake and risk of cancer, particularly colorectal cancer. This association is stronger for meats cooked until well done or grilled. Heterocyclic amines (HCA) are formed by pyrolysis of aminoacids and creatinine during cooking of meats, and the production of HCA increases with increasing temperature and duration of cooking. HCAs are potent mutagens and carcinogens in laboratory animals, and thus could be one of the agents responsible for the association between cancer risk and meat intake. We have recently developed a diet questionnaire with detailed information on meat cooking practices that may be related to the production of HCA and other food carcinogens. In conjunction with the questionnaire, we have also developed a database with HCA levels found in meat dishes cooked to different degrees. The combination of this questionnaire and database allows the estimation of usual intake of HCA, and it is currently being used in several case-control and cohort studies.

Food born HCAs need to be metabolized before they can exert their genotoxic or carcinogenic activity. The first step for the activation of HCAs is *N*-oxidation by several P450 enzymes, primarily CYP 1A2. The *N*-hydroxyl metabolites from this reaction can be further activated by *N*-acetylation catalized by NAT1/NAT2 enzymes. Reactive metabolites are detoxified by conjugation reactions such as glucuronidation, sulfatation and GST-conjugation. Genetic variations in these metabolizing enzymes could be responsible for differences in enzyme activity and thus could modulate individual susceptibility to the intake of HCAs present in meats. Studies to date provide insufficient evidence for a role of metabolic polymorphisms on susceptibility to HCAs. These studies suffer from several methodological limitations such as exposure misclassification and small sample sizes to evaluate interactions. Finally, it is unlikely that a single chemical will be responsible for the observed associations between red meat intake and cancer, or that a single genetic variant will be responsible for genetic susceptibility to HCAs or other carcinogens present in meats. Therefore, future studies should consider complex mixtures of carcinogenic and anti-carcinogenic factors in foods and combinations of functionally relevant genetic variants, rather than taking a "reductionist" approach of looking at single substances and genes. Large case-control and cohort studies using state-of-the-art methods of exposure assessment to meat consumption and HCAs and collecting DNA samples for genotype determination are currently underway. These studies will provide valuable contributions to this field in the near future.

10.2 Introduction

There is increasing evidence for an association between red meat intake and risk of cancer, particularly colorectal cancer. This association is stronger for meats cooked until well done or grilled. Heterocyclic amines (HCA) is a family of compounds predominantly formed by pyrolysis of aminoacids and creatinine during cooking of meats. The formation of these compounds increases with increasing surface temperature and duration of cooking, especially with cooking methods involving direct transfer of heat from source to the food such a barbecue or grilling (see paper by Dr. W. Pfau in this volume and references thereby).

These substances can form DNA adducts in exposed animals, are potent mutagens in the Ames/*Salmonella* assay, and can induce tumors in laboratory animals exposed at high doses. Thus, HCAs might be one of the factors contributing to the increased risk of cancer observed with high consumption of red meat.

As most xenobiotics, HCAs need to be metabolized before they can exert their mutagenic or carcinogenic activity [1]. Therefore, polymorphic variation in the activity of metabolizing enzymes could modulate host susceptibility to these compounds. In view of this, consideration of different sources of host susceptibility to HCAs is crucial to advance our understanding on the potential role of HCAs in human cancer.

The first half of this paper reviews new developments in methods to assess dietary exposure to HCAs in epidemiologic studies and the epidemiologic evidence for an association between HCA exposure and cancer risk. The second half of the paper reviews metabolic polymorphisms that may affect host susceptibility to HCA exposure, and the current epidemiologic evidence for genetic susceptibility to meat intake or dietary HCAs. The paper concludes with a discussion on the main limitations of these studies and suggestions for future directions.

10.3 Heterocyclic aromatic amines and cancer in human populations

10.3.1 Exposure assessment of HCAs in epidemiologic studies

To investigate cancer risk posed by HCAs to humans in epidemiologic studies, accurate estimation of exposure is needed. Exposure to HCAs can be estimated using measures of external exposure (i. e., amount of HCAs consumed) and/or measures of internal exposure (i. e., actual amount of HCAs in the body) that reflect not only intake frequency but also other factors such as rate of absorption, activation and elimination from the body. This section focuses on the assessment to external exposure to dietary HCAs and will not cover research on biomarkers of internal exposure to HCAs.

Since humans are exposed to HCAs mainly through cooked meat consumed in the diet, accurate information on usual level of meat consumption is needed to estimate external exposure to HCAs. Epidemiological studies to date have used crude surrogates for HCAs exposure such as total level of red meat intake, level of doneness, browness, frying and gravy consumption. Questionnaires used in these studies have focused on doneness level of "red meat" which may be inadequate to assess an individual's exposure to HCAs since substantial heterogeneity of HCA levels is present in different types of red meats cooked at the same level of doneness [2, 3]. Furthermore, different types of meat such as chicken and fish also contain HCAs, so questions limited to doneness of red meat may not provide adequate information on total HCA consumption.

To capture the total amount of HCAs consumed by an individual, we have developed a meat cooking practice module that can be added to a food frequency questionnaire (FFQ), the primary method used to estimate long-term "usual intake" of foods in epidemiological studies of chronic diseases [4]. This module obtains information from the subjects on usual intake of different types of meat with the portion size, method of cooking, degree of internal doneness and external browning. Photographs of meats cooked to different degrees of doneness are used to standardize the responses. The meat cooking practice module was validated using 12 day dietary records and one 24-hour recall with detailed information on meat cooking from over 150 people who previously completed the meat module (data not published). Furthermore, we are also planning to use biomarkers of internal exposure, such as urinary HCA parent compounds and metabolites, in conjunction with the 24-hour recall or dietary records in a triad approach to validate dietary questionnaires.

In addition to the meat module, we have developed a database for HCAs present in the most commonly consumed meat items in the US that allows the estimation of levels of HCA intake [5]. Multiple meat samples were cooked using different methods and levels of doneness, and multiple samples of cooked meats were obtained from several restaurants and fast-food chains. All samples for one particular cooking method and doneness level were finely ground to form a composite sample. Levels of 2-amino-3,8-dimethylimidazo[4,5-*f*]quinoxaline (MeIQx), 2-amino-3,4,8-trimethylimidazo[4,5-*f*]quinoxaline (DiMeIQx), 2-amino-1-methyl-6-phenylimidazo[4,5-*b*]pyridine (PhIP), 2-amino-3-methylimidazo[4,5-*f*]quinoline (IQ), 2-amino-3,4-dimethylimidazo[4,5-*f*]quinoline (MeIQ) were measured in each of the extracts from composite samples by HPLC [2, 3, 6, 7]. In general, HCA content of different meats increased by doneness level but the individual HCAs measured were not produced to the same extent by each cooking method and doneness level.

In summary, we are integrating type of meat, cooking method, and doneness levels into a questionnaire which can be used in an epidemiologic study to assess HCA exposure. The generated database contains HCA values for a matrix of type of meat, cooking method, and doneness that can be linked to questions in the meat module of the food frequency questionnaire.

10.3.2 Cancer risk and HCA intake in human populations

10.3.2.1 Studies of meat intake and risk of cancer

A large number of epidemiological studies have evaluated the relationship between meat intake and risk of cancer. There is considerable heterogeneity in the characterization of meat intake across studies, however most of the evidence for an association with cancer risk comes from studies that considered red meat intake defined as beef, pork and lamb. Studies of red meat intake and cancer risk were recently reviewed by an international panel of experts from the World Cancer Research Fund (WCPF) and the American Institute for Cancer Research (AICR). This panel of experts concluded that high intake of red meat *probably* increases the risk of developing colorectal cancer, and *possibly* increases the risk of pancreas, breast, prostate and kidney cancer [8].

The formation of heterocyclic amines depends on the temperature and duration of cooking and other factors such as the use of direct flame and the use of fats and oils. However, only a limited number of studies have evaluated the effects of cooking methods. The interpretation of results from theses studies is difficult due to the use of different terminology for cooking methods and the consideration of confounding factors such as meat itself, animal fat and animal protein. The WCRF panel of experts evaluated the epidemiologic evidence from these studies and concluded that: "there is no convincing evidence that any method of cooking modifies the risk of any cancer, nor there is evidence of any probably causal relationship" [9]. However, they indicated that high intake of grilled or barbecued meat *possibly* increases the risk of stomach and colorectal cancer, and that there is *insufficient* evidence from a few studies that consumption of fried foods increases the risk of bladder cancer. A few studies have evaluated the level of meat doneness and the degree of external browning, providing some evidence for an association with an increased risk of colorectal cancer and adenomas [10–14]. However, data from these studies is sparse and no final conclusions can be drawn.

Two case-control studies in Uruguay have reported an increased risk of breast and gastric cancer associated with high intake of HCAs [15, 16]. To date, these are the only two published studies that used estimates of HCA levels. Estimates of HCA levels were obtained from previously published data from heterogeneous sources that often used information on the HCA content of meat samples cooked to maximize the production of these chemicals ([5] and references cited thereby). Therefore, the values used may not be representative of the levels in meats usually cooked by the study population. A number of case-control and cohort studies that have incorporated new methodologies to estimate levels of HCA intake are currently being conducted and will be publishing results in the near future.

We are currently evaluating the relationship between meat intake, cooking method, level of doneness and level of HCA intake estimated using our newly developed database in three case-control studies of colorectal adenomas, lung cancer and breast cancer. In a case-control study of colorectal adenomas at the National Navy Medical Center, we found an increased risk of adenomas associated with red meat consumption but not with total or white meat intake [17]. The increased risk was mainly associated with the consumption of well done/ very well done red meat compared to rare/medium red meat. High temperature cooking methods were also associated with increased risk, particularly with red meat that was grilled. When we considered levels of HCA intake, we found a significant association between MeIQx and risk of adenomas that remained significant after adjustment for red meat but not vice versa. The odds ratios for MeIQx and grilled meats did not substantially change when adjusted for each other. Analyses are currently underway to explore these preliminary results in more detail. In the case-control study of lung cancer among women in Missouri, we found evidence for an increased risk of lung cancer after high intake of red meat, especially fried and/or well-done [18]. Finally, in the case-control study of breast cancer, nested among cohort members of the Iowa Women's Health Study, high consumption of well-done meats was associated with an increased risk of breast cancer (manuscript in press). Analyses using estimates of HCA intake from the last two studies are also underway.

10.4 Genetic susceptibility to heterocyclic aromatic amines

As many xenobiotics, HCAs need to be metabolized before becoming mutagenic or carcinogenic [1]. Individual differences in the ability to metabolize HCAs may thus influence host susceptibility to these compounds [19, 20]. Metabolic variation can arise from both inherited or environmental factors which can be assessed in epidemiologic studies using phenotype and genotype assays. In this section, we review the epidemiologic evidence for individual susceptibility to dietary HCAs due to metabolic polymorphisms. Other factors such as polymorphisms in DNA repair genes could contribute to genetic susceptibility to HCAs [21], however consideration of these factors is outside the scope of this paper.

10.4.1 Polymorphisms for heterocyclic amine metabolism in humans

The first and obligatory step in the activation of heterocyclic amines is *N*-hydroxylation. This reaction is mediated by both hepatic and extrahepatic cytochrome P450 enzymes, however hepatic metabolism by CYP 1A2 seems to be the major pathway in humans [1, 19]. CYP 1A1 is involved in *N*-hydroxylation of HCAs to a lesser extent, however a role of this enzyme in detoxification through ring-hydroxylation seems more likely [22]. In the presence of flavonoids, both CYP 3A4 and CYP 3A5 can also activate HCAs both in the liver and extrahepatic tissues including the colon mucosa [1]. More recently, CYP 1B1, expressed primarily in extrahepatic tissues, has been shown to activate HCAs [23].

N-hydroxylamines can be further activated by phase II enzymes involved in both *N,O*-acetylation (*N*-acetyltransferases) and *N,O*-sulfonation (phenol sulfotransferases). NAT1 and NAT2 enzymes mediate the *N*-acetyltransferase activity in humans. Only NAT2 is expressed in liver, however, both NAT1 and NAT2 have high activities in the colon mucosa [24]. There are three types of phenol sulfotransferases in humans, two thermostable forms (SULT 1A1 and SULT 1A2) and one thermolabile form (SULT 1A3). SULT 1A1 seems to be the main enzyme involved in the sulfate conjugation of *N*-hydroxylamines [25]. Sulfotransferases are probably expressed at highest levels in the liver; however, the levels can vary widely in different tissues from different individuals [26, 27]. Other phase II enzymes such as glucuronosyltransferases and glutathiontransferases are involved in reactions that generally result in detoxification of HCA metabolites [1, 19].

All of these enzymes exhibit polymorphic variations in human populations and can be induced by drugs, environmental and dietary factors [19, 20]. The CYP 1A2 and NAT2 polymorphism can be detected by phenotype assays that measure the ratio of caffeine metabolites in urine after caffeine consumption [28]. The NAT2 phenotype has a bimodal distribution that can clearly distinguish between two distinct groups of subjects, slow and rapid acetylators. The CYP 1A2 phenotype seems to follow a unimodal distribution and thus, there is no clear distinction between "slow" and "rapid" *N*-oxidizers, but rather a continuous gradient from slow to rapid *N*-oxidizers. The reasons for the variability of these two enzymes are different. For NAT2, the variability is primarily due to genetic variants that determine function [29, 30], with environmental factors playing a minor role in the phenotype. In contrast, both genetic and environmental factors are likely to be responsible for the variability in CYP 1A2 [29, 31]. Various environmental factors, such as smoking, or certain dietary components, e.g. cruciferous vegetables and polycyclic aromatic hydrocarbons (PAHs), as well as HCAs in high-temperature cooked meats are known to induce the enzyme activity [31].

In spite of intense study and the identification of a few polymorphic variants of the CYP 1A2 gene, a clear genetic source of the variability remains to be demonstrated [32]. On the other hand, the acetylation phenotype is explained by a genetic polymorphism in the NAT2 gene locus [30]. At least, fifteen variant alleles at the NAT2 gene locus have been linked with the acetylation phenotype [33]. Expression of the NAT1 enzyme also presents phenotypic variation that is partly explained by allelic variants in the NAT1 gene locus. Several NAT1 variant alleles have been described, however, the functional significance of many of these variants is unknown. Different allelic variants might have very different functional consequences. For instance, NAT1*14 and NAT1*15 alleles produce defective proteins that lead to a reduced metabolism of NAT1 substrates, whereas the NAT1*10 allele has been associated with an elevated NAT1 activity [30].

Different phenotype and genotype assays for other enzymes involved in HCA metabolism have been recently described and are currently being investigated. For instance, polymorphisms in SULT 1A1 and CYP 1B1 gene loci have been reported recently, however their functional significance is still largely unknown.

10.4.2 Epidemiologic studies on metabolic polymorphisms, meat/HCA intake and cancer risk

10.4.2.1 Colorectal adenomas and cancer

Several epidemiologic studies have evaluated the relationship between variations in the CYP 1A2 and NAT2 enzyme activity and the risk of colorectal adenomas and cancer [34], but only a few studies took into consideration meat intake [13, 35, 36, 39]. A metanalysis including all studies published before 1996, concluded that subjects with the rapid NAT2 acetylator phenotype or genotype have an about 20% increased risk of developing colorectal cancer [34]. A study by Lang *et al.* including 75 cases of colorectal cancer or polyps and 205 controls in the U.S. evaluated both the CYP 1A2 and NAT2 phenotypes, as well as the frequency of meat consumption and cooked meat preference [13]. The authors concluded that rapid metabolizers for both enzymes have increased susceptibility to colon cancer or adenomas, and that this risk is further increased among subjects that prefer well done meat. However, the small number of subjects included in this study precludes any conclusions about a possible interaction between these metabolic polymorphisms and meat intake or meat cooking preference. A study of 174 cases of colorectal cancer and 174 controls in the U.K., failed to observe differences in the frequency of NAT2 fast genotype among cases and controls [36]. However, they observed an increased risk of cancer associated with high intake of fried meat only among subjects with the NAT2 fast genotype.

The NAT1 genotype has been evaluated only in a few studies. A case-control study reported a significant association between the NAT1 rapid acetylator genotype and an increase in risk of colorectal cancer [37], whereas another case-control study found no association with colorectal adenomas [38]. A recent case-control study of colorectal cancer nested in the Physician's Health Study failed to observe a significant increase in risk among subjects with the slow acetylator genotype for NAT1 or NAT2 [39]. However, the authors reported an association between high red meat intake and risk of colorectal cancer only among slow acetylators for both the NAT1 and NAT2 genes. This increase in risk was statistically significant only among men older than 60 years of age. Interactions between meat/HCA intake and metabolic polymorphisms are currently being investigated in the case-control study of colorectal adenomas at the Navy Medical Center [17].

10.4.2.2 Bladder cancer

The slow acetylation phenotype has long been associated with increased risk of arylamine-related urinary bladder cancer [40, 41]. Subjects that are slow acetylators are thought to be at increased risk of bladder cancer due to an impairment in their ability to detoxify arylamines by *N*-acetylation. The evidence for a role of the NAT2 slow genotype or phenotype on smoking-related bladder cancer, and the evidence for a NAT1 role is weaker [42–45]. No current studies have evaluated a possible interaction between meat intake and the acetylator genotype in relation to bladder cancer risk. We are planning to evaluate this hypothesis in an undergoing multicenter case-control study in Spain that will include about 1,500 cases of bladder cancer and 1,500 hospital controls with detailed information of meat intake and cooking methods.

10.4.2.3 Breast cancer

Studies to date have not found significant differences in the frequency of the *NAT2* slow acetylation genotype among cases of breast cancer and controls [46–49]. A recent report by Ambrosone *et al.* considered frequency of meat intake and NAT2 genotype in relation to breast cancer risk among pre- and postmenopausal women [50]. This study failed to observe a relationship between any of these variables and breast cancer risk. Several ongoing studies using dietary questionnaires with detailed information on meat intake will be contribute more information of the potential role of genetic susceptibility to HAC exposure in breast cancer.

10.5 Limitations and future directions

The study of genetic susceptibility to dietary substances such as HCA intake in epidemiologic studies presents several limitations of note. First, assessment of dietary exposure to carcinogenic factors in the population is prone to measurement error from several sources. Relevant sources of error include: difficulties of subjects to recall past dietary exposures, failure to consider all relevant sources of HCA exposure in dietary questionnaires (e.g. method of meat cooking, level of meat doneness, restaurant versus home cooking etc.), and difficulties in obtaining comprehensive and accurate databases to estimate carcinogenic levels in foods. In addition, there are also sources of error in measuring genetic susceptibility due to failure to measure functionally important genetic variants and to a lesser extend due to laboratory errors. Second, studies of interactions between genetic susceptibility markers and exposure to dietary carcinogenic factors require large sample sizes in order to achieve adequate statistical power. Third, given that diet is a complex mixture of carcinogens, co-carcinogens and anti-carcinogens, a "reductionist" approach of evaluating only a single compound or family of compounds such as HCAs may not be appropriate. This is especially true with cooked meats that may also contain other mutagens/carcinogens such as polycyclic aromatic hydrocarbons, as well as other factors that have been found to increase the risk of cancer such as saturated fats. Similarly, metabolizing enzymes are part of complex metabolic pathways involving a large number of enzymes with overlapping substrates and therefore, genetic susceptibility is likely to be the result of a combination of genetic variants rather than a single genetic variant.

In order to advance our knowledge on the role of HCA in cancer and a possible genetic susceptibility to these compounds, it is critical to develop valid and accurate methods to assess both external and internal HCA exposure. Newly developed methods will contribute to reduce measurement error in the assessment of HCA intake. In addition, there is a need to develop reliable biomarkers of long-term internal exposure that can be measured in easily accessible tissues. To clarify complex relationships between genetic variants, HCA intake and risk of cancer will also require large studies. A new generation of large case-control and cohort studies using state of the art methods to assess HCA exposure are currently underway and will provide important clues to these relationships in the near future. Better knowledge on metabolic pathways for HCAs and the functional importance of metabolic polymorphisms is also needed to identify combinations of genetic variants that could modify individual susceptibility to HCAs.

References

[1] Windmill, K. F.; McKinnon, R. A.; Zhu, X.; Gaedigk, A.; Grant, D. M.; McManus, M. E. (1997) The role of xenobiotic metabolizing enzymes in arylamine toxicity and carcinogenesis: Functional and localization studies. *Mutat. Res.* **376**, 153–160.

[2] Sinha, R.; Knize, M. G.; Salmon, C. P.; *et al.* (1998) Heterocyclic amine content of pork products cooked by different methods and to varying degrees of doneness. *Food Chem. Toxicol.* **36**, 289–297.

[3] Sinha, R.; Rothman, N.; Salmon, C. P.; *et al.* (1998) Heterocyclic amine content in beef cooked by different methods to varying degrees of doneness and gravy made from meat drippings. *Food Chem. Toxicol.* **36**, 279–287.

[4] Tompson, F. E.; Byers, T. (1994) Dietary assessment resource manual. *J. Nutr.* **124**, 2245S–2317S.

[5] Sinha, R.; Rothman, N. (1997) Exposure assessment of heterocyclic amines (HCAs) in epidemiologic studies. *Mutat. Research* **376**, 195–202.

[6] Knize, M. G.; Sinha, R.; Rothman, N.; *et al.* (1995) Heterocyclic amine content in fast-food meat products. *Food Chem Toxicol.* **33**, 545–551.

[7] Sinha, R.; Rothman, N.; Brown, E. D.; *et al.* (1995) High concentrations of the carcinogen 2-amino-1-methyl-6-phenylimidazo-[4,5-*b*]pyridine (PhIP) occur in chicken but are dependent on the cooking method. *Cancer Res.* **55**, 4516–4519.

[8] Nutrition food and cancer: Meat, poultry, fish and eggs (1997). In: *Food, nutrition and the prevention of cancer: A global perspective.* World Cancer Research Fund and American Institute for Cancer Research, Washington DC, USA, p. 452–459.

[9] Nutrition food and cancer: Cooking (1997). In: *Food, nutrition and the prevention of cancer: A global perspective.* World Cancer Research Fund and American Institute for Cancer Research, Washington DC, USA, p. 497–500.

[10] Schiffman, M. H.; Felton, J. S. (1990) Re: Fried foods and the risk of colon cancer. *Am. J. Epidemiol.* **131**, 76–78.

[11] Gerhardsson, de Verdier, M.; Hagman, U.; Peters, P. K.; Steineck, G.; Overvik, E. (1991) Meat, cooking methods and colorectal cancer: a case-referent study in Stockholm. *Int. J. Cancer* **49**, 520–525.

[12] Steineck, G.; Gerhardsson de Verdier, M.; Overik, E. (1993) The epidemiological evidence concerning intake of mutagenic activity from fried surface and the risk of cancer cannot justify preventive measures. *Eur. J. Cancer Prev.* **2**, 293–300.

[13] Lang, N. P.; Butler, M. A.; Massengill, J.; Lawson, M.; Craig Stotts, R.; Haurer-Jensen, M.; Kadlubar, F. F. (1994) Rapid metabolic phenotypes for acetyltransferase and cytochrome P450 1A2 and putative exposure to food-borne heterocyclic amines increase the risk for colorectal cancer or polyps. *Cancer Epidemiol. Biomarkers Prev.* **3**, 675–682.

[14] Probst-Hensch, N. M.; Sinha, R.; Lin, H. J.; Longnecker, M. P.; Witte, J. S.; Ingles, S. A.; *et al.* (1997) Meat preparation and colorectal adenomas in a large sigmoidoscopy-based case-control study. *Cancer Causes Control* **8**, 175–183.

[15] De-Stefani, E.; Ronco, A.; Mendilaharsu, M.; Guidobono, M.; Deneo-Pellegrini, H. (1997) Meat intake, heterocyclic amines, and risk of breast cancer: a case-control study in Uruguay. *Cancer Epidemiol. Biomarkers Prev.* **6**, 573–581.

[16] De Stefani, E.; Boffetta, P.; Mendilaharsu, M.; Carzoglio, J.; Deneo-Pellegrini, H. (1998) Dietary nitrosamines, heterocyclic amines and risk of gastric cancer: A case-control study in Uruguay. *Nutr. Cancer* **30**, 158–162.

[17] Sinha, R.; Chow, W. H.; Kulldorf, M.; *et al.* (1998) Well done, grilled red meat, 1-amino-3,8-dimethylimidazo[4,5-*f*]quinoxaline (MeIQx), metabolizing enzymes and risk of colorectal adenomas. *Proc. Am. Assoc. Cancer Res.* **39**, 364.

[18] Sinha, R.; Kulldorf, M.; Crandall, J.; Brown, C. C.; Alevanja, M. C. R.; Swanson, C. A. (1998) Fried, well done red meat and risk of lung cancer in women (United States). *Cancer Causes Control* **9**(6), 621–630.

[19] Kadlubar, F. F.; Butler, M. A.; Kaderlik, K. R.; Chou, H. C.; Lang, N. P. (1992) Polymorphisms for aromatic amine metabolism in humans: relevance for human carcinogenesis. *Environ. Health Perspect.* **98**, 69–74.

[20] Kadlubar. F. F. (1994) Biochemical individuality and its implications for drug and carcinogen metabolisms: recent insights from acetyltransferase and cytochrome P450 1A2 phenotyping and genotyping in humans. *Drug Metab. Rev.* **26**, 37–46.

[21] Mohrenweiser, H. W.; Jones, I. M. (1998) Variation in DNA repair is a factor in cancer susceptibility: a paradigm for the promises and perils of individual and population risk estimation? *Mutat. Res.* **400**, 15–24.

[22] Kadlubar, F. F.; Hammons, G. J. (1987) The role of cytochrome P450 in metabolism of chemical carcinogens. In: Guengerich, F. P. (Ed.) *Mammalian cytochromes P450.* Boca Raton FL, CRC Press, p. 81–130.

[23] Shimada, T.; Hayes, C. L.; Yamazaki, H.; *et al.* (1996) Activation of chemically diverse procarcinogens by human cytochrome P450 1B1. *Cancer Res.* **56**, 2979–2984.
[24] Turesky R. J.; Lang N. P.; Butler M. A.; Teitel C. H.; Kadlubar F. F.; (1991) Metabolic activation of carcinogenic heterocyclic aromatic amines by human liver and colon. *Carcinogenesis (Lond.)* **12**, 1839–1845.
[25] Glissen, R. A.; Bamforth, K. L.; Stavenuiter, J. H.; Coughtrie, M. W. H.; Meermen, J. H. (1994) Sulfation of aromatic hydroxamic acids and hydroxylamines by multiple forms of human live sulfotransferases. *Carcinogenesis.* **15**, 39–45.
[26] Rein, G.; Glover, V.; Sandler, M. (1982). Multiple forms of phenolsulfotransferase in human tissues. *Biochem. Pharmacol.* **31**, 1893–1897.
[27] Chou, H. C.; Lang, N. P.; Kadlubar, F. F. (1995) Metabolic activation of *N*-hydroxy arylamines and *N*-hydroxy heterocyclic amines by human sulfotransferase(s). *Cancer Res.* **55**, 525–529.
[28] Butler, M. A.; Lang, N. P.; Young, J.; *et al.* (1992) Determination of CYP 1A2 and NAT2 phenotypes in human population by analysis of caffeine urinary metabolites. *Pharmacogenetics* **2**, 116–127.
[29] Sinha, R.; Caporaso, N. (1997) Role of genetic susceptibility in activation of heterocyclic amines. *Ann. Epidemiol.* **7**, 350–356.
[30] Grant, D M.; Hughes, N. C.; Janezic, S. A.; *et al.* (1997) Human acetyltransferase polymorphisms. *Mutat. Res.* **376**, 61–70.
[31] Sinha, R.; Rothman N.; Brown ED.; *et al.* (1994) Pan-fried meat containing high levels of heterocyclic aromatic amines but low levels of polycyclic aromatic hydrocarbons induces cytochrome P450 1A2 activity in humans. *Cancer Res.* **54**, 6154–6159.
[32] Nakajima, M. T.; Yokoi, M.; Mizutami, S. S.; Kadlubar, F. F.; Kamataki, T. (1994) Phenotyping of CYP 1A2 in Japanese population by analysis of caffeine urinary metabolites: absence of mutation prescribing the phenotype in the CYP 1A2 gene. *Cancer Epidemiol. Biomarkers Prev.* **3**, 413–421.
[33] Vatsis, K. P.; Weber, W. W.; Bell, D. A.; *et al.* (1995) Nomenclature for *N*-acetyltransferases. *Pharmacogenetics* **5**(1), 1–17.
[34] D'Errico, A.; Taioli, E.; Chen, X.; Vineis, P. (1996) Genetic metabolic polymorphisms and the risk of cancer: a review of the literature. *Biomarkers* **1**, 149–173.
[35] Vineis, P.; McMichael, A. (1996) Interplay between heterocyclic amines in cooked meat and metabolic phenotype in the etiology of colon cancer. *Cancer Causes Control* **7**, 479–486.
[36] Welfare, M. R.; Cooper, J.; Bassendine, M. F.; Daly, A. K. (1997) Relationship between acetylator status, smoking, diet and colorectal cancer risk in the north-east of England. *Carcinogenesis (Lond.)* **18**, 1351–1354.
[37] Bell, D.; Stephens, E. A.; Castranio, T.; *et al.* (1995) Polyadenylation polymorphisms in the acetyltransferase 1 gene (*NAT1*) increases risk of colorectal cancer. *Cancer Res.* **55**, 3537–3542.
[38] Probst-Hensch, N. M.; Haile, R. W.; Li, D. S.; *et al.* (1996) Lack of association between the polyadenylation polymorphisms in the NAT1 (acetyltransferase 1) gene and colorectal adenomas. *Cancinogenesis (Lond.)* **17**, 2125–2129.
[39] Chen, J.; Stampfer, M. J.; Hough, H. L.; *et al.* (1998) A prospective study of *N*-acetyltransferase genotype, red meat intake, and risk of colorectal cancer. *Cancer Res.* **58**, 3307–3311.
[40] Cartwright, R. A.; Glashan, R. W.; Rogers, H. J.; *et al.* (1982) Role of *N*-acetyltransferase phenotypes in bladder carcinogenesis: a pharmacogenetic epidemiological approach to bladder cancer. *Lancet* **2**, 842–845.
[41] Weber, W. W.; Mattano, S. S.; Levy, G. N. (1988) Acetylation pharmacogenetics and aromatic amine-induced cancer. In: King, C. M.; Romano, L. J.; Schuetzle, D. (Eds.) *Carcinogenic and mutagenic responses to aromatic amines and nitrosamines*, Elsevier Science Publishing Co. Inc., New York, USA, p. 115–123.

[42] Risch, A.; Wallace, D. M.; Bathers, S.; Sim, E. (1995) Slow *N*-acetylator genotype is a susceptibility factor in occupational and smoking related bladder cancer. *Hum. Mol. Genet.* **4**, 231–235.
[43] Brockmoller, J.; Cascorbi, I.; Kerb, R.; Roots, I (1996) Combined analysis of inherited polymorphisms in arylamine *N*-acetyltransferase 2, glutathione S-transferase M1 and T1, microsomal epoxide hydrolase and cytochrome P450 enzymes as modulators of bladder cancer risk. *Cancer Res.* **56**, 3915–3925.
[44] Okkels, H.; Sigsgaard, T.; Wolf, H.; Autrup, H. (1997) Arylamine *N*-acetyltransferase 1 (NAT1) and 2 (NAT2) polymorphisms in susceptibility to bladder cancer: influence of smoking. *Cancer Epidemiol. Biomarkers Prev.* **2**, 225–231.
[45] Taylor, J. A.; Umbach, D. M.; Stephens, E.; *et al.* (1998) The role of *N*-acetylation polymorphisms in smoking-associated bladder cancer: evidence of a gene-gene-exposure three-way interaction. *Cancer Res.* **58**, 3603–3610.
[46] Agundez, J. A.; Ladero, J. M.; Olivera, M.; *et al.* (1995) Genetic analysis of the arylamine *N*-acetyltransferase polymorphism in breast cancer patients. *Oncology* **52**(1), 7–11.
[47] Ambrosone, C. B.; Freudenheim, J. L.; Graham, S.; *et al.* (1996) Cigarette smoking, *N*-acetyltransferase 2 genetic polymorphisms, and breast cancer risk. *J. Am. Med. Assoc.* **276**(18), 1494–1501.
[48] Hunter, D. J.; Hankinson, S. E.; Hough, H.; *et al.* (1997) A prospective study of *NAT2* acetylation genotype, cigarette smoking, and risk of breast cancer. *Carcinogenesis* **18**(11), 2127–2132.
[49] Millikan, R. C.; Pittman, G. S.; Newman, B.; *et al.* (1998) Cigarette smoking, *N*-acetyltransferases 1 and 2, and breast cancer risk. *Cancer Epidemiol. Biomarkers Prev.* **7**(5), 371–378.
[50] Ambrosone, C. B.; Freudenheim, J. L.; Sinha, R.; *et al.* (1998) Breast cancer risk, meat consumption and *N*-acetyltransferase (*NAT2*) genetic polymorphisms. *Int. J. Cancer* **75**, 825–830.

11 A Review of Recent Advances in the Genotoxicity of Carcinogenic Mycotoxins

Guy Dirheimer

11.1 Abstract

Mycotoxins are produced by microscopic toxigenic fungi which grow on human food and animal feed. These mycotoxins are highly stable to heat treatments and, as the range of their biological activity is large, their presence in commodities constitutes sometimes a serious health problem. Several of them are carcinogenic either by their genotoxic or by their epigenetic effect. It is important to know by which mechanism a mycotoxin is carcinogenic in order to define the safety factor which will be used in the calculation of its acceptable daily intake. This review describes the genotoxic effects of several mycotoxins: aflatoxins, sterigmatocystin, ochratoxin A, citrinin, zearalenone, fusarin C and griseofulvin, in particular the DNA adducts caused by these mycotoxins and, when they are known, the mutations which ensue. The often contradictory results obtained in the genotoxicity tests with other mycotoxins: patulin, trichothecenes, fumonisins are also presented.

11.2 Introduction

Mycotoxins constitute a heterogeneous group of secondary metabolites produced by microscopic toxigenic fungi which can appear at each stage of food production from the plant growth to the processing of foods and feeds. Due to their various toxic effects and their high stability to heat treatments their presence is potentially hazardous to the health of both humans and animals. Their clinical toxicological syndromes are very broad ranging from acute toxicity, which is rarely encountered, to a wide spectrum of long term toxic effects that include target organ toxicities, impaired immunity, reduced reproductive efficiency, and cancer. We shall focus in this paper on this last but important aspect of the mycotoxin toxicity. It is known that chemical carcinogens can be either genotoxic (DNA-reactive carcinogens) or act by epigenetic mechanisms (promoters). For toxicological evaluation and the estimation of safety factors that are used in the calculation of its acceptable daily intake, it is important to know by what mechanism a mycotoxin is genotoxic, because the safety factors used for genotoxic compounds are much higher than for non-genotoxic ones.

11.3 Aflatoxins

Aflatoxin B_1 (AFB_1) is the most intensively studied example of a mycotoxin. It was shown to be a potent genotoxic agent in many tests. It is also carcinogenic in several animal species. In the rat, development of hepatocellular carcinoma is preceded by the formation of liver foci and hyperplastic nodules [1].

The strength and consistency of the epidemiological data (reviewed recently in [2, 3]) indicated that there is sufficient evidence for the carcinogenicity to humans (Group 1) in whom it causes hepatocellular carcinoma.

AFB_1 is activated by the cytochrome P450 (CYP) mixed function oxidase system to the carcinogenic intermediate AFB_1-*exo*-8,9-epoxide [4], a process mediated in man liver mainly by CYP 3A4 and 1A2 [5, 6]. The relative contribution of these two CYP at low substrate concentrations encountered in the diet was recently studied by Gallagher *et al.* [7] by *in vitro* methods. They showed that CYP 1A2 dominates the activation of AFB_1 at submicromolar concentrations. However mechanisms of AFB_1 activation by prostaglandin H-synthase or lipoxygenases have also been described [8–11]. Genetic variability in the expression of these CYP and enzymes may result in substantial inter-species and intra-strain differences in susceptibility to the carcinogenic action of AFB_1 [3].

Formation of DNA adducts of AFB_1-epoxide is well-known for 20 years. Treatment of rats or mice with a single dose of AFB_1 at the LD_{50} resulted in 2,300 adducts/10^9 nucleotides in the mouse liver DNA and of 126,000 adducts/10^9 nucleotides in the rat liver DNA [12]. The primary site of adduct is the N_7 position of a guanine nucleotide (AFB_1-N7-Gua) [13, 14] (Figure 11.1).

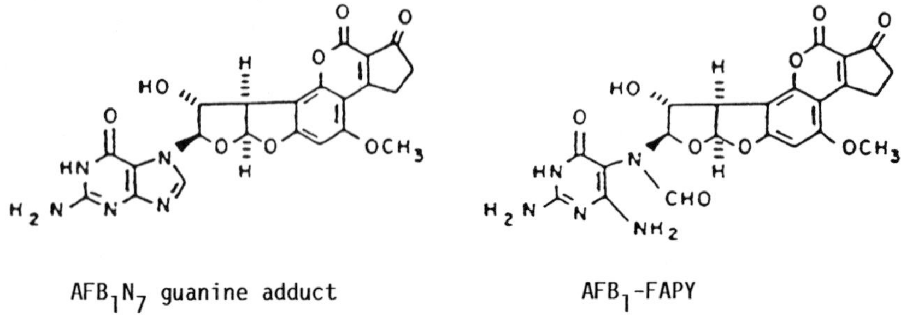

Figure 11.1: Structure of AFB_1-N7-guanine (left) and of AFB_1-FAPY (right).

A recent advance in chemical synthesis has facilitated the production of AFB_1-epoxide, which has enabled investigations of the mechanisms of the reaction of this epoxide with DNA [15]. Given that the half-life of AFB_1 epoxide in water is approximately one second, it has been proposed that AFB_1 epoxide is first intercalated on the 5′ side of the target guanine [16]. This hypothesis is supported by the observation that the reactivity of AFB_1 epoxide with double-stranded B-form DNA is greatly enhanced as compared with single-stranded DNA or alternative duplex structures [17, 18]. From NMR structural studies on the adduct, it is speculated that 5′ intercalation facilitates adducts formation by positioning the epoxide for in-line nucleophilic attack by the guanine N_7 [19, 20]. Furthermore, modifications of the ring systems in AFB_1 that decrease planarity and therefore intercalation ability, a situation that exists with aflatoxin G_1, decrease affinity for DNA and result in lower reactivity [21]. Finally, addition of the intercalating agent ethidium bromide to target DNA or formation of an intercalation inhibitor *cis*-syn thymidine benzofurane [22] before treatment with AFB_1 epoxide greatly reduces guanine reactivity. These experiments support a model in which the intercalated intermediate provides a kinetic and entropic advantage for productive reaction with DNA over hydrolysis [23]. Moreover, in the absence of this favorable interaction, AFB_1 epoxide is readily hydrolyzed to the inactive AFB_1 diol. Thus intercalation of AFB_1 epoxide precedes covalent bond formation.

As a consequence of this modification of guanines, the predominant mutation in AFB_1-induced rat tumors has been identified as $G:C \rightarrow A:T$ base transition, with a low frequency of $G:C \rightarrow T:A$ transversions [24]. It has been proposed that the $G:C \rightarrow T:A$ transversion mutation is generated by depurination of the guanine residue promoted by the positively charged imidazole ring and preferential insertion of adenine opposite the apurinic (AP) sites [25]. However, under slightly basic conditions, the imidazole ring of AFB_1-N7-guanine opens to the chemically and biologically stable AFB_1 formamidopyrimidine (AFB_1-FAPY) (Figure 11.1) [26]. The initial AFB_1-N7-guanine adduct, the AFB_1-FAPY and the AP site individually or collectively represent the likely chemical precursors to the genetic effects of AFB_1. Recently, Bailey *et al.* [27] compared the mutation frequencies of AFB_1-N7-guanine and the AP site (by replication of a modified bacteriophage M13 in SOS induced *E. coli*). Both types of DNA modifications gave mainly $G \rightarrow T$ transversions. The mutation frequency of AFB_1-N7-guanine was 4% including about 3% $G \rightarrow T$. They concluded that AFB_1-N7-guanine adducts and not the AP sites are responsible for the mutations.

In human hepatocelular carcinomas, a high frequency of $G:C \rightarrow T:A$ transversions was reported in cases of persons living in areas with high levels of food contamination with AFB_1 [28], particularly in the third position of codon 249 of the p53 tumor suppressor gene. This leads to a AGG to AGT transversion (arginine to serine) and the inactivation of the gene.

In a recent paper, Autrup *et al.* [29] used Big Blue transgenic mice treated prior to AFB_1 with phorone to deplete the level of glutathione which normally detoxifies the epoxides. Again the major mutations (18 out of 23) were $G:C \rightarrow T:A$ transversions, two were $G:C \rightarrow C:G$ transversions, two were A insertions and one a T insertion. Although, the authors did not find mutational hotspots, three mutations occurred at location 86 and location 928.

In addition to AFB_1, three other natural aflatoxins are known: B_2, G_1, G_2. Table 11.1 summarizes their genetic and related effects [2]. Aflatoxin B_2 becomes bound to DNA of rats treated *in vivo* after metabolic conversion to AFB_1. Aflatoxin G_1 binds to DNA and produces chromosomal aberrations in rodents treated *in vivo* (summarized in [2]).

Finally, several metabolites of aflatoxin B_1 have been characterized (for recent reviews, see [3, 6]). Aflatoxicol, the reduced metabolite of AFB_1 must be reoxidized to AFB_1 to give covalent adducts to DNA [30]. Aflatoxin Q_1, a 3-hydroxylated metabolite, is less mutagenic and gives less DNA adducts than AFB_1 [4, 31, 32]. Aflatoxin M_1, the 9a-hydroxylated metabolite gives also DNA adducts and is carcinogenic to animals. Thus it is classified 2B by IARC [2].

Table 11.1: Genetic and related effects of AFB_1 analogues.

	Non mammalian systems				Mammalian systems									
					In vitro								*In vivo*	
	Procaryotes		Lower eucaryotes		Animal cells					Human cells			Animals	
	DNA dam.	Gene mut.	Mitot. recomb.	Gene mut.	DNA dam.	Gene mut.	SCE	Trans-form.	Chrom. aberr.	DNA dam.	SCE	Chrom. aberr.	DNA dam.	Chrom. aberr.
Afla B_2		+	–[1]	–[1]	+	–[1]	+[1]	+[1]		–[1]			+	
Afla G_1	+	+	–[1]	+	+		+[1]		+[1]	+	+[1]	+[1]	+	+
Afla G_2	–			–[1]	+	–	+			–[1]				
Afla M_1		+			+									

+[1] considered to be positive, but only one valid study is available; –[1] considered to be negative, but only one valid study available.

DNA dam. = DNA damage
Gene mut. = Gene mutation
Mitot. recomb. = Mitotic recombination
SCE = Sister chromatid exchange
Transform. = Transformation
Chrom. aberr. = Chromosomal aberrations

11.4 Sterigmatocystin (STC)

This compound has a close structural resemblance to the aflatoxins (Figure 11.2) and is produced by *Aspergillus versicolor*, *Aspergillus nidulans* and other fungi. Like AFB_1 it is metabolized to an epoxide which binds preferably to the N7 of guanine [33].

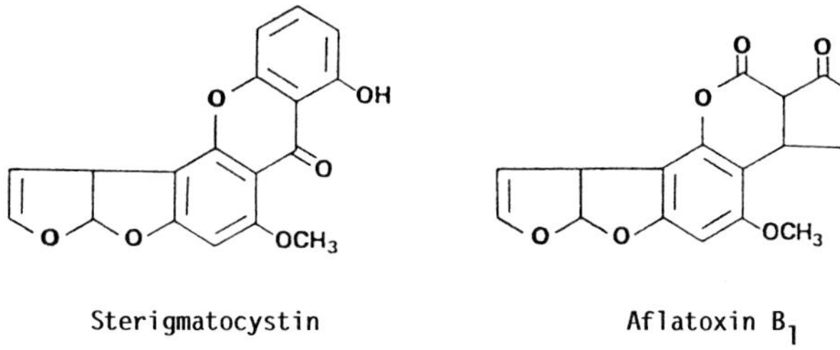

Figure 11.2: Structure of sterigmatocystin (left) and aflatoxin B_1 (right).

In the Ames test STC exhibited a maximum activity at 10 µg/plate with *S. thyphimurium* TA98, without S-9 and at 100 µg/plate with S-9 [34].

Using the SOS chromotest (PQ37), STC was genotoxic with and without metabolic activation already with 1 µg/ml [35]. Using the SOS microplate assay in the presence of hepatic S-9 mix, STC was positive at doses of 0.1–1 µg/ml, but caused a slight depression of SOS response at higher dose [36].

STC was active in the induction of umu gene response in the chimeric plasmid pSK 1002, carried in *Salmonella typhimurium* TA 1535, after incubation with human liver preparation [37].

In cultured cells, STC induced chromosomal aberrations and enhanced 8-azaguanine-resistant mutations [38] similar to AFB_1.

On cultured V79 Chinese hamster lung cells, STC induced 8-azaguanine-resistant mutations in a time and concentration dependant manner [39].

In vitro micronucleus assays in the V79 cultures in the presence of S-9 gave elevated frequences with STG [40].

Thus all the tests performed either with bacteria or with eucaryotic cells were positive and confirmed the carcinogenic activity of STC observed by several authors.

11.5 Ochratoxin A (OTA)

The second best studied mycotoxin is ochratoxin A (OTA). For a long time, OTA was not considered to be genotoxic as it gave negative results in most short-term assays used to detect gene mutations and produced controversial results in assays for unscheduled DNA synthesis (UDS) and sister chromatid exchange (SCE) (for reviews, see [2, 41–44]). OTA did not induce chromosome aberrations in CHO cells [45]. However, as early as 1985, Creppy *et al.* [46] showed that OTA causes DNA single-strand breaks in mouse spleen cells *in vitro*. These breaks were also found *in vivo* in kidney, liver and spleen of Balb/c mice having received a single *i.p.* dose of OTA and in rat kidney and liver, after gavage treatment for 12 weeks at a level equivalent to low dietary concentrations (2 mg/kg in food) [47, 48]. This was confirmed by Stetina and Votava [49].

Subsequently, we could show that OTA treatment induces the formation of DNA adducts in a dose- and time-dependent manner, both in mice and rats [50–52]. After a single oral dose of 2.5 mg/kg bw to mice, the adducts reached a maximum at 48 h when 103, 42 and 2.2 adducts per 10^9 nucleotides were found respectively in kidney, liver and spleen DNA. More recently, a high level of DNA-adducts was also found in the urinary bladder of mice [53]. The decrease of the adducts, most probably due to repair, is more rapid in liver and spleen than in kidney (Figure 11.3) [54]. The level of adducts was reduced by 70 %

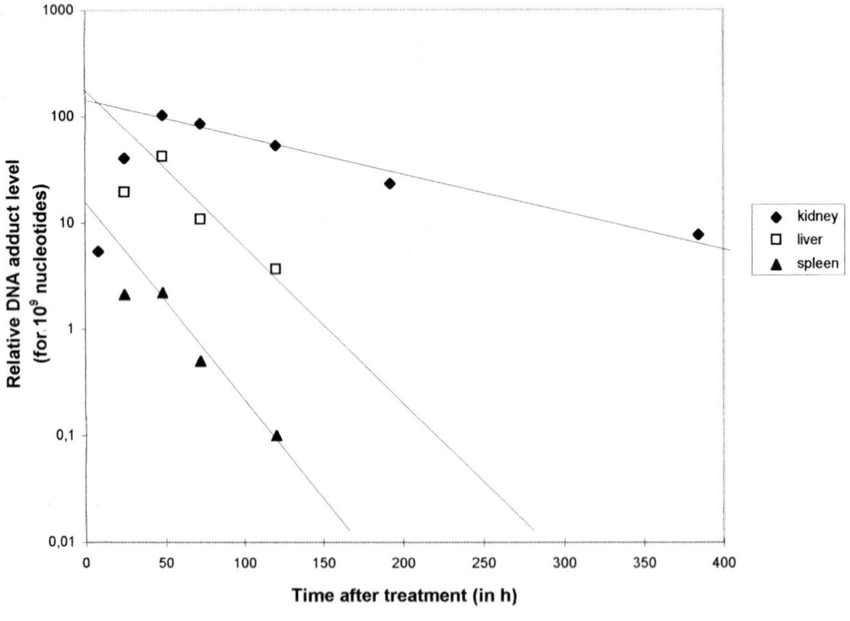

Figure 11.3: Decrease of DNA adduct levels in mice kidney, liver and spleen treated with a single oral dose of 2.5 mg/kg bw of OTA [92].

after 8 days in kidney DNA, but, after 16 days, there were still 8 adducts per 10^9 nucleotides in this DNA, whereas in liver DNA the adducts were no longer measurable. These OTA-induced DNA adducts were not exactly the same in the different organs, indicating differences in the metabolism and/or import of the metabolites in the different organs.

DNA adducts were still found in kidney of Balb/c male mice given an oral dose as low as 20 μg OTA/kg bw [55]. This corresponds to 0.25 ppm in feed. In male rats kidney DNA adducts could also be visualized after an oral administration of 18 μg OTA/kg bw [55].

OTA also induced SOS response in *E. coli* [56]. It increased SCE rate in porcine urinary bladder epithelial cell culture [57] and chromosomal aberrations, particularly X-trisomy in cultured human lymphocytes [58]. It also induced micronuclei formation in ovine seminal vesicle cell cultures as we have shown recently [59] and gave clastogenic and division disruptive changes in metaphase chromosomes [60]. Hennig *et al.* [61] showed that the Ames test, which was negative in presence or absence of S9 mix, became positive when conditioned medium from OTA-exposed hepatocytes were used. In another paper from the same laboratory, OTA was shown to be mutagenic in NIH-3T3 cells [62]. The analysis of these mutations revealed that predominantly large deletions could be found. DNA adducts were recently observed in monkey kidney cells [63], in human bronchial epithelial cells [64–66] and in porcine urinary bladder epithelial cells [67] (Figure 11.4).

However, in spite of all these results, the genotoxic effect of OTA is still questioned [68], because the OTA metabolites(s) reacting on DNA have not yet been found and because DNA adducts could be formed by secondary effects. *In vitro* assays can exclude such a mechanism. As stated above, the level of DNA adducts has been found higher in the kidney than in the liver. Moreover, pretreatment with indomethacin, an inhibitor of prostaglandin H-synthase (PGHS), dramatically reduces the DNA adducts formation in mice kidney and urinary bladder. These results suggest that PGHS may play an important role in OTA adducts formation [53].

In view of the prospective role of PGHS, we performed *in vitro* experiments using kidney microsomes, incubated with DNA, OTA and arachidonic acid (AA) [67, 69, 70]. Our results showed that OTA gives rise to DNA adducts *in vitro*. This is in agreement with the *in vivo* studies and provides evidence for direct genotoxic activity of the mycotoxin after its metabolism.

In addition, when NADPH was used as cosubstrate (to explore the role of cytochromes P450), the adduct levels were lower than when arachidonic acid was used. These results lend support to the hypothesis of the activation of OTA by the peroxidase activity of PGHS.

To identify the nucleotides of DNA modified by the OTA metabolite(s), we have used dAMP, dCMP, dGMP and dTMP as substrates, under the same conditions as used with DNA. The adducts were only found in dGMP, in higher amounts when arachidonic acid was used compared to NADPH. We have also used dG-p-dC and polydG-polydC as alternative substrates and similar results were obtained. Thus the guanine residues of DNA are the target for the genotoxic effect of OTA, which is now obvious.

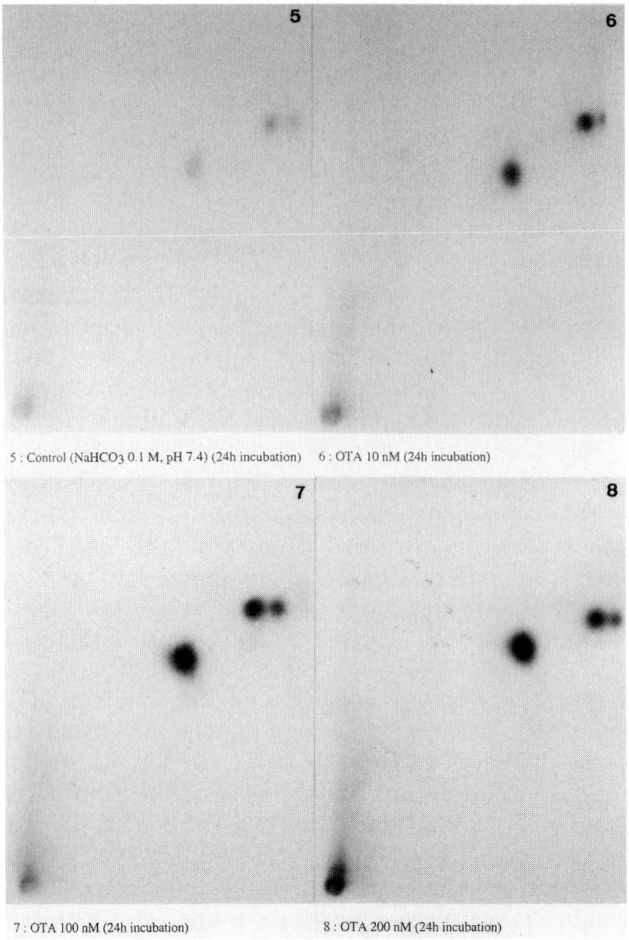

5
6
5 : Control (NaHCO3 0.1 M, pH 7.4) (24h incubation)
6 : OTA 10 nM (24h incubation)
7
8
7 : OTA 100 nM (24h incubation)
8 : OTA 200 nM (24h incubation)

Figure 11.4: DNA adducts of porcine urinary bladder epithelial cells incubated with increasing amounts of OTA for 6 h.

Horseradish peroxidase has an enzymatic action very close to the peroxidase activity of the PGHS. Thus we have conducted the same experiments substituting the kidney microsomes with horseradish peroxidase and using cumene-hydroperoxide in place of arachidonic acid as cosubstrate. DNA, dGMP, dG-p-dC or polydG-polydC served as substrates in presence of OTA. Consequently, we have obtained similar patterns of adducts as those with kidney microsomes. The results of these investigations support the implication that the activation of OTA to genotoxic compound(s) occurs by a peroxidase mechanism leading to adducts in DNA guanylic residues.

Finally, tumorous tissues from three kidneys and five bladders of Bulgarian patients, coming from the area of Balkan endemic nephropathy and undergoing surgery for cancer, and tissues from three non-malignant kidneys were analyzed for DNA adducts [71]. Several adducts, with the same R_F values as those obtained from mouse kidney after treatment with OTA (one major and some minor adducts), were detected, mainly in kidney, but also in bladder tissues from Bulgarian patients. No DNA adducts were detected in non-malignant human kidney tissues. These results provide new evidence of the possible role of OTA in the development of tumors of the urinary tract.

Let us finally remark that OTA was tested for carcinogenicity by oral administration in mice and rats (for review, see [2]). It increased the incidence of hepatocellular tumors in mice of each sex and produced renal-cell adenoma and carcinoma in male mice and Fischer 344 rats of each sex. In this last study, the incidence of renal tubular cell adenomas and carcinoma observed in male rats, after treatment with OTA, was the highest of any of the National Cancer Institute NTP studies to date. In addition, respectively 12 or 43 total metastases from renal carcinoma occurred in the 50 rats treated respectively with 70 µg or 210 µg/kg bw/day [45]. These results, in addition to the genotoxic effects of OTA, should lead the authorities to classify OTA as 2A as to its carcinogenic risk to human.

11.6 Citrinin

Citrinin has some structural analogy with ochratoxin A (Figure 11.5). It is produced by at least 26 fungal species, mainly *Penicillium* and some *Aspergilli*. Of these, several *Penicillium* produce also ochratoxin A (*P. viridicatum, P. cyclopium, P. palitans, P. purpurescens*). Citrinin was found to contaminate some food and foodstuffs (for reviews, see [72–74]). The genotoxicity of citrinin was tested by Ueno and Kubota [75] who found a positive reaction in the *Bacillus subtilis* rec+/–assay when tested at 20 and 100 µg/disc. Martin *et al.* [76] confirmed this result and showed that even with 20 µg/ml a positive reaction was observed at pH 6. However, citrinin was not mutagenic in the Ames test (reviewed in [77]). Using permeabilized *E. coli* cells, Martin *et al.* [78] showed that citrinin caused single-strand breaks in the DNA of *E. coli* at a concentration of

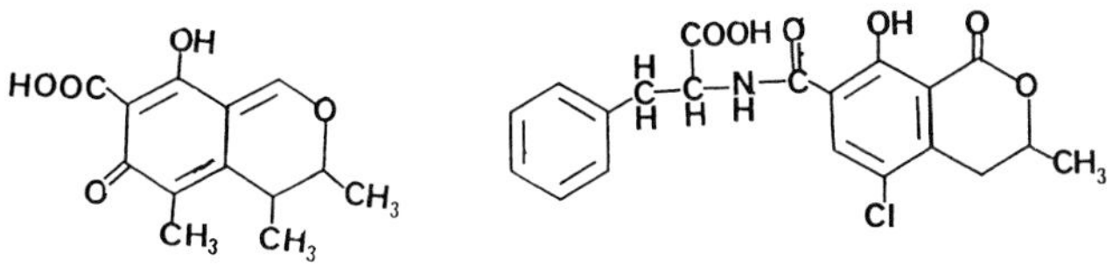

Citrinin Ochratoxin A

Figure 11.5: Structure of citrinin (left) and ochratoxin A (right).

100 μg/ml and that it induced DNA repair synthesis at a concentration of 5 μg/ml. At 300 μg/ml *lambda* ts prophage was induced in a lysogenic *E. coli* strain.

In order to investigate whether such DNA damage led to mutagenicity, Brakhage *et al.* [79] employed a reversion test using an amber mutant of phage M_{13}. They could show, with 50 μg citrinin/ml at pH 6, a low but significant increase of reversion of the phage. DNA sequence of 112 revertants showed in all cases single base substitution mutations in the former amber codon (positions 3066–3068). In some cases, a second T:A → C:G mutation was observed about 50 bases upstream of the amber codon (position 3118). All these results clearly showed a genotoxic effect of citrinin on bacteria.

However, with eucaryotic cells, the following short-term tests were negative: (i) mitogenic gene conversion in *Saccharomyces cerevisiae*; (ii) VDS in hepatocytes (at a concentration of 0.2 g/l); (iii) SCE in Chinese hamster V79 cells in the presence of an exogenous metabolic system from rat or human liver. Citrinin was however a potent inducer of chromosomal aberrations in the clastogenicity assay on V79-E cells in presence of liver microsomes, but only in a narrow range and at high concentration (5×10^{-4} M). It was inactive at 5×10^{-5} M [80].

Several carcinogenic studies were done with mice and rats. However, as noted in the IARC Monograph 1986 [77], two of them which were negative suffered either from the small number of animals used or from short duration. However, a long-term feeding study with 1 g citrinin per kg of diet, done with male Fisher 344 rats for 80 weeks, led to massive tumors in the kidneys of all treated rats. These tumors were benign and were classified as clear-cell adenomas [81]. Thus the IARC experts [77] concluded that there is limited evidence for the carcinogenicity of citrinin to experimental animals.

11.7 Zearalenone

Zearalenone is an estrogenic mycotoxin produced by several *Fusarium* species, particularly *F. graminearum*. It has been found in several cereals (maize, oats, wheat, rye [2]).

The genotoxic effects of zearalenone and some of its derivatives have been reviewed [2]. Contradictory results were obtained using several systems. The Ames test was negative as well as a test with *Saccharomyces cerevisiae*, whereas zearalenone showed a positive DNA damaging effect in recombination tests with *Bacillus subtilis* [75]. However, positive results for SCE were found in Chinese hamster ovary cells as well as chromosomal aberrations and polyploidy [82].

In 1989, Keith *et al.* [83] using the ^{32}P-postlabelling assay of Reddy and Randerath [84] found DNA adducts in the liver of Balb/c female and male mice after oral administration of zearalenone (2 mg/kg body weight) (Figure 11.6). The amount and number of adducts were much higher in female than in male

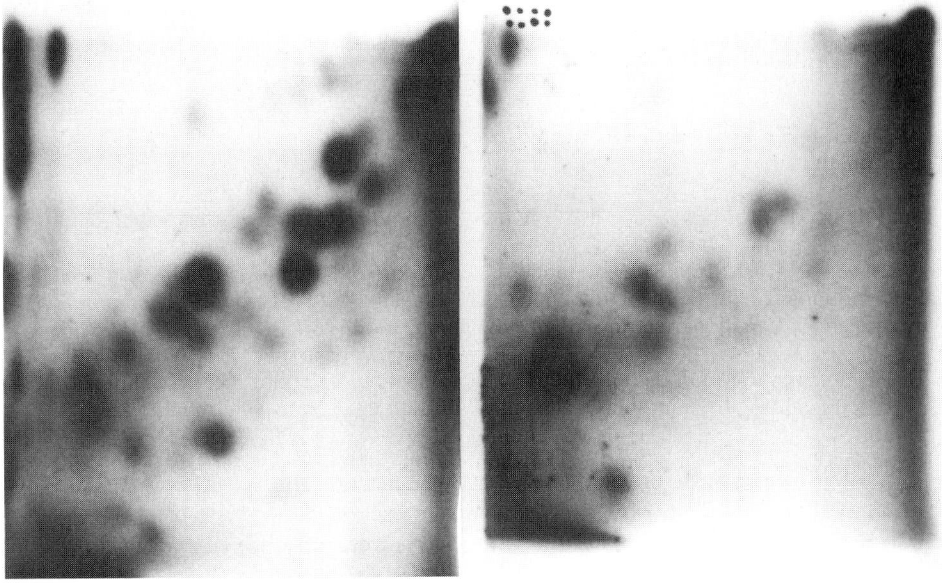

Figure 11.6: Liver DNA adducts of female (left) and male (right) Balb/c mice treated with a single oral dose of 2 mg/kg bw of zearalenone [83].

mice. These results were confirmed by Pfohl-Leszkowicz *et al.* [85, 86] who systematically studied the kidney and liver DNA adducts of female Balb/c mice also treated with 2 mg/kg bw *i.p.* or orally. The total DNA adduct levels reached about 1390 and 114 per 10^9 nucleotides in liver and kidney respectively, after *i.p.* treatment. After oral treatment about 5-times lower adduct levels were found in liver and 12-times lower in kidney. Between 12 to 15 different adducts were detected, some existing in both tissues, others being specific for one or the other organ. The major adduct was not the same in the two organs. No DNA adducts were detected in the genital organs after a single dose of zearalenone, but five repeated administrations of 1 mg/kg in 10 days led to respectively 1000, 61 and 17 adducts/10^9 nucleotides, in liver, kidney and genital organs. Surprisingly no DNA adducts could be detected in the same organs of Sprague-Dawley rats.

The higher adduct level found in liver is in agreement with the liver neoplastic lesions found in B6C3F1 mice after a 2-year oral carcinogenicity study [2, 87]. Pituirary adenomas and carcinomas were also found in both sexes.

By *in vitro* incubation of hepatic mice C57 microsomes with DNA and zearalenone, Steiblen [88] obtained 47 adducts/10^9 nucleotides confirming the *in vivo* results.

These results show clearly the genotoxicity of zearalenone, but only for one species: mice, whereas in rats no DNA adducts were found. This confirms the result of Li *et al.* [89] which did not found liver, kidney and uterus DNA adducts in female Sprague-Dawley rats after 3 weeks treatment with 0.05 mg/kg zearalenone in the diet. This difference between the two animal species remains to be explained.

11.8 Patulin

Patulin is an unsaturated α,β-lactone, mainly found in moldy fruits, especially apples, but it occurs also in other fruits, peaches, tomatoes, etc. (for a review, see [77]). The use of such fruits in juice or cider manufacture can result in quite high concentrations of patulin in the resultant juice [72]. Contradictory results were obtained in short-term tests (Table 11.2). Several authors reported negative results in the Ames test, but positive results were obtained with a mutant of *Bacillus subtilis* deficient in recombination. Lee and Röschenthaler [90] showed that patulin causes DNA single-strand breaks in *E. coli* at a concentration of 10 µg/ml. At 50 µg/ml double-strand breaks were observed also.

Table 11.2: Results of the short-term tests for patulin.

		DOSE
DNA damage		
E. coli	–	3– 50 µg/disk
B. subtilis mutant	+	20–100 µg/disk
SOS assay		
E. coli	–	24 µg/ml
Mutagenic		
S. typhimurium	–	250 µg/plate
Host mediated assay	–	20 mg/kg bw
S. cerevisia		
Petite mutants	+	20–70 µg/ml
Mitotic recombination	–	100 µg/ml
DNA strand breaks		
HeLa cells	+	32 µg/ml
FM3A cells	+	3.2 mg/ml
Mutations		
FM3A cells	+	100 µg/ml
UDS		
Liver cells	–	246 mg/l
Somatic mosaicism		
Drosophila	+	493 mg/l

Table 11.2 (continued)

		DOSE
SCE		
Human lymphocytes	+	0.1 mg/l
Chinese hamster V79	–	1.54 mg/l
Chromosomal aberrations		
Human peripheral leucocytes	+	532 µg/l
FM3A cells	+	32 µg/l
in vivo		
SCE		
Chinese hamster bone-marrow cells	–	20 mg/kg bw
Chromosomal aberrations		
Chinese hamster bone-marrow cells	+	10–20 mg/kg bw
Dominant mutations	–	10 mg/kg bw

Single- and double-strand breaks of DNA caused by patulin have also been reported in eucaryotic cell culture: in FM3A cells [38] and in HeLa cells [91]. A clastogenic activity of patulin in Chinese hamster V79-E cells was also detected, but only in a very narrow dose range (5×10^{-6} M). It was lost when S9 was added. It was suggested that the clastogenicity is caused by interaction with chromosomal proteins and not with DNA [92]. SCE and UDS tests were negative *in vitro* and *in vivo* (Table 11.2). Some of these differences may be due to the doses used (Table 11.2), but the blood cells (leucocytes, lymphocytes) seem to be the most sensitive. The chromosomal aberrations were found both in cell culture and *in vivo*.

As there is inadequate evidence for the carcinogenicity in experimental animals, patulin was classified in Group 3 by IARC [77].

11.9 Trichothecenes

Only a few studies concern the genotoxic effects of deoxynivalenol, nivalenol and fusarenone X. They have been reviewed by IARC [2].

Deoxynivalenol (vomitoxin) was not mutagenic in the Ames test. In mammalian systems it inhibited gap-junctional intercellular communication in the Chinese hamster V79 cells *in vitro* (dose: 0.3 mg/l) and induced cell transformation in BALB/3T3 mouse embryo cells *in vitro* (dose: 0.2 mg/l). However, it did not produce gene mutation at the hprt locus in V79 cells (dose: 3 mg/l), nor did it induce UDS in rat hepatocytes *in vitro* even at very high doses (1 g/l). Very recently, it was shown to have a clastogenic effect with primary rat hepatocytes (after 3 h exposure at 1 μg/ml) [93]. Thus more work should be done concerning this compound.

For nivalenol there are even less results: in V79 cells *in vivo* it slightly increased the frequencies of chromosomal aberrations and SCE was found.

Finally, fusarenone X did not give a positive Ames test, but increased the frequency of petite mutations in yeast. In Chinese hamster V79 cells *in vitro* it gave chromosomal aberrations and caused a weakly positive sister chromatid exchange (dose: 3 mg/l). With a concentration of 32 mg/l it weakly induced DNA single strand breaks in human HeLa cells *in vitro*.

In conclusion, these three trichothecenes should be more carefully studied with recent techniques to get a clearer opinion about their genotoxicity.

11.10 Fumonisins

Fumonisins represent a new class of mycotoxins produced mainly by *Fusarium moniliforme* and other *Fusarium* which occur worldwide and grow mainly on maize. Numerous fumonisins have been described so far from which the four B-type fumonisins are the most abundant.

Fumonisins B_1 (FB_1) was considered by the IARC monograph 56 working group [2] and a recent review by Dutton describes its toxic and carcinogenic effects [94].

FB_1, as well as FB_2 and FB_3, were non-mutagenic in the Ames test using TA 97a, TA 98, TA 100 and TA 102, at concentrations up to 10 mg/plate, either in absence or presence of S9 [95, 96]. FB_1 was also devoid of activity in the SOS chromotest with *E. coli* PQ37 in presence or absence of S9 and gave negative results in differential DNA repair assay with *E. coli* K12 strain [93].

In contrast, Sun and Star [97] using a commercial bioluminescent bacterial (*Vibrio fischeri*) genotoxicity test, claimed that FB_1 showed genotoxicity without S9 at a concentration range 5–20 μg/ml.

Contradictory, results were obtained in cell cultures. On one hand, Gelderblom *et al.* [98, 99] showed that FB_1 and FB_2 do not induce UDS in rat hepatocytes both *in vitro* (concentration up to 40–80 μM/plate) (result confirmed by Norred *et al.* [100, 101] at 250 μM, and Yoo *et al.* [102], and *in vivo* (100 mg/kg)). On the other hand, moderate increases of the micronucleus (MN) frequencies were shown at low concentrations of FB_1 ($\leq$ 1 μg/ml), but no clear dose-response effects were seen; in addition at higher exposure levels, the MN frequencies declined. In the chromosomal aberration experiments with primary rat hepatocytes pronounced dose-dependent effects were observed. The increase was 6-fold after exposure of the cells to 1 μg/ml for 3 h [93].

FB_1 has been shown to be hepatocarcinogenic in rats fed chronically over a period of 18–26 month [103, 104]. It induces leukoencephalomalacia in horses (reviewed in [2, 94]). In addition, epidemiologic data suggest that ingestion of *F. moliniforme* contaminated corn is linked to human esophagial cancer in areas of South Africa and China [2]. Thus both IARC [2] and CEPA (California Environmental Protection Agency) [105] concluded that toxins derived from *F. moniliforme* are possibly carcinogenic to humans (Group 2B).

The question of the mechanism of action of FB_1 is still not clarified. It might well be that its carcinogenicity is linked to its strong promoting activity [98, 106, 107]. This could be explained by its action on protein kinase C [108].

11.11 Fusarin C

Fusarin C is also produced by *Fusarium moniliforme*. It gives DNA strand breaks in *Salmonella typhimurium* in presence of S9 and reverse mutations in several Ames strains (for references, see [2]). It induced also gene mutation, SCE, chromosomal aberrations and micronucleus formation in Chinese hamster V79 lung cells [109], but attempts to detect DNA adducts in *S. typhimurium* or calf thymus were unsuccessfull [110, 111].

Fusarin C induced papillomas and carcinomas of the esophagus and forestomach in one study done with female mice and rats after oral gavage [112]. Thus more studies should be performed with this mycotoxin to obtain sufficient evidence of its genotoxic and carcinogenic potential.

11.12 Griseofulvin

This natural antibiotic produced by *Penicillium griseofulvum* is commonly used in human for the treatment of dermatomycoses. Its toxic effects have been extensively reviewed recently [113]. A number of studies indicate that griseofulvin causes hepatic cancer in mice.

Let us summarize its genotoxic effects (references in [113]). Griseofulvin does not induce point mutation and DNA repair in the usual bacterial cell test systems, but seems to interfere with excision repair processes and induces SCE *in vivo*. Griseofulvin causes disturbance of microtubuli formation. This can explain several studies which indicate that griseofulvin induces structural and numerical chromosomal aberrations and micronuclei. Thus, there is sufficient evidence for carcinogenicity and genotoxicity of griseofulvin to animals [114].

11.13 Conclusion

We did not, in this review, describe all the other genotoxic mycotoxins. Only a few publications dealing with them have appeared (for general reviews, see [115, 116]).

Versicolorin A and austocystin A and D which are structurally related to aflatoxins, are mutagenic together with the structurally unrelated austdiol, kojic acid and viridicatumtoxin [117]. Versicolorin A gave also a positive SOS chromotest [118] and elicited DNA repair synthesis. Austocystin B, C and H are potent mutagens with frameshifting activity after metabolic activation (cited in [115]). Viomellein and xanthomegnin are, like ochratoxin A, dihydroisocoumarin containing mycotoxins. Viomellein is a weak SOS inducing agent [119] and xanthomegnin is positive in the hepatocyte primary culture/DNA repair test.

Penicillic acid, a small lactone like patulin, was the first mycotoxin described. It was suspected by Dickens and Jones [120] to be carcinogenic and was found to have DNA-attacking ability in the rec-assay by Ueno and Kubota [75]. It also induced azaguanine-resistant mutations as well as DNA breaks in mammalian cell lines [38, 91]. Chromosome aberrations were also found by the same authors (reviewed in [77]).

Rubratoxins have a complex structure (for review see [115]) with an unsaturated lactone ring. Natori *et al.* [121] observed chromosome damage ranging from breaks to complete destruction in HeLa cells cultures.

Several mycotoxins belong to hydroxyanthraquinones (emodin) or to modified bianthraquinones (luteoskyrin, rugulosin and rubroskyrin). Emodin is mutagenic to *S. typhimurium* TA 1537 in presence of S9 [122] and on cultured mouse carcinoma FM3A cells [123]. Luteoskyrin was reviewed in IARC [74]. It was positive in the hepatocyte primary culture/DNA repair test [124]. It had also a DNA-attacking ability in the rec-assay [75] and it induced respiratory deficient mutants in *S. cerevisiae* [125]. When cultured cells were exposed to luteoskyrin, the 8-hydroxydeoxyguanine content of DNA was increased. Thus, this anthraquinone may induce the formation of hydroxy radicals [126]. It has been reported to be carcinogenic in liver [127]. Rugulosin interacts *in vitro* with DNA in absence of metabolic activating systems [115, 116]. Rugulosin was also positive in the rec-assay with *B. subtilis* [75]. Rubratoxin B caused extensive chromosomal damage in exposed HeLa cells [121, 128]. It was also mutagenic in the dominant lethal assay [129]. Rubroskyrin and (+)rugulosin induced forward mutations to 8-azaguanine resistance in *S. typhimurium* strain TM 677 [130].

Penicillium roqueforti toxin (PR toxin) at a concentration of 2.5×10^{-5} M produced single strand breaks in DNA [131]. It gave a positive Ames test after preincubation with the TA98 strain and S9 fraction before incorporation into the agar overlay at an amount of 100 µg/plate [34].

Viomellein was a weak SOS inducing agent [119].

Mutagenic activity of mycophenolic acid has been reported [117, 132].

Averufin, luteosporin and chrysacin were positive in the hepatocyte primary culture/DNA repair test [124].

This list is certainly not complete and a lot of work remains to be done in this field.

Finally, it must be emphasized that contaminated food and feed rarely contain only one mycotoxin [133]. In fact, they very often occur in combinations and can have synergistic effects. Only a few studies deal with multiple mycotoxin exposure and their effects. It is obvious that there is a need for research on the effect of the genotoxicity of these combinations.

Acknowledgements

This research was supported by the Association de la Recherche contre le Cancer (ARC) and the Ministère de l'Environnement (Action Santé-Environnement).

References

[1] Wogan, G. N.; Poglialungo, S.; Newberne, P. M. (1974) Carcinogenic effects of low dietary levels of aflatoxin B1 in rats. *Food Cosmet. Toxicol.* **12**, 681–865.

[2] IARC (1993) *Some naturally occurring substances, food items and constituents, heterocyclic, aromatic amines and mycotoxins* (IARC, Ed.), Lyon.

[3] Eaton, D. L.; Gallagher, E. P. (1994) Mechanisms of aflatoxin carcinogenesis. *Annu. Rev. Pharmacol. Toxicol.* **34**, 135–172.

[4] Gurtoo, H. L.; Dahms, R. P.; Paigen, B. (1978) Metabolic activation of aflatoxins related to their mutagenicity. *Biochem. Biophys. Res. Commun.* **81**, 965–972.

[5] Guengerich, F. P.; Johnson, W. W.; Ueng, Y.-F.; Yamazaki, H.; Shimada, T. (1996) Involvement of cytochrome P450, glutathione S-transferase and epoxide hydrolase in the metabolism of aflatoxin B1 and relevance to risk of human liver cancer. *Environ. Health Perspect.* **104** (Suppl. 3), 557–562.

[6] Guerre, P.; Galtier, P.; Burgat, V. (1996) Le métabolisme un facteur de susceptibilité à la toxicité des aflatoxines. *Revue Méd. Vét.* **147**, 879–892.

[7] Gallagher, E. P.; Kunze, K. L.; Stapleton, P. L.; Eaton, D. L. (1996) The kinetics of aflatoxin B1 oxidation by human cDNA-expressed and human liver microsomal cytochromes P450 1A2 and 3A4. *Toxicol. Appl. Pharmacol.* **141**, 595–606.

[8] Battista, J. R.; Marnett, L. J. (1985) Prostaglandin H synthase-dependent epoxidation of aflatoxin B1. *Carcinogenesis* **6**, 1227–1229.

[9] Datta, K.; Kulkarni, A. P. (1994) Oxidative metabolism of aflatoxin B1 by lipoxygenase purified from human term placenta and intrauterine conceptal tissues. *Teratology* **50**, 311–317.

[10] Liu, L.; Massey, T. E. (1992) Bioactivation of aflatoxin B1 by lipooxygenases, prostaglandin H synthase and cytochrome P7 monooxygenase in guinea-pig tissues. *Carcinogenesis* **13**, 533–539.

[11] Liu, L.; Daniels, J. M.; Stewart, R. K.; Massey, T. E. (1990) *In vitro* prostaglandin H synthase and monooxygenase-mediated binding of aflatoxin B1 to DNA in guinea pig tissue microsomes. *Carcinogenesis* **11**, 1915–1919.

[12] Croy, R. G.; Wogan, G. N. (1981) Quantitative comparison of covalent aflatoxin-DNA adducts formed in rat and mouse livers and kidneys. *J. Natl. Cancer Inst.* **66**, 761–768.

[13] Essigmann, J. M.; Croy, R. G.; Nadzan, A. M.; Busby, W. F. J.; Reinhold, V. N.; Buchi, G.; Wogan, G. N. (1977) Structural identification of the major DNA adduct formed by aflatoxin B1 *in vitro*. *Proc. Natl. Acad. Sci. USA* **74**, 1870–1874.

[14] Essigmann, J. M.; Green, C. L.; Croy, R. G.; Fowler, K. W.; Buchi, G. H.; Wogan, G. N. (1983) Interactions of aflatoxin B1 and alkylating agents with DNA: Structural and functional studies. *Cold Spring Harbor Symp. Quant. Biol.* **47**, 327–337.

[15] Baertschi, S. W.; Raney, K. D.; Stone, M. P.; Harris, T. M. (1988) Preparation of aflatoxin B1-8.9-epoxide; the ultimate carcinogen of aflatoxin B1. *J. Am. Chem. Soc.* **110**, 7929–7931.

[16] Gopalakrishnan, S.; Byrd, S.; Stone, M. P.; Harris, T. M. (1989) Carcinogen-nucleic acid interactions: equilibrium binding studies of aflatoxin B1 with the oligodeoxynucleotide d(ATGCAT)$_2$ and with plasmid pBR322 support intercalative association with the B-DNA helix. *Biochemistry* **28**, 726–734.

[17] Misra, R. P.; Muench, K. F.; Humayun, M. Z. (1983) Covalent and non-covalent interactions of aflatoxin with defined deoxyribonucleic acid sequences. *Biochemistry* **22**, 3351–3359.

[18] Raney, V. M.; Harris, T. M.; Stone, M. P. (1993) DNA conformation mediated aflatoxin B1-DNA binding and formation of guanine N7 adducts by aflatoxin B1 8.9-exo-epoxide. *Chem. Res. Toxicol.* **6**, 64–68.

[19] Iyer, R. S.; Coles, B. F.; Raney, K. D.; Thier, R.; Guengerich, F. P.; Harris, T. M. (1994) DNA adduction by the potent carcinogen aflatoxin B1, mechanistic studies. *J. Am. Chem. Soc.* **116**, 1603–1609.

[20] Gopalakrishnan, S.; Harris, T. M.; Stone, M. P. (1990) Intercalation of aflatoxin B1 in two oligonucleotide adducts: comparative ^{1}H-NMR analysis of d(ATCAFBGAT): d(ATCGAT) and d(ATAFBGCAT)2. *Biochemistry* **29**, 10438–10448.

[21] Raney, K. D.; Gopalakrishnan, S.; Byrd, S.; Stone, M. P.; Harris, T. M. (1990) Alteration of the aflatoxin cyclopentanone ring to a δ-lactone reduces intercalation with DNA and decreases formation of guanine N7 adducts by aflatoxin epoxides. *Chem. Res. Toxicol.* **3**, 254–261.

[22] Kobertz, W. R.; Wang, D.; Wogan, G. N.; Essigmann, J. M. (1997) An intercalator inhibitor altering the target specificity of DNA damaging agents: Synthesis of site-specific aflatoxin B1 adducts in a p53 mutation hotspot. *Proc. Natl. Acad. Sci. USA* **94**, 9579–9584.

[23] Johnson, W. W.; Guengerich, F. P. (1997) Reaction of aflatoxin B1 exo-8.9-epoxide with DNA: Kinetic analysis of covalent binding and DNA-induced hydrolysis. *Proc. Natl. Acad. Sci. USA* **94**, 6121–6125.

[24] Soman, N. R.; Wogan, G. N. (1993) Activation of the c-Ki-*ras* oncogene in aflatoxin B1-induced hepatocellular carcinoma and adenoma in the rat: Detection by denaturing gradient gel electrophoresis. *Proc. Natl. Acad. Sci. USA* **90**, 2045–2049.

[25] Foster, P. L.; Eisenstadt, E.; Miller, J. H. (1983) Base substitution mutations induced by metabolically activated aflatoxin B1. *Proc. Natl. Acad. Sci. USA* **80**, 2695–2698.

[26] Busby, W. F.; Wogan, G. N. (1984) Aflatoxins. *Chemical Carcinogens* **2**, 945–1136. Am. Chem. Soc., Washington DC.

[27] Bailey, E. A.; Iyer, R. S.; Stone, M. P., Harris, T. M.; Essigmann, J. M. (1996) Mutational properties of the primary aflatoxin B1-DNA adduct. *Proc. Natl. Acad. Sci. USA* **93**, 1535–1539.

[28] Hsu, J. C.; Metcalf, R. A.; Sun, T.; Welch, J. A.; Wang, N. J.; Harris, C. C. (1991) Mutation hotspot in the p53 gene in human hepatocellular carcinomas. *Nature* **350**, 427–428.

[29] Autrup, H.; Jørgensen, E. C. B.; Jensen, O. (1996) Aflatoxin B1-induced lacI mutation in liver and kidney of transgenic mice C57BL/6N: effect of phorone. *Mutagenesis* **11**, 69–73.

[30] Loveland, P. M.; Wilcox, J. S.; Bailey, G. S. (1987) Metabolism and DNA binding of aflatoxicol and aflatoxin B1 *in vivo* and in isolated hepatocytes from rainbow trouth (Salmo gairdneri). *Carcinogenesis* **8**, 1065–1079.

[31] Raney, K. D.; Shimada, T.; Kim, D. H.; Groopman, J. D.; Harris, T. M.; Guengerich, F. P. (1992) Oxidation of aflatoxins and sterigmatocystin by human liver microsomes: Significance of aflatoxin Q1 as a detoxication product of aflatoxin B1. *Chem. Res. Toxicol.* **5**, 202–219.

[32] Yourtee, D. M.; Rohrig, T. M. (1985) The *in vivo* metabolism of aflatoxin Q1 by mouse and rabbit liver preparations. *Res. Com. Chem. Pathol. Pharmacol.* **50**, 103–123.

[33] Essigmann, M.; Barker, L. J.; Fowler, K. W.; Francisco, M. A.; Reinhold, V. N.; Wogan, G. N. (1979) Sterigmatocystin-DNA interaction: Identification of a major adduct formed after metabolic activation *in vitro*. *Proc. Natl. Acad. Sci. USA* **76**, 179–183.

[34] Ueno, Y.; Kubota, K.; Ito, T.; Nakamura, Y. (1978) Mutagenicity of carcinogenic mycotoxins in *Salmonella typhimurium*. *Cancer Res.* **38**, 536–542.

[35] Krivobock, S.; Olivier, P.; Marzin, D. R.; Seigle-Murandi, F.; Steiman, R. (1987) Study of the genotoxic potential of 17 mycotoxins with the SOS chromotest. *Mutagenesis* **2**, 433–439.

[36] Sakai, M.; Abe, K.; Okumura, H.; Kawamura, O.; Sugiura, Y.; Horie, Y.; Ueno, Y. (1992) Genotoxicity of fungi evaluated by SOS microplate assay. *Nat. Toxins* **1**, 27–34.

[37] Shimada, T.; Iwasaki, M.; Martin, M. V.; Guengerich, F. P. (1989) Human liver microsomal cytochrome P450 enzymes involved in the bioactivation of procarcinogens detected by umu gene response in *Salmonella typhimurium* TA 1535/pSK1002. *Cancer Res.* **49**, 3218–3228.

[38] Umeda, M.; Tsutsui, T.; Saito, M. (1977) Mutagenicity and inducibility of DNA single-strand breaks and chromosome aberrations by various mycotoxins. *Gann* **68**, 619–625.

[39] Noda, K.; Umeda, M.; Ueno, Y. (1981) Cytotoxic and mutagenic effects of sterigmatocystin on cultured Chinese hamster cells. *Carcinogenesis* **2**, 945–949.

[40] Ellard, S.; Mohammed, Y.; Dogra, S.; Wolfel, C.; Doehmer, J.; Parry, J. M. (1991) The use of genetically engineered V79 Chinese hamster cultures expressing rat liver CYP 1A1, 1A2 and 2B1 cDNAs in micronucleus assays. *Mutagenesis* **6**, 461–479.

[41] Dirheimer, G. (1996) Mechanistic approaches to ochratoxin toxicity. *Food Add. Contam.* **13**, 43–44.

[42] Committee on Food Additives (1991) Ochratoxin A. Toxicological evaluation of certain food additives and contaminants. FAO/WHO. 37th Report of the Joint FAO/ WHO Expert, 365–417.

[43] Committee on Food Additives (1996) Ochratoxin A. Toxicological evaluation of certain food additives and contaminants. FAO/WHO. 44th Report of the Joint FAO/WHO Expert, 363–376.

[44] Kuiper-Goodman, T.; Scott, P. M. (1989) Risk assessment of the mycotoxin ochratoxin. *Biomed. Environ. Sci.* **2**, 433–439.

[45] US Department of Health and Human Services, National Institutes of Health, Research Triangle Park, NC. (1988) Technical Report on the Toxicology and Carcinogenesis Studies of Ochratoxin. G. Boorman (Ed.) NTP. NIH Publications N° 88-2813.

[46] Creppy, E.-E.; Kane, A.; Dirheimer, G.; Lafarge-Frayssinet, C.; Mousset, S.; Frayssinet, C. (1985) Genotoxicity of ochratoxin A in mice: DNA single-strand break evaluation in spleen, liver and kidney. *Toxicol. Lett.* **28**, 29–85.

[47] Kane, A.; Creppy, E.-E.; Roth, A.; Röschenthaler, R.; Dirheimer, G. (1986) Distribution of the [3H]-label from low doses of radioactive ochratoxine A ingested by rats, and evidence for DNA single-strand breaks caused in liver and kidneys. *Arch. Toxicol.* **58**, 219–224.

[48] Kane, A. (1986) Intoxication subchronique par l'ochratoxine A, mycotoxine contaminant les aliments: Effets néphrotoxiques et génotoxiques. Ph. D., Université Louis Pasteur, Strasbourg.

[49] Stetina, R.; Votava, M. (1986) Induction of DNA single-strand breaks and DNA synthesis inhibition by patulin, ochratoxin A, citrinin and aflatoxin B1 in cell lines CHO and AWRF. *Folia Biologica (Praha)* **32**, 128–144.

[50] Pfohl-Leszkowicz, A.; Chakor, K.; Creppy, E.-E.; Dirheimer, G. (1991) DNA adduct formation in mice treated with ochratoxin A. In: Castegnaro, M.; Plestina, R.; Dirheimer, G.; Chernozemsky, I. N.; Bartsch, H. (Eds.) *Mycotoxins, endemic nephropathy and urinary tract tumours.* IARC Scientific Publications N° 115, Lyon, p. 245–253.

[51] Pfohl-Leszkowicz, A.; Grosse, Y.; Kane, A.; Castegnaro, M.; Creppy, E.-E.; Dirheimer, G. (1993) Preponderance of DNA-adducts in kidney after ochratoxin A exposure. In: *Human Ochratoxicosis and its Pathologies* (Creppy, E.-E.; Castegnaro, M.; Dirheimer, G.; Eds.), John Libbey, Eurotext N° 231, Montrouge, p. 199–207.

[52] Pfohl-Leszkowicz, A.; Grosse, Y.; Kane, A.; Creppy, E. E.; Dirheimer, G. (1993) Differential DNA adduct formation and disappearance in the mouse tissues after treatment with the mycotoxin ochratoxin A. *Mutat. Res.* **289**, 265–273.

[53] Obrecht-Pflumio, S.; Grosse, Y.; Pfohl-Leszkowicz, A.; Dirheimer, G. (1996) Protection by indomethacin and aspirin against genotoxictiy of ochratoxin A, particularly in the urinary bladder and kidney. *Arch. Toxicol.* **70**, 244–248.

[54] Obrecht-Pflumio, S. (1998) Détermination des mécanismes de la génotoxicité et du métabolisme de l'ochratoxine A, une mycotoxine cancérogène. Ph. D., University Louis Pasteur, Strasbourg.

[55] Grosse, Y. (1996) Détermination des voies métaboliques impliquées dans la génotoxicité de la mycotoxine ochratoxine A, par la mise en évidence d'adduits à l'ADN. Ph. D., Institut National Polytechnique de Toulouse.

[56] Malaveille, C.; Brun, C.; Bartsch, H. (1994) Structure-activity studies in *E. coli* strains on ochratoxin A (OTA) and its analogues implicate a genotoxic free radical and a cytotoxic thiol derivative as reactive metabolites. *Mutat. Res.* **307**, 141–147.

[57] Föllmann, W.; Hillebrand, I. E.; Creppy, E.-E.; Bolt, H. M. (1995) Sister chromatid exchange frequency in cultured isolated porcine urinary bladder epithelial cells (PUBEC) treated with ochratoxin A and alpha. *Arch. Toxicol.* **69**, 280–286.

[58] Manolova, Y.; Manolov, G.; Parvanova, L.; Petkova-Bocharova, T.; Castegnaro, M.; Chernozemsky, I. N. (1990) Induction of characteristic chromosomal aberrations, particularly x-trisomy, in cultured human lymphocytes treated by ochratoxin A, a mycotoxin implicated in Balkan endemic nephropathy. *Mutat. Res.* **231**, 143–149.

[59] Degen, G. H.; Gerber, M. M.; Obrecht-Pflumio, S.; Dirheimer, G. (1997) Induction of micronuclei with ochratoxin A in ovine seminal vesicle cell culture. *Arch. Toxicol.* **71**, 365–371.

[60] Dharmshila, K.; Sinha, S. P. (1994) Effect of retinol on ochratoxin-produced genotoxicity in mice. *Food Chem. Toxicol.* **32**, 471–475.

[61] Hennig, A.; Fink-Gremmels, J.; Leistner, L. (1991) Mutagenicity and effects of ochratoxin A on the frequency of sister chromatid exchange after metabolic activation. In: Castegnaro, M.; Plestina, R.; Dirheimer, G.; Chernozemsky, I. N.; Bartsch, H. (Eds.) *Mycotoxins, endemic nephropathy and urinary tract tumours.* IARC Scientific Publications N° 115, Lyon, P. 255–269.

[62] De Groene, E. M.; Hassing, I. G.; Blom, M. J.; Seinen, W.; Fink-Gremmels, J.; Horbach, G. J. (1997) Development of human cytochrome P450 expressing cell lines: Application in mutagenicity testing of ochratoxin A. *Cancer Res.* **56**, 299–304.

[63] Grosse, Y.; Baudrimont, I.; Castegnaro, M.; Betbeder, A.-M.; Creppy, E.-E.; Dirheimer, G.; Pfohl-Leszkowicz, A. (1995) Formation of ochratoxin A metabolites and DNA-adducts in monkey kidney cells. *Chem. Biol. Interactions* **95**, 175–187.

[64] Grosse, Y.; Castegnaro, M.; Mace, K.; Bartsch, H.; Mohr, U.; Dirheimer, G.; Pinelli, E.; Pfeifer, A.; Pfohl-Leszkowicz, A. (1995) Cytochrome P450 isoforms implicated in ochratoxin A genotoxicity determined by DNA adduct formation. *Clin. Chem.* **41**, 93–95.

[65] Grosse, Y.; Monje, M. C.; Mace, K.; Pfeifer, A.; Pfohl-Leszkowicz, A. (1997) Use of bronchial epithelial cells expressing human cytochrome P450 for study on the metabolism and genotoxicity of ochratoxin A. *In vitro Toxicol.* **10**, 97–106.

[66] Grosse, Y.; Pfeifer, A.; Mace, K.; Harris, C. C.; Dirheimer, G.; Pfohl-Leszkowicz, A. (1994) Biotransformation of ochratoxin A by human bronchial epithelial cells (BEAS-2B) expressing human cytochrome P450s and implication of glutathione conjugation. *Toxicol. Lett.* (Suppl. 74), 33.

[67] Obrecht-Pflumio, S.; Föllmann, W.; Dirheimer, G. (1998) unpublished results.

[68] Schlatter, C.; Studer-Rohr, J.; Rasonyi, T. H. (1996) Carcinogenicity and kinetic aspects of ochratoxin A. *Food Add. Contam.* **13**, 43–44.

[69] Obrecht-Pflumio, S.; Dirheimer, G. (1997) *In vitro* DNA and nucleotides adducts formation caused by ochratoxin A. In: Märtlbauer, E. U. (Ed.) *Proceedings of the 19. Mykotoxin-Workshop*, p. 73–77.

[70] Obrecht-Pflumio, S.; Dirheimer, G. (1997) *In vitro* DNA and nucleotides adducts formation caused by ochratoxin A. *Mutat. Res.* **379** (No. 1, Suppl. 1), S156.

[71] Pfohl-Leszkowicz, A.; Grosse, Y.; Castegnaro, M.; Nicolov, I. G.; Chernozemski, I. N.; Bartsch, H.; Betbeder, A. M.; Creppy, E.-E.; Dirheimer, G. (1993) Ochratoxin A-related DNA adducts in urinary tract tumours of Bulgarian subjects. In: Phillips, D. H.; Castegnaro, M.; Bartsch, H. (Eds.) *Postlabelling methods for detection of DNA adducts. IARC scientific publication N° 124.* Lyon, p. 141–148.

[72] Frank, H. K. (1977) Occurence of patulin in fruits and vegetables. *Ann. Nutr. Alim.* **31**, 459–465.

[73] Frank, H. K. (1992) Citrinin. *Z. Ernährungswiss.* **31**, 164–177.

[74] IARC (1983) Vol. 5 Some mycotoxins. In: *Environmental carcinogens. Selected methods of analysis.* IARC Scientific Publication N° 44, Lyon.

[75] Ueno, Y.; Kubota, K. (1976) DNA-attacking ability of carcinogenic mycotoxins in recombination deficient mutant cells of Bacillus subtilis. *Cancer Res.* **36**, 445–451.

[76] Martin, W.; Lorkowsi, G.; Creppy, E.-E.; Muller, K.; Dirheimer, G.; Röschenthaler, R. (1982) Citrinin: some aspects on the mode of action in yeast and bacteria. In: *Mycotoxins and phycotoxins* (Vienna, T. U. O., Ed.), p. 305–308, Vienna.

[77] IARC (1986) Some naturally occurring food components furocoumarins and ultraviolet radiation. In: *IARC monographs on the evaluation of carcinogenic risks of chemicals to humans* (IARC, Ed.), Vol. 40, Lyon.

[78] Martin, W.; Lorkowski, G.; Creppy, E.-E.; Dirheimer, G.; Röschenthaler, R. (1986) Action of citrinin on bacterial chromosomal and plasmid DNA *in vitro* and *in vivo*. *Appl. Environ. Microbiol.* **52**, 1273–1279.
[79] Brakhage, A. A.; Bürger, M. G.; Creppy, E. E.; Dirheimer, G.; Röschenthaler, R. J. (1988) Base substitution mutations induced by the mycotoxin citrinin. *Arch. Toxicol.* Suppl. **12**, 341–346.
[80] Thust, R.; Kneist, S. (1979) Activity of citrinin metabolized by rat and human microsome fractions in clastogenicity and SCE assay on Chinese hamster V79-E cells. *Mutat. Res.* **67**, 321–339.
[81] Arai, M.; Hibino, T. (1983) Tumorigenicity of citrinin in male F344 rats. *Cancer Lett.* **17**, 281–287.
[82] Galloway, S. M.; Armstrong, M. J.; Reuben, C.; Colman, S.; Brown, B.; Cannon, C.; Bloom, A. D.; Nakamura, F.; Ahmed, M.; Duk, S.; Rimpo, J.; Margolin, B. H.; Resnick, M. A.; Anderson, B.; Zeiger, E. (1987) Chromosome aberrations and sister chromatid exchanges in Chinese hamster ovary cells: Evaluation of 108 chemicals. *Environ. Mol. Mutagens* (suppl. 10) **10**, 1–175.
[83] Keith, G.; Creppy, E.-E.; Dirheimer, G.; Randerath, K. (1989) unpublished results.
[84] Reddy, M. V.; Randerath, K. (1986) Nuclease P1 mediated enhancement of sensitivity of ^{32}P-postlabeling test for structurally diverse DNA adducts. *Carcinogenesis* **7**, 1543–1551.
[85] Pfohl-Leszkowicz, A.; Chekir, L.; Creppy, E.-E.; Dirheimer, G.; Bacha, H. (1994) Genotoxicity of zearalenone. In: *Proceedings of the 16th Mycotoxin Workshop, Hohenheim, May 16–18, 1994.*
[86] Pfohl-Leszkowicz, A.; Chekir-Ghedira, L.; Bacha, H. (1995) Genotoxicity of zearalenone, an estrogenic mycotoxin: DNA adduct formation in female mouse tissues. *Carcinogenesis* **16**, 2315–2329.
[87] Galloway, S. M. (1982) Carcinogenesis bioassay of zearalenone in F344/N rats and B6C3F1 mice. National Toxicology Program Technical report series, Dept. Health and Human Services, Research Triangle Park, NC, N° 235.
[88] Steiblen, G. (1996) Genotoxicité de la zearalenone: Implication du récepteur Ah. Mémoire de Diplôme d'Etat de Docteur en Pharmacie, Université Louis Pasteur, Strasbourg.
[89] Li, D.; Chen, S.; Randerath, K. (1992) Natural dietary ingredients (Oats and Alfalfa) induce covalent DNA modifications (I-compounds) in rat liver and kidney. *Nutr. Cancer* **17**, 205–216.
[90] Lee, K.-S.; Röschenthaler, R. J. (1986) DNA-damaging activity of patulin in *Escherichia coli. Appl. Environ. Microbiol.* **52**, 1046–1054.
[91] Umeda, M.; Yamamoto, T.; Saito, M. (1972) DNA-strand breakage of HeLa cells induced by several mycotoxins. *J. Exp. Med.* **42**, 527–539.
[92] Thust, R.; Kneist, S.; Mendel, J. (1982) Patulin, a further clastogenic mycotoxin, is negative in the SCE-assay in Chinese hamster V79-E cells *in vitro. Mutat. Res.* **103**, 91–97.
[93] Knasmuller, S.; Bresgen, N.; Kassie, F.; Mersch-Sundermann, V.; Gelderblom, W.; Zohrer, E.; Eckl, P. M. (1997) Genotoxic effects of three Fusarium mycotoxins, fumonisin B1, moliniformin and vomitoxin in bacteria and in primary cultures of rat hepatocytes. *Mutat. Res.* **391**, 39–48.
[94] Dutton, M. F. (1996) Fumonisins, mycotoxins of increasing importance. Their nature and their effects. *Pharmacol. Ther.* **70**, 137–161.
[95] Gelderblom, W. C. A.; Snyman, S. D. (1991) Mutagenicity of potentially carcinogenic mycotoxins produced by *Fusarium moniliforme. Mycotoxin Res.* **7**, 46–52.
[96] Park, D. L.; Rua, S. M.; Mirocha, C. J.; Abd-Alla, E. A. M.; Weng, C. Y. (1992) Mutagenic potentials of fumonisin contaminated corn following ammonia decontamination procedure. *Mycopathologia* **117**, 105–108.
[97] Sun, T. S.; Star, H. M. (1993) Evaluation and application of a bioluminescent bacterial genotoxicity test. *J. Assoc. Off. Anal. Chem. Int.* **76**, 893–893.

[98] Gelderblom, W.; Marasas, W. F. O.; Thiel, P.; Semple, E.; Farber, E. (1989) Possible non-genotoxic nature of active carcinogenic components produced by *Fusarium moniliforme*. *Proc. Am. Assoc. Cancer Res.* **30**, 144.

[99] Gelderblom, W. C. A.; Semple, E.; Marasas, W. F. O.; Farber, E. (1992) The cancer-initiating potential of the fumonisin B mycotoxins. *Carcinogenesis* **13**, 433–437.

[100] Norred, W. P.; Plattner, R. D.; Vesonder, R. F.; Hayes, P. M.; Bacon, C. W.; Voss, K. A. (1990) Effect of *Fusarium moniliforme* metabolites on unscheduled DNA synthesis (UDS) in rat primary hepatocytes. *The Toxicologists* **10**, 165.

[101] Norred, W. P.; Plattner, R. D.; Vesonder, R. F.; Bacon, C. W.; Voss, K. A. (1992) Effects of selected secondary metabolites of *Fusarium moniliforme* on unscheduled synthesis of DNA by rat primary hepatocytes. *Food Chem. Toxicol.* **30**, 233–237.

[102] Yoo, H.-S.; Norred, W. P.; Wang, E.; Merril, A. H. J.; Riley, R. T. (1992) Fumonisin inhibition of de novo sphingolipid biosynthesis and cytotoxicity are correlated in LLC-PK cells. *Toxicol. Appl. Pharmacol.* **114**, 9–15.

[103] Gelderblom, W. C. A.; Kriek, N. P. J.; Marasas, W. F. O.; Thiel, P. G. (1991) Toxicity and carcinogenicity of the *Fusarium moniliforme* metabolite, fumonisin B1, in rats. *Carcinogenesis* **12**, 1247–1251.

[104] Sydenham, E. W.; Shephard, G. S.; Thiel, P. G.; Marasas, W. F. O.; Stockenström, S. (1991) Fumonisin contamination of commercial corn-based human foodstuffs. *J. Agric. Food Chem.* **39**, 2014–2018.

[105] California Environmental Protection Agency (CEPA); Reproductive and Cancer Hazard Assessment Section; Office of Environmental Health Hazard Assessment (1995). *Evidence on the carcinogenicity of fumosin B1 and fusarin C (draft).* p. 32.

[106] Gelderblom, W. C. A.; Snyman, S. D.; Abel, S.; Lebepe-Mazur, S.; Smuts, C. M.; Van Der Westhuizen, L.; Marasas, W. F. O. (1995) Hepatotoxicity and carcinogenicity of the fumonisins in rats. A review regarding mechanistic implications for establishing risk in humans. In: Jackson, L.; De Vries, J. W.; Bullerman, L. B. (Eds.) *Fumonisins in food.* Plenum Press, New York, p. 279–296.

[107] Gelderblom, W. C.; Snyman, S. D.; Lebepe-Mazur, S.; Van Der Westhnizen, L.; Kriek, N. P.; Marasas, W. F. O. (1996) The cancer promoting potential of fumonisin B1 in rat liver using diethylnitrosamine as a cancer initiator. *Cancer Lett.* **109**, 101–108.

[108] Yeung, J.; Wang, H.-Y.; Prelusky, D. B. (1996) Fumosin B1 induces protein kinase C translocation via direct interaction with diacylglycerol binding site. *Toxicol. Appl. Pharmacol.* **141**, 178–184.

[109] Cheng, S. J.; Jiang, Y. Z.; Li, M. H.; Lo, H. Z. (1985) A mutagenic metabolite produced by *Fusarium moniliforme* isolated from Linxian Country, China. *Carcinogenesis* **6**, 903–905.

[110] Lu, S.-J.; Ronai, Z. A.; Li, M. H.; Jeffrey, A. M. (1988) *Fusarium moniliforme* metabolites: Genotoxicity of culture extracts. *Carcinogenesis* **9**, 1523–1527.

[111] Lu, F.-X.; Lee, M.-X.; Shen, D.-C. (1990) The metabolism and DNA binding of 3H-fusarin C in rats (Chin.). *Acta acad. med. sin.* **12**, 231–235.

[112] Li, M.-X.; Jian, Y.-Z.; Han, N.-J.; Fan, W.-G.; Ma, J.-L.; Bjeldanes, L. E. (1992) Fusarin C induced esophageal and forestomach carcinoma in mice and rats (Chin.). *Chin. J. Oncol.* **14**, 27–29.

[113] Knasmuller, S.; Parzefall, W.; Helma, C. F. K.; Ecker, S.; Schulte-Herman, R. (1997) Toxic effects of griseofulvin. Disease models, mechanisms and risk assessment. *Crit. Rev. Toxicol.* **27**, 495–537.

[114] IARC (1987) *Overall evaluations of carcinogenicity. An updating of IARC Monographs on the evaluation of carcinogenic risks to human* (IARC, Ed.), Suppl. No. 7 Vol. 1.42, Lyon.

[115] Betina, V. (1989) Structure-activity relationships among mycotoxins. *Chem.-Biol. Interactions* **71**, 105–146.

[116] Stark, A. A. (1980) Mutagenicity and carcinogenicity of mycotoxins. DNA binding as a possible mode of action. *Ann. Rev. Microbiol.* **34**, 235–262.

[117] Wehner, F. C.; Thiel, P. G.; Van Rensburg, S. J.; Demasius, I. P. (1978) Mutagenicity to *Salmonella typhimurium* of some Aspergillus and Penicillium mycotoxins. *Mutat. Res.* **58**, 193–203.

[118] Krirobock, S.; Olivier, P.; Marzin, D. R.; Seigle-Murandi, F.; Steiman, R. (1987) Study of the genotoxic potential of 17 mycotoxins with the SOS chromotest. *Mutagenesis* **2**, 433–439.

[119] Auffray, Y.; Boutibonnes, P. (1987) Genotoxic activity of some mycotoxins using the SOS chromotest. *Mycopathologia* **100**, 49–53.

[120] Dickens, F.; Jones, H. E. H. (1961) Carcinogenic activity of a series of reactive lactones and related substances. *Br. J. Cancer* **15**, 85–109.

[121] Natori, S.; Sakaki, S.; Kurata, H.; Undagawa, S.; Ichinoe, M.; Saito, M.; Umeda, M.; Ohtsueo, K. (1970) Production of rubratoxin B by *Penicillium purpurogenum. Appl. Microbiol.* **19**, 613–617.

[122] Masuda, T.; Haraikawa, K.; Morooka, N.; Nakamo, S.; Ueno, Y. (1985) 2-Hydroxyemodin, an active metabolite of emodin in the hepatic microsomes of rats. *Mutat. Res.* **149**, 327–332.

[123] Morita, H.; Umeda, M.; Masuda, T.; Ueno, Y. (1988) Cytotoxic and mutagenic effects of emodin on cultured mouse carcinoma FM3A cells. *Mutat. Res.* **204**, 329–332.

[124] Mori, H.; Kawai, K.; Ohbayashi, F.; Kuniyasu, T.; Williams, G. M. (1984) Genotoxicity of a variety of mycotoxins in the hepatocyte primary culture. DNA repair test using rat and mouse hepatocytes. *Cancer Res.* **44**, 2918–2923.

[125] Ueno, Y.; Nakajima, M. (1974) Production of respiratory-deficient mutants of *Saccharomyces cerevisiae* by (–)-luteoskyrin and (+)-rugulosin. *Chem. Pharm. Bull.* **22**, 2258.

[126] Akuzawa, S.; Yamaguchi, H.; Masuda, T.; Ueno, Y. (1992) Radical mediated modification of deoxyguanine and deoxyribose by luteoskyrin and related anthraquinones. *Mutat. Res.* **266**, 63–69.

[127] Uraguchi, K.; Saito, M.; Noguchi, Y.; Takahashi, K.; Enomoto, M.; Tatsuno, T. (1972) Chronic toxicity and carcinogenicity in mice of the purified mycotoxins, luteoskyrin and cyclochlorotine. *Food Cosmet. Toxicol.* **10**, 193–207.

[128] Umeda, M.; Saito, A.; Saito, M. (1970) Cytotoxic effects of toxic culture filtrate of *Penicillium purpurogenum* and its toxic metabolite rubratoxin B on HeLa cells. Comparative study of the effect of rubratoxin B, colcemid, and vinblastine. *Jpn. J. Exp. Med.* **40**, 409–423.

[129] Evans, M. A.; Harbison, R. D. (1977) Prenatal toxicity of rubratoxin B and its halogenated analogue. *Toxicol. Appl. Pharmacol.* **39**, 13–22.

[130] Stark, A. A.; Townsend, J. M.; Wogan, G. N.; Demain, A. L.; Manmade, A.; Ghosh, A. C. (1978) Mutagenicity and antibacterial activity of mycotoxins produced by *Penicillium islandicum Sopp* and *Penicillium rugulosum. J. Environ. Pathol. Toxicol.* **2**, 313.

[131] Aujard, C.; Moule, Y.; Moreau, S.; Darraq, N. (1979) Persistence of DNA damage induced in cultured liver cells by PR toxin, a mycotoxin from *Penicillium roqueforti. Toxicol. Eur. Res.* **2**, 273–278.

[132] Umeda, M. (1977) Cytotoxicity of mycotoxins. In: Rodrichs, J. V.; Hesseltine, C. W.; Mehlman, M. A. (Eds.) *Mycotoxins in human and animal health.* Pathotox Publ. Inc., Park Forest South Ill., p. 712–729.

[133] Huff, W. E.; Kubena, L. F.; Harwey, R. B.; Doerr, J. A. (1988) Mycotoxin interactions in poultry and swine. *J. Animal Sci.* **56**, 2351–2355.

12 The Role of Nitrosation: Exogenous *vs* Endogenous Exposure to *N*-Nitroso Compounds

David E. G. Shuker

12.1 Abstract

For more than 40 years since the discovery of the carcinogenicity of dimethylnitrosamine we have steadily accumulated a large store of knowledge about the mode(s) of action of *N*-nitroso compounds (NOC). However, despite this knowledge there is still great uncertainty about the role of NOC in human cancer. For many years the main route of exposure to NOC was thought to be external (exogenous) and much effort was expended in identifying and minimising these exposures. However, the discovery of endogenous synthesis of NOC in animals by Sander and Bürkle in 1969 [1] and in humans by Ohshima and Bartsch in 1981 [2] opened up the possibility of a completely different and possibly major source of NOC. Evidence has since accumulated that body fluids, particularly, gastric juice, do contain NOC and that certain pathologies (e.g., chronic inflammation or bacterial overgrowth) increase the levels of NOC. More recently, evidence has emerged to suggest that endogenous nitrosation does not necessarily lead to stable, measurable NOC but instead to direct acting products of nitrosation. However the question still remains – to what extent does nitrosation from whatever source contribute to human cancer? Recent improvements in analytical methods for measurement of low levels of DNA damage in humans have provided evidence for NOC exposures and several lines of work suggest that nitrosation of amino acids accounts for much of the alkyl-DNA damage which has been observed. Weisburger and colleagues [3] showed that the main mutagen in Chinese pickled fish was a derivative of nitrosated methionine. Sedgwick [4] found that nitrosation of amino acids may be responsible for background levels of DNA damage in bacteria and mammalian cells in culture. We have recently shown that the simplest α-amino acid, glycine, is converted by nitrosation to both carboxymethylating and methylating agents [5]. This latter observation is consistent with a number of reports of detection of O^6-methylguanine in human DNA [6]. In conclusion, endogenous nitrosation does appear to be a contributor to the human burden of carcinogens and the precursors are probably predominantly dietary in origin.

12.2 Introduction

N-Nitroso compounds (NOC) are a class of carcinogens that have attracted much interest since *N*-nitrosodimethylamine (NDMA) was found to be a liver carcinogen in rats by Magee and Barnes in the mid 1950's [7]. The extensive literature on NOC has been reviewed in several volumes [8, 9] and much original work can be found in the publications arising from a series of international meetings organised by the International Agency for Research on Cancer [10]. The tobacco specific nitrosamines constitute a distinct group of NOC for which there is increasing evidence of their pivotal role in the induction of lung cancer [11].

N-Nitrosamines and *N*-nitrosamides are, for the most part, stable compounds derived from the parent nitrogen precursors by replacement of a proton by the nitroso group [12]. However, this is only part of the spectrum of *N*-nitrosation chemistry, most of which is relevant to humans (Figure 12.1). The significance of the pathways resulting from *N*-nitrosation of primary amines, in relation to dietary exposures, will be considered in detail later. The observation that NOC, in particular, the *N*-nitrosamines, could induce cancers at most sites in many species of animal [13] supported the view that the presence of many different NOC in the environment contributed to the burden of human cancer. It is interesting to note, however, that Barnes felt the value of NOC was primarily as experimental carcinogens, only indirectly contributing to our knowledge of cancer [14]. However, in the light of the great expansion of work on NOC, Magee came to a different conclusion in 1989 [15]. Nonetheless, the lack of an obvious connection between NOC in the environment, particularly in food, and human cancer risk at any site has lead to a view that NOC are perhaps "experimental carcinogens looking for human cancers to be the cause of". NOC are potent carcinogens in many (>40) species that have been tested and the evidence that humans are not especially resistant to NOC carcinogenicity comes from several quarters. Several *N*-nitrosochloroethylureas are used in the treatment of various cancers. The incidence of treatment-related second

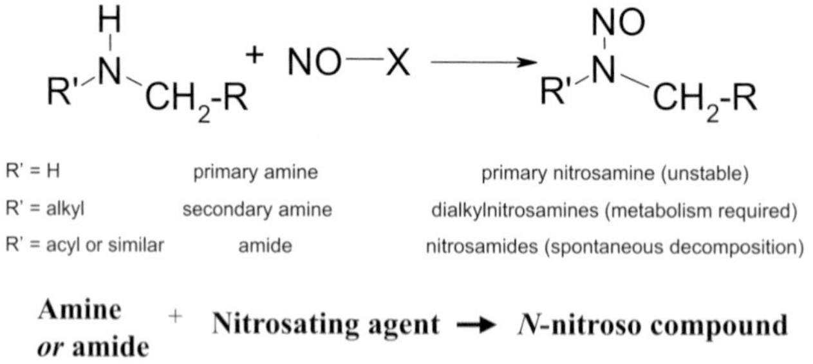

Figure 12.1: Summary of the main *N*-nitrosation pathways giving rise to *N*-nitroso compounds.

cancers is elevated in patients treated with MeCCNU showing that it is carcinogenic to humans and it has been classified as a group 1 carcinogen by IARC [16]. As indicated above, in the case of tobacco smoking there is accumulating evidence that tobacco-specific nitrosamines (TSNA) account for a major part of the pulmonary tumor burden [11]. Furthermore, it has been calculated that a typical tobacco-smoker in the US is exposed to a total life time dose of TSNA equivalent to that which induces tumors in experimental animals [17]. The purpose of this review is to briefly summarise the evidence for human exposure to preformed NOC from dietary sources and to evaluate the contribution this exposure makes to the overall burden of NOC when endogenous processes are considered.

12.3 Exogenous exposures to NOC

Improvements in analytical methods for low levels of NOC in the environment (notably, air, occupational settings, consumer products and foodstuffs) in the early 1970's resulted in realisation that NOC were ubiquitous. However, except in the case of certain occupational exposures and tobacco smoking, the levels of NOC involved were very low (<1 μg/kg/day). NOC have been detected in several human foods and beverages including nitrite-cured meat, smoked or salt-dried fish and beer giving rise to exposures up to 2.3 μg/day/person of volatile NOC and 10–100 μg/day/person of non-volatile NOC [18, 19]. One area which attracted much attention was the presence of NOC in alcoholic beverages, with some beers containing high levels of dialkylnitrosamines (up to 20 μg/kg). In some respects this situation represents a success story in controlling exogenous exposure to carcinogens since changes in the various malting processes were found to dramatically reduce NOC levels by 20–100-fold [18]. However, surveillance is still carried out on alcoholic beverages [20], with a particular concern being the possible synergism between NOC and alcohol for esophageal cancer [21].

NOC have been detected in blood and other physiological fluids and early studies attempted to relate the levels to external exposures [22]. However, with possible exceptions of passive exposure to environmental tobacco smoke, which does appear to give rise to detectable levels of TSNA – specific DNA adducts in non-smokers [23] and some occupational exposures [22], current evidence suggests that such NOC are probably present as a result of endogenous synthesis.

12.4 Endogenous synthesis of NOC

The discovery of endogenous synthesis of NOC firstly by Sander and Bürkle in 1969 [1] and then in humans by Ohshima and Bartsch in 1981 [2] opened up a whole new area of research which has irrevocably changed our view of human exposure to NOC and their role in cancer [24]. *N*-Nitrosoproline (NPRO) is an unusual nitrosamine in that it is resistant to metabolic activation and is excreted unchanged in urine (Figure 12.2). Furthermore, the precursor L-proline (PRO), a natural amino acid, is capable of being efficiently nitrosated under mildly acidic conditions whereas the more basic dialkylamines are protonated and are thus much less reactive. Interestingly, early in the work it was recognised that background levels of NPRO were present in human urine, and in an elegant series of animal experiments with stable isotope labelled PRO and nitrite, Hotchkiss and colleagues [25] showed that urinary NPRO was derived from either wholly exogenous or wholly endogenous sources as well as the intermediate mixed routes. Because of the continuing interest in the role of NOC in gastric cancer arising out of a hypothesis initially suggested, and consequently elaborated, by Correa [26] several groups tackled the difficult problem of quantifying NOC levels in the dynamic system of the gastric contents [27]. Eventually, a well-conducted and extensive series of observations by Xu and Reed [28] demonstrated that total NOC concentrations varied as a function of intragastric pH with maximal values being found at either end of the scale from pH~2 to pH~8 (Figure 12.3). The methodology used for quantification of NOC, as total NOC, was such that the identity of the individual compounds could not be ascertained. However, it is known that precursor nitrogenous compounds can be converted to NOC under acidic conditions (*via* a nitrous acid-mediated pathway) or at neutral or mildly alkaline conditions (*via* reaction with gaseous oxides of nitrogen of bacterial from other sources) [12].

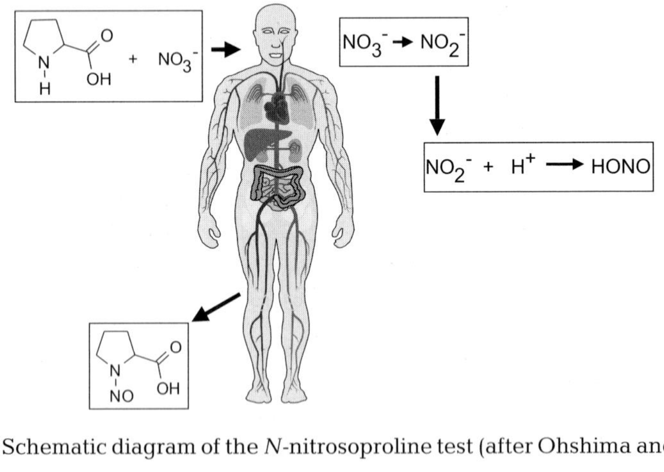

Figure 12.2: Schematic diagram of the *N*-nitrosoproline test (after Ohshima and Bartsch [2]).

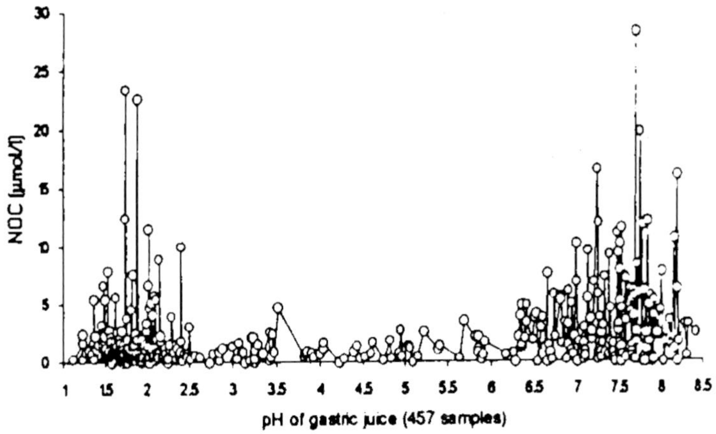

Figure 12.3: Levels of total *N*-nitroso compounds (NOC) in fasting gastric juice at a wide range of pH values (from Xu and Reed (1993) *Carcinogenesis* **14**, 2546–2551).

12.5 Alkyl-DNA adducts as indicators of NOC exposure

NOC form reactive DNA-alkylating agents upon spontaneous decomposition (diazoalkanes, nitrosoureas, nitrosamides and related compounds) or after oxidative metabolism (dialkylnitrosamines). The intermediate, probably an alkyldiazonium ion, reacts with all of the nucleophilic centres in DNA (Figure 12.4) [29]. Much interest has focused on the formation of O^6-alkylguanine, and to a lesser extent O^4-alkylthymine, adducts since these are promutagenic modifications leading to misincorporation of opposing bases on DNA replication [30]. Efficient repair processes exist for most forms of alkyl-DNA damage and levels of any particular DNA adduct in humans reflect a dynamic equilibrium between exposure, leading to adduct formation, and repair, leading to removal of adduct. There have been a number of reports of the methyl adduct, O^6-methyldeoxyguanosine (O^6-MedG), being present in DNA from different organs (Table 12.1) and, in some cases with elevated levels being associated with increased risk of cancer [31, 32]. A number of methylating NOC such as *N*-methyl-*N*-nitro-*N*-nitrosoguanidine (MNNG) and *N*-nitroso-*N*-methylurea (MNU), have been used to induce experimental gastric tumors [13]. It is unlikely however that nitroso compounds such as MNNG or MNU occur naturally in the human GI tract and, in fact, the endogenous nitrosation of dietary amino acids and peptides is a more likely reaction [33], although the bile acid conjugates also constitute a large endogenous source of nitrosatable substrates [34]. Indirect evidence for endogenous nitrosation has come from *in vitro* systems where background levels of DNA alkylation appear to be related to nitrosation of amino acids, includ-

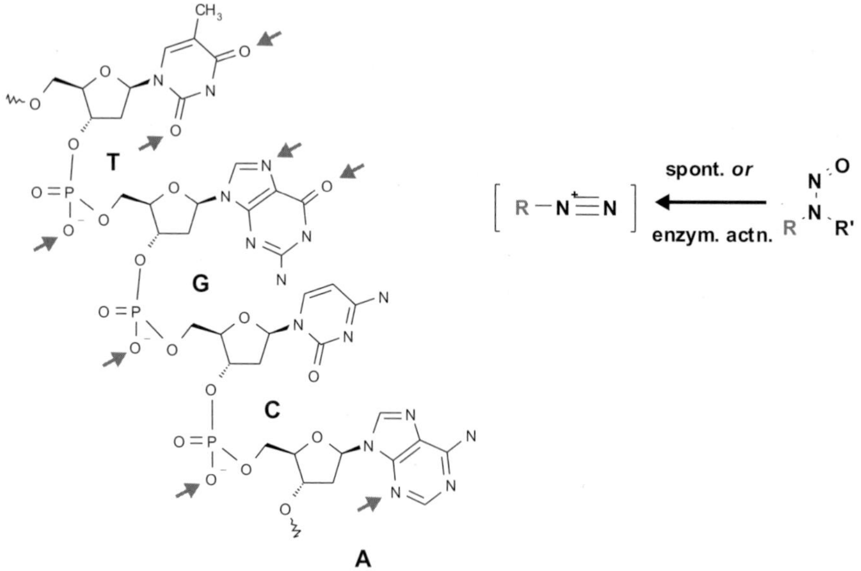

Figure 12.4: The main sites of DNA alkylation by *N*-nitroso compounds.

ing glycine [4]. As glycine is one of the most abundant amino acids in nature, it would seem likely that its nitrosation products would constitute a major source of alkylating agents.

12.6 Nitrosation of glycine and other amino acids as a source of endogenous DNA damage

Glycine is the simplest α-amino acid and is among the most abundant amino acids in dietary proteins. Free glycine occurs in biological fluids at millimolar concentrations [35]. Nitrosation of glycine esters to give stable diazoacetic esters was first reported in 1904 [36] and salts of diazoacetic acid itself were reported in 1908 [37]. Remarkably, apart from kinetic studies on decomposition [38], there has been no examination of the potential toxicity or carcinogenicity of this simple compound. However there has been a recent resurgence of interest in the chemistry and biology of diazo- and *N*-nitrosopeptides and this was reviewed by Challis [39]. Diazo- and *N*-nitrosopeptides are consistently mutagenic in many test systems and several are potent carcinogens in animal models [e.g. 40]. To date, there have been no studies directed at evaluating the pos-

Table 12.1: Summary of studies reporting the presence of O^6-MedG in human tissues arising from environmental (including lifestyle) or endogenous exposures.

Exposure	Population	Tissue	Method	Proportion alkylated[a]	O^6-MedG Range (μmol/mol dG/dA)	Ref.
Environmental[c]	Lin Xian (China)	Esophagus and stomach	RIA	18/26 9/11	0.016–026 0.024–0.14	32
	Singapore (S.E. Asia)	Esophagus and stomach	RIA	27/53	0.016–0.08	47
	Shanghai (China)	Stomach	RIA	15/53	0.053–0.718	48
	Nile delta (Egypt)	Bladder	RIA	44/46	0.012–048	49
General survey/controls [d]	European (France and Germany	Esophagus, stomach and colon	RIA	5/12	0.032–0.064	32
	Manchester (UK)	Stomach and colon	RIA	27/53	0.011–>0.3	50
	Manchester (UK)	Bladder	RIA	4/12	0.034–0.225	49
	Athens (Greece)	Stomach	CRA	1/20	0.083	51
	Japan	Liver Leukocytes	HPLC/ ^{32}P PL	13/15 15/15	0.11 –0.67 0.007–0.46	52
	17 populations	Leukocytes	CRA	21/413	0.085–0.42	31
	Shanghai (China) Manchester (UK)	Stomach Stomach	IAC/ ^{32}P PL	7/7 5/5	0.021–0.041 0.028–0.07	6

[a] Number of samples containing O^6-MedG/total number of samples
[b] Amount of O^6-MedG detected in positive samples
[c] Tissues taken from individuals at increased risk for developing cancer in specific tissues due to environmental factors
[d] Tissues taken from individuals not known to be at an increased risk for developing cancer. RIA, radioimmunoassay; HPLC, high performance liquid chromatography; PL, postlabelling; ELISA, enzyme-linked immunosorbent assay; IAC, immunoaffinity column; CRA, Competitive repair assay.

sible contribution of these compounds to human exposure although Shephard and Lutz [33] calculated that nitrosation of dietary amino acids was a major source of carcinogen exposure in humans. As indicated above, current methods for analysis of total NOC do not permit identification of individual compounds. Furthermore, the DNA alkylation products of even simple nitrosated peptides are likely to be very complex because of the presence of 20 normal amino acids. Thus, there are 20 different possible dipeptides N-terminal in glycine alone and the numbers of possible peptides of higher number and N-terminal in all the different amino acids increases rapidly according to strictly combinatorial arithmetic (i.e. there are $20^2 = 400$ different possible dipeptides, and $20^3 = 8{,}000$ different tripeptides, etc.). Therefore, not least for reasons of sheer practicality, it seems reasonable to focus in the first instance on nitrosation of simple amino acids.

In several studies of DNA damage by nitrosated glycine derivatives the expected formation of carboxymethyl adducts (N^7-carboxymethylguanine [7CMG], N^3-carboxymethyladenine and O^6-carboxymethylguanine [O^6-CMG]) had been observed [41–43]. However, on closer examination it became clear that methylation was also occurring [44]. This latter observation is particularly interesting in view of the number of observations of elevated O^6-MeG in DNA from the human gastrointestinal tract (Table 12.1). The mechanism of the production of a methylating agent from nitrosated glycine derivatives has not been fully elucidated but it almost certainly involves decarboxylation of the intermediate diazoacetate or diazoniumacetate species.

The formation of characteristic carboxymethyl adducts by nitrosated glycine derivatives affords an approach to evaluating their contribution to the overall burden of diet related NOC exposures in man. For example, 7CMG is excreted unchanged in urine and could be used as a non-invasive marker [42]. O^6-CMG appears to be resistant to the action of O^6-alkylguanine transferase (AGT) and would be expected to persist, therefore constituting a good candidate biomarker in DNA. A sensitive immunoslot blot (ISB) method has been developed for O^6-CMG in intact DNA [43] and examination of human DNA extracted from blood or gastric biopsies revealed the presence of O^6-CMG at relatively high levels (2–10 μmol/mol dG [blood], 1–20 μmol/mol dG [gastric biopsies]) (Harrison *et al.*, unpublished observations). Previous *in vitro* studies with a range of nitrosated glycine derivatives had established that the ratio of O^6-CMG to O^6-MeG ranged from 10:1 to 40:1 [5] so it was not unexpected that, if such compounds were responsible for the background levels of DNA methylation in man, more O^6-CMG would be found than O^6-MeG. The ratio of O^6-CMG to O^6-MeG appears to be ca. 100:1 in humans and this would be consistent with the hypothesis that nitrosated glycine derivatives may be the source of this damage, since O^6-MeG would be repaired by AGT and therefore be present at lower than initial adduct ratios.

Further evidence for this hypothesis is the observation in human volunteers that increased protein intake results in elevated levels of intra-intestinal total NOC [45] and a number of epidemiological studies that link high protein intake with increased risk of GI tract, particularly colorectal, cancers [46].

Nitrosation of other natural amino acids may be relevant for human cancer risk, as indicated by the discovery that nitrosation of methionine in the presence of high chloride ion concentrations leads to the production of a highly mutagenic derivative, 2-chloro-4-methylthio-butanoic acid [3].

12.7 Conclusions

There is a large body of evidence to indicate that humans are exposed to NOC from exogenous and endogenous sources. In order to evaluate the role of nitrosation in the etiology of human cancers as well as the relative contributions of the two major routes of exposure, it appears that methods which measure the overall effect of exposure (such as DNA or protein adducts) are likely to be informative. However, unlike many other carcinogenic exposures which give rise to many characteristic adducts (such as aflatoxin B_1), many NOC result in "generic" damage, such as DNA methylation, which cannot be unambiguously assigned to any single agent. An exception to this situation which may be particularly productive in future, is the use of accelerator mass spectrometry (AMS) which quantifies radio-isotopes, particularly ^{14}C, at very low levels allowing labelled precursors to be safely administered to humans. However, this approach is reminiscent of the early experimental studies on DNA alkylation using labelled compounds and which resulted in the background levels of DNA damage being undetected, until analytical methods of sufficient sensitivity were developed.

Acknowledgements

Support from the Medical Research Council and the Ministry of Agriculture Fisheries and Food is gratefully acknowledged.

References

[1] Sander, J.; Bürkle, G. (1969) Induktion maligner Tumoren bei Ratten durch gleichzeitige Verfütterung von Nitrit und sekundären Aminen. *Zeit. Krebs.* **76**, 93–96.
[2] Ohshima, H.; Bartsch, H. (1981) Quantitative estimation of endogenous nitrosation in humans by monitoring *N*-nitrosoproline excreted in the urine. *Cancer Res.* **41**, 3658–3662.
[3] Chen, W.; Weisburger, J. H.; Fiala, E. S.; Carmella, S. G.; Chen, D.; Spratt, T. E.; Hecht, S. S. (1995) Unexpected mutagen in fish. *Nature* **374**, 599.
[4] Sedgwick, B. (1997) Nitrosated peptides and polyamines as endogenous mutagens in O^6-alkylguanine–DNA alkyltransferase deficient cells. *Carcinogenesis* **18**, 1561–1567.
[5] Harrison, K. L.; Jukes, R.; Cooper, D. P.; Shuker, D. E. G. (1999) Concomitant formation of O^6-carboxymethyl- and O^6-methyl-deoxyguanosine in DNA exposed to nitrosated glycine derivatives. *Chem. Res. Toxicol.* **12**, 106–111.
[6] Povey, A. C.; Cooper, D. P. (1995) The development, validation and application of a ^{32}P-postlabelling assay to quantity O^6-methylguanine in human DNA. *Carcinogenesis* **16**, 1665–1669.
[7] Magee, P. N.; Barnes, J. M. (1967) Carcinogenic nitroso compounds. *Adv. Cancer Res.* **10**, 163–246.
[8] Preussmann, R.; Stewart, B. W. (1984) *N*-Nitroso carcinogens. In: Searle, C. E (Ed.) *Chemical carcinogens, Vol. 2 (ACS Monograph 182).* American Chemical Society, Washington DC, p. 643–828.
[9] Forman, D.; Shuker, D. E. G. (1989) Introduction: Nitrate, nitrite and nitroso compounds, *Cancer Surv.* **8**, 205–206.
[10] IARC (1990) Relevance to human cancer of *N*-nitroso compounds, tobacco smoke and mycotoxins. In: O'Neill, I. K.; Chen, J.; Lu, S. H.; Bartsch, H. (Eds.). *IARC Scientific publication No. 105.* International Agency for Research on Cancer, Lyon, France, p. 1–600; and previous volumes in the series.
[11] Hecht, S. S. (1999) Tobacco smoke carcinogens and lung cancer. *J. Natl. Cancer Inst.* **91**, 1194–1210.
[12] Shuker, D. E. G. (1988) The chemistry of *N*-nitrosation. In: Hill, M. J. (Ed.) *Nitrosamines – Toxicology and microbiology.* Ellis Harwood, Chichester, England, Ch. 3.
[13] Lijinsky, W. (1992) *Chemistry and biology of* N-*nitroso compounds.* Cambridge University Press, Cambridge, England.
[14] Barnes J. M. (1974) Nitrosamines. *Essays Toxicol.* **5**, 1–15.
[15] Magee, P. N. (1989) The experimental basis for the role of nitroso compounds in human cancer. *Cancer Surv.* **8**, 207.
[16] IARC (1987) *IARC Monographs on the evaluation of carcinogenic risks to humans (Supplement 7).* International Agency for Research on Cancer, Lyon, France, p. 150.
[17] Hecht, S. S.; Hoffman, D. (1989) The relevance of tobacco-specific nitrosamines to human cancer. *Cancer Surv.* **8**, 273–294.
[18] Hotchkiss, J. H. (1989) Preformed *N*-nitroso compounds in foods and beverages. *Cancer Surv.* **8**, 295–321.
[19] Challis, B. C. (1996) Environmental exposures to *N*-nitroso compounds and precursors: general review of methods and current status. *Eur. J. Cancer Prev.* **5**, 19–26.
[20] MAFF (1995) NDMA and ATNC in retail malt whisky. *MAFF UK Food surveillance information sheets, No. 57* (http://www.maff.gov.uk/food/infsheet/1995/no57/57whisky.htm).
[21] Craddock, V. M. (1993) *Cancer of the esophagous.* Cambridge University Press, Cambridge, England.

[22] Tricker, A. R. (1989) Environmental exposures to preformed nitroso compounds. *Cancer Surv.* **8**, 251–272.

[23] Hecht, S. S. (1998) Biochemistry, biology and carcinogencity of the tobacco-specific *N*-nitrosamines. *Chem. Res. Toxicol.* **11**, 560–603.

[24] Bartsch, H.; Ohshima, H.; Pignatelli, B.; Calmels, S. (1988) Human exposure to endogenous *N*-nitroso compounds: quantitative estimates in subjects at high risk of cancer of the oral cavity, oesophagus, stomach and urinary bladder. *Cancer Surv.* **8**, 335–362.

[25] Perciballi, M.; Conboy, J. J.; Hotchkiss, J. H. (1989) Nitrite cured meats as a source of endogenously and exogenously formed *N*-nitrosoproline in the ferret. *Food Chem. Toxicol.* **27**, 111–116.

[26] Correa, P. (1992) Human gastric carcinogenesis a multistep and multifactorial process. *Cancer Res.* **52**, 6735–6740.

[27] Kyrtopoulos, S. A. (1989) *N*-nitroso compound formation in human gastric juice. *Cancer Surv.* **8**, 423–442.

[28] Xu, G. P.; Reed, P. L. (1993) *N*-Nitroso compounds in fresh gastric juice and their relation to intragastric pH and nitrite employing an improved analytical method. *Carcinogenesis* **14**, 2547–2551.

[29] Singer, B.; Grunberger, D. (1983) *Molecular biology of mutagens and carcinogens.* Plenum Press, New York, p. 45–96.

[30] Singer, B.; Essigman, J. M. (1991) Site specific mutagenesis, retrospective and prospective. *Carcinogenesis* **12**, 949–955.

[31] The Eurogast Study Group (1994) O^6-Methylguanine in blood leucocyte DNA: an association with the geographic prevalence of gastric cancer and with low levels of serum pepsinogen A, a marker of severe chronic atrophic gastritis. *Carcinogenesis* **15**, 1815–1820.

[32] Umbenhauer, D.; Wild, C. P.; Montesano, R.; Saffhill, R.; Boyle, J. M.; Huh, N.; Kirstein, U.; Thomale, J.; Rajewsky, M. F.; Lu, S. H. (1985) O^6-Ethyldeoxyguanosine in oesophageal DNA among individuals at high risk of oesophageal cancer. *Int. J. Cancer* **36**, 661–665.

[33] Shephard, S. E.; Lutz W. K. (1989) Nitrosation of dietary precursors. *Cancer Surv.* **8**, 401–421.

[34] Shuker, D. E. G.; Tannenbaum, S. R.; Wishnok, J. S. (1981) *N*-Nitrosobile acid conjugates. I. Synthesis, chemical reactivity and mutagenic activity. *J. Org. Chem.* **46**, 2092–2096.

[35] Komorowska, M.; Szafran, H.; Popiela, T.; Szafran, Z. (1981) Free amino acids of human gastric juice. *Acta Physiol. Pol.* **32**, 559–567.

[36] Curtius, T. (1904) Über die freiwillige Zersetzung des Glykocollesters. *Chem. Ber.* **43**, 1285–1300.

[37] Müller, E. (1908) Über pseudo-Diazoessigsäure. *Chem. Ber.* **41**, 3116–3139.

[38] Kreevoy, M. M.; Konasewich, D. E. (1970) The mechanism of hydrolysis of diazoacetate ion. *J. Phys. Chem.* **74**, 4464–4472.

[39] Challis, B. C. (1989) Chemistry and biology of nitrosated peptides. *Cancer Surv.* **8**, 363–384.

[40] Anderson, D.; Blowers, S. D. (1994) Limited cancer bioassay to test a potential food chemical. *Lancet* **344**, 343–344.

[41] Zurlo, J.; Curphey, T. J.; Hiley, R.; Longnecker, D. S. (1982) Identification of 7-carboxymethylguanine in DNA pancreatic acinar cells exposed to azaserine. *Cancer Res.* **42**, 1286–1288.

[42] Shuker, D. E. G.; Howell, J. R.; Street, B. W. (1987) Formation and fate of nucleic acid and protein adducts from *N*-nitroso bile acid conjugates. In: Bartsch, H.; O'Neill, I. K.; Schülte-Hermann, R. (Eds.). *Relevance of* N-*nitroso compounds for human cancer: Exposures and mechanisms, IARC scientific publications No. 84.* International Agency for Research on Cancer, Lyon, p. 187–190.

[43] Harrison, K. L.; Fairhurst, N.; Challis, B. C.; Shuker, D. E. G. (1997) Synthesis, characterisation, and immunochemical detection of O^6-(carboxymethyl)-2′-deoxyguanosine: A DNA adduct formed by nitrosated glycine derivatives. *Chem. Res. Toxicol.* **10**, 652–659.
[44] Shuker, D. E. G.; Margison, G. P.; (1997) Nitrosated glycine derivatives as a potential source of O^6-methylguanine in DNA. *Cancer Res.* **57**, 366–369.
[45] Bingham, S.; Pignatelli, B.; Pollock, J.; Ellul, A.; Mallaveille, C.; Gross, G.; Runswick, S.; Cummings, J. H.; O'Neill, I. K. (1996) Does increased formation of endogenous *N*-nitroso compounds in the human colon explain the association between red meat and colon cancer? *Carcinogenesis* **17**, 515–523.
[46] UK Department of Health (1998) *Nutritional aspects of the development of cancer.* Report on Health and Social Subjects No. 48, Her Majesty's Stationary Office, London.
[47] Saffhill, R.; Badawi, A. F.; Hall, C. N. (1988) Detection of O^6-methylguanine in human DNA. In: Bartsch, H.; Hemminki, K.; O'Neil, I. K. *Methods for detecting DNA damage in humans.* IARC Scientific Publication No. 89, IARC, Lyon, France, p. 301–305.
[48] Cooper, D. P.; Yao, G. F.; Qu, U. H.; O'Connor, P. J. (1991) DNA methylation in individuals at high risk for stomach cancer. *Brit. J. Cancer* **63**, 65–68.
[49] Badawi, A. F.; Mostafa, M. H.; Aboul-Azm, T.; Haboubi, N. Y.; O'Connor, P. J.; Cooper, D. P. (1992) Promutagenic methylation damage in DNA from patients with bladder cancer associated with schistosomiasis and from normal individuals. *Carcinogenesis* **13**, 877–881.
[50] Hall, C. N.; Badawi, A. F.; O'Connor, P. J.; Saffhill, R. (1991) The detection of alkylation damage in the DNA of human gastrointestinal tissues. *Brit. J. Cancer* **64**, 59–63.
[51] Kyrtopoulos, S. A.; Ampatzi, P.; Davaris, P.; Hartiopoulos, N.; Golematis, B. (1990) Studies in gastric carcinogenesis, IV. O^6-Methylguanine and its repair in normal and atrophic biopsy specimens of human gastric mucosa. Correlation of O^6-alkyl-guanine – DNA alkyl transferase activities in gastric mucosa and circulating lymphocytes. *Carcinogenesis* **11**, 431–436.
[52] Kang, H.; Komishi, C.; Kuroki, T.; Huh, N. (1995) Detection of O^6-methylguanine, O^4-methylthymine and O^4-ethylthymine in human liver and peripheral blood leukocyte DNA. *Carcinogenesis* **16**, 1277–1280.

13 Meat, Other Dietary Factors and Intestinal Ammonia and *N*-Nitro Compound Formation in Relation to Colorectal Cancer

Sheila A. Bingham

13.1 Abstract

The amount of nitrogen entering the large bowel, mainly in the form of protein, peptides and amino acids, can be increased by increasing protein intake. In humans, the increase in nitrogen entering the colon, as a result of consuming high meat diets and bacterial activity, increases faecal ammonia concentration. Ammonia is a promoter of carcinogenesis induced by NOC in rodent models. In studies in which high amounts (300 to 600 g per day) of meat have been fed, bran, resistant starch, and vegetables have not reduced faecal ammonia levels.

N-nitroso compounds are also found in the colon and are formed endogenously because the amines and amides produced primarily by bacterial decarboxylation of amino acids can be *N*-nitrosated in the presence of a nitrosating agent. In the anaerobic large bowel, nitrate, entering the body partly in food and water, or from endogenous synthesis via iNOS, is reduced to nitrite in the colon during dissimilatory nitrate metabolism by the colonic flora. Supplements of nitrate have therefore been shown to elevate faecal NOC levels. A number of facultive and anaerobic colonic bacteria are able to catalyse the formation of *N*-nitroso compounds at an optimum pH of 7.5.

We have been investigating the hypothesis that high meat diets increase faecal NOC levels, and that an increase in fermentable carbohydrate reaching the colon could be expected to reduce them. We first studied the effect of red meat consumption on faecal NOC levels in eight male volunteers who consumed diets low or high in meat (60 or 600 g/day) as beef, lamb or pork whilst living in a metabolic suite. Increased intake of red meat induced a significant ($p<0.024$) three-fold increase from 40 (se 7) to an average of 113 (se 25) μg per day of NOC, mainly as acidic and basic nitrosamines. Subsequent studies have confirmed this effect of red meat, and shown that there is a dose response to 0, 60, 240 and 420 g meat per day. Further studies are in progress but so far there are inconclusive differences in faecal NOC concentration in response to red and white meat.

Contrary to expectation that NOC would be reduced when high red meat diets are supplemented with bran, there was no reduction in NOC levels. Later studies have also shown no reduction with resistant starch on faecal NOC levels, nor with vegetables, but similar effects on faecal weight, transit time, and hence contact of NOC with the large bowel mucosa. These effects would suggest an indirect mechanism for any protective effect of plant foods against cancer, but this has not been investigated epidemiologically.

13.2 Introduction

The strong evidence that the majority of large bowel cancers are attributable to environmental factors means that it is a potentially preventable disease. Up to 80 % of bowel cancer in western populations are currently attributed to diet [1].

However, the extent to which diet is capable of causing somatic alterations in genes known to be involved in the causation of bowel cancer, or is able to prevent or mitigate these alterations, is an emerging area of research. For example, the kras mutations most commonly involved in colon cancer are G → A transitions at the second G of a GG pair at codon 12 or 13 of K-*ras* and these are characteristic of alkylating agents such as *N*-nitroso compounds (NOC) [2]. Butyrate, formed during fermentation of carbohydrates in the large bowel, has well known effects on gene expression and DNA repair enzymes.

For the time being, evidence linking diet with colorectal cancer is limited to epidemiological associations with diet with support from experimental studies and plausible hypotheses. A major problem in epidemiological studies of large bowel cancer is the absence of an easily accessible intermediate risk marker, known to alter in response to diet in metabolic studies, that can be used to link dietary intake and the presence of the disease in either intervention or prospective studies. Information being collected from all participants in the European prospective Investigation of Cancer (EPIC) includes not only estimates of diet but also the collection of biological specimens which will be used to link diet and cancer registrations with intermediate markers of risk such as hormonal status, DNA adducts, biomarkers of diet and genotypic risk factors [3, 4].

13.3 Epidemiology of diet and colorectal cancer

Armstrong and Doll attributed much of the international variation in large bowel cancer incidence between countries to dietary differences, especially meat and fat consumption [5]. In a study of individual surveys of food consumption in 12 countries, a positive association with protein and fat ($r = 0.60$ and 0.62 respectively) was confirmed, and weak negative associations with NSP ($r = -0.23$). There were however strong ($r = -0.70$) inverse associations between colorectal cancer incidence and starch intake. These were maintained after partial correlation controlling for meat and fat consumption [6]. International comparisons of colorectal cancer rates in relation to vegetable consumption have not been conducted, but in the UK, regional differences in colorectal cancer mortality are strongly related to consumption of vegetables excluding potatoes ($r = -0.94$) [7].

Results of cohort studies set up several years ago are now beginning to appear in the literature, but the need for accuracy in dietary assessment has only recently been recognised, hence very crude assessments of dietary intake, mainly based on short lists of food, food frequency questionnaires, were used in most of these. These assessments would have been associated with a substantial degree of measurement error, not amenable to correction. For example, if meat is associated with increased risk, lower rates for cancer would be expected in vegetarians. In a recent meta analysis of five cohorts, meat eaters were not at greater risk than non meat eaters, although the standardised mortality ratios for all cohorts was low, and the amount of meat consumed by meat eaters was unable to be established [8].

Various indices of meat consumption have been measured in eight prospective studies [9–16] although few studies have reported risks from total meat consumption in colorectal cancer. Two studies [9, 12] have shown significant evidence of increased trends of colorectal cancer risk with "red" meat consumption and one a significant reduction in risk from white meat [9], and a significant elevated risk for processed meat consumption and large bowel cancer was shown in two studies [9, 13]. Though the majority of studies have wide confidence intervals and therefore non significant results, there is a trend for red and processed meats to elevate colon cancer risks, and some evidence that white meat is associated with either no effect, or reduction in risk.

Seven studies have assessed fibre or fibre containing foods, and two [10, 16] detected a significant inverse protective trend between dietary fibre and colorectal incidence. No study demonstrated an enhanced risk. Five did not detect significant trends, but in one [9] fibre was protective in high fat consumers and in two a non significant inverse trend [12, 16] was shown. The lack of association with trends in bowel cancer in the Male Health Professionals Study [12] occurred despite higher fibre consumption in the lowest risk group. Relative risks for studies which report them tend to be less than 1.00 and range from 0.9 to 0.5 in individuals classified in the upper ends of the distribution of fibre intake. Six prospective studies have investigated vegetable consumption. None have reported increased trends in risk with increased consumption, and three have shown significant protective effects [10, 16, 17]. In one, significant protective effects were shown in women, and a non significant higher risk was shown in men [18]. A non significant higher risk was also shown in the study of Hirayama [14] and no significant findings were shown in the study of Giovannucci *et al.* [12]. Generally risks are in the direction of lower risk with increased consumption and relative risks in the order of 0.9 to 0.5.

13.4 Starch, non starch polysaccharides and fermentation

Carbohydrate entering the large bowel stimulates anaerobic fermentation, leading to the production of short chain fatty acids (SCFA), acetate, propionate, and butyrate, gas, and an increase in microbial cell mass (biomass). The SCFA are absorbed by the intestinal mucosa where they stimulate sodium absorption and bicarbonate production [19]. Sugars such as lactose in lactase deficient individuals, and oligosaccharides such as fructooligosaccharides and inulin are substrates for fermentation, but normally it is the polysaccharides that are quantitatively most important. Of these, only 12 g of NSP ("fibre") is available in western diets, an amount that is insufficient to account for the known amounts of SCFA produced [19]. However, studies in man have shown that a significant amount of starch escapes digestion in the small gut, depending on the physical form of the food eaten, the granule type, and how it is cooked and processed [20]. This starch, resistant starch (RS), reaches the large bowel and is also a substrate for fermentation.

Carbohydrate fermentation has a number of implications for protection against large bowel cancer. Burkit [21] and Higginson and Oettle [22] were amongst the first to attribute differences in colorectal cancer rates to lack of dietary fibre, long transit time and low stool weight. The stimulation of bacterial growth, together with water binding to residual unfermented NSP, leads to an increase in stool weight, dilution of colonic contents and faster transit time through the large gut [19, 21]. Long transit time has not been related to large bowel cancer risk, but there is a strong inverse association between high stool weight and colorectal cancer incidence [23]. Low stool weight leads to constipation, which together with use of cathartics are risk factors for colorectal cancer. Odds ratios are 1.48 (1.32–1.66) and 1.46 (1.33–1.61) respectively, with attributable risks for colon cancer of 4.4 % in the US population [24]. The association between low stool weight and bowel disease and the linear relationship between NSP consumption and stool weight, with a 5 g increase for every 1 g of NSP consumed, is the basis for UK and WHO recommendations for an 18 g population average intake of NSP, a 50 % increase for the UK and most Western populations [25, 26].

During fermentation, approximately 60, 20 and 20 % molar ratios of acetate, propionate and butyrate are formed in the large gut [19]. Molar ratios can however be varied according to the substrate. In *in vitro* batch cultures, 29 % of butyrate can be produced from starch, compared with 2–8 % from NSP sources [27], and in humans molar ratios of butyrate can be made to increase by 50 % by feeding the glucosidase inhibitor acarbose [28]. Starch may therefore be a better source of butyrate than NSP.

Butyrate was suggested as a protective agent in colon cancer in 1981 [29]. In cultured cell lines, it is a well recognised anti-proliferative agent, arresting cell growth in G_1, and inducing differentiation. Histone deacetylase is inhibited and other SCFA are much less active in this respect [30]. Alterations in gene expression occur and chromatin accessibility to DNA repair enzymes is altered [31]. Rodent studies have shown that luminal butyrate levels are inversely associated with colonic cell proliferation, and positively associated with histone acetylation [32]. High starch diets, substrates for increased amounts of butyrate in the lumen, have been shown to reduce proliferative activity in the colon of mice and in humans [33, 34]. Paraskeva and colleagues [35] have suggested that butyrate induces apoptosis which may account for its role in reducing proliferation.

Despite the effects of starch and butyrate in cell lines and in humans *in vivo*, studies using direct acting carcinogens in rodents have not shown a protective effect against carcinogenesis of large amounts of butyrate added to drinking water or food [36, 37]. The effect of starch on chemical carcinogenesis has not been investigated intensively, and there are few reports in the literature as yet. Using potato starch granules as the source of resistant starch, tumorigenesis induced by dimethylhydrazine was increased in rodents, but suppressed when wheat bran was added. The increase in tumorigenicity was not due to increased proliferation since bran did not suppress proliferation [38]. One other study has shown enhancement with high amylose starch, and another reduction with corn starch compared with a high sucrose diet [39, 40]. A large number of studies have investigated the effect of purified sources of dietary fibre on chemically initiated colorectal cancer in animals. A collation of overall findings of experimental studies by the Federation of American Societies for Experimental Biology [41] showed that bran appeared to have a consistently protective effect against chemical carcinogenesis. However, "soluble fibres" are associated with tumor enhancement [42].

Fermentation may be important in prevention of colorectal cancer via other effects as well [29]. The production of SCFA reduces luminal pH, and bacterial 7α-dehydroxylase activity, and hence conversion of primary to secondary bile acids deoxycholic and lithocholic acids, is also inhibited [34]. Production of phosphatidylcholine diacylglycerol (DAG), one of two intracellular messengers formed from phosphatidyl inositol, is enhanced in human fermentation systems by the presence of deoxycholic acid [43]. DAG increases the affinity of protein kinase C for calcium and renders it active at physiological levels of this ion, phosphorylating serine and threonine residues in many target organs. Phorbol esters are well known promotors because they resemble DAG but are not degraded. Increased levels of PKC have been reported in colonic tumor tissue [44]. Total faecal DAG levels have been shown to be reduced by a supplement of 15 g wheat bran in women [45].

13.5 Meat and nitrogen metabolism in the colon

The association between meat consumption and colorectal cancer is usually attributed to the formation of heterocyclic amines in meat when it is cooked. However, the possibility that meat alters nitrogen metabolism and enhances the production of endogenous promotors and carcinogens within the colon is attracting increasing attention.

The amount of nitrogen entering the large bowel, mainly in the form of protein, peptides and amino acids, can be increased by increasing protein intake [46]. There are many different types of proteolytic bacteria found in the large gut which, depending on pH and substrate availability, may respond to active carbohydrate fermentation in the right colon, or to protein released from bacterial cell lysis in the left colon when readily fermented carbohydrates, such as pectin, are exhausted [47]. Some versatile bacteria deaminate to form ammonia, SCFA, and a variety of other products including phenols and branched chain fatty acids [47]. When carbohydrate fermentation is active, ammonia is assimilated into glutamine or glutamate and the amino group is distributed to other amino acids as required [47].

In humans, the increase in nitrogen entering the colon as a result of consuming high meat diets increases faecal ammonia concentration. Ammonia is a promoter of carcinogenesis in rodent models [48] and patients with uterosigmoidostomies who have very high luminal ammonia concentrations have a greatly increased risk of developing tumors distal to the site of ureteric implantation [49]. In *in vitro* fermentation studies, and in some *in vivo* studies, carbohydrate in the form of starch and non starch polysaccharides reportedly reduces faecal ammonia concentration [47]. However, in studies in which high amounts (300 to 600 g per day) of meat have been fed, bran, resistant starch, and vegetables have not reduced faecal ammonia levels [50–53].

N-nitroso compounds are also found in the colon and are formed endogenously because the amines and amides produced primarily by bacterial decarboxylation of amino acids can be *N*-nitrosated in the presence of a nitrosating agent. Chemical *N*-nitrosation may occur under neutral or alkaline conditions (as in the small and large intestine). In the anaerobic large bowel, nitrate, entering the body partly in food and water, is reduced to nitrite in the colon during dissimilatory nitrate metabolism by the colonic flora [54]. Supplements of nitrate have therefore been shown to elevate faecal NOC levels [55]. A number of facultive and anaerobic colonic bacteria are able to catalyse the formation of *N*-nitroso compounds at an optimum pH of 7.5 [56].

There is another significant source of endogenous nitrate production, which has been deduced for some time, since nitrate excretion exceeds that consumed in food and water [57]. Wagner *et al.* [58] showed that nitrate synthesis is enhanced during immunostimulation and Stuehr and Marletta [59] showed that nitrite and nitrate are produced from macrophages. Studies with N15 established that the source is dietary arginine, used to produce NO, which together with superoxide causes oxidative injury and cell death [60]. Other fields of research established that NO accounted for the biological activity of endothelium-derived relaxing factor and inducible nitric oxide synthase produces continuous amounts of NO from arginine [61]. Increased arginine, from protein, might be expected to increase urine nitrate excretion, an effect which has been shown in animals [62, 63]. Thus, NO from stimulated macrophages in the large bowel mucosa, together with nitrite produced from reduced nitrate diffusing into the gut, are therefore available for *N*-nitroso compound formation.

In the 1980s it was found that faecal samples contain negligible amounts of volatile *N*-nitroso compounds [64], but since that time newer methods to measure total NOC by chemiluminescence have been developed and the presence of NOC in faeces in animals and man is now well established [51, 52, 55, 65, 66].

In humans, we have been investigating the hypothesis that high meat diets increase faecal NOC levels, and that an increase in fermentable carbohydrate reaching the colon could be expected to reduce them [67]. All of these studies are carried out in a metabolic suite, where diet can be carefully controlled and all specimens collected over prolonged periods. All diets are isoenergetic and contain equal amounts of fat, and are matched to each individual's energy expenditure. We first studied the effect of red meat consumption on faecal NOC levels in eight male volunteers who consumed diets low or high in meat (60 or 600 g/day) as beef, lamb or pork whilst living in a metabolic suite. Increased intake of red meat induced a significant ($p<0.024$) three-fold increase from 40 (se 7) to an average of 113 (se 25) µg per day of NOC, mainly as acidic and basic nitrosamines [51]. Subsequent studies have confirmed this effect of red meat, and shown that there is a dose response to 0, 60, 240 and 420 g meat per day [52, 66].

In two volunteers, there was no effect of 600 g white meat and fish on faecal NOC; mean low white meat diet, 68 (se 10) µg per day, high white meat diet 56 (se 6) µg per day. This suggests that the increase in faecal NOC and nitrosating products is brought about by a specific effect of red meat not seen with white meat. A major difference between red and white meat is in their content of iron, which is poorly absorbed from the small intestine. This may be due to the fact that iron and molybdenum are integral components of nitrate reductase and are essential for enzyme activity [68]. Faecal nitrate reductase is a key step in determining the levels of production of nitrosating agents such as nitrite from nitrate [56]. Further studies are in progress but so far there are inconclusive differences in faecal NOC concentration in response to red and white meat [69].

In *in vitro* systems, the presence of starch alters nitrogen metabolism so that more is incorporated into bacterial cell walls, when faecal ammonia is reduced [47]. In humans, increased bran also reduces faecal ammonia [29]. Contrary to expectation that NOC would also be reduced when high red meat diets are supplemented with 20 g phytate-free wheat bran in six volunteers, there was no reduction in NOC levels. However, faecal weight increased and hence the contents of the lumen were diluted. Transit time is inversely related to faecal weight. The net result would have been less contact between NOC arising from a high red meat diet and the colonic mucosa with the high bran diet. Later studies have also shown no reduction with resistant starch on faecal NOC levels, nor with vegetables, but similar effects on faecal weight, transit time, and hence contact of NOC with the large bowel mucosa [52, 69]. The lack of effect on faecal ammonia levels has been noted above. These effects would suggest an indirect mechanism for any protective effect of plant foods against cancer, but this has not been investigated epidemiologically.

13.6 Conclusion

Despite interesting possibilities, a direct link between the epidemiology of most dietary factors, intermediate risk markers and the end points of cancer in humans has yet to be established. One preliminary study has shown an increase in relative risks for mutations in the kras gene and meat consumption in colon cancer cases [70]. It is likely that more studies of this type, linking somatic mutations, diet and cancer will be published, allowing a much stronger case to be made for or against meat being related to cancer. Meanwhile, there are public health recommendations to increase vegetable and NSP consumption in order to increase stool weight and reduce the risk of large bowel cancer. There are also recommendations, based on statistical concepts, to reduce meat intake [71, 72]. However, the type, amount, processing, cooking, dose responses of meat or protein increasing risk of cancer are uncertain.

References

[1] Willett, W. C. (1995) Diet, nutrition and avoidable cancer. *Environ. Health Perspect.* **103** (Suppl. 8), 165–170.

[2] Bos, J. L. (1989) *ras* Oncogenes in human cancer: a review. *Cancer Res.* **49**, 4682–4689.

[3] Riboli, E. (1992) Background and rational of EPIC. *Ann. Oncol.* **3**, 783–791.

[4] Day, N.; Oakes, S.; Luben, R.; Khaw, K-T.; Bingham, S.; Welch, A.; Wareham, N. (1998) EPIC in Norfolk: Study design and characteristics of the cohort. *Br. J. Cancer*, in press.

[5] Armstrong B.; Doll, R. (1975) Environmental factors and cancer incidence in different countries *Int. J. Cancer* **15**, 617–631.

[6] Cassidy, A.; Bingham, S.; Cummings, J. H. (1994) Starch Intake and Colorectal Cancer Risk: An international comparison. *Br. J. Cancer* **69**, 937–942.

[7] Bingham, S.; Williams, D. R. R.; Cole, T. J.; James, W. P. T. (1979) Dietary fibre and regional large bowel cancer mortality. *Br. J. Cancer* **40**, 456–463.

[8] Key, T. J.; Fraser, G. E.; Thorogood, M.; Appleby, P. N.; Beral, V.; Reeves, G.; Burr, M. L.; Chang-Claud, J.; Frentzel-Beyme, R.; Kuzma, J. W.; Mann, J.; McPherson, K. (1998) Mortality in vegetarians and non-vegetarians: a collaborative analysis of 8300 deaths among 76,000 men and women in five prospective studies. *Public Health Nutr.* **1**, 33–4114.

[9] Willett, W. C.; Stampher, M. J.; Colditz, G. A.; Rosner, B. A.; Spiezer, F. E. (1990) Relation of meat, fat and fibre intake to risk of colon cancer in a prospective study among women. *New Engl. J. Med.* **323**, 1664–1672.

[10] Thun, M.; Calle, E. E.; Namboodiri, M. M.; Flanders, W. D.; Coates, R. J.; Byers, T.; Boffeta, P.; Garfinkel, L.; Heath, C. W. (1992) Risk factors in fatal colon cancer in a large prospective study. *J. Natl. Cancer Inst.* **84**, 1491–1500.

[11] Bostick, R. M.; Potter, J. D.; Kushi, L. H.; Sellers, T. A.; Steinmetz, K. A.; McKenzie, D. R.; Gapstyr, S. M.; Folsom, A. R. (1994) Sugar, meat, and fat intake and non-dietary risk factors for colon cancer incidence in Iowa women. *Cancer Causes Control* **5**, 38–52.

[12] Giovannucci, E.; Rimm, E.; Stampfer, M. J .; Colditz, G. A.; Ascherio, A.; Willett, W. C. (1994) Intake of fat, meat, and fibre in relation to risk of colon cancer in men. *Cancer Res.* **54**, 2390–2397.

[13] Goldbohm, R. A.; Van Den Brandt, P.; Van't Veer, P.; Brants, H. A. M.; Dorant, E.; Sturmans, F.; Hermus, R. J. J. (1994) A prospective study on the relation between meat consumption and the risk of colon cancer. *Cancer Res.* **54**, 718–723.

[14] Hirayama, T. (1981) A large scale cohort study on the relationship between diet and selected cancer of digestive organs. In: Bruce, W. R. *et al.* (Eds.) *Banbury Report 7*, Cold Spring Harbour Laboratory, USA, p. 409–429.

[15] Phillips, R. L.; Snowdon, D. A. (1985) Dietary relationships with fatal colorectal cancer among SDA. *J. Natl. Cancer Inst.* 74, 307–317.

[16] Heilbrun, L. K.; Nomura, A.; Hankin, J.; Stemmerman, G. N. (1989) Diet and colorectal cancer with special reference to fiber intake. *Int. J. Cancer* **44**, 1–6.

[17] Morgan, J. W.; Frazer, G. F.; Phillips, R. L.; Andress, M. H. (1988) Dietary factors and colon cancer incidence among seventh day adventists. *Am. J. Epidemiol.* **128**, 918(A).

[18] Shibata, A.; Paganini-Hill, R. K.; Henderson, R.; Henderson, B. E. (1992) Intake of vegetables, fruits, β-carotene, vitamin C and supplements and cancer among the elderly: a prospective study. *Br. J. Cancer* **66**, 673–679.

[19] Cummings, J. H. (1981) Short chain fatty acids in the human colon. *Gut* **22**, 763–779.

[20] Englyst, H. N.; Kingman, S. M.; Cummings, J. H. (1992). Classification and measurement of nutritionally important starch fractions. *Eur. J. Clin. Nutr.* **46**, S33–S50.

[21] Burkitt, D. P. (1969) Related disease – related cause? *Lancet* **2**, 1229–1231.

[22] Higginson, J.; Oettle, A. G. (1960) Cancer incidence in the Bantu and Cape coloured race of South Africa. *J. Natl. Cancer Inst.* **24**, 584–671.

[23] Cummings, J. H.; Bingham, S. A.; Heaton, K. W.; Eastwood, M. A. (1992) Faecal weight, colon cancer and dietary intake of NSP (dietary fibre). *Gastroenterology* **103**, 1783–1789.

[24] Sonnenberg, A.; Muller, A. (1993) Constipation and cathartics as risk factors in colorectal cancer. *Pharmacology* **47**, 224–233.

[25] Department of Health (1990) *Dietary reference values for the UK Rep. Health Soc. Subj. 41.* HMSO, London.

[26] WHO (1990) Diet, nutrition and chronic disease. *Technical Report Series* 797, WHO, Geneva.

[27] Englyst, H. N.; Hay, S.; Macfarlane, G. T. (1987). Polysaccharide breakdown by mixed populations of human faecal bacteria. *FEMS Microbiol. Ecol.* **96**, 163–171.

[28] Scheppach, W.; Fabian, C.; Sachs, M.; Kasper, H. (1988) Effect of starch malabsorption on faecal SCFA excretion in man. *Sci. J. Gastroenterol.* **23**, 755–759.

[29] Cummings, J. H.; Stephen, A. M.; Branch, W. J. (1981) Implications of dietary fibre breakdown in the human colon. In: Bruce, W. R.; Correa, P.; Lipkin, M.; Tannenbaum, S.; Wilkins, T. D. (Eds.) *Banbury Report 7, Gastrointestinal Cancer.* Cold Spring Harbor Laboratory, p. 71–81.

[30] Kruh, J.; Defer, N.; Tichonicky L. (1994) Effects of butyrate on cell proliferation and gene expression Chap 18. In: Cummings, J. H. *et al.* (Eds.) *Physiological and clinical aspects of SCFA.* Cambridge University Press.

[31] Smith, P. J. (1986) Butyrate alters chromatin accessibility to DNA repair enzyme. *Carcinogenesis* **7**, 423–429.

[32] Boffa, L. C.; Luption, J. R.; Mariani, M. R.; Ceppi, M.; Newmark, H.; Scalmati, A.; Lipkin, M. (1992) Modulation of colonic cell proliferation, histone acetylation and luminal short chain fatty acids by variation of dietary fibre (wheat bran) in rats. *Cancer Res.* **52**, 5906–5912.

[33] Caderni, G.; Bianchini, F.; Dolora, P.; Kreibel, D. (1989) Proliferative activity in the colon of the mouse and its modulation by dietary starch, fat, and cellulose. *Cancer Res.* **49**, 1655–1659.

[34] Van Munster, I. P.; Tangerman, A.; Nagengast, F. M. (1994) The effect of resistant starch on colonic fermentation, bile acid metabolism, and mucosal proliferation. *Dig. Dis. Sci.* **39**, 834–842.

[35] Hague, A.; Manning, A. M.; Hanlon, K. A.; Hueschtchav, L.; Hart, D.; Paraskeva, C. (1993) Sodium butyrate induces apoptosis in human colonic tumor cell lines. *Int. J. Cancer* **55**, 498–505.

[36] Freeman, H. J. (1986) Effects of differing concentrations of sodium butyrate in DMH induced rat intestinal neoplasia. *Gastroenterology* **91**, 596–602.

[37] Deschner, E. E.; Ruperto, J. F.; Lupton, J. R.; Newmark, H. L. (1990) Butyrate does not enhance AOM induced colon tumorigenesis. *Cancer Lett.* **52**, 79–8.

[38] Young, G. P.; McIntyre, A.; Albert, V.; Folino, M.; Muir, J.; Gibson, P. R. (1996) Wheat bran suppresses potato starch-potentiated colorectal tumorigenesis at the aberrrant-crypt stage in a rat model. *Gastroenterology* **110**, 508–514.

[39] Sakamoto, J.; Nakaji, S.; Sugawara, K.; Iwane, S.; Munakata, A. (1996) Comparison of resistant starch with cellulose diet on 1,2-dimethylhydrazine-induced colonic carcinogenesis in rats. *Gastroenterology* **110**, 116–120.

[40] Luceri, C.; Caderni, G.; Lancioni, L.; Aiolli, S.; Dolara, P.; Mastrandea, V.; Scardazzq, F.; Morozzi, G. (1996) Effects of repeated boluses of sucrose on proliferation and on AOM-induced aberrrant crypt foci in rat colon. *Nutr. Cancer* **25**, 187–196.

[41] Pilch, S. (Ed.) (1987) *Physiological effects and health consequences of dietary fibre*. Bethesda, Maryland, FASEB, USA.

[42] Jacobs, L. R. (1990) Influence of soluble fibres on experimental colon carcinogenesis. In: Kritchevsky, D.; Bonfield, C.; Anderson, J. W. (Eds.) *Dietary Fibre*. Plenum, New York, p. 389–420.

[43] Morotomi, M.; Giullem, J. G.; Logerfo, P.; Weistein, B. (1990) Production of DAG by human intestinal microflora. *Cancer Res.* **50**, 3595–3599.

[44] Guillem, J. G.; Obrian, C. A.; Fitzer, C. J.; Johnson, M. D.; Forde, K. A.; Logerfo, P.; Weinstein, B. (1987) Studies on PKC and colon carcinogenesis. *Arch. Surg.* **122**, 1475–1478.

[45] Reddy, B. S.; Simi, B.; Engle, A. (1994) Effect of types of fibre on colonic DAG in women. *Gastroenterology* **106**, 883–889.

[46] Silvester, K. R.; Cummings, J. H. (1995) Does digestibility of meat protein help to explain large bowel cancer risk? *Nutr. Cancer* **24**, 279–288.

[47] Macfarlane, G.; Cummings, J. H. (1991) The colonic flora, fermentation, and large bowel digestive function. In: Phillips, S.; Pemberton, J. H, Shorter, R. G (Eds.) *The large intestine: physiology, pathophysiology and disease*. Raven, New York, p. 51–92.

[48] Clinton, S. K.; Bostwick, D. G.; Olson, L. M.; Mangian, H. J.; Visek, W. J. (1988) Effects of ammonium acetate and sodium cholate on *N*-methyl-*N'*-nitro-*N*-nitrosoguanidine-induced colon carcinogenesis of rats. *Cancer Res.* **48**, 3035–3039.

[49] Tank, E. S.; Krausch, D. N.; Lapides, J. (1973) Adenocarcinoma of the colon associated with ureterosigmoidoscopy. *Dis. Colon Rectum* **16**, 300–304.

[50] Cummings, J. H.; Hill, M. J.; Bone, E. S.; Branch, W. J.; Jenkins, D. J. A. (1979) The effect of meat protein and dietary fiber on colonic function and metabolism. II. Bacterial metabolites in faeces and urine. *Am. J. Clin. Nutr.* **32**, 2094–2101.

[51] Bingham, S.; Pignatelli, B.; Pollock, J.; Ellul, A.; Mallaveille, C.; Gross, G.; Runswick, S.; Cummings, J. H.; O'Neill, I. K. (1996) Does increased formation of endogenous *N*-nitroso compounds in the human colon explain the association between red meat and colon cancer? *Carcinogenesis* **17**, 515–523.

[52] Silvester, K. R.; Bingham, S. A.; Cummings, J. H.; O'Neill, I. K. (1997) The effect of meat and resistant starch on faecal excretion of *N*-nitroso compounds from the human large bowel *Nutr. Cancer* **29**, 13–23.

[53] Murphy, C. (1998) The effects of red meat, vegetables and tea on faecal nitrogen and ammonia excretion in healthy male volunteers. *University of Coleraine Undergraduate Report*, p. 1–36.

[54] Allison, C.; Macfarlane, G. T. (1988) Effect of nitrate on methane production and fermentation by slurries of human faecal bacteria. *J. Gen. Microbiol.* **134**, 1397–1405.

[55] Rowland, I. R.; Granli, T.; Bockman, O. C.; Key, P. E.; Massey, R. C. (1991) Endogenous *N*-nitrosation in man assessed by measurement of apparent total *N*-nitroso compounds in faeces. *Carcinogenesis* **12**, 1359–1401.

[56] Calmels, S.; OShima, H.; Vincent, P.; Gounot, A.; Bartsch, H. (1985) Screening of micro-organisms for nitrosation catalysis at pH 7. *Carcinogenesis* **6**, 911–915.

[57] Witter, J. P.; Balish, E.; Gatley, S. J. (1979) Origin of excess urinary nitrate in the rat. *Cancer Res.h* **42**, 3645–3648.

[58] Wagner, D. A.; Young, V. R.; Tannerbaum, S. R. (1983) Mammalian nitrate synthesis. *Proc. Natl. Acad. Sci.* **80**, 4518–4521.

[59] Stuehr, D. J.; Marletta, M. A. (1985) Mammalian nitrate synthesis: mouse macrophages produce nitrate and nitrite in response to *E. coli* lipopolysaccharide. *Proc. Natl. Acad. Sci.* **82**, 7738–7742.

[60] Iyengar, R.; Stuehr, D. J.; Marletta, M. A. (1987) Macrophage synthesis of nitrate, nitrite, and *N*-nitrosamines, precursors and the role of the respiratory burst. *Proc. Natl. Acad. Sci.* **84**, 6369–6373.

[61] Anggard, E. (1994) Nitric oxide: mediator, murderer and medicine. *Lancet* **343**, 1199–1206.

[62] Mallett, A. K.; Walters, D. G.; Rowland, I. R. (1988) Protein related differences in the excretion of nitrosoproline and nitrate by the rat. *Food Chem. Toxicol.* **26**, 831–835.

[63] Ward, J. M.; Aanjo, T.; Ohannesian, L.; Keefer, L. K.; Devor, D. E.; *et al.* (1988) Inactivity of fecapentaene 12 as a rodent carcinogen or tumor initiator. *Cancer Lett.* **42**, 49–59.

[64] Archer, M. C.; Saul, R. L.; Lyang-Ja, L.; Bruce, W. B. (1981) Analysis of nitrate, nitrite and nitrosamines in human faeces. In: Bruce, W. B. *et al.* (Eds.) *Banbury Report 7*, Cold Spring Harbor, NY, p. 321–7.

[65] Massey, R.; Key, P.; Mallett, A.; Rowland, I. (1988) An investigation of the endogenous formation of ATNC in conventional and germ free rats. *Food Chem. Toxicol.* **26**, 595–600.

[66] Hughes, R.; O'Neill, I. K.; Pollock, J. R. A.; Cummings, J. H.; Bingham, S. (1997) Dose response effect with dietary meat on faecal *N*-nitroso compound levels in humans. *Gastrointest. Oncol.* **112**, A581.

[67] Bingham, S. (1988) Meat, starch and NSP and large bowel cancer. *Am. J. Clin. Nutr.* **48**, 762–767.

[68] Stouthammer, A. H. (1976) Biochemistry and genetics of nitrate reductase in bacteria. *Adv. Microb. Physiol.* **14**, 315–375.

[69] Hughes, R. (1998) *The effects of diet on colonic* N-*nitrosation and biomarkers of DNA damage*, PhD Thesis, Cambridge.

[70] Freedman, A. N.; Michalek, A. M.; Muro, K.; Mettlin, C. J.; Brooks, J. S.; Petrelli, N. J.; Caporaso, N. E.; Hamilton, S. R. (1997) Meat consumption is associated K-*ras* mutations in tumors of the distal colorectum. *Proc. Am. Ass. Cancer Res.* **38**, 457A.

[71] COMA (1998) *Nutritional aspects of the development of cancer.* Department of Health Report on Health and Social Subjects, London.

[72] WCRF (1997) *Food, nutrition and the prevention of cancer.* American Institute for Cancer Research, Washington DC.

C Carcinogenic Factors: Endogenous

14 DNA Damage by Nitrogen and Oxygen Free Radicals and its Modulation by Dietary Constituents

Steven R. Tannenbaum

14.1 Abstract

The process of carcinogenesis is currently viewed as a series of steps involving the inactivation of tumor suppressor genes and the activation of oncogenes. Both types of genetic change involve damage to DNA in the form of either mutations or deletions. In addition, the sequence of steps for transformation of a somatic cell requires alternation of DNA damage and cell division, resulting in expansion of the new target cell population.

The overall hypothesis to be presented in this lecture is that DNA damage, mutation, and cytotoxicity will arise as a result of nitrosative deamination, $NO^{\bullet}$ radical reactions, and oxygen radical damage when target cells are exposed to generator cells that produce $NO^{\bullet}$. Depending upon the dose rate, total dose, types of cells, and other circumstances, $NO^{\bullet}$ may drive cells into apoptosis through multiple pathways, or inhibit apoptosis and enhance mutation through damage to bases, strand breaks, and cross-links.

14.2 Chemistry of nitric oxide

The chemistry of nitric oxide in oxygenated biological systems is extremely complex due to the large number of chemical species formed and the numerous parallel reactions to consider. The first pathway involves direct reaction of $NO^{\bullet}$ with cellular targets after simple diffusion of $NO^{\bullet}$.

Nitric oxide also reacts to form additional reactive species that can participate in other types of chemistry. One major fate of nitric oxide is reaction with superoxide anion ($O_2^{-\bullet}$), to yield peroxynitrite, $ONOO^-$. This is an extremely fast reaction due to the fact that both species are radicals. The rate of the nitric oxide/superoxide reaction is near the diffusion limit with a rate constant of $6.7 \times 10^9\ M^{-1}s^{-1}$. This rate constant is approximately 3.5-times larger than that for the superoxide dismutase (SOD)-catalyzed decomposition of $O_2^{-\bullet}$ indicating that the nitric oxide/superoxide reaction may predominate over the superoxide/SOD reaction. The formation of both nitric oxide and superoxide does indeed occur simultaneously in cells such as macrophages, neutrophils, Kupffer cells, and endothelial cells. In the vicinity of these cells, peroxynitrite may be present at high concentrations, although the mechanism and extent of $ONOO^-$ formation are strongly influenced by the relative fluxes of $O_2^{-\bullet}$ and $NO^{\bullet}$.

14.3 Nitric oxide-induced DNA damage via the N_2O_3 pathway

The formation of N_2O_3 can cause either direct or indirect damage. Direct damage results from the nitrosation of primary amines on DNA bases ultimately leading to deamination. Nitrosative deamination is also a well known consequence of the reaction of primary amines with acidic nitrite. In fact, the chemistry of N_2O_3 formation is nearly identical to that from nitrous acid, i. e. nitrite at an acidic pH, because it is the anhydride of HNO_2. The overall reaction rates are actually higher based upon N_2O_3 at neutral or basic pH than at acidic pH because of the increased concentration of free amine under these conditions [1]. There are a number of other $NO^{\bullet}$-derived nitrosating agents that may be important under other conditions, however at physiological pH, N_2O_3 formation from nitric oxide has been demonstrated to be most important [2]. Direct attack of N_2O_3 on DNA can lead to DNA deamination via diazonium ion formation [1]. Hydrolysis of the diazonium ion completes the deamination. The end result of this process is the net replacement of an amino group by a hydroxyl group. Any DNA base containing an exocyclic amino group can undergo deamination upon reaction with N_2O_3. Therefore, adenine, cytosine, 5-methylcytosine, and guanine can all be deaminated forming hypoxanthine, uracil, thymine and xanthine respectively. The potential consequences of these reactions vary from nucleoside to nucleoside. Deamination of guanine leads to xanthine formation. Mispairing of xanthine can cause a G:C → A:T transition. Xanthine is unstable in DNA and can depurinate readily leaving an abasic site. The cell may replicate past the abasic site following the "A" rule involving insertion of an adenine opposite the abasic site resulting in a G:C → T:A transversion mutation [3]. More likely, however, the abasic site may be cleaved by endonucleases resulting in the formation of single-strand breaks [4]. Deamination of cytosine forms uracil which can give rise to a G:C → A:T transaction mutation through mispairing. Given the high levels of uracil glycosylase in a cell, uracil can be easily repaired [5]. Methylation of cytosine to form 5-methylcytosine and subsequent deamination to thymine could also result in a G:C → A:T transition. In addition, the deamination of adenine to hypoxanthine needs to be considered because mispairing of hypoxanthine with cytosine can lead to a A:T → G:C transition.

Caulfield *et al.* [6], have studied the relative rates of $NO^{\bullet}$-induced deamination of dG and dC in different environments in order to assess the importance of DNA structure in determining the reactivity of the different bases towards nitric oxide. Relative reactivity varied between deoxynucleosides, single and double stranded DNA. In all cases xanthine formation was twice that of uracil formation.

Recent evidence indicated that the reaction of deoxyguanidine with nitric oxide might actually be more complex than previously expected. Multiple products besides xanthine may be formed [7]. Oxanosine, the ribonucleoside of this compound, had previously been isolated as a novel antibiotic in 1981 from a bacterial culture. The compound 2′-deoxyoxanosine was synthesized from oxanosine and exhibited a stronger antineoplastic activity than oxanosine. Whether 2′-deoxyoxanosine is actually formed from nitric oxide is somewhat questionable because in the studies carried out, the pH dropped to 2.9 and relatively high concentrations of nitrite were employed.

$NO^{\bullet}$-treatment can also lead to the formation of single strand breaks in DNA. The formation of $NO^{\bullet}$-induced strand breaks has been examined both *in vitro* and *in vivo*. Nguyen *et al.* demonstrated the $NO^{\bullet}$ causes both dose- and time-dependent DNA single-strand breaks in TK6 cells [8]. However, when supercoiled plasmid DNA was treated extracellularly with $NO^{\bullet}$ no nicking of DNA was observed [9]. Surprisingly, when the same plasmid was transfected into CHO cells after treatment it was found to contain strand breaks. The most likely explanation for these results is that $NO^{\bullet}$-treatment can indirectly lead to single strand breaks via formation of xanthine which can depurinate leaving an abasic site. Intracellularly, these abasic sites are recognized by endonucleases which readily cleave them leading to formation of single strand breads. This hypothesis is supported by time course experiments measuring formation of abasic sites *vs* strand breaks [9]. CHO cells were treated with $NO^{\bullet}$ (85 nmol/ml min) for one hour. Directly after treatment only abasic sites were detected. Abasic sites were reduced to background 12 h later but a high percentage of single-strand breaks were found suggesting that abasic sites may be cleaved to form strand breaks. DNA double-strand breaks were detected 24 h later indicating that unrepaired single-strand breaks may be converted to double-strand breaks with may be toxic to a cell.

DNA intrastrand, DNA interstrand, and DNA-protein cross-link formation have been demonstrated upon treatment of DNA with nitrous acid. This raises the possibility that these products may be formed in $NO^•$-treated DNA via the intermediacy of N_2O_3. We have confirmed this possibility by quantitative analysis of the cross-links.

14.4 DNA damage from peroxynitrite

In contrast to nitric oxide, which is involved primarily in the deamination chemistry of DNA, most of the damage inflicted on DNA by peroxynitrite is oxidative in nature. DNA-treatment with peroxynitrite generally leads to much more damage than treatment with an equivalent dose of nitric oxide. In addition to the higher levels of damage present in DNA after $ONOO^-$-treatment, the spectrum of damage also tends to be much more complex. All of this makes sense given that peroxynitrite is intrinsically much more reactive than nitric oxide.

Evidence that peroxynitrite is formed *in vivo* is steadily accumulating. As is the case with nitric oxide, peroxynitrite's transient nature required that its production in a cell be monitored indirectly. Beckman and co-workers were among the first to examine the formation of peroxynitrite from activated macrophages. They detected significant formation of peroxynitrite utilizing the nitration of 4-hydroxyphenylacetate as a marker of peroxynitrite activity. Lewis *et al.* [10] performed a kinetic analysis on the fate of $NO^•$ synthesized by activated macrophages using endproduct measurements of nitrite and nitrate. Hydrolysis of N_2O_3 forms nitrite while $ONOO^-$ decay leads to nitrate formation. Their results indicate that approximately half of the released nitric oxide forms N_2O_3 while the remainder combines with O_2 to form $ONOO^-$. Interestingly, it appears from the Lewis study that $ONOO^-$ formation occurs partially extracellularly and not exclusively inside the macrophage. This can be inferred from the observation that adding SOD to the media significantly reduced $ONOO^-$ formation. If peroxynitrite formed only within the cell it should not have been affected by the extracellular addition of SOD. deRojas-Walker *et al.* [11] identified DNA deamination (xanthine) and oxidation (8-oxo-dG, FAPY-G, and 5-hydroxymethyl-uracil) base products in activated macrophage DNA. The formation of these products was inhibited by NG-methyl-L-arginine, a nitric oxide synthase inhibitor. Since both deamination and oxidation products were observed, their results demonstrate that effects of $NO^•$ produced by a macrophage must be mediated not just by N_2O_3 but also by $ONOO^-$. Immunohistochemical techniques using antibodies against 3-nitrotyrosine have also been used for $ONOO^-$ detection in animal models. For example, SJL mice, especially those bearing the RcsX tumor, suffer from chronic inflammation with activation of macrophages in the spleen and lymph nodes. Nitrotyrosine staining performed on SJL mice tissue sections showed positive results in cells adjacent to *i*NOS-expressing macrophages and also within the macrophages themselves.

The two main types of chemistries attributed to $ONOO^-$ are oxidations and nitrations. Therefore it is not very surprising that the two main products that have been identified so far from the reaction of dG with $ONOO^-$ are 8-oxo-dG and 8-nitro-dG. 8-oxo-dG has long been considered an attractive biomarker for monitoring DNA damage in studies with various oxidizing agents. The role of 8-oxo-dG in mutagenesis and carcinogenesis has been widely investigated and several studies have shown a correlation between the formation of 8-oxo-dG and carcinogenesis. The identification of 8-nitro-dG is especially significant because it provides an analytical tool for the measurement of $ONOO^-$ specific DNA damage. Unlike 8-oxo-dG which is known to cause $G:C \rightarrow A:T$ transitions no mutagenicity data are available as yet of 8-nitro-dG. Recent work on the further reactions of 8-oxo-dG with peroxynitrite will be reported.

14.5 NO-induced mutagenesis

The discovery that phagocytic monocytes and neutrophils can produce NO in addition to reactive oxygen species has led to recent studies of DNA damage and mutations induced by NO and its reactive metabolites under a variety of experimental conditions. Initial investigations involved exposure of target DNA to NO gas, with the following consequences. Exposure of *S. typhimurium* induced C to T transition mutations, while treatment of nucleosides produced several deamination products [12]. Treatment of TK6 human lymphoblastoid cells induced *HPRT* mutations as well as DNA deamination (xanthine, hypoxanthine) and strand breaks [8]. NO gas mutagenized the plasmid pSP189, causing primarily A:T to G:C and G:C to A:T mutations in the *supF* gene [13]. Treatment of Chinese hamster ovary (CHO) cells with NO gas or with peroxynitrite was followed by formation of AP sites and single-strand DNA breaks [9]. NO donor drugs have also been used to treat DNA in a variety of experimental models. Treatment of pSP189 with DEA/NO or spermine/NO induced predominantly G:C to A:T mutations [14]. Treatment of bronchial epithelial cells with NO donor drugs failed to induce detectable levels of mutation in the *HPRT* or p53 genes [15]. Treatment of pSP189 with peroxynitrite induced strand breaks and *supF* mutations, predominantly G:C to T:A and G:C to A:T [16].

Study of the DNA damage and mutagenicity of macrophage-derived NO *in vivo* is essential to assess its potential significance as a genotoxic hazard. In response to this need, we have developed the SJL-*lac*Z transgenic animal model, and have produced preliminary evidence of mutagenicity associated with NO overproduction. Similar results have been achieved for mutations in the endogenous HPRT gene in RAW 264.7 macrophages. The potential for DNA damage and mutation by NO and its products is indisputable at the present time.

References

[1] Tannenbaum, S. R.; *et al.* (1994) In: *Nitrosamines and related* N-*nitroso compounds.* American Chemical Society, Washington, DC, p. 120–135.
[2] Lewis, R. S.; *et al.* (1995) *J. Am. Chem. Soc.* **117**, 3933–3939.
[3] Loeb, L. A.; Preston, B. D (1986) *Annu. Rev. Genet.* **20**, 201–230.
[4] Lindahl, T.; Anderson, A. (1972) *Biochem.* **11**, 3618–3623.
[5] Domena, J. D.; *et al.* (1988) *Biochem.* **27**, 6742–6751.
[6] Caulfield, J.; *et al.* (1998) *J. Biol. Chem.* **273**, 12 689–12 695.
[7] Suzuki, T. R.; *et al.* (1996) *J. Am. Chem. Soc.* **118**, 25.15–25.16.
[8] Nguyen, T.; *et al.* (1992) *Proc. Natl. Acad. Sci. USA* **89**, 3030–3024.
[9] Tamir, S.; *et al.* (1996) *Chem. Res. Toxicol.* **9**, 821–827
[10] Lewis, R.; *et al.* (1995) *J. Biol. Chem.* **270**, 29 350–29 355.
[11] DeRojas-Walker, T.; *et al.* (1995) *Chem. Res. Toxicol.* **8**, 473–477.
[12] Wink, D. A.; *et al.* (1991) *Science* **254**, 1001–1003.
[13] Routlege, M. N.; *et al.* (1993) *Carcinogenesis* **14**, 1251–1254.
[14] Inoue, S.; Kawanishi, S. (1995) *FEBS Lett.* **371**, 86–88.
[15] Felley-Bosco, E.; *et al.* (1995) *Carcinogenesis* **16**, 2069–2074.
[16] Juedes, M. J.; Wogan, G. N. (1996) *Mutat. Res.* **349**, 51–61.

15 Oxidative DNA Damage and its Cellular Consequences

Bernd Epe, D. Ballmaier, O. Will, S. Hollenbach, H.-C. Mahler, M. I. Homburg and J. P. Radicella

15.1 Abstract

Oxidative DNA damage is endogenously generated in apparently all types of cells, most probably via reactive oxygen species formed as by-products of the cellular oxygen metabolism. As a result of the balance between generation and removal of the oxidative DNA modifications, steady-state (background) levels of the lesions can be detected in the cellular DNA. Although at least some of the oxidative DNA modifications, e.g. 8-hydroxyguanine, are premutagenic, there is only limited knowledge about the mutagenic and carcinogenic risk associated with the endogenous oxidative damage. Here we report on attempts (i) to quantify the steady-state (background) levels of oxidative DNA modifications in various mammalian cells, (ii) to identify the cellular factors that have major impact on these levels and (iii) to assess the mutagenicity associated with the oxidative DNA damage. The results indicate that repair endonucleases allow a sensitive quantification of oxidative DNA modifications in human lymphocytes and cultured mammalian cells. Among the factors shown to modulate the steady-state levels were glutathione, ascorbic acid and elevated temperature. To assess the mutagenic potential of oxidative DNA damage, we determined the mutation frequencies induced by several types of oxidants in the *gpt* locus of AS52 cells and bacteria (*gpt* gene in a plasmid) and normalized for the numbers of DNA modifications induced. Per base modification sensitive to Fpg protein (8-oxoG), several oxidants that induced the same type of damage profile were found to also induce the same number of mutations. This number was as high as the number of mutations induced by UVB per pyrimidine dimer in the same system. The data support the assumption that endogenous oxidative DNA damage is an important risk factor for gene mutations and its consequences in mammalian cells.

15.2 Introduction

Oxidative DNA damage is endogenously generated in apparently all types of cells under physiological conditions and constitutes a natural challenge to the integrity of the genome. Reactive oxygen species (ROS) generated in the oxygen metabolism of the cells are assumed to be responsibe, although the relevant mechanisms (nature of the enzymes and species involved) remain to be established.

All cells are protected from oxidative DNA damage and its consequences by two lines of defence: firstly by an efficient scavenging of ROS by several enzymes (superoxide dismutase, catalase, glutathione peroxidase) and antioxidants (e.g. glutathione, ascorbic acid) and secondly by specific DNA repair enzymes, which remove oxidative DNA modifications from the genome. The defence systems, however, are not perfect and the balance between generation and removal of the oxidative DNA modifications results in steady-state (background) levels, which have been detected in the nuclear and mitochondrial DNA of all cells by various techniques.

There is no doubt that several types of oxidative DNA modifications, e.g. 8-hydroxyguanine (8-oxoG) are pre-mutagenic [1–5]. Therefore, the steady-state levels of oxidative DNA modifications are expected to contribute to the spontaneous mutation rates in all cells and could play a role in carcinogenesis, the development of several age-correlated degenerative diseases and the process of aging itself [6–8]. Support for this assumption comes from recent findings that defects in the specific repair of 8-oxoG in bacteria and yeast result in mutator phenotypes, i.e. in several-fold increases of the spontaneous mutation rates [9, 10]. Acquisition of a mutator phenotype has been proposed to play a pivotal role in the process of carcinogenesis, since it can explain the high number of somatic mutations found in cancer cells [11]. Recent results indicate that the human DNA repair gene responsible for the removal of 8-oxoG, *OGG1*, is indeed mutated in several human tumors [12].

To better define the role of oxidative DNA damage as a natural risk factor for the development of cancer and other diseases, we have addressed in our studies the following questions:

- what are the steady-state (background) levels of oxidative DNA modifications in various mammalian cells?
- which endogenous and exogenous factors have major impact on these levels? and
- which mutagenicity is associated with the oxidative DNA damage?

15.3 Steady-state levels of oxidative DNA modifications in mammalian cells

Only a limited number of techniques is sensitive enough to allow quantifications of the numbers of oxidative DNA modifications in untreated cells (steady-state levels). These include HPLC with electrochemical detection (HPLC/ECD) [13], gas chromatography/mass spectrometry (GC/MS) [14], ^{32}P-postlabelling [15] and immunochemical methods [16]. In addition, the recognition of oxidative DNA modifications by specific repair endonucleases has been exploited for a determination of endonuclease-sensitive modifications. The assays make use of the fact that the enzymes incise the DNA at the substrate modifications generating single-strand breaks (SSB), which can be very sensitively determined by various techniques such as alkaline elution [17], alkaline unwinding [18], single cell gel electrophoresis (COMET assay) [19] and the relaxation assay [20]. Several types of DNA modifications can be determined in parallel, depending on the repair endonucleases available. A certain limitation of the assays is the limited and not always fully established substrate specificity of the enzymes.

Surprisingly, the various detection methods gave conflicting results for the steady-state levels of oxidative modifications such as 8-oxoG in the same type of cell (reviewed by Ref. [21, 22]). For HPLC/ECD, GC/MS and ^{32}P-postlabelling, an oxidation of guanine during isolation and derivatisation was shown to be responsible for at least some of the discrepancies [23–30]. Its avoidance resulted in steady-state levels that were similarly low as the steady-state level of DNA modifications sensitive to the repair endonuclease Fpg protein, which include 8-oxoG [26, 30]. We recently summarized our evidence that an artifactual oxidation of guanine is indeed avoided in the determination of endonuclease-sensitive modifications by alkaline elution and that on the other hand all substrate modifications in the cellular DNA are recognized under the assay conditions [31, 32]. Nevertheless, there is an obvious need for a better (and comparative) validation of the various assays.

Steady-state levels of various endonuclease-sensitive modifications in human lymphocytes and AS52 (Chinese hamster ovary) cells are shown in Figure 15.1. The recognition spectrum of the repair endonucleases is listed in Table 15.1. The results indicate that DNA base modifications sensitive to Fpg protein, which according to the present knowledge comprise the purine modifications 8-oxoG and formamidopyrimidines (imidazol ring-opened purines), are much more frequent than oxidative pyrimidine modifications sensitive to endonuclease III and sites of base loss, which are specifically recognized by endonuclease IV but are also substrates for all other repair endonucleases (Table 15.1). The dominance of Fpg-sensitive modifications was also observed in all other types of cultured mammalian cells that have been analyzed so far by the same technique [32]. A recent quantification of thymine glycols, a well-known substrate of endonuclease III, by an immunochemical method yielded steady-state

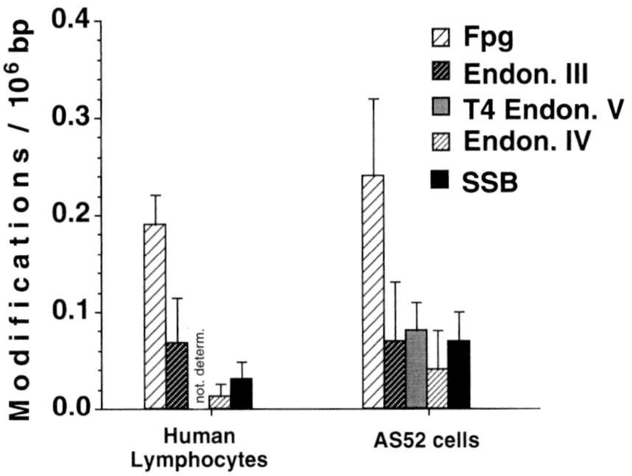

Figure 15.1: Steady-state (background) levels of DNA modifications recognized by various repair endonucleases (Table 15.1) in chromosomal DNA from human lymphocytes (averaged from 3 donors) and AS52 Chinese hamster cells determined by alkaline elution. Data are taken from Ref. [32].

Table 15.1: Recognition of DNA modifications by selected repair endonucleases.

Repair endonuclease	**Recognition spectrum**[a]	
	Sites of base loss	**Base modifications**
Fpg protein	+	8-oxoG[b]; Fapy[c]
Endonuclease III	+	5,6-dihydropyrimidines; hyd[d]
T4 endonuclease V	+	CPD[e]
Endonuclease IV	+	–

[a] See Ref. [33–37].
[b] 7,8-Dihydro-8-oxoguanine (8-hydroxyguanine).
[c] Formamidopyrimidines (imidazole ring-opened purines), e. g. 4,6-diamino-5-formamidopyrimidine (Fapy-A) and 2,6-diamino-4-hydroxy-5-formamidopyrimidine (Fapy-G).
[d] 5-Hydroxy-5-methylhydantoin and other ring-contracted and fragmented pyrimidines.
[e] Cyclobutane pyrimidine photodimers.

levels of approx. 0.05 modifications per 10^6 bp [16], which is in reasonable agreement with the low level of endonuclease-III-sensitive base modifications observed in the enzymic assays (Figure 15.1).

The absolute level of Fpg-sensitive modifications showed considerable variation with the cell type [31]. The differences between the cell types are experimental evidence that the steady-state levels determined are not predominantly experimental artifacts.

15.4 Factors that influence the steady-state levels of oxidative DNA damage

A reduction of the steady-state levels of oxidative modification in somatic cells is expected to reduce the adverse consequences mentioned above. Therefore, the identification of the exogenous and endogenous factors that have major influence on the levels of oxidative damage in the cells is of high interest. Cellular processes that may be modulated include the metabolic generation of ROS, the scavenging of ROS and the repair of the oxidative DNA modifications.

The generation of ROS is increased under conditions of oxidative stress. An example is the growth of cells at elevated temperatures (thermal stress), which causes an augmented generation of ROS possibly via mitochondrial uncoupling [38]. As shown in Figure 15.2, the steady-state level of Fpg-sensitive modifications in AS52 Chinese hamster cells was indeed higher at 41 °C than at 37 °C. As may be expected, the effect was more pronounced in cells depleted of glutathione [39]. Other groups have shown that autoimmune disease [40] and hyperbaric oxygen therapy [41] are associated with increased levels of oxidative base modifications in lymphocytes, possibly via the induction of oxidative stress.

The modulation of the antioxidant defence is of particular interest with respect to potential carcinogenic/anti-carcinogenic factors in food. While habitual alcohol intake increased the level of 8-oxoG in lymphocytes [42], long-term supplementation with ascorbic acid was shown to decrease the steady-state levels of oxidative base modifications in some (but not all) cases [43–45]. Surprisingly, some of the effects were observed by means of a conventional GC/MS technique which yields very high absolute steady-state levels (150 8-oxoG residues

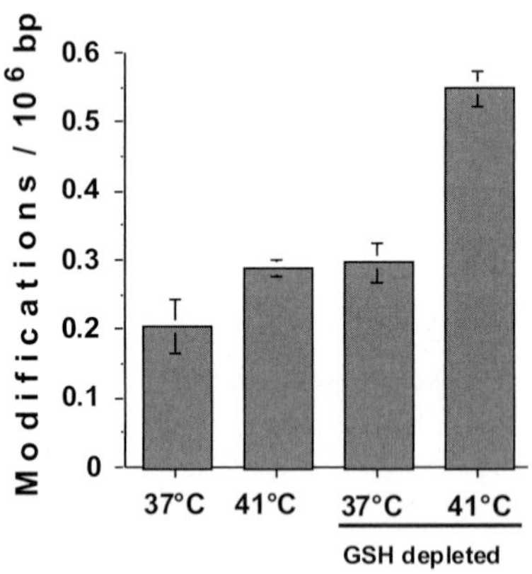

Figure 15.2: Influence of thermal stress on the steady-state levels of oxidative DNA modifications sensitive to Fpg protein in AS52 cells. Cells were cultured at 37 °C and 41 °C, with or without a preceeding depletion of cellular glutathione by incubation with buthionine-sulfoximine (1 mM) for 24 h.

per 10^6 bp) suspected to be caused by artifactual guanine oxidation (see above). In addition, an increase of 8-hydroxyadenine observed under the same conditions [44] is difficult to explain.

In our own experiments, the steady-state level of Fpg-sensitive modifications was determined before and 2 h after oral application of a single dose (1 g) of ascorbic acid. The results revealed a reproducible and significant drop of the basal level of oxidative damage to 73±14% in all volunteers (unpublished results) and thus confirmed a pronounced effect of dietary ascorbic acid.

Interestingly, the steady-state levels of Fpg-sensitive modifications in various types of human and rodent cells were to a significant extent inversely correlated with the intracellular glutathione levels [39]. The finding supports the assumption that glutathione is a major cellular factor for the inhibition of endogenous oxidative DNA damage. However, supplementation of the cell culture media with additional thiols such as *N*-acetylcysteine and cysteine ethylester, a precursor of glutathione, did not reduce the levels of Fpg-sensitive modifications in the cells, but at high concentration even increased the oxidative damage. The results may be explained by the well-known pro-oxidant effect of many reductants, which results from their autoxidation or from the reduction of cellular transition metals, which then act as Fenton catalysts. It could also indicate that the protection by GSH and other thiols against endogenous ROS requires adequate concentrations of other factors of the cellular antioxidant defence system and therefore is not significantly improved if only the thiol concentrations are raised.

To study the influence of DNA repair on the steady-state levels of oxidative DNA damage, AS52 Chinese hamster cells were transfected with the human *OGG1* gene coding for the 8-oxoG endonuclease. While the rate of repair of additionally induced Fpg-sensitive modifications was two-fold increased in the tranfectants, the steady-state levels of Fpg-sensitive modifications in untreated cells were unchanged (unpublished results). The result could indicate that the 8-oxoG glycosylase is not rate-limiting for the removal of 8-oxoG from the genome under normal growth conditions.

15.5 Mutagenicity associated with oxidative DNA damage

An estimation of the mutagenicity associated with endogenous oxidative DNA damage is of high interest: on the one hand, it should help to assess the significance of the oxidative damage for the (spontaneous) cancer incidence and thus help to estimate the benefit that can be achieved by a reduction of the oxidative damage in humans, e. g. by supplementation of dietary antioxidants (see above). On the other hand, the mutagenicity data could be compared with the mutagenicities associated with low doses of xenobiotics (in particular oxidants) and thus allow risk assessments based on the relative extents of exogenous and endogenous mutagenic risks.

Two types of studies appear promising. In a suitable cell line, the effect of a modulation of the steady-state level of oxidative damage (by manipulation of one of the cellular factors discussed above) can be compared with the resulting modulation of the spontaneous mutation rate. Experiments with AS52 cells, which carry the bacterial *gpt* gene for mutation analysis [46] and appropriate transfectants are in progress. Secondly, the mutation frequencies induced by various types of exogenous oxidants may be normalized for the numbers of a suitable marker modification induced (e.g. 8-oxoG) and used to calculate the mutation frequency that should arise from the steady-state level of the marker modification if a linear dose response is assumed. The approach is based on the assumption that the number of different oxidative DNA damage profiles induced in mammalian cells under various conditions is limited since the number of different types of oxidative stress in the cells is probably relatively low. Indeed, ROS such as hydroxyl radicals, peroxynitrite or singlet oxygen have been shown to induce characteristic DNA damage profiles both under cell-free conditions and in cells which were independent of the mechanism by which the ROS were generated [47–50]. It was particularly interesting to note that several mild (less reactive) oxidants such as singlet oxygen, alkoxyl radicals and triplet-exited photosensitizers (acridine orange, Ro19-8022) gave rise to rather similar DNA damage profiles which were dominated by a high relative number of purine modifications sensitive to Fpg protein and only low numbers of SSB, sites of base loss and oxidative pyrimidine modifications (Figure 15.3). As shown in

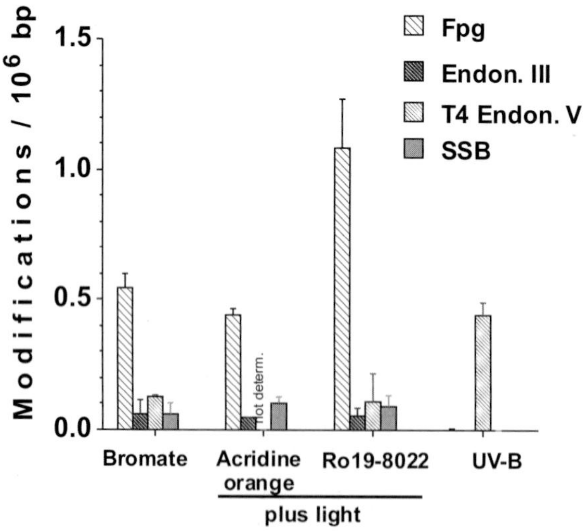

Figure 15.3: DNA damage profiles induced in AS52 by exposure to (a) potassium bromate (7.5 mM; 15 min; 37 °C), (b) acridine orange (6.6 μM) plus visible light (0.57 kJ/m^2), (c) Ro19-8022 (0.05 μM) plus visible light (166 kJ/m^2) and (d) UV-B (10 J/m^2). Background levels of modifications (see Figure 15.1) were substracted.

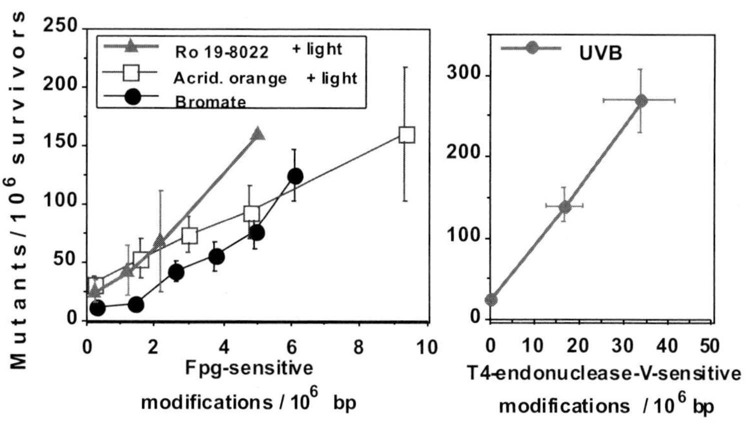

Figure 15.4: Numbers of mutants induced in the *gpt* locus of AS52 cells by various agents (see Figure 15.3), plotted against the numbers of Fpg-sensitive or T4-endonuclease-V-sensitive modifications induced. The numbers of modifications were extrapolated from data determined at lower concentrations or light doses assuming a linear dose response.

Figure 15.4, mutation frequencies induced by these agents at the same level of oxidative DNA damage in the *gpt* locus of AS52 cells were rather similar. The number of mutations per Fpg-sensitive modification was as high as the number of mutations induced by UVB per pyrimidine dimer in the same system (Figure 15.4, left panel). (The damage profile by UVB consists of more than 99 % pyrimidine dimers, see Figure 15.3 and Ref. [51].) The result suggests that oxidative DNA damage is not generally less mutagenic than other types of damage.

15.6 Conclusions

The results outlined above support the assumption that oxidative DNA damage plays an important role as an endogenous mutagen. However, a quantification of the mutagenicity associated with the steady-state levels of oxidative DNA modifications observed under physiological conditions is still a difficult problem. Recent results indicate that the levels of oxidative base modifications such as 8-oxoG have been highly overestimated in many previous studies. On the other hand, the data presented above indicate that the mutagenicity associated with oxidative DNA damage profiles is relatively high when calculated per 8-oxoG residue induced.

A better understanding of the correlation between the steady-state levels of oxidative DNA damage in the cells and its direct consequences (such as the induction of mutations) is of high interest since (i) the level of oxidative DNA damage is an easily accessible end-point *in vivo*, despite the technical problems discussed above, and (ii) there are good indications that the cellular levels of oxidative DNA damage are influenced by various endogenous and exogenous factors, at least some of which are subject to simple nutritional and medical intervention.

References

[1] Breimer, L. H. (1990) Molecular mechanisms of oxygen radical carcinogenesis and mutagenesis: the role of base damage. *Mol. Carcinog.* **3**, 188–197.

[2] Feig, D. I.; Sowers, L. C.; Loeb, L. A. (1994) Reverse chemical mutagenesis: Identification of the mutagenic lesions resulting from reactive oxygen species-mediated damage to DNA. *Proc. Natl. Acad. Sci. USA* **91**, 6609–6613.

[3] Wood, M. L.; Dizdaroglu, M.; Gajewski, E.; Essigmann, J. M. (1990) Mechanistic studies of ionizing radiation and oxidative mutagenesis: genetic effects of a single 8-hydroxyguanine (7-hydro-8-oxoguanine) residue inserted at a unique site in a viral genome. *Biochemistry* **29**, 7024–7032.

[4] Cheng, K. C.; Cahill, D. S.; Kasai, H.; Nishimura, S.; Loeb, L. A. (1992) 8-Hydroxyguanine, an abundant form of oxidative DNA damage, causes G → T and A → C substitutions. *J. Biol. Chem.* **267**, 166–172.

[5] Moriya, M. (1993) Single-stranded shuttle phagemid for mutagenesis studies in mammalian cells: 8-oxoguanine in DNA induces targeted G:C → T:A transversions in simian kidney cells. *Proc. Natl. Acad. Sci. U.S.A.* **90**, 1122–1126.

[6] Wallace, D. G. (1992) Mitochondrial genetics: a paradigm for ageing and degenerative diseases? *Science* **256**, 628–632.

[7] Gutteridge, J. M. C. (1993) Free radicals in disease processes: a compilation of cause and consequences. *Free Radical Res. Commun.* **19**, 141–158.

[8] Beckman, K. B.; Ames, B. N. (1998) The free radical theory of ageing matures. *Physiol. Rev.* **78**, 547–581.

[9] Michaels, M. L.; Cruz, C.; Grollman, A. P.; Miller, J. H. (1992) Evidence that MutY and MutM combine to prevent mutations by an oxidative damaged form of guanine. *Proc. Natl. Acad. Sci. USA* **89**, 7022–7025.

[10] Thomas, D.; Scot, A. D.; Barbey, R.; Padula, M.; Boiteux, S. (1997) Inactivation of OGG1 increases the incidence of G:C → T:A transversions in *Saccharomyces cerevisiae*: evidence for endogenous oxidative damage to DNA in eukaryotic cells. *Mol. Gen. Genet.* **254**, 171–178.

[11] Loeb, L. A. (1991) Mutator phenotype may be required for multistage carcinogenesis. *Cancer Res.* **51**, 3075–3079.

[12] Chevillard, S.; Radicella, J. P.; Levalois, C.; Lebeau, J.; Poupon, M.-F.; Oudard, S.; Dutrillaux, B.; Boiteux, S. (1998) Mutations in OGG1, a gene involved in the repair of oxidative DNA damage, are found in human lung and kidney tumors. *Oncogene* **16**, 3083–3086.

[13] Kasai, H.; Crain, P. F.; Kuchino, Y.; Nishimura, S.; Ootsuyama, A.; Tanooka, H. (1986) Formation of 8-hydroxyguanine moiety in cellular DNA by agents producing oxygen radicals and evidence for its repair. *Carcinogenesis* **7**, 1849–1851.

[14] Nackerdien, Z.; Olinski, R.; Dizdaroglu, M. (1992) DNA base damage in chromatin of γ-irradiated cultured human cells. *Free Radical Res. Commun.* **16**, 259–273.
[15] Devanaboyina, U.; Gupta, R. C. (1996) Sensitive detection of 8-hydroxy-2′-deoxyguanosine in DNA by ^{32}P-postlabelling assay and the basal levels in rat tissues. *Carcinogenesis* **17**, 917–924.
[16] Le, X. C.; Xing, J. Z.; Lee, J.; Leadon, S. A.; Weinfeld, M. (1998) Inducible repair of thymine glycol detected by an ultrasensitive assay for DNA damage. *Science* **280**, 1066–1069.
[17] Epe, B.; Hegler, J. (1994) Oxidative DNA damage: endonuclease fingerprinting. *Methods Enzymol.* **234**, 122–131.
[18] Hartwig, A.; Dally, H.; Schlepegrell, R. (1996) Sensitive analysis of oxidative DNA damage in mammalian cells: use of the bacterial Fpg protein in combination with alkaline unwinding. *Toxicology* **110**, 1–6.
[19] Collins, A. R.; Duthie, S. J.; Dobson, V. L. (1993) Direct enzymatic detection of endogenous oxidative base damage in human lymphocyte DNA. *Carcinogenesis* **14**, 1733–1735.
[20] Hegler, J.; Bittner, D.; Boiteux, S.; Epe, B. (1993) Quantification of oxidative DNA modifications in mitochondria. *Carcinogenesis* **14**, 2309–2312.
[21] Halliwell, B.; Dizdaroglu, M. (1992) The measurement of oxidative damage to DNA by HPLC and GC/MS techniques. *Free Radical Res. Commun.* **16**, 75–87.
[22] Collins, A. R.; Cadet, J.; Epe, B.; Gedik, C. (1997) Problems in the measurement of 8-oxoguanine in human DNA. Report of a workshop, DNA oxidation, held in Aberdeen, UK, 19–21 January, 1997. *Carcinogenesis* **18**, 1833–1836.
[23] Ravanat, J. L.; Turesky, R. J.; Gremaud, E.; Trudel, L. J.; Stadler, R. (1995) Determination of 8-oxoguanine in DNA by gas chromatography-mass spectrometry and HPLC-electrochemical detection: overestimation of the background level of the oxidized base by the gas chromatography-mass spectrometry assay. *Chem. Res. Toxicol.* **8**, 1039–1045.
[24] Hamberg, M.; Zhang, L.-Y. (1995) Quantitative determination of 8-hydroxyguanine by isotope dilution mass spectrometry. *Anal. Biochem.* **229**, 336–344.
[25] Douki, T.; Delatour, T.; Bianchini, F.; Cadet, J. (1996) Observation and prevention of an artifactual formation of oxidised DNA bases and nucleosides in the GC-EIMS method. *Carcinogenesis* **17**, 347–353.
[26] Nakajima, M.; Takeuchi, T.; Morimoto, K. (1996) Determination of 8-hydroxydeoxyguanosine in human cells under oxygen-free conditions. *Carcinogenesis* **17**, 787–791.
[27] Finnegan, M. T. V.; Herbert, K. E.; Evans, M. D.; Griffiths, H. R.; Lunec, J. (1996) Evidence for sensitization of DNA to oxidative damage during isolation. *Free Radical Biol. Med.* **20**, 93–98.
[28] Kvam, E.; Tyrrell, R. M. (1997) Induction of oxidative DNA base damage in human skin cells by UV and near visible radiation. *Carcinogenesis* **18**, 2379–2384.
[29] Schuler, D.; Otteneder, M.; Sagelsdorff, P.; Eder, E.; Gupka, R. C.; Lutz, W. K. (1997) Comparative analysis of 8-oxo-2′-deoxyguanosine in DNA by ^{32}P- and ^{33}P-postlabelling and electrochemical detection. *Carcinogenesis* **18**, 2367–2371.
[30] Helbock, H. J.; Beckman, K. B.; Shigenaga, M. K.; Walter, P. B.; Woodall, A. A.; Yeo, H. C.; Ames, B. (1998) DNA oxidation matters: The HPLC-electrochemical detection assay of 8-oxo-deoxyguanosine and 8-oxo-guanine. *Proc. Natl. Acad. Sci. USA* **95**, 288–293.
[31] Pflaum, M.; Will, O.; Epe, B. (1997) Determination of steady-state levels of oxidative DNA base modifications in mammalian cells by means of repair endonucleases. *Carcinogenesis* **18**, 2225–2231.
[32] Pflaum, M.; Will, O.; Mahler, H.-C.; Epe, B. (1998) DNA oxidation products determined with repair endonucleases in mammalian cells: types, basal levels and influence of cell proliferation. *Free Radical Res.* **29**, 585–594.

[33] Boiteux, S. (1993) Properties and biological functions of the NTH and FPG proteins of *Escherichia coli*: two DNA glycosylases that repair oxidative damage in DNA. *Photochem. Photobiol. B.* **19**, 87–96.
[34] Demple, B.; Harrison, L. (1994) Repair of oxidative damage to DNA: enzymology and biology. *Ann. Rev. Biochem.* **63**, 915–948.
[35] Seeberg, E.; Eide, L.; Bjoras, M. (1995) The base excision repair pathway. *Trends Biol. Sci.* **20**, 391–397.
[36] Karakaya, A.; Jaruga, P.; Bohr, V. A.; Grollman, A. P.; Dizdaroglu, M. (1997) Kinetics of excision of purine lesions from DNA by *Escherichia coli* Fpg protein. *Nucleic Acids Res.* **25**, 474–479.
[37] Cunningham, R. (1997) DNA glycosylases. *Mutat. Res.* **383**, 189–196.
[38] Skibba, J. L.; Powers, R. H.; Stadnicka, A.; Cullinane, D. W.; Almagro, U. A.; Kalbfleisch, J. H. (1991) Oxidative stress is a precursor to the irreversible hepatocellular injury caused by hyperthermia. *Int. J. Hyperthermia* **7**, 749–761.
[39] Will, O.; Mahler, H.-C.; Arrigo, A.-P.; Epe, B. (1998) Influence of glutathione levels and heat-shock on the steady-state levels of oxidative DNA base modifications in mammalian cells. (submitted for publication).
[40] Bashir, S.; Harris, G.; Denman, M. A.; Blake, D. R.; Winyard, P. G. (1993) Oxidative DNA damage and cellular sensitivity to oxidative stress in human autoimmune diseases. *Ann. Rheum. Dis.* **52**, 659–666.
[41] Speit, G.; Dennog, C.; Lampl, L. (1998) Biological significance of DNA damage induced by hyperbaric oxygen. *Mutagenesis* **13**, 85–87.
[42] Nakajima, M.; Takeuchi, T.; Takeshita, T.; Morimoto, K. (1996) 8-hydroxydeoxyguanosine in human leukocyte DNA and daily health practice factors: effects of individual alkohol sensitivity. *Environ. Health Perspect.* **104**, 1336–1338.
[43] Duthie, S. J.; Ma, A.; Ross, M. A.; Collins, A. R. (1996) Antioxidant supplementation decreases oxidative DNA damage in human lymphocytes. *Cancer Res.* **56**, 1291–1295.
[44] Podmore, I. D.; Griffiths, H. R.; Herbert, K. E.; Mistry, N.; Mistry, P.; Lunec, J. (1998) Vitamin C exhibits pro-oxidant properties. *Nature* **392**, 559.
[45] Rehman, A.; Collis, C. S.; Yang, M; Kelly, M.; Diplock, A. T.; Halliwell, B.; Rice-Evans, C. (1998) The effect of iron and vitamin C co-supplementation on oxidative damage to DNA in human volunteers. *Biochem. Biophys. Res. Commun.* **246**, 293–298.
[46] Tindall, K. R.; Stankowski Jr., L. F. (1989) Molecular analysis of spontaneous mutations at the gpt locus in Chinese hamster ovary (AS52) cells. *Mutat. Res.* **220**, 241–253.
[47] Epe, B.; Pflaum, M.; Boiteux, S. (1993) DNA damage induced by photosensitizers in cellular and cell-free systems. *Mutat. Res.* **299**, 135–145.
[48] Epe, B.; Häring, M.; Ramaiah, D.; Stopper, H.; Abou-Elzahab, M. M.; Adam, W.; Saha-Möller, C. R. (1993) DNA damage induced by furocoumarin hydroperoxides plus UV (360 nm). *Carcinogenesis* **14**, 2271–2276.
[49] Epe, B.; Ballmaier, D.; Adam, W.; Grimm, G. N.; Saha-Möller, C. R. (1996) Photolysis of *N*-hydroxypyridinethiones: a new source of hydroxyl radicals for the direct damage of cell-free and cellular DNA. *Nucleic Acids Res.* **24**, 1625–1631.
[50] Epe, B.; Ballmaier, D.; Roussyn, I.; Briviba, K.; Sies, H. (1996) DNA damage by peroxynitrite characterized with DNA repair enzymes. *Nucleic Acids Res.* **24**, 4105–4110.
[51] Kielbassa, C.; Roza, L.; Epe, B. (1997) Wavelength dependence of oxidative DNA damage induced by UV and visible light. *Carcinogenesis* **18**, 811–816.

16 Fermentation Profiles in the Gut and Functional Effects of Food Components

Wolfgang Scheppach

16.1 Abstract

Polysaccharides and proteins/peptides which escape enzymatic digestion in the small bowel undergo bacterial fermentation in the colon. The dominating anaerobic bacteria (bacteriodes, bifidobacteria, eubacteria) exhibit a substrate preference for polysaccharides. This results in a saccharolytic type of fermentation in the proximal colon (caecum, ascending and transverse colon). The yield of short-chain fatty acids (SCFAs: acetate, propionate, *n*-butyrate), the major end products of this process, varies among polysaccharide substrates. Starch and fructooligosaccharides are especially good sources of *n*-butyrate. Only after exhaustion of polysaccharide sources proteins and peptides are broken down (mainly in the descending and sigmoid colon) resulting in a higher yield of branched-chain fatty acids. It is not known whether this finding is of relevance to the question of carcinogenesis.

There is good evidence for a protective role of SCFAs, especially *n*-butyrate, in the adenoma-carcinoma sequence. Butyrate inhibits hyperproliferation of colonic crypt cells which is regarded as an early step in carcinogenesis. *n*-Butyrate and, to a lesser degree, propionate inhibit growth and favour differentiation in colon cancer cell lines. Recently it has been shown that SCFAs induce apoptosis in carcinoma and adenoma cell lines. At the molecular level, *n*-butyrate causes acetylation of DNA-histone complexes and regulates the expression of various genes relevant to cancer. This includes the bcl-2/bax gene family involved in the regulation of apoptosis and cyclins/cyclin-dependent kinases important for cell cycle progression. There are plausible mechanisms whereby complex carbohydrates may affect carcinogenesis in the human colorectum.

16.2 Introduction

Carbohydrates escaping enzymatic digestion in the small intestine are delivered to the colon where they are fermented by the anaerobic microflora. These include non-starch polysaccharides (the major fraction of dietary fibre), resistant starch, and endogenous polysaccharides from mucus and shed epithelial cells. The most important end products of bacterial carbohydrate breakdown are short-chain fatty acids (SCFAs: acetate, propionate and *n*-butyrate). SCFAs are rapidly and efficiently taken up by the epithelial cells which line the colonic lumen. Butyrate serves as an energy yielding substrate in the colonocytes and, additionally, affects several cellular functions (proliferation, membrane synthesis, sodium absorption). Propionate and acetate are released by the basolateral membrane to the portal circulation and may have effects far from their production site. The physiology of the fermentation process including SCFA production has been reviewed in depth by Macfarlane and Cummings [1].

16.3 Colonic SCFA concentrations and carcinogenesis

SCFAs are produced in the proximal colon of hindgut fermenters (including man) in an average molar ration of acetate:propionate:butyrate equivalent to 60 : 25 : 10 mmol/l. This ratio, however, is not constant but is determined by the kind of substrate fermented. Assuming that butyrate protects against cancer formation, fermentable substrates should be examined which are primarily broken down to butyrate.

It has been shown *in vitro* and *in vivo* that the fermentation of starch yields high levels of butyrate. To study the impact of starch malabsorption on faecal SCFAs, 11 healthy volunteers consumed a controlled diet rich in starch for two 4-week periods. They received the glucosidase inhibitor acarbose in one of the study periods and placebo in the other. The faecal concentration (µmol/g wet weight) on *n*-butyrate (+58 %) rose significantly when acarbose was added to the diet. The faecal excretion (mmol/day) of total SCFAs (+95 %) and of their constituents acetate (+97 %) and *n*-butyrate (+182 %) was significantly higher when starch malabsorption was induced by acarbose.

Another approach was used by McIntyre *et al.* [3]. These authors argued that butyrate concentrations show a falling gradient along the large bowel with highly fermented dietary fibre (guar, oat bran), being low in the distal colon, where benign and malignant tumors are most prevalent. They suggested a "lente" fermented fibre (wheat bran) to maintain faecal butyrate concentration at caecal values. In the rat, they induced tumors with dimethylhydrazine, and assessed the impact of different fibre-containing diets on the number and size of tumors. Significantly fewer tumors were seen in the rats fed wheat bran compared with those fed guar or oat bran, and the total tumor mass was lowest in rats fed wheat bran. The concentration of butyrate in stools correlated significantly and negatively with tumor mass. Thus, the type of fibre which is associated with high butyrate concentrations in the distal large bowel is protective against colorectal cancer in this animal model. In this context, it is interesting that a low butyrate to acetate ratio has been found in enema samples from patients with adenomatous polyps and colon cancer [4].

16.4 SCFAs and proliferation of normal colonocytes

Data on the interaction of SCFAs with colonic epithelial proliferation are mainly derived from *in vitro* work employing intact colonic tissue or permanent cell lines. When the impact of SCFAs on colonic carcinogenesis is addressed, a paradoxical effect on normal and neoplastic epithelial cells becomes evident. While butyrate and, to a lesser degree, propionate reduce proliferation of cancer cells *in vitro*, all three major SCFAs stimulate proliferation in normal colonic epithelium.

Scheppach *et al.* [5] assessed the effect of SCFAs on proliferation of normal caecal mucosa in biopsy specimens obtained from 45 individuals at routine colonoscopy. Tissues were incubated for 4 h with sodium salts of SCFAs at physiological concentrations or equimolar NaCl (control). Cell proliferation was measured autoradiographically by pulse labelling with [^{3}H]thymidine for 1 h. Caecal crypt proliferation was raised significantly in all incubation experiments with SCFAs. Butyrate (10 mmol/l, increase +89%) and propionate (25 mmol/l, increase +70%) were as effective in stimulating proliferation as the combination of the 3 SCFAs (acetate 60 mmol/l + propionate 25 mmol/l + butyrate 10 mmol/l, increase +103%), although the effect of acetate (increase +31%) was minor. SCFAs stimulated DNA synthesis only in the basal 3 of 5 crypt compartments, which is considered the physiological proliferation zone. Other authors have obtained similar results in rat experiments [6, 7].

As demonstrated by Aghidassi *et al.* [8], the suppression of normal crypt cell proliferation may impair nutritional recovery in rats with small bowel resection. Animals with 80% small bowel resection were fed a liquid diet enterally for 16 days, with or without metronidazole to reduce fermentation. This antibiotic significantly lowered the total amount of SCFAs in the caecum. Resected rats receiving metronidazole had a significantly lower weight gain, carcass protein, nitrogen balance and mucosal dry weight, protein and DNA, compared with resected rats without metronidazole. This study provides evidence that SCFAs are important luminal trophic factors that may favour intestinal adaptation in rats with massive small bowel resection. It is unlikely that the stimulation of a physiological pattern of proliferation can be related to the process of carcinogenesis.

16.5 Butyrate and colonic "hyperproliferation"

"Normal" proliferation of colonocytes occurs in the basal 60 % of the crypts. In the upper 40 % of the crypts, however, proliferation stops and colonocytes become fully differentiated, to be extruded after approximately 7 days of upward migration. The expansion of the proliferative compartment to the crypt surface (hyperproliferation, involving the upper 40 % of the crypt length) is considered a preneoplastic biomarker. SCFAs stimulate proliferation in the basal crypt compartments *in vitro* without increasing cell labelling in the upper crypt [5]. The effect of butyrate on hyperproliferating colonic epithelium differs clearly from the effect on normal epithelium.

Bartram *et al.* [9] induced hyperproliferation *in vitro* by incubating biopsy specimens from the human ascending colon with deoxycholic acid. This secondary bile acid has been found to be co-carcinogenic in many animal models. Deoxycholic acid significantly raised the upper crypt labelling index, measured by bromodeoxyuridine immunohistochemistry. When the tissue was co-incubated with deoxycholic acid and *n*-butyrate, the increase of DNA synthesis in the upper crypt was no longer observed. Similar data were obtained in biopsies from the human rectosigmoid colon [10]. In conclusion, butyrate antagonised hyperproliferation induced by deoxycholic acid.

16.6 Butyrate and adenoma cell lines

Limited data are available concerning potential effects of SCFAs on cells at intermediate stages of the adenoma-carcinoma sequence. To date, it is unknown where the "switch" from stimulation to suppression of proliferation occurs.

In a paper by Hague *et al.* [11], the induction of apoptosis by butyrate (1–4 mmol/l) was studied in two adenoma cell lines (RG/C2 and AA/C1). Butyrate increased the number of apoptotic cells characterised by internucleosomal DNA fragmentation. In contrast, transforming growth factor β_1, which is thought to have an important role in the control of growth in colonic epithelium, did not induce apoptosis. The adenoma cell line RG/C2 did not contain wildtype *TP53*, therefore this tumor suppressor gene is not required to mediate signals for the induction of apoptosis in colonic tumor cells. Butyrate may induce apoptosis either by suppressing anti-apoptotic bcl-2 expression or stimulating pro-apoptotic bak expression, depending on the cell line studied [12]. Butyrate also induced apoptosis in colon carcinoma cells (HT-29, SW-620), which has been confirmed by Heerdt *et al.* [13]. A disruption of the balance between cell gain through mitosis and cell loss through programmed cell death (apoptosis) is thought to be an important event in carcinogenesis.

16.7 Butyrate, propionate, and proliferation of colon cancer cell lines

Abundant literature has accumulated on butyrate inhibition of the growth of human colon cancer cell lines. In these isolated tumor cells, butyrate suppresses proliferation at concentrations between 1 and 5 mmol/l, without impairing cell viability; at higher levels cytotoxic effects occur. In contrast with these findings, normal colonic tissue tolerates butyrate concentrations of 10–60 mmol/l. The growth-limiting action of butyrate is observed in lines from many colon carcinomas (SW-620, SW-480, CaCo-2, HT-29, HRT-18, LIM-1215, SK-CO-1) and malignancies from other organs (e. g., breast, pancreas).

In the original paper by Kim *et al.* [14] human colonic adenocarcinoma cell lines (SW-620, SW-480) were incubated with sodium butyrate (mmol/l) for 8 days. Doubling times were increased between 1.18 and 7.6-fold while cell viability was unaffected. The removal of butyrate from the medium resulted in the resumption of rapid growth. Gross morphological alterations including cell enlargement, process formation, and cellular flattening occurred during culture with butyrate.

Gamet *et al.* [14] demonstrated in the HT-29 adenocarinoma cell line that, similar to butyrate (2–5 mmol/l), propionate (2–10 mmol/l) also inhibited growth. The addition of acetate (2–10 mmol/l) to the medium had no effect on cell proliferation. The antiproliferative action of butyrate and propionate was associated with an inhibition of ornithine decarboxylase, a key enzyme in polyamine synthesis. These data suggest that propionate, together with butyrate, may partially account for the protective effect of some dietary fibres with regard to carcinogenesis.

The results of Kim *et al.* [14] and Gamet *et al.* [15] were confirmed in our laboratory [16] when HT-29 cells were incubated with SCFAs. Butyrate (2 mmol/l) and, to a lesser extent, propionate (5 mmol/l) inhibited proliferation. Branched-chain fatty acids (*iso*-butyrate, *iso*-valerate, 1–2 mmol/l), which arise during the fermentation of proteins or peptides, had no effect on growth of this colon cancer cell line. These data support the hypothesis that protein and carbohydrate fermentation may affect the colonic environment in different ways.

16.8 Butyrate and markers of differentiation

There is good evidence that butyrate does not inhibit growth of colon cancer cells simply by cytotoxic action. On the contrary, this fatty acid induces markers of differentiation at the same time as it inhibits proliferation. In the presence of butyrate, tumor cells assume a phenotype more like the original non-neoplastic tissue. Other markers of differentiation induced by butyrate include alkaline phosphatase and other hydrolases, carcinoembryonic antigen [17] and membrane-associated glycoproteins and glycolipids [18].

Gum *et al.* [19] studied the effect of butyrate on the induction of alkaline phosphatase in the human colonic tumor cell line LS-174-T. Culture of these cells in the presence of butyrate (2 mmol/l) caused the activity to increase from <0.0001 unit/mg of protein to >0.7 unit/mg of protein over an 8-day period. This induction proceeded in a non-linear manner, with a lag time of 2–3 days occurring before enzymatic activities began to rise. Northern blot analysis indicated that treatment of these cells with butyrate caused >20-fold induction of a 2700-base messenger RNA that hybridised to a cDNA probe for placental-like alkaline phosphatase (PLAP). These results indicate that a placental-like protein and messenger RNA are induced by butyrate in LS-174-T cells, with a time course consistent with cellular differentiation preceeding induction.

More recently, the same group [20] examined the PLAP gene promotor in LS-174-T cells, using transient transfection experiments. Chimaeras from various lengths of the PLAP promotor clone were made, and transfected cells were studied in the presence or absence of butyrate (2 mmol/l). The region between nucleotides −363 and −170 was found to represent a strong negative control element within the PLAP promotor (total length 512 nucleotides). The effects of this negative control element were significantly reduced in the presence of sodium butyrate. Thus, the region which is functionally involved in PLAP gene transcriptional regulation, as judged by promotor deletion experiments, appears to be crucial for the regulation of the PLAP gene and its induction by butyrate.

It should be emphasised that the differentiating action of butyrate is observed in tumor cells, but not in non-neoplastic colonocytes cultured under indentical conditions [21]. This paradox could possibly be solved by investigating intermediate stages of the adenoma-carcinoma sequence.

16.9 Effect of butyrate on histones

Attempts have been made to investigate the effects of butyrate on proliferation and differentiation of neoplastic cells at the molecular level. Early trials have focused on acetylation, phosphorylation and methylation [22] of DNA-histone complexes in various cell lines. Witlock *et al.* [23] incubated HeLa cells for 16 h in the presence of sodium butyrate (1–10 mmol/l). The extent of histone acetylation was estimated by quantitative densitometry of the H4 region of the stained gel. The amount of ^{32}P in H3 was determined by autoradiography and quantitative densitometry. At a concentration of 1–5 mmol/l, butyrate increased the extent of histone acetylation dose-dependently by inhibiting histone deacetylase activity. The changes in H3 phosphorylation closely paralleled changes in histone acetylation, suggesting a relationship between the two modifications.

The consequence of hyperacetylation could be a release of bonds between DNA and histones. This may result in an increased accessibility of DNA, not only to nucleases, but also to various factors involved in the control of gene expression. At present, the importance of these butyrate effects in uncertain.

16.10 Effects of butyrate on gene expression

While the action of butyrate on histones is considered non-specific, there is recent evidence that this fatty acid may affect gene expression in a highly specific manner. This subject has recently been reviewed by Kruh *et al.* [24]. The butyrate effects may be summarised as follows:

- The synthesis of a limited number of proteins is induced, which includes alkaline phosphatase [19], glycoproteins, hormone receptors and ionbinding metallothioneins.
- Butyrate suppresses cancer-specific properties in tumor cells, which recover normal molecular characteristics; little is known about the genes that may be involved.
- Butyrate inhibits proliferation of colon cancer cells, probably by causing an arrest at the early G1 phase. This could, at least partly, result from the effect of butyrate on the expression of genes involved in the control of the cell cycle, including oncogenes. Butyrate has been shown to reduce the expression of cMYC (3T3 fibroblasts, CaCo-2 cells), cSRC (SW-620 cells), cMYB (LIM-1215 cells) and cRAS (HT-29 cells), while inducing the expression of cFOS (3T3 fibroblasts). Generally, the regulation of these oncogenes in the described manner is associated with a higher degree of differentiation and, inversely, with cell growth [25].

Foss *et al.* [26] assessed the association of butyrate-induced (2 mmol/l) differentiation of human colon carcinoma cells (SW-620), with the expression of src-related tyrosine protein kinase activities and the abundance of $pp60^{c\text{-}src}$ and $p56^{lck}$ were found to parallel the butyrate-induced phenotypic alterations. These data indicate that a higher degree of differentiation of SW-620 cells is associated with a down regulation of src-related kinases.

Recent approaches include effects of *n*-butyrate on cell cycle inhibitors [27] and interactions with overexpressed cyclooxygenase-2 [28]. Another target for butyrate may be the *ras*-dependent signal transduction cascade by which extracellular growth factors affect DNA transcription.

16.11 Butyrate and tumor invasiveness

Another target of butyrate to affect carcinogenesis is the extracellular matrix. One factor controlling tumor invasion may be cell surface bound urokinase, which activates plasminogen to plasmin. There is a wide variety of substrates for plasmin including fibrin (plasmatic fibrinolysis), laminin, fibronectin and proteoglycan (extracellular matrix). Urokinase is secreted by normal [29] and neoplastic [30] colonocytes, and remains at the cell surface bound to receptors. The penetration of malignant cells to the substratum may be facilitated by urokinase.

Gibson *et al.* [31] measured secreted and cell-associated levels of urokinase and plasminogen activator inhibitor 1 in colonic crypt cells. Butyrate (0.001–4 mmol/l) caused a concentration-dependent inhibition of both secreted and cell-associated urokinase content. Acetate and propionate had minimal effects. Butyrate also stimulated plasminogen activator inhibitor 1 secretion. Levels of transcripts for urokinase and the inhibitor changed with butyrate exposure in parallel with the levels of secretion of the respective proteins. Cells from the cancer group showed significantly reduced inhibitor secretion and abnormal responses to butyrate (greater inhibition of urokinase secretion and no stimulation of inhibitor secretion).

16.12 Butyrate as a treatment for metastatic cancer in experimental animals

In a remarkable *in vivo* study by Perrin *et al.* [32] peritoneal carcinomatosis was induced in the rat using PROb colon cancer cells. Established carcinomatosis was treated with intraperitoneal injections of interleukin 2, sodium butyrate and a combination of the two. While both monotherapies were ineffective in reducing tumor mass, the combination resulted in a 60 % overall survival rate, and included cases of complete cure. Accompanying *in vitro* work revealed that sodium butyrate enhanced the expression of major histocompatibility complex class I molecules. After butyrate treatment, PROb cells became more sensitive to lymphokine-activated killer cells. The authors concluded that butyrate may act by increasing immunogenicity of colon cancer cells. In another study by Velazquez *et al.* [33] butyrate inhibited the seeding and growth of colorectal metastases to the liver in mice.

16.13 Conclusion

It is likely that colonic carcinogenesis is an example of how endogenous (genetic) and exogenous (nutritional) factors interact. According to Vogelstein's model [34], genetic alterations in the colonic mucosa accumulate over one or two decades which lead up to the formation of a malignant tumor. There is, however, evidence that nutritional factors determine the speed of progression in the adenoma-carcinoma sequence. Protective factors (SCFAs and others) may outweigh the detrimental effects of accelerating factors (e.g., secondary bile acids). Future research should be focused on the molecular mechanisms whereby nutrition affects carcinogenesis in the human large bowel. This approach could form the basis for prevention strategies.

References

[1] Macfarlane, G. T.; Cummings, J. H. (1991) The colonic flora, fermentation, and large bowel digestive function. In: Philips, S. F.; Pemberton, J. H.; Shorter, R. G. (Eds.) *The Large Intestine: Physiology, Pathophysiology, and Disease.* New York, Raven Press, p. 51–91.

[2] Scheppach, W.; Fabian, C.; Sachs, M.; Kasper, H. (1988) The effect of starch malapsorption on faecal short-chain fatty acid excretion in man. *Scand. J. Gastroenterol.* **23**, 755–759.

[3] McIntyre, A.; Gibson, P. R.; Young, G. P. (1993) Butyrate production from dietary fibre and protection against large bowel cancer in a rat model. *Gut* **34**, 386–391.

[4] Weaver, G. A.; Krause, J. A.; Miller, T. L.; Wolin, M. J. (1988) Short chain fatty acid distribution of enema samples from a sigmoidoscopy population: an association of high acetate and low butyrate ratios with adenomatous polyps and colon cancer. *Gut* **29**, 1539–1543.

[5] Scheppach, W.; Bartram, P.; Richter, A.; *et al.* (1992) Effect of short-chain fatty acids on the human colonic mucosa *in vitro. J. Parenter. Enteral Nutr.* **16**, 43–48.

[6] Sakata, T. (1987) Stimulatory effect of short-chain fatty acids on epithelial cell proliferation in the rat intestine: a possible explanation for trophic effects of fermentable fibre, gut microbes and luminal trophic factors. *Br. J. Nutr.* **58**, 95–103.

[7] Frankel, W. L.; Zhang, W.; Singh, A.; *et al.* (1994) Mediation of the trophic effects of short-chain fatty acids on the rat jejunum and colon. *Gastroenterology* **106**, 375–380.

[8] Aghdassi, E.; Plapler, H.; Kurian, R.; *et al.* (1994) Colonic fermentation and nutritional recovery in rats with massive small bowel resection. *Gastroenterology* **107**, 637–642.

[9] Bartram, H. P.; Scheppach, W.; Schmid, H.; *et al.* (1993) Proliferation of human colonic mucosa as an intermediate biomarker of carcinogenesis: effects of butyrate, deoxycholate, calcium, ammonia, and pH. *Cancer Res.* **53**, 3283–3288.

[10] Bartram, H. P.; Englert, S.; Scheppach, W.; *et al.* (1994) Antagonistic effects of deoxycholic acid and butyrate on epithelial cell proliferation in the proximal and distal human colon. *Z. Gastroenterol.* **32**, 389–392.
[11] Hague, A.; Manning, A. M.; Hanlon, K. A.; Hutschtscha, L. I.; Hart, D.; Paraskeva, C. (1993) Sodium butyrate induces apoptosis in human colonic tumor cell lines in a p53-independent pathway: implications for the possible role of dietary fibre in the prevention of large-bowel cancer. *Int. J. Cancer* **55**, 498–505.
[12] Hague, A.; Diaz, G. D.; Hicks, D. J.; Krajewski, S.; Reed, J. C.; Paraskeva, C. (1997) bcl-2 and bak play a pivotal role in sodium butyrate-induced apoptosis in colonic epithelial cells; however overexpression of bcl-2 does not protect against bak-mediated apoptosis. *Int. J. Cancer* **72**, 989–905.
[13] Heerdt, B. G.; Houston, M. A.; Augenlicht, L. H. (1994) Potentiation by specific short-chain fatty acids of differentiation and apoptosis in human colonic carcinoma cell lines. *Cancer Res.* **54**, 3288–3294.
[14] Kim, Y. S.; Tsao, D.; Siddiqui, B.; *et al.* (1980) Effects of sodium butyrate and dimethylsulfoxide on biochemical properties of human colon cancer cells. *Cancer* **45**, 1185–1192.
[15] Gamet, L.; Daviaud, D.; Denis-Pouxviel, C.; Remesy, C.; Murat, J.-C. (1992) Effects of short-chain fatty acids on growth and differentiation of the human colon cancer cell line HT29. *Int. J. Cancer* **52**, 286–289.
[16] Richter, F.; Weimer, A.; Stahl, A.; Richter, A.; Scheppach, W.; Bartram, H. P.; Kasper, H. (1996) Divergent effects of short chain fatty acids on growth and differentiation of HT29 colon cancer cells. *Gastroenterology* **110**, A 583.
[17] Saini, K.; Steele, G.; Thomas, P. (1990) Induction of carcinoembryonic antigen-gene expression in human colorectal carcinoma by sodium butyrate. *Biochem. J.* **272**, 541–544.
[18] Siddiqui, B.; Kim, Y. S. (1984) Effects of sodium butyrate, dimethyl sulfoxide, and retinoic acid on glycolipids of human rectal adenocarcinoma cells. *Cancer Res.* **44**, 1648–1652.
[19] Gum, J. R.; Kam, J. C.; Hicks, J. W.; Sleisenger, M. H.; Kim, Y. S. (1987) Effects of sodium butyrate on human colonic adenocarcinoma cells. Induction of placental-like alkaline phosphatase. *J. Biol. Chem.* **262**, 1092–1097.
[20] Deng, G., Liu, G., Hu, L.; Gum, J. R.; Kim, Y. S. (1992) Transcriptional regulation of the human placental-like alkaline phosphatase gene and mechanisms involved in its induction by sodium butyrate. *Cancer Res.* **52**, 3378–3383.
[21] Gibson, P. R.; Moeller, I.; Kagelari, O.; Folino, M.; Young, G. P. (1992) Contrasting effects of butyrate on the expression of phenotypic markers of differentiation in neoplastic and non-neoplastic colonic epithelial cells *in vitro*. *J. Gatroneterol. Hepatol.* **7**, 165–172.
[22] De Haan, J. B.; Gevers, W.; Parker, M. I. (1986) Effects of sodium butyrate on the synthesis and methylation of DNA in normal cells and their transformed counterparts. *Cancer Res.* **46**, 713–716.
[23] Whitlock, J. P.; Galeazzi, D.; Schulman, H. (1983) Acetylation and calcium dependent phosphorylation of histone H3 in nuclei from butyrate treated HeLa cells. *J. Biol. Chem.* **258**, 1299–1304.
[24] Kruh, J.; Tichonicky, L.; Defer, N. (1994) Effect of butyrate on gene expression. In: Binder, H. J.; Cummings, J. H.; Soergel, K. (Eds.) *Short chain fatty acids*. Dordrecht, Kluwer, p. 135–147.
[25] Toscani, A.; Soprano, D. R.; Soprano, K. J. (1988) Molecular analysis of sodium butyrate-induced growth arrest. *Onocogene Res.* **3**, 223–238.
[26] Foss, F. M.; Veillette, A.; Sartor, O.; Rosen, N.; Bolen, J. B. (1989) Alterations in the expression of $pp60^{c\text{-}src}$ and $p56^{lck}$ associated with butyrate-induced differentiation of human colon carcinoma cells. *Oncogene Res.* **5**, 13–23.

[27] Wang, J.; Friedman, E. A. (1998) Short-chain fatty acids induce cell cycle inhibitors in colonocytes. *Gastroenterology* **114**, 940–946.
[28] Tsujii, M.; DuBois, R. N. (1995) Alterations in cellular adhesion and apoptosis in epithelial cells overexpressing prostaglandin endoperoxidase synthase 2. *Cell* **83**, 493–501.
[29] Gibson, P.; Rosella, O.; Rosella, G.; Young, G. (1994) Secretion of urokinase and plasminogen activator inhibitor 1 by normal colonic epithelium *in vitro*. *Gut* **35**, 969–975.
[30] Sier, C. F. M.; Vloedgraven, H. J. M.; Ganesh, S.; *et al.* (1994) Inactive urokinase and increased levels of its inhibitor type 1 in colorectal cancer liver metastasis. *Gastroenterology* **107**, 1449–1456.
[31] Gibson, P. R.; Rosella, O.; Rosella, G.; Young, G. P. (1994) Butyrate is a potent inhibitor of urokinase secretion by normal colonic epithelium *in vitro*. *Gastroenterology* **107**, 410–419.
[32] Perrin, P.; Cassagnau, E.; Burg, C.; *et al.* (1994) An interleukin 2/sodium butyrate combination as immunotherapy for rat colon cancer peritoneal carcinomatosis. *Gastroenterology* **107**, 1679–1708.
[33] Velazquez, O. C.; Jabbar, A.; DeMatteo, R. P.; Rombeau, J. L. (1996) Butyrate inhibits seeding and growth of colorectal metastases to the liver in mice. *Surgery* **120**, 440–448.
[34] Fearon, E. R.; Vogelstein, B. (1990) A genetic model for colorectal tumorigenesis. *Cell* **61**, 759–767.

D Anticarcinogenic Factors

17 Involvement of Free Radicals in Carcinogenesis and Modulation by Antioxidants

Anthony T. Diplock †

17.1 Abstract

Substantial epidemiological evidence links high antioxidant status in human populations with a low risk of degenerative disease. The consumption of large amounts of fruit and vegetables has been shown unequivocally in a considerable number of studies to be associated with a lowered risk of several kinds of cancer in a number of different body sites, and this effect is present even after adjusting for heavy smoking. It remains to be proven with certainty that this effect is due to the content in the fruits and vegetables of antioxidant substances which are the protective agents, but substantial circumstantial evidence points to the likelihood that it may prove to be correct that antioxidant nutrient and non-nutrient substances in fruit and vegetables are the most important anti-carcinogenic factors in these foods. In addition to the classical antioxidants vitamin C and vitamin E, β-carotene may be functioning as an antioxidant or by some other means; also present are many phenolic substances, among which are the flavonoids, which may also contribute to the total antioxidant potential of the diet and thus may lower the risk of cancer. Similar considerations apply to atherosclerosis, with its cardiovascular and cerebrovascular risk connotations; it has been clearly shown that antioxidant nutrients can lower the oxidation of low density lipoproteins in the vascular wall, which is thought to be a major early factor in the development of arteriosclerosis. It is also possible that antioxidants can play an important role in delaying the thrombotic events associated with the onset of heart attack or stroke, and there is thus a possibility that antioxidants have a dual effect in lowering the risk of vascular accidents. If it is proven that antioxidants can lower risk of these degenerative diseases it is necessary to be assured that enhancement of the diet with antioxidants is safe and free from risk of undesirable side effects. The literature on vitamin C and vitamin E shows that their toxicity is very low. High levels of intake of vitamin E are contraindicated in subjects with vitamin K-associated coagulation disorders. Anxieties in earlier literature about association between high levels of intake of vitamin C and renal oxalate stone formation, uricosuria, vitamin B_{12} destruction, mutagenicity and iron overload, have proved to be unfounded. Toxicity of β-carotene has also been classically considered to be very low and there is evidence that intakes of 15–50 mg/day are without side-effects, except hypercarotenaemia in some subject at the higher levels of intake. The finding of an enhanced risk of lung cancer in heavy smokers given high doses of β-carotene requires further investigation, although it is likely that the cancer in these subjects was already established as a pre-cancerous lesion before the β-carotene was given.

17.2 Introduction

Carcinogenesis is a multi-factorial process that takes place over a long period of time. The well-recognized stages of initiation and promotion lead to a truly transformed cell type; this involves many stages at which free radical-involving processes might be implicated, and free radical processes may also have an indirect effect on carcinogenesis. The role of such free radical-involving or -derived processes is of critical importance to the question of whether reduction of risk of cancer by antioxidants is a likely premise.

17.3 Free radical generation and DNA as target for attack

Primary radical species differ markedly in their chemical reactivity, and potential for damage to living cells. Aside from the question of whether it is formed at all in the biological environment, of great potential significance as an initiator of damage is the hydroxyl radical, $OH^•$, since it reacts with most biological macromolecules in its immediate vicinity with rate constants in the order of 10^{-10} or more. This is both an advantage and a disadvantage; the primary damage caused by $OH^•$ is likely to be very local and more probably containable, although it must be appreciated that secondary radicals, or radical products, may be produced and that these may be able to move within the intracellular or even intercellular environment, and may be severely detrimental at sites far removed from the point of the initial attack.

With respect to damage specifically to DNA, which may be of significance in carcinogenesis, measurement of the production *in vivo* of modified purine and pyrimidine bases, which are probably derived from DNA excision and repair, has given rise to the estimate that oxygen radicals cause about 10,000 DNA base modifications per cell per day [1]. Such damage by radical species may lead to mutagenesis and carcinogenesis. An account of the effect of free radicals on DNA is that edited by Halliwell and Aruoma (1993) [2]. It is clear that damage to DNA may be of primary significance in the etiology of cancer, by causing direct mutagenic effects, by being involved in the promotion of transformation of mutated cells, as well as in more diffuse effects through expression of genes that may be important in the same context. Both DNA damage and mitogenesis, by agents that increase the rate of mitosis, are thought to be of significance in the cancer process. Four endogenous processes that lead to significant DNA damage are oxidation, methylation, deamination and depurination, and repair mechanisms exist in most cellular systems for these processes [3]. The measurement of DNA adducts have demonstrated that oxidation is likely to be the most significant endogenous damage. Oxidative DNA damage is prevented, *inter alia*, by glutathione and in glutathione-depleted cells oxidative DNA damage increases. Rapidly dividing cells are much more likely to mutate than non-dividing cells and mitogen-induced increases in mitosis expose the cells to greater risk. Lowering the rate of mitogenesis causes a greatly lowered incidence of cancer.

The chemistry of attack by free radicals on DNA is very complex; lesions in chromatin include damage to bases, sugar lesions, DNA-nucleoprotein cross-links, single strand-breaks, and abasic lesions [4]. Hydroxyl radicals, e^-_{aq} and H atoms react with DNA bases by addition and abstraction. Reaction of $OH^\bullet$ with both pyrimidines and purines gives several addition products and the detection of specific degradation products of DNA metabolites is probably the best evidence available that $OH^\bullet$ formation occurs *in vivo*. Sugar radicals can be formed by the reaction of $OH^\bullet$ with deoxyribose and further reaction of the sugar radicals formed can lead to DNA strand breakage. The further reactions of free radical-altered DNA bases are beyond the scope of this short review. A number of methodologies have been developed for measuring and characterizing these products and the application of these techniques to body fluids offers the possibility of non-invasive assessment of free radical damage to DNA *in vivo* [5]. Detection of altered sugars may also be measured to provide further information about this kind of DNA damage. Photochemical reactions induced by photosensitizers may also lead to cellular DNA modification [6]; DNA base modifications are the most common kinds of such damage and strand breaks are of lesser significance. Singlet oxygen is likely to be involved in these processes. Oxidative modification of guanine and thymine residues, and covalent adduct formation are among the types of reactions that occur. The damage may cause cytotoxicity by blocking DNA replication which can also give rise to mutation by misreading of modified sites by DNA polymerases or by induction of DNA recombinatorial events; both may be counteracted by the normal cellular repair mechanisms but the possibility exists that, in particular where mitosis is occurring, the repair mechanisms may not be adequate to prevent progression of the damaged cell to a cancerous one.

The nature of the free radical species that is involved in these processes, and their origin, has not been clearly resolved. It is clear that hydroxyl radicals and singlet oxygen produce several different types of damage to DNA, and alkoxyl radicals have also been shown to cause DNA damage [7]. There is however no information available as to the formation of singlet oxygen in the cell nucleus. The reactivity of the oxygen species implies that they must be generated in the nucleus and it is unlikely that migration of such reactive species can account for direct damage to DNA. If it is true that free iron and copper exist in the nucleus [8], then the relatively stable compound hydrogen peroxide, which can migrate freely within cellular structures, may give rise to hydroxyl radicals *in situ* in the nucleus by Fenton-type reactions. A source of hydrogen peroxide that may be of particular significance in the mutagenic and carcinogenic process are activated phagocytes which are in close proximity to the site of initiation of tumorigenesis, or in sites of inflammation. The oxygen burst associated with the phagocytic function is a considerable source of both hydrogen peroxide and the superoxide anion radical.

There is thus a considerable likelihood that relatively straightforward molecular alterations to DNA structure could lead to mutagenesis and eventual carcinogenesis. However, this is an overly simplistic view and a number of other potential mechanisms of carcinogenesis must also be considered. The involvement of free radicals in cell proliferation has been reviewed by Burdon [9], and the question of the involvement of lipid peroxides and their products in the carcinogenic process has been considered by Cheeseman [10] and by Morrero and Marnett [11]. The mechanisms underlying the regulation of mammalian cell proliferation are being clarified and the primary controlling agents are specific serum growth factors that provide external signals. They interact with cell surface receptors setting in motion intracellular second messenger systems which are able to activate genes, including proto-oncogenes, to provide key biochemical timing devices. These events are of great complexity but there is no doubt that free radical-related events are important as modulators of growth regulation.

17.4 Lipid peroxidation and carcinogenesis

An important potential relationship also may exist between lipid peroxidation and the causation of mutagenesis and carcinogenesis involving damage to DNA. This is the interaction of aldehyde metabolites of lipid hydroperoxides with DNA which may have consequences for the cell cycle and in carcinogenesis.

Many observations show an inverse relationship between levels of lipid peroxidation and rates of cellular proliferation (for example [12]). Bursts of DNA synthesis in the S-phase of regenerating rat liver were shown to be accompanied by lowered lipid peroxidation and a rise in vitamin E level [13]. However, supplementation of growth media of cultured cells with physiological concentrations of α-tocopherol leads to variable results; Balb/3T3 cells were shown to undergo stimulation by the vitamin [14] wheras in other experiments α-tocopherol was without effect [15]. The reason for this difference may lie in the variability of the very small levels of α-tocopherol to be found in different batches of the foetal calf sera used in the culture medium. In experiments with (BHK-21) cells in culture, addition of α-tocopherol led to significant growth enhancement which was larger in such cells derived by transformation with the DNA tumor polyoma virus (BHK-21/PyY cells) [16]; this was taken to indicate that α-tocopherol does indeed stimulate cellular proliferation. In the same experiments it was also shown that the growth enhancing effect was accompanied by lowered malonyldialdehyde formation, an index of lipid peroxidation.

Lipid peroxides may also affect the rate of cell proliferation through aldehyde products of lipid hydroperoxide breakdown; a variety of carbonyls, such particularly as hydroxyalkenals like 4-hydroxynonenal, may affect enzymic activities and other functions of proteins by their reaction with thiol and amino groups of the protein side-chain. This occurs at high concentrations of the aldehyde but this may however have little physiological relevance because, at lower concentrations, the aldehydes appear to have different effects on cell proliferation, which for example are mediated through effects on adenyl cyclase and phospholipase C activities, and on proto-oncogene expression. These effects at low, perhaps physiological, concentrations may have considerable significance in the cancer process [9].

The multi-factorial nature of carcinogenesis necessitates consideration also of the role of free radicals, and of lipid peroxides in particular, in chemical carcinogenesis. In a review by Morrero and Marnett [11], the metabolic activation of carcinogens by peroxyl radicals is considered and its effect in the metabolic activation of polycyclic hydrocarbons and their DNA conjugates. Phorbol ester and non-phorbol ester tumor promoters also increase oxygen radical production in mouse skin, which may be relevant to the process of tumor promotion. The mechanisms by which peroxyl radicals may enhance the initiation of tumors is clear, for example, epoxidation of benzo[*a*]pyrene-7,8-diol by peroxyl radicals generates a potent mutagen which leads to tumor initiation. But measurement of the relative amounts of the DNA adduct formed by epoxidation indicates that peroxyl radicals play only a minor role in the process. On the other hand there is strong evidence that generation of peroxyl radicals occurs during tumor promotion by phorbol esters.

17.5 Modulation by dietary antioxidants of free radical attack: epidemiological studies

It is clear from the foregoing that the many possible ways in which free radical-related events in cellular metabolism may impinge on the cancer process, must provide an excellent rationale, which has not as yet been explained in detail, from which the potential role of antioxidants as agents that reduce the risk of cancer will eventually emerge. Antioxidants are at present divided into the accepted nutrient substances, which include the minerals copper, zinc, manganese and selenium, and vitamins C and E, and the non-nutrient antioxidants which include the major carotenoids, which may or may not act as antioxidants *in vivo*, and the large range of polyphenolic substances present in diet, notably the flavonoids. Of particular significance to this discussion are selenium, vitamin C and vitamin E, and β-carotene.

A detailed study of the then available epidemiological evidence regarding what was perceived to be the vexed question of the possible anti-carcinogenic effects of selenium was made by the WHO Task Group on selenium in human health [17] (in particular pages 214–221 and 235–236). Negative correlations had been drawn between cancer incidence, or death rates, and some general population characteristics such as blood selenium levels, or the average level of selenium intake in populations, in specified geographical zones. In the years since the publication of the WHO study many papers have appeared which address the question of the role of selenium in cancer prevention. The publications of particular significance are those of the prospective and interventional epidemiological kind that appear to support a role for selenium as an anti-carcinogen. For the present, however, the question must remain an open one, although it seems likely that, as more carefully controlled clinical epidemiological trials are conducted, it will be found that selenium is indeed a risk factor in the prevention of cancer in some, but not in all, sites in human subjects.

There is considerable interest in the possibility that dietary vitamin E, vitamin C and β-carotene may be able to lower the risk of a wide range of human cancers. The search for a single protective agent has confused the issue of the link between a high intake of a specified nutrient with lowered risk of a particular cancer on the one hand, and the much more extensive evidence that links a high intake of fresh fruit and vegetables with a lower risk on the other. Reviews of the epidemiological literature have suggested that there is a protective role for vitamin C [18, 19] and β-carotene [20, 21] against the incidence of cancer. In all the studies, where correspondents were asked about their fruit and vegetable consumption, it is not certain that the effect that was reported was indeed due to the nutrient in question, and it must be recognized that no allowance was made for the effect of other factors in the foods on the risk under investigation. Although the evidence linking low intake of any specified individual antioxidant with elevated risk of cancer is not very impressive there is nevertheless overwhelming evidence that links low incidence of cancer in many body sites with a high intake of fresh fruits and vegetables. In a large meta-analysis of many pieces of evidence [22] this relationship was studied in detail. It was accepted for the purposes of this review that the anticancer activity of carotenoids derives from the action of the carotenoid itself rather than from its first being converted in the body into vitamin A. The methodology of studies in the literature differs somewhat, and the meaning of relative risk, in particular when it is given a numerical value, may differ also. The review [22] used a method by which results can be compared between different studies. In all such studies, information about intake of a nutrient through diet was obtained by a questionnaire on frequency of consumption of named foods. Respondents were then grouped into those with low, moderate or high intake of individual foods, groups of foods, or of a nutrient contained in them which was calculated. Risk of cancer is expressed as Relative Risk (RR) and the risk of cancer in the group exposed to a factor (such as low fruit and vegetable intake, or low carotenoid intake) is expressed as a ratio of the risk in the group not so exposed (those with the highest intake). Thus an RR of >1.0 indicates an increased risk of disease, and an RR of 2.0 in a low-consuming group indicates twice the risk of cancer compared with the high-consuming group. Similarly an RR of <1.0 indicates a lowered risk of disease where the same comparison is being made. Table 17.1 is derived from the review [22]. The data concern established invasive cancer, and precancerous conditions are omitted. The statistical significance of the results was based on the results reported by the individual authors: $p<0.05$ or more, the lower level of the 95 % confidence interval at an RR ≥ 1.0. In almost all the studies reviewed adjustment for smoking was made or the effects in smokers were reported separately.

Table 17.1: Summary of human epidemiological studies up to 1992 of fruit and vegetable intake and reduction of cancer risk (from: Block, Patterson, Subar (1992) [22]).

Site	No. of studies	Protective (p<0.05)	Harmful (p<0.05)	Relative Risk (Range)
All	170	132	6	
All except prostate	156	128	4	
Lung	25	24	0	2.2 (1.2–7.0)
Oral cavity/pharynx	9	9	0	2.3 (1.7–2.5)
Larynx	4	4	0	2.0 (2.1–2.8)
Esophagus	16	15	0	2.0 (0.7–4.8)
Stomach	19	17	1	2.5 (0.5–5.8)
Colorectal	27	20	3	1.9 (0.3–3.3)
Bladder	5	3	0	2.1 (1.6–2.1)
Pancreas	11	9	0	2.8 (1.4–6.4)
Cervix	8	7	0	2.0 (1.2–4.7)
Ovary	4	3	0	1.8 (1.1–2.3)
Breast	14	8	2	1.3 (1.1–2.8)
Prostate	14	4	2	1.3 (0.6–3.5)
Miscellaneous	8	6	0	

Note: in many studies significance levels of $p<0.01$ or stronger were reported. Significance was defined here as $p<0.05$ for comparison purposes only.

It is clear from the results presented in Table 17.1 that there is strong consistency in the data that link a low level of intake of fresh fruits and vegetables with a higher risk of cancer. However, at around the time that this work was published, there was a tendency to assume that the effective agents in dietary fruit and vegetables were the antioxidants. This may be far from the true case. Intervention studies that involve specific nutrients have given mixed results. In the study in Linxian in which supplementation with specific vitamin and mineral combinations was undertaken [23], it was hoped to reveal whether supplementation with specific vitamins or minerals might lead to lower cancer incidence or mortality. Four combinations of nutrients were tested and doses ranged from one- to two-times the US RDA's. A total of 2127 deaths occurred among trial participants during the intervention period. Cancer was the leading cause of death with 32% of all deaths being due to esophageal or gastric cancer. This population is unusual because this is a very high rate of incidence of gastric and esophageal cancer. Lower total mortality occurred among those subjects receiving supplementation with β-carotene, vitamin E and selenium together ($p = 0.03$). This reduction was mainly due to lower cancer rates (RR = 0.87, 95% CI = 0.75–1.00) with stomach cancer being especially significant (RR = 0.79, 5% CI = 0.64–0.99), with the reduction in the risk becoming apparent 1–2 years after the supplementation began. The authors concluded that these results were "not definitive but that the promising findings should stimulate further research to clarify the potential benefits of micronutrient supplements".

The Finnish α-Tocopherol β-Carotene (ATBC) study of 29,133 heavy chronic smokers tested the effects of 20 mg β-carotene, either alone or in combination with 50 IU vitamin E (33.5 α-TE), for an average of 6 years. There was a significant *increase* of lung cancer incidence (16%) in the groups which received β-carotene [24]. A more detailed analysis of the results revealed that the increased risk of lung cancer appeared to be restricted to participants who had smoked more than 20 cigarettes per day over an average period of 30 years [25].

The β-Carotene and Retinol Efficacy Trial (CARET) with 18,314 subjects at high risk for lung cancer (heavy smokers, and asbestos-exposed workers) evaluated the combination of 30 mg β-carotene and 25,000 IU vitamin A over an average of 4 years. The intervention group had a significantly increased risk of lung cancer (relative risk = 1.36) [26]. A reduced risk of lung cancer (relative risk = 0.80) was seen in subjects who were former smokers at the beginning of the study. Interestingly, participants with high initial serum β-carotene concentrations had a 31% reduction in risk of lung cancer ($p = 0.003$), regardless of which group they were randomized to. This effect was also seen in the ATBC study and Physicians' Health Study which was conducted over 12 years in 22,071 male physicians who consumed 50 mg β-carotene every second day. There was no beneficial influence of β-carotene on overall cancer incidence. However due to the long duration of the trial, it is important to note that there were no adverse effects reported [27].

It is thus clear from the foregoing that mixed messages emerge at present from the literature concerning intervention with individual antioxidants in the cancer process. A major factor that contributes to this confusion is undoubtedly the timing of the intervention; in both the ATBC and CARET studies, intervention was made following many years of high risk behaviour, smoking, at a time when it would be expected that many of the subjects would have been in a precancerous condition. It is clear that the intervention was apparently instrumental in turning this precancerous state in some individuals into overt cancer, and the explanation for this is not immediately apparent. However it nevertheless seems likely that intervention with β-carotene at a much earlier stage of the cancer process would have been more likely to have had beneficial results, because it is particularly at those early stages that free radicals are thought to have their major impact.

17.6 Enhancement of antioxidant intake and safety

If it becomes likely that an increase in the human dietary intake of antioxidants is to be recommended as a means designed to lower the risk of cancer, then it must be certain that this proposed intervention will be without risk due to some toxic effect. From the point of view of safety, the effect of antioxidants must strictly be concerned with their effects *in vivo*; however, the possibility that a given compound may be converted by chemical reaction, or bacterial action, into a toxic substance within the gastrointestinal tract, or during storage of the food that contains it, must also be borne in mind because such products may themselves be absorbed and exert their toxicity *in vivo*. With respect to vitamin C, the conclusion from an exhaustive survey of the literature is that oral intake of high (up to 200 mg/day) levels of vitamin C are safe and entirely free from side-effects; even very high levels (up to 2000 mg/day) have not been consistently reported to result in side effects although some reports, of low reliability, may suggest that minor side-effects may occur. With regard to vitamin E, the following conclusions which were reached [28, 29] with respect to the safety of oral intake of vitamin E by human subjects, can be endorsed here. (i) The toxicity of vitamin E is very low. (ii) Animal studies show vitamin E is not mutagenic, carcinogenic or teratogenic. (iii) Reported increases in serum lipids in human subjects following high oral dosage were found to be inconsistent and of little significance when careful studies were conducted. (iv) In double blind human studies, oral dosage resulted in few side-effects, even at a dosage as high as 3.2 g per day. (v) Dosage up to 1000 mg per day is considered to be entirely safe and without side-effects. (vi) Oral intake of high levels of vitamin E can exacerbate the blood coagulation defect of vitamin K deficiency: high vitamin E intake is thus contra-indicated in these subjects.

The question has been raised of the safety [30] of long-term ingestion of amounts of vitamin E in doses of 100–400 mg per day which were found [31, 32] to confer significant protection against the risk of coronary artery disease. Higher doses than this (about 500 mg daily) have also been shown to confer benefit in apparently causing improvement in subjects with angiographically proven cardiovascular disease [33]. Further, results [34] of a trial in which 100 mg D-α-tocopheryl acetate was administered for six years showed that there were no adverse effects during clinical follow-up. It is not possible at present to state categorically that oral ingestion of large amounts of vitamin E is entirely safe for long periods, because no one has been able to test the toxicity of the vitamin for such periods of time. The usual criteria of safety applied to any new drug suggests that it should be found that long-term supplementation will prove to be free from harmful side-effects and that this may also be beneficial to health.

With regard to β-carotene, supplementation of normal individuals in the population with moderate supplements is safe. The safety of this for heavy smokers, who are at high risk of developing lung cancer, has been put in question by the two recent studies [24, 26] described above. The increase in incidence of lung cancer (18% and 28% in the two studies) has no easy explanation but it is necessary to caution that, until more work is done, administration of more than 10 mg/day β-carotene to smokers is contra-indicated. Observational epidemiological evidence suggests that subjects who are not heavy smokers may benefit from β-carotene administration which is likely to exert a protective role at an early stage of the cancer process. β-Carotene supplementation at levels up to 10–12 mg per day is entirely safe.

References

[1] Ames, B. N.; Shingenaga, M. K.; Park, E. M. (1991) In: Davies, K. J. A. (Ed.) *Oxidation damage and repair: Chemical, biological and medical aspects.* Pergamon, New York, p. 181–187.

[2] Halliwell, B.; Aruoma, O. I. (1993) *DNA and free radicals.* Ellis Horwood, New York, London, p. 1–332.

[3] Ames, B. N.; Shigenaga, M. K. (1993) In: Halliwell, B.; Aruoma, O. I. (Eds.) *DNA and free radicals.* Ellis Horwood, New York, London, p. 1–15.

[4] Dizdaroglou, M. (1993) In: Halliwell, B.; Aruoma, O. I. (Eds.) *DNA and free radicals.* Ellis Horwood, New York, London, p. 19–39.

[5] Halliwell, B. (1993) In: Halliwell, B.; Aruoma, O. I. (Eds.) *DNA and free radicals.* Ellis Horwood, New York, London.

[6] Epe, B. (1993) In: Halliwell, B.; Aruoma, O. I. (Eds.) *DNA and free radicals.* Ellis Horwood, New York, London.

[7] Meneghini, R.; Martins, E. L. (1993) In: Halliwell, B.; Aruoma, O. I. (Eds.) *DNA and free radicals.* Ellis Horwood, New York, London, p. 83–93.

[8] Thorsten, K.; Romslo, I. (1984) Uptake of iron from transferrin by isolated hepatocytes. *Biochim. Biophys. Acta* **804**, 200–208.

[9] Burdon, R. (1994) In: Rice-Evans, C. A.; Burdon, R. H. (Eds.) *Free radical damage and its control.* Elsevier, Amsterdam, London, New York, Tokyo, p. 153–183.
[10] Cheeseman, K. H. (1993) In: Halliwell, B.; Aruoma, O. I. (Eds.) *DNA and free radicals.* Ellis Horwood, New York, London, p. 109–144.
[11] Morrero, R.; Marnett, L. J. (1993) In: Halliwell, B.; Aruoma, O. I. (Eds.) *DNA and free radicals.* Ellis Horwood, New York, London, p. 145–161.
[12] Barrera, G.; *et al.* (1991) *Exp. Cell. Res.* **197**, 148–152.
[13] Slater, T. F.; *et al.* (1986) In: Rice-Evans, C. (Ed.) *Free radicals, cell damage and disease.* Reichlieu Press, London, 57–72.
[14] Giasuddin, A. S. M.; Diplock, A. T. (1979) The influence of vitamin E and selenium on the growth and plasma membrane permeability of mouse fibroblasts in culture. *Arch. Biochem. Biophys.* **196**, 270–280.
[15] Boscoboinik, D.; Szewczyk, A.; Hensy, C.; Azzi, A. (1991) *J. Biol. Chem.* **268**, 6188–6194.
[16] Goldring, C. E. P.; Rice-Evans, C. A.; Burdon, R. H.; Rao, R.; Diplock, A. T. (1993) - Tocopherol uptake and its influence on cell proliferation and lipid peroxidation in transformed and non-transformed baby hamster kidney cells. *Arch. Biochem. Biophys.* **303**, 429–435.
[17] WHO (1987) *Environmental Health Criteria. 58. Selenium.* WHO, Geneva, p. 1–306.
[18] Block, G.; Menkes, M. (1989) In: Moon, T. E.; Micozzi, M. S. (Eds.) *Diet and cancer prevention: investigating the role of micronutrients.* Marcel Dekker, New York, p. 341–388.
[19] Block, G. (1991) Vitamin C and cancer prevention: the epidemiologic evidence. *Am. J. Clin. Nutr.* **53**, 270S–282 S.
[20] Ziegler, R. G. (1989) A review of epidemiologic evidence that carotenoids reduce the risk of cancer. *J. Nutr.* **119**, 116–122.
[21] Basu, T. K.; Temple, N. J.; Hodgson, A. M. (1988) Vitamin A, β-carotene and cancer. *Prog. Clin. Biol. Res.* **259**, 255–267.
[22] Block, G.; Patterson, B.; Subar, A. (1992) Fruit, vegetables and cancer prevention: a review of the epidemiological evidence. *Nutr. Cancer* **18**, 1–29.
[23] Blot, W. J.; *et al.* (1993) Nutrition intervention trials in Linxian, China: Supplementation with specific vitamin? mineral combinations, cancer incidence and disease-specific mortality in the general population. *J. Natl. Cancer Inst.* **85**, 1483–1492.
[24] ATBC* (1994) The effect of vitamin E and β-carotene on the incidence of lung cancer and other cancers in male smokers. *New Engl. J. Med.* **330**, 1029–1035.
[25] Albanes, D.; *et al.* α-Tocopherol and β-carotene supplements and lung cancer incidence in the α-tocopherol, β-carotene cancer prevention study: effects of base-line characteristics and study compliance [see comments]. *J. Natl. Cancer Inst.* **88**, 1560–1570 (1996).
[26] Omenn, G. S.; *et al.* (1996) Effects of combination of β-carotene and vitamin A on lung cancer and cardiovascular disease. *New Engl. J. Med.* **334**, 1150–1155.
[27] Taylor-Mayne, S. (1996) β-Carotene, carotenoids and disease prevention in humans. *FASEB J.* **10**, 690–701.
[28] Kappus, H.; Diplock, A. T. (1992) Tolerance and safety of vitamin E: a toxicological position report. *Free Radical Biol. Med.* **13**, 55–74.
[29] Diplock, A. T. (1995) Safety of antioxidant vitamins and β-carotene. *Am. J. Clin. Nutr.* **62**, 1510S–1516S.
[30] Steinberg, D. (1993) Antioxidant vitamins and coronary heart disease [editorial; comment]. *New Engl. J. Med.* **328**, 1487–1489.
[31] Stampfer, M. J.; *et al.* (1993) Vitamin E consumption and the risk of coronary disease in women. *New Engl J Med* **328**, 1444–1449.
[32] Rimm, E. B.; *et al.* (1993) Vitamin E consumption and the risk of coronary heart disease in men. *New Engl. J. Med.* **328**, 1450–1456.

[33] Stephens, N. G.; *et al.* (1996) Randomised controlled trial of vitamin E in patients with coronary disease: Cambridge Heart Antioxidant Study (CHAOS). *Lancet* **347**, 781–786.
[34] Takamatsu, S.; *et al.* (1995) Effects on health of dietary supplementation with 100 mg D-α-tocopheryl acetate daily for 6 years. *J. Int. Med. Res.* **23**, 342–357.

18 Anticarcinogenic Factors in Plant Foods and Novel Techniques to Elucidate Their Potential Chemopreventive Activities

Beatrix L. Pool-Zobel

18.1 Abstract

High vegetable and fruit consumption is associated with the reduction of cancer in multiple tissues. Phytoprotectants (micronutrients, secondary plant ingredients) and products formed during the fermentation of non-digestible components in the gut have been implicated as acting anticarcinogenic. The phytochemicals comprise of at least 10 different groups, each containing several hundred or thousand individual compounds. Gut fermentation products are multiple and heterogeneous as well, and include soluble compounds, enzymes, and insoluble components. The major mechanisms of action are to prevent etiological risk factors from causing cancer initiation (primary prevention, blocking activity) or progression (secondary prevention, suppressing activity). There are numerous studies with isolated compounds which show that they exert a large diversity of protective mechanisms in various model systems. However, it has been very difficult to actually pinpoint individual substances as being responsible for the anti-carcinogenic effects of plant foods. Epidemiological evidence indicates that only carotenoids, vitamins C and E, selenium and *Allium* compounds may possibly decrease tumor risks. Also, in several chemoprevention trials where natural substances are given as supplements to high risk subjects, the evidence of chemoprotective activity is not conclusive or lacking. Clearly more research is needed to identify phytoprotectants, their properties and mechanisms by which they contribute to reducing risks in multiple tissues. This review will first address general aspects on how to comprehensively evaluate chemopreventive food ingredients. Additionally, using examples of own research, novel techniques (detection of DNA damage and detection of oxidized DNA-bases with the COMET assay, receptor/ligand interactions with a biosensor, glutathione S-transferase induction with ELISA), will be presented with which the activities of phytoprotectants (carotenoids, anthocyanidins, isoflavonoids, lignans, butyrate) have been monitored in target cells of tumorigenesis *in vitro* and are being developed as biomarker techniques. Together with the *in vitro* studies, results of human dietary intervention trials with foods containing the individual compounds will thus provide valuable information on mechanisms of protective effects by plant foods.

18.2 Introduction

The mortality rates of the 10 most frequent cancers in males and females from European countries are presented in Table 18.1. Neoplasms of the lung, in tissues of the gastrointestinal tract and in hormone dependent tissues are most frequent in both sexes [1]. Different causative agents have been identified for different tumor types. The most important avoidable life style factors are tobacco smoke and diet, each causing approximately 30–35 % of all human tumors in western countries [2]. Individual dietary factors may contribute to enhancing risks for the various cancers. These factors are heterogeneous. In many cases individual compounds have been suggested as being involved, but little definitive evidence is actually available. The individual risk factors are specific for different tissues and the state of current knowledge is limited on the bases of individual compounds. For the assessment of food groups or dietary regimens, however, diets high in total fat or in saturated/animal fat are considered possibly causative for most of the frequently afflicted tissues.

18.3 Mechanisms of carcinogenesis

The mechanisms of cancer development include the induction of alterations (mutations, amplification, recombination) in proto-oncogenes, in tumor suppressor genes or in DNA repair genes. The alterations may be of different quantity and quality in the various tumor target tissues (Table 18.1). However, the common denominator is that the accumulation of the alterations produces a clonal selection of cells with aggressive and invasive growth properties. Furthermore an increased stimulation of cell proliferation by growth factors at an early stage of tumor development additionally potentates the progression of the carcinogen-

Table 18.1: Some characteristica of the most frequent, dietary related tumors in Europe.

	Colon	Breast	Prostate	Stomach	Lung	Esophagus	Pancreas	Bladder
Mortality (Europe) males	21.5	/	16.7	12.2	56.4	7.8	7.3	7.3
(per 100.000) females	14.3	28.4	/	4.9	20.3	3.8	5.2	/
genetic alterations[c] proto-oncogenes & tumor suppressor genes (alterations may be *germline*)	*APC*, MCC, *DCC*, KRAS2, P53, TGFBR2 *MSH2, MLH1*	MYB,MYC, cyclin D1, D3, EGFR, HER2, HRAS, *P53*, RB1, *BRCA1*, *BRCA2*, HSTF1, INT2, YES1	MET, HER2, P53, RB1, *HPC-1*	HER2, MET, K-SAM, BCL2, APC, P-cadherin, P53	MYC, MYCN, MYCL, HRAS, RB1, P53, RAF1, JUN	P53, EGF, EGFR, $TGF\alpha$, (for adeno-ca)	KRAS2, DPC4, EGFR, P53	MYC, HRAS, EGFR, P53
etiologic risk factors[d]	diet, IFBD genetic disposition (FAP, HNPCC)	hormonal factors, diet, ionising radiation genetic disposition (BRCA1/2, LFS, AT)	diet	diet infection	smoking diet	smoking diet	diet, smoking	smoking, occupational exposure to carcinogens, infection
link with diet[e]	+	(+)	(+)	+	(+)	+	+	–
enhancement	+	+	(+)	+	+	+	+	+
protection								

Table 18.1 (continued)

	Colon	Breast	Prostate	Stomach	Lung	Esophagus	Pancreas	Bladder
enhancing factors dietary constituents, food processing, other factors	red meat, alcohol (BMI, adult height, frequent eating, total fat, saturated/ animal fat, processed and heavily cooked meat, eggs)	**Energy and related factors**, rapid growth and great adult height BMI, adult weight gain, alcohol, (total fat, saturated/ animal fat, meat)	(total fat, saturated/ animal fat, meat, milk and dairy products)	salting (starch, grilled/ barbecued meat and fish) H. pylori	tobacco (total fat, saturated/ animal fat, alcohol)	**alcohol** (cereals, very hot drinks)	(high energy intake, meat)	(coffee) S. haema-tobium
preventive factors dietary constituents food processing, other factors	**Energy and related factors,** physical activity, **vegetables** (NSP/fibre, starch)	vegetables and fruits (physical activity, NSP/fibre)	(vegetables)	**vegetables and fruits**, **refrigeration** (whole grain cereals, green tea)	**vegetables and fruits**	**vegetables and fruits**	vegetables and fruits (NSP/fibre)	vegetables and fruits
individual food compounds	(carotenoids)	(carotenoids)		vitamin C (carotenoids allium com-pounds)	carotenoids (vitamin C, E, sele-nium)	(carotenoids vitamin C)	(vitamin C)	

1[a] data from [1]; /, not specified, since tumor is gender-specific or not among 10 most frequent tumors in Europe;
1[b] types of diseases
1[c] Most frequent genetic changes, usually arising in somatic tissues; mutations for which germline origin has been identified are *italic*. For references see text and [5];
1[d] FAP, familial adenomatosis polyposis; HNPCC, hereditary non polyposis colon cancer; BRCA, Breast Cancer gene; LFS, Li-Fraumeni Syndrome; AT, Atexia telangiectasia Syndrome; IFBD, inflammatory bowel diseases;
1[e] summary of link with diet [11]: **+**, (bold lettering) convincing evidence, consistent associations according to epidemiological studies; +, (normal lettering) some evidence, probable associations according to epidemiological studies; (+), (lettering in parenthesis) possible associations, supportive epidemiological or experimental studies are available, but are limited in quantity or quality; – no evidence; BMI, body mass index; NSP, non-starch polysaccharides.

esis process [3]. Only approximately 1% of all cancers are due to inheritance of genetic alterations [4]. Most other cancers carry multiple mutations which are acquired in the respective somatic tissues during life time. These may be induced by reactive intermediates which result to a larger extent from a high risk diet than from a low risk diet. Life long dietary regimens, e.g. with high energy food and meat intake, but with low vegetable and fruit consumption is expected to increase the exposure to reactive oxygen intermediates (ROS), products of lipid peroxidation, protein and fat pyrolysis products, which may induce genotoxic lesions and act as growth factors. Figure 18.1 depicts the mechanisms by which the different factors can increase cell growth. Genotoxic agents may lead to alterations in the cancer relevant genes, thus resulting in an impaired control of normal cell growth. These alterations are mutations, amplification and recombination of proto-oncogene sequences which lead to increases of normal protein or to synthesis of altered proteins which enhance proliferation stimuli. Also, mutations, deletions, recombinations of tumor suppressor genes will lead to loss of proteins or to malfunctioning proteins which fail to suppress abnormal growth processes. Alterations in repair genes enhance mutation-induction and result in genetic instability. Endogenous and exogenous dietary factors may interact with membrane associated (e.g. receptor tyrosine kinases) or intracellular receptors (e.g. estrogen receptors) and enhance growth stimuli. An increase of apoptosis induced by nutritional agents can aid mutational altered cells to survive and

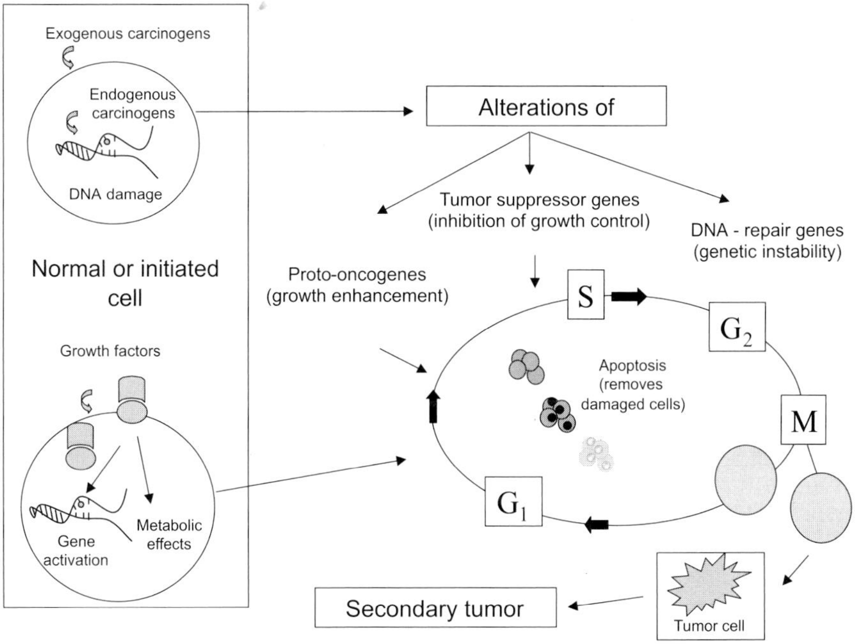

Figure 18.1: Schematic presentation of mechanisms by which risk factors may enhance cell proliferation (modified according to [5], with permission).

further transform to yield malignant cell growth [5]. The result of the combined processes is a higher probability that initiated cells will be transformed, survive and serve as origin for clonal cell expansion and growth of malignant tumors. It is estimated that approximately 20–60 % of human tumors are dietary related because exogenous agents in food and endogenous genotoxic agents formed during nutritionally related metabolism may induce these effects [2, 6].

18.4 The concept of chemoprevention/dietary prevention

Conversely, these dietary related cancers are also considered to be preventable by high vegetable and fruit consumption. Plant foods are protective in many tissues. The numerous findings are exemplified by studies of Block *et al.* [7], or Steinmetz [8] reviewed by Hill [9] and more recently by Potter and Steinmetz [10]. A very comprehensive review has been recently published [11]. The general conclusion is that cancer prevention is also one of the final expected benefits of a healthy diet.

This type of dietary prevention can be considered to be a complex form of chemoprevention, which is the *"use of pharmacological or natural agents that inhibit the development of invasive cancer either by blocking the DNA damage that initiates carcinogenesis or by arresting or reversing the progression of premalignant cells in which such damage has already occurred"* [12]. A difference is that dietary intervention mainly aims at preventing cancer in healthy "normal" humans of the general population (primary prevention), whereas chemoprevention (with single pharmaceutical agents) stresses more the active intervention in high risk subjects or in already diseased individuals (secondary prevention).

The dietary ingredients that are implicated as being the actual chemopreventive factors are phytoprotectants (micronutrients, secondary plant ingredients) and products formed during the fermentation of non-digestible components in the gut. According to Wattenberg [13], the major mechanisms are to prevent carcinogens from exerting their mutagenic effects (blocking agent activity) or to prevent the initiated cell from developing into a further transformed or cancerous cell (suppressing activity). As is shown in Figure 18.2, the blocking agent activity may begin at early stages, already by preventing carcinogen formation [14]. This has been shown for the example of vitamin C which prevents the formation of *N*-nitrosamines in the gastric environment by inhibiting nitrosation reactions. The consequences are a reduced risk for the formation of gastric tumors [15, 16].

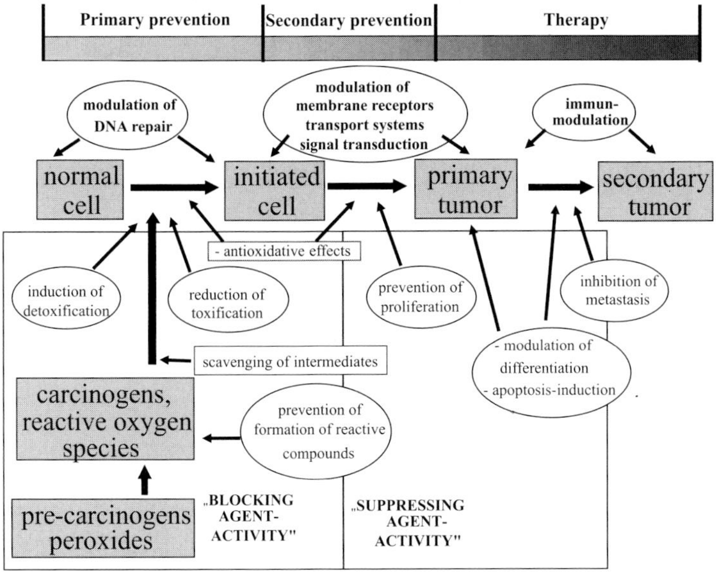

Figure 18.2: Potential mechanisms during which chemoprotective agents and dietary ingredients may interfere with processes of carcinogenesis (modified from [14], with permission).

18.5 Phytoprotectants

On the basis of these activities, some of the many possible phytoprotectants can be identified which, from a mechanistic point of view, may contribute to the reduction of cancer risks from fruits and vegetables. These phytoprotectants belong to a very wide range of compounds. Some of them are essential for the human, as is known for vitamins and trace elements (Table 18.2). In addition to the ones mentioned here, also vitamins D, E and several types of vitamin B have anticarcinogenic potential. Especially vitamin C is probably associated with a decreased risk in the stomach, and possibly in the lung, esophagus and pancreas [11]. In addition to the inhibition of nitrosation, discussed above, another mode of activity is its antioxidative potential. Also selenium and sulphide compounds contained in garlic, onion and leek (Table 18.2) are possibly associated with a decrease of tumors in stomach and in the lung [11]. Selenium is an antioxidative agent, partially due to its being a cofactor of the important antioxidative enzyme glutathione peroxidase. Recently, a strong inverse association between prediagnostic selenium levels in toe nail clippings and risk of advanced prostate cancer has been reported [17]. In contrast, the activity of the *allium* compounds also shown in

Table 18.2: Micronutrients and organosulfur compounds.

Compound	Structure	Occurence in plants and human data
Vitamin A (Retinoic acid)	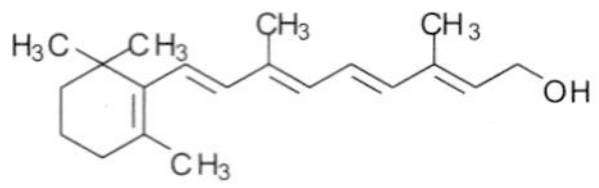	Occurs only in animals, precursor in plants is β-carotenoid, RDI is 100 µg retinol equivalents. Important for functions of vision, systemic role in maintenance of life. Plasma levels range from 1.5 to 1.9 µM.
Vitamin C (Ascorbic acid)	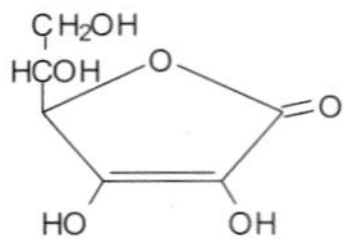	Is a water-soluble micronutrient and ubiquitously present in all plants. The RDI is 80 mg. Functions as a reducer of transition metals, participates in hydroxylation reactions; and is a cofactor for biosynthesis of collagen. The plasma levels range from 50 to 100 µM.
Organselenium	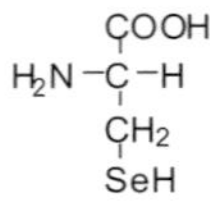	Onions contain the selenoamino acids selenocystine and selenomethionine. Garlic is one of several vegetables with elevated levels of selenium (>3 mg/g fresh weight $\times$ 10.000). Selenium is an essential micronutrient whose absence causes skeletal and cardiac muscle dysfunction. Is required for proper function of the immunsystem and defense against oxidative damage. Plasma levels are at 120 µg/l.
Isothiocyanates	$R-N=C=S$	Occur as glucosinolate conjugates in cruciferous vegetables. Hydrolysis is catalysed by myrosinase, which is released when the vegetables are damaged. These compounds are responsible for the sharp taste of the plant.

Table 18.2 is probably more a result of scavenging activity which inactivates electrophilic intermediates of genotoxic carcinogens [18]. The monoterpene limonene, shown in Table 18.3, is an aromatic compound found in citrus fruits. Its special mode of chemopreventive activity is to inhibit cell growth by inhibiting farnesyl transferases, the consequence of which is a retardation of cell proliferation (see below). Carrots and tomatoes contain the related tetraterpenes, β-carotene and lycopene. These are responsible for giving the vegetables the typical orange and red colour, respectively. The compounds are antioxidants by quenching ROS [19, 20]. They possibly may reduce risks for the development of colon, breast, stomach and esophageal cancers. They are probably associated with a reduced risk for lung cancer [11]. However, in three lung cancer chemoprevention trials with β-carotene, the daily application as supplement either failed to show an effectiveness of this carotenoid or even enhanced the lung tumor incidence for smokers in two of the studies [21, 22]. The reasons for this are still under current investigation. However, possible explanations for the results of the ATBC trial are that an incorrect dose was used, the duration of the study was too short or the study population was inappropriate [23]. Another feasible explanation is that fruits and vegetables are protective not only because of their β-carotene content but on account of the combined action of many carotenoids, other micronutrients and phytoprotectants contained in the whole plant foods. A recent intervention study with tomato and carrot juice, at least has shown pronounced antioxidative and protective properties (decrease in oxidised bases in lymphocyte DNA) in normal, healthy, non-smoking human subjects [24]. Other phytoprotectants, shown in Table 18.4, with antioxidative potential include the large group of polyphenols, which may scavenge ROS or protect against H_2O_2 [25–28]. Their mechanisms are different than those of selenium or carotenoids and additionally include their capacity to chelate transition metal ions [29]. Polyphenols include the large class of flavonoids, chemicals which are widely distributed in plant foods and therefore ingested in significant quantities [30, 31]. One additional protective mechanism which they exert is the induction of antioxidative or other chemopreventive enzyme systems [32–34]. Also inhibition of cell proliferation and modulation of protooncogene and tumor suppressor gene-expression are possible cellular effects exerted by flavonoids [35–39]. Some of these properties have also partially been shown for a special subgroup of the polyphenols, namely the phytohormones (Table 18.5) [40, 41]. These ingredients of soy products (isoflavonoids) and of flaxseed or other grains (lignans), moreover may also interact with estrogen receptors and thus act antiestrogenic and estrogenic as well [42]. The consequence of this interaction is the inhibition or stimulation of cell proliferation [43]. Finally, the induction of apoptosis, also a 'clearing mechanism of tissues', is considered to be protective when occurring in initiated cells [44, 45]. One of the compounds detected at a higher rate of a so called positive fermentation profile is butyrate (Table 18.6). This is formed in the gut by the microflora, upon ingestion of dietary fibre, resistant starch, complex carbohydrates [46, 47]. It has the interesting property of inducing apoptosis in transformed colon cells and inhibiting programmed cell death in normal cells [48–50]. Thus, it should act favourably in both normal epithelium and in colon tissue with the first neoplastic lesions.

Table 18.3: Terpenes.

Compound	Structure	Occurence in plants and human data
Limonene	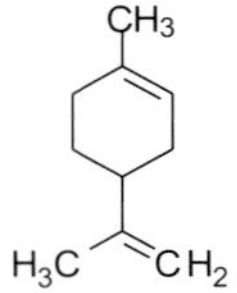	Monoterpene; found in the essential oils of citrus fruits, cherry, spearmint, dill, caraway and other plants. In plants they function as chemoattractants or chemorepellents and are responsible for the plants' pleasant fragrance. Limonene is abundant in orange peel oil. They are derived from the mevalonate pathway in plants and are not produced in mammals. Perillyl alcohol is the hydroxymetabolite of limonene: Plasma levels range from 1.5 to 10 or 20 μM after clinical application [127].
Limonin	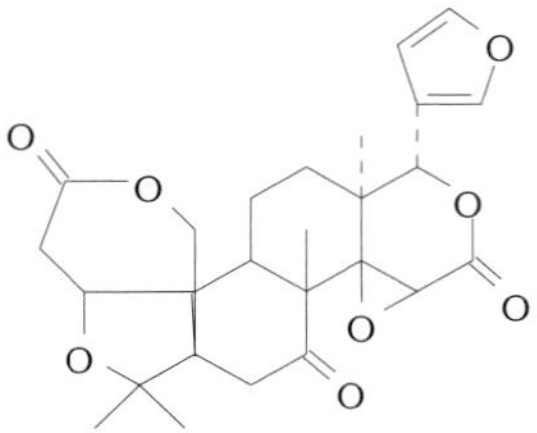	Triterpene, found in grapefruit, responsible for bitter taste.
β-Carotene	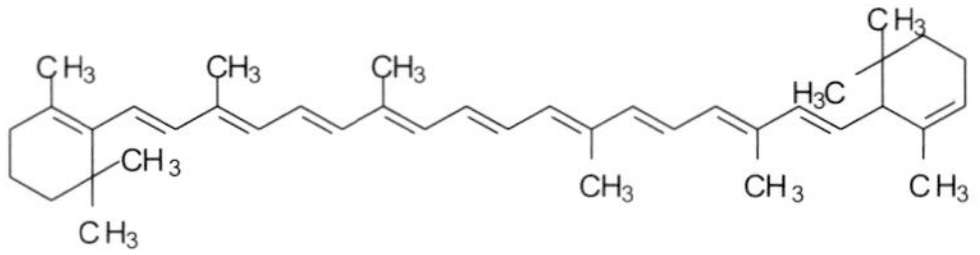	Carotinoid, provitamin A, orange pigment, found in carrots, green leafy vegetables, yellow melons, mangos. Plasma levels increase after daily carrot juice consumption [129]. Increased vegetable consumption increases plasma levels from 6733 to 22376 μg/l [128]. Normal range ist 0.3–0.6 μM [20].
Lycopene	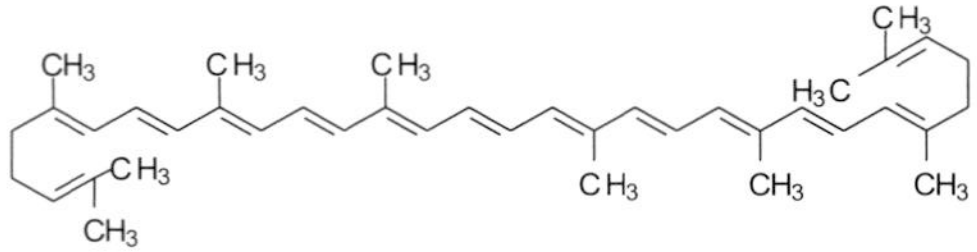	Non-provitamin A carotenoid, red pigment, found in tomatos, watermelon. Plasma levels increase after daily tomato juice consumption [129]. Increased vegetable consumption increases plasma levels from 2288 to 3334 μg/l [128]. Normal range is 0.5–1 μM [20].

Table 18.4: Polyphenolics.

Compound	Structure	Occurence in plants and human data
Cyanidin	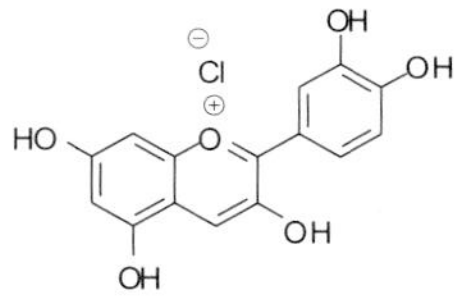	Anthocyanidin, blue red colored pigment found in Elderberry, *Aronia melanocarpa* [130].
Malvidin	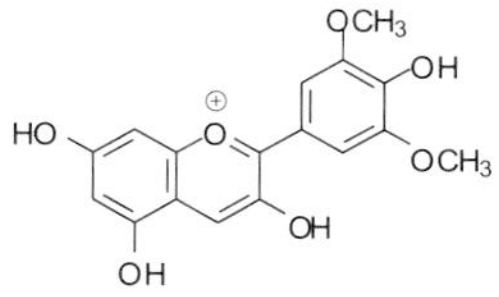	Anthocyanidin, blue colored pigment found in grapes [131].
Quercetin	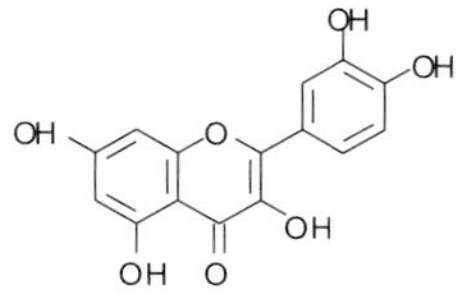	Total daily quercetin intake (17–19 mg) is mainly derived from tea (35–39 %) and from onions (35–38 %) in The Netherlands [132]. Ingestion of quercetin rich meals leads to plasma levels of 0.15 to 0.65 µM 3–4 hours after the meal [133].
Epigallocatechin	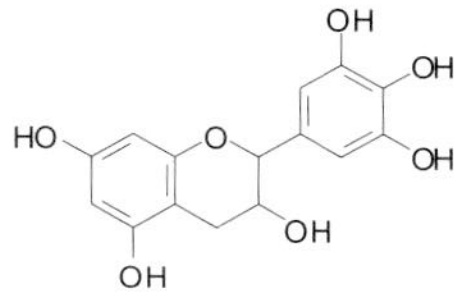	EC and related compounds are in green tea, black tea contains the theaflavins and thearubigins. Tea consumption accounts for half the total daily intake of flavonoids (27–29 mg) in The Netherlands [132]. 1.4–2.4 h after ingestion of 1.5–4.5 of green tea solids, plasma levels of EC and related compounds reach appr. 1–2 µM [134].

Table 18.5: Phytohormones.

Compound	Structure	Occurence in plants and human data
Genistein	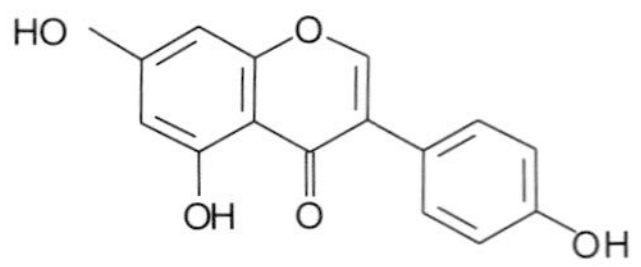	Present in soya plants as gycosides, in fermented soya products also as the agylcone, is formed also from biochanin A. 0.4–0.65 μM were found in 3 males and 1 female after consumption of soya meal with 78 μmol [115].
Daidzein	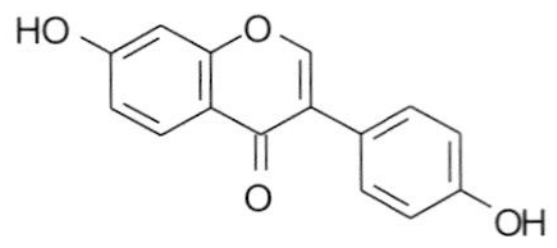	Present in soya products, in clover. Is formed from formononetin and metabolised by the gut flora to equol, o-desmethylangolensin and other metabolites [135]. Peak plasma levels reached appr. 0.5 μM after high consumption of soya or clover [136]. 0.3–0.4 μM were found after consumption of a soya meal containing 53 μmoles [115].
Enterolactone	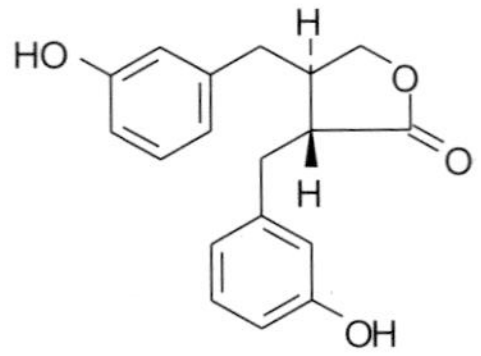	Is formed by gut flora metabolism from the precursor matairesinol in plants (flaxseed, grains, vegetables, coffee and tea) or indirectly by gut flora metabolism from secoisolariciresinol and enterodiol. Approximately 0.14–0.85 μM were found in the plasma after high linseed consumption by women [136]. 0.4–0.65 μM were found by Barnes *et al.* [115].
Tamoxifen	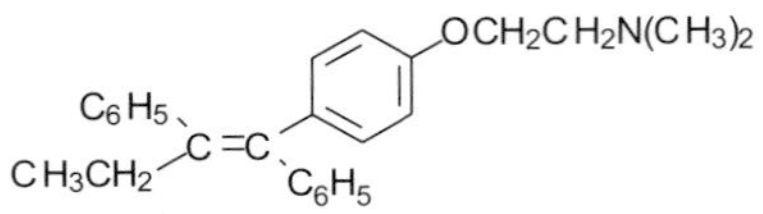	Chemotherapeutic, chemoprotective agent, used for therapy or prevention of breast cancer [137].

Table 18.6: Gut-fermentation profile.

Ingestion of	May lead to	Which is associated with decreased risk
Probiotics Oligosaccharides Non starch polysaccharides Fibre	Increased stool bulk Increased water content in stool An altered composition of gut flora (usually increase of lactic acid- and butyrate-producing bacteria) [139, 140]	By increasing fecal output, lowering exposure time and concentration of carcinogens in faeces, scavenging carcinogens through reaction with unsoluble components
		By lowering concentration of genotoxic compounds associated with foods of a vegetable-poor diet (high fat, meat consumption) [138]
		By decreasing pH, which in turn may inactivate carcinogens or carcinogen producing enzymes
		By reducing formation of carcinogen producing enzymes (β-glucuronidase, nitrate reductase)
		By increasing "favorable" enzymes (glycosidases which enhance liberation of aglycones from plant glycosides [141]
		By increasing production of butyrate, which has several types of protective activities [127]

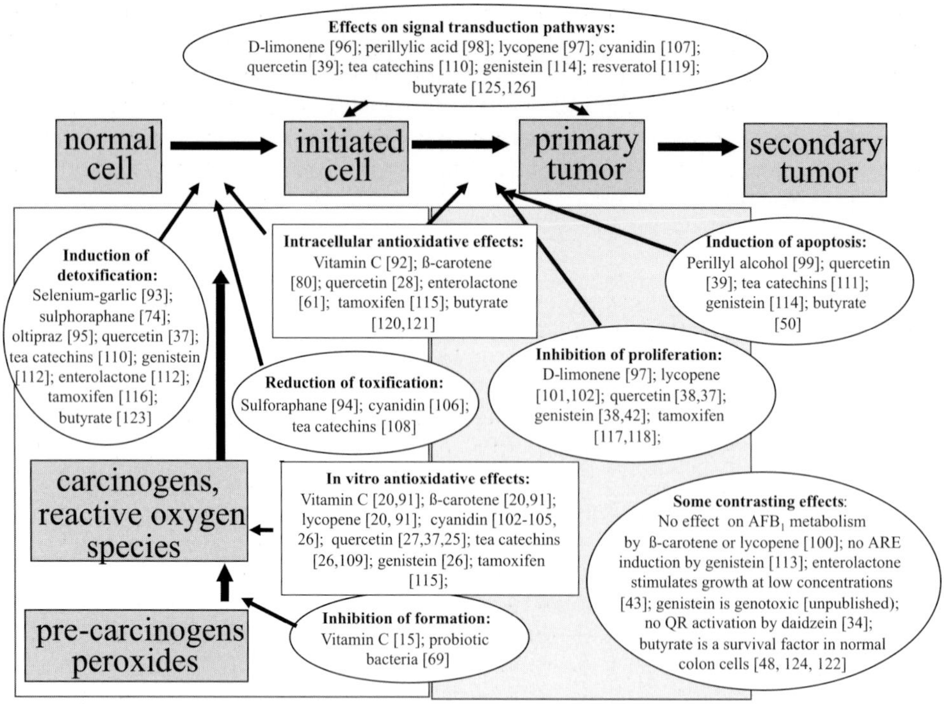

Figure 18.3: Summary of individual chemopreventive activities by selected phytoprotectants.

In summary, as is shown in Figure 18.3, the diverse, few selected compounds of this review, already span a large range of activities, including blocking and suppressing agent properties.

18.6 Mechanisms of protective activities

The compounds exemplified in the Figure 18.3 more or less have blocking and suppressing activities. However, they have been shown to inhibit carcinogen-induced tumor formation in animal experiments. Moreover, epidemiological data exist implicating that there is a reduced risk after high consumption of foods containing these compounds. Therefore these mechanistic data also point to the modes of chemopreventive activities by which vegetables and fruit may be cancer preventive. For each of these compounds and many related compounds of the same groups, numerous studies are available with *in vitro* assays, with cultured cells, in animal studies and in human trials. Only a few examples of these diverse activities for individual compounds are presented. The detailed presentation of these and other manifold investigations are of course beyond the scope of this review. Therefore, the following will focus on some major modes of protective activities presented in the figures, describe in which manner the selected compounds can act accordingly, and discuss the meaning of the findings for overall protection by foods which contain these substances.

18.6.1 Antigenotoxic activities

Genetic damage (DNA breaks, oxidative DNA damage, genome instability) may be indicative of an increased risk. This is based on the assumption that increased DNA damage will enhance the probability of mutations occurring in critical target genes and cells. It is also based on the probability that increased DNA damage is the result of a higher load of genotoxic agents which will enhance the several steps of the complex process of carcinogenesis. These genotoxic agents are the actual etiological risk factors of dietary related cancers. Based upon the positive associations of dietary factors and tumor risks in different tissues (Table 18.1), it is apparent that energy and related factors, animal fat, red meat, and alcohol are important causative dietary factors. It is assumed that carcinogens found in the foods of such diets, such as pyrolysis products (heterocyclic amines, polycyclic aromatic hydrocarbons) and others contribute to cancer induction [51]. Also endogenously formed metabolites and substances like nitroso compounds [52, 53], reactive oxygen species (ROS), and peroxides [54, 55], hormones (e.g. estradiol which is associated with ROS formation during oxidative metabolism) [56, 57], products of lipid peroxidation (4-hydroxyalkenals, aldehydes) [58] additionally contribute. Contaminants have also been implicated as risk factors such as mycotoxins (e.g. aflatoxin B_1), or metal contaminants [59]. It has, however, not been possible to actually pinpoint individual compounds as being particularly responsible for the overall cancer load. Some exceptions may be the special role of aflatoxin B_1 for liver cancer in developing countries, or the role of hormones in breast cancer. Therefore, the relevant determination of "antigenotoxicity" of phytoprotectants should involve their capacity to inhibit a variety of possible risk factors. This type of assessment has been done by several groups and the most frequently employed indicator system was the *Salmonella typhimurium* assay [13]. Studies like this may point to a direct inactivation of the genotoxins by phytoprotectants (e.g. scavenging reactions). In approaches additionally utilising external microsomal systems, modulation of specific modes of genotoxin activation or deactivation, however, may also be assessed. A new, and probably more relevant approach is to study the potential risk factors in cells of the target tissues and then assess the additional modulator capacity of phytoprotectants *in vitro* and *in vivo* [60, 61].

One approach enabling detection of genotoxic effects in remote tissues of experimental animals, but also in the minute samples of human biopsies is the COMET assay. This technique is also known as the single cell gel electrophoresis assay (SCGE). For this, cells are isolated by enzyme digestion from the rat tissue or human biopsies and incubated with the risk factors [62]. Subsequently they are suspended into 75 µl low melting point agarose (0.7 % (w/v) in PBS at pH 7.4, 37 °C) and distributed onto microscopical slides precoated with 0.5 % (w/v) normal melting point agarose. After solidification of the agar, slides are submersed into a lysis solution (1 % (w/v) *N*-laurolyl sarcosin (Na-salt), 10 % (v/v) DMSO, 100 mM Na_2EDTA, 1 % (v/v) Triton X-100, 2.5 mM NaCl and 10 mM Tris) for 60 minutes to remove proteins. The slides are then placed into an electrophoresis chamber containing alkaline buffer (1 mM Na_2EDTA in 300 mM NaOH) for DNA unwinding. After 20 minutes the current is switched on and electrophoresis is carried out at 25 V, 300 mA for 20 minutes. The slides are removed from the alkaline buffer and washed 3-times, 5 minutes with neutralisation buffer (0.4 M Tris-HCl, pH 7.5). Slides are stained with 20 µg/ml ethidium bromide (100 µl/slide). All steps, beginning with the isolation of the cells, are conducted under red light. Using an image analysis system COMET images are evaluated by recording image lengths, % fluorescence in tail (tail intensity) and other parameters [63].

The assay has been used for determining damage by risk factors in rat and human cells of colon, gastric, nasal, lung and breast epithelial cells [60, 64, 65]. Vegetable consumption by humans, lactulose in human gut flora associated rats, and probiotics prevent DNA damage *in vivo* and *in vitro*, in peripheral lymphocytes and in colon cells [24, 66–69].

18.6.2 Antioxidative activities

Many phytoprotectants have the specific capacity to prevent the formation of reactive oxygen species (ROS) or to inactivate them. This particular type of "antigenotoxicity" is better known as antioxidative activity. Antioxidants protect against free radical damage by scavenging ROS or by ending radical chain reactions. Moreover, polyphenolic compounds are able to act as antioxidants by virtue of the hydrogen-donating capacity of their phenolic groups or by chelating transition metals, which catalyse the ROS formation from peroxides. Antioxidative activities are expected to contribute to cancer prevention since both initiation and progression is enhanced by oxidative stress [54, 55, 70]. Oxidants cause DNA damage which may be converted to mutations and induce proto-oncogenes [71, 72]. If this antioxidative effect occurs outside of the cells (e.g. in plasma, colon lumen, or extracellular compartments) there will be a reduced exposure load to ROS. The antioxidative effect within the cell will also protect the DNA by removing the ROS. Using the COMET assay, it is now possible to specifically monitor oxidative DNA damage even in minute tissue samples [73]. Consumption of carrot juice caused a reduction of oxidated DNA bases *in vivo* in peripheral lymphocytes [24]. This may have been due to the action of all-*trans*-β-carotene, which *in vitro* also inhibits oxidised DNA bases in human lymphocytes and in colon tumor cells (manuscript in preparation).

There is, however, the additional effect that the oxidative/reductive environment within a cell may be altered by antioxidants. The consequence of this is an important stimulant for gene expression. Genes which may e.g. be induced by reducing conditions are quinoline reductase or glutathione S-transferase (GST), both important chemopreventive enzymes [32, 74]. The process of activation may proceed over the activation of signal transduction pathways from an antioxidant/redox sensor via an antioxidant responsive element (ARE) and/or be the consequence of altered binding capacities of transcription factors to the DNA [75–77]. The *in vivo* reduction of genetic damage (see above) is connected to induction of GST as well [66]. Additional examples are presently being found in our laboratory.

18.6.3 Induction of chemopreventive enzymes

Protective, antioxidative dietary ingredients stimulate gene activation (e.g. for the chemopreventive, phase II enzymes) and thus render the cell to be less vulnerable and more resistant to risk factors. Therefore antioxidants have a multiple role in protection. Firstly, by direct interaction with ROS, and secondly, indirectly, by induction of the endogenous defence systems. A third mechanism may be the inhibition of enzymes which activate carcinogens to their reactive species. This type of interaction is assessed by measuring e.g. cytochrome P450 metabolism using biochemical substrate conversions. Inactivation may be due to a direct effect of the phytochemicals on the protein or the catalytic centre of the enzyme. Therefore the dietary induction of GSTs may be considered a protective mechanism. In contrast, some of the phytoprotectants may also induce cytochrome P450-enzymes and this is then associated with increased risk, if carcinogens are metabolically activated at a higher rate [78]. However, not all metabolic conversions of cytochrome P450 involve the generation of carcinogens, and many phytoprotectants do not necessarily selectively stimulate only phase II enzyme synthesis. The complexity of effects of chemoprotective agents on phase I and phase II enzymes makes it very difficult to generalise individual findings made in single cell types, after selected doses and times of treatment. Moreover, the extrapolation of results to the diversity of the human exposure situation is difficult. However, more and more data are being generated, showing that fruit and vegetable associated with a reduced risk in tumor formation, are also actually increasing levels of protective enzymes in humans *in vivo* [66, 79]. At the moment only single or few genes are being monitored at the time. Future developments as with cDNA arrays, will make a monitoring of many different genes of the metabolism systems possible. In any case, more mechanism-based research in this area will be necessary in order to interpret unequivocally the meaning of these altered patterns of gene expression for the potential of cancer prevention by individual phytoprotectants, or by complex foods containing the compounds. *In vitro*, all-*trans*-β-carotene, but not lycopene, efficiently reduced DNA breaks and oxidised pyrimidine DNA bases while simultaneously inducing GST [80].

18.6.4 Antiproliferative activity

Additional cellular processes which may serve as biomarkers are cell proliferation or apoptosis (intermediate endpoints). Their detection will indicate early steps of carcinogenesis and could be modulated unfavourably by risk factors or more positively by protective factors. Especially the arrest of proliferation in cells which have already been initiated and are undergoing the processes of transformation are particularly attractive points for prevention by phytoprotectants. The most straightforward approach is the cytotoxicity mediated cell death, including apoptosis, which will prevent proliferation of cultured cells. More indirectly, one mechanism which has been of interest is the property of certain hormone-like plant ingredients to bind to endogenous estrogen receptors. This interaction has been observed for phytohormones such as isoflavonoids and lignans, but also for therapeutically active compounds such as tamoxifen and resveratrol (also a natural phytohormone, occurring in grapes). The consequence of the estrogen receptor binding may be an inhibition of cell proliferation, if the interaction is "static" and has no further physiological consequences. However, also proestrogenic, growth promoting effects have been observed, when the phytohormone acts as an estrogen and it's binding stimulates growth of the affected cell. The question of whether a phytohormone will act estrogenic or antiestrogenic is dependent on the chemical, it's concentration and on the receptors within the target cells. Interestingly, the phytohormones (and polyphenols in general) may also inhibit growth of cells not expressing the estrogen receptor. This may be due to interference with other growth factor receptors, non-identified but similar the ER, or maybe to other mechanisms, such as antioxidative effects.

Another important mode of chemoprevention leading to decreased cell proliferation is the inhibition of farnesyl protein transferase. This activity, initially detected for the examples of monoterpenes counteracts signal transduction pathways involving *ras* proteins. These serve as central connectors between signals generated at the plasma membrane and nuclear effectors [81]. They are responsible for growth control. The proteins require posttranslational modification with farnesyl moiety for anchorage into the plasma membrane and thus for functionality [82]. Inhibition of this step by monoterpenes will inhibit *ras*-dependent proliferative activity in cancerous and precancerous lesions [83]. Much effort is being directed at developing drugs with this type of activity. Meanwhile D-limonene and it's oxidised metabolite perillyl alcohol (also found naturally in cherries) is being studied in the first clinical therapy trials [84].

We are using a novel approach, monitoring extracellular acidification using a microphysiometer [85]. The method utilises the property of cells to excrete protons during energy metabolism, but also as a result of receptor/ligand interactions. Thus epidermal growth factor and insulin induce acidification in human colon tumor cell line HT29 clone 19A. They are known to have the respective receptors. 17β-estradiol does not elicit a response and it is also known, that these cells are devoid of an estrogen receptor. The interesting finding we have, however, is that lignans and equol yield acidification. This may be due to "orphaned receptors" and the consequence could also be enhanced proliferation, which is presently being investigated in our cell culture system.

18.6.5 Induction of apoptosis

Apoptosis, or programmed cell death, is the result of a systematic intracellular cascade of events centred on the activation of cysteine proteases, largely caspases [86]. This genetically determined cascade leads to the death of the cells, which are subsequently removed from the tissue by phagocytosis. It is a normal physiological process to eliminate extraneous or dangerous cells. A malfunctioning of apoptosis is expected to be critical for the development of cancer. Thus the effect of phytoprotectants on the processes of apoptosis is gaining more and more attention. Agents which damage DNA such as chemotherapeutic drugs, X-irradiation, and UV-irradiation can initiate apoptosis [87]. It may also be triggered by proteins involved in the regulatory process of this programmed cell death. Phytoprotectants which induce apoptosis in transformed cells and inhibit apoptosis in normal cells are expected to be of relevant protective activity. Free radical species may modulate apoptosis, independent of the tumor suppressor p53, and chain breaking antioxidants or green tea phenols can provoke this type of apoptosis [88]. Reactive oxygen species are not always required to induce apoptosis, however. In transformed cells, sulfur containing antioxidants have also been used as inducers of apoptosis. They were able to activate p53, probably via a redox sensor as described above for the chemoprotective enzymes. In contrast, chain breaking antioxidants such as vitamin E did not induce p53 dependent apoptosis [89]. These types of activities by antioxidants have potential for secondary cancer prevention, in that transformed cells will be excluded from the tissue.

18.6.6 Modulation of biomarkers

The biomarkers of dietary chemoprevention may be divided into several major categories: In order to assess efficacy of dietary intervention, the biomarkers needed will be a combination of those to determine two types of *effect* (reduction of damage and induction of protective processes), those to determine two types of *exposure* (increase in protective factors and decrease in risk factors) and those which acknowledge several types of susceptibility properties (age, sex, predisposing diseases, immunological status, predetermining and predisposing genetic alterations). A review on biomarker application for cancer prevention is in progress. On the basis of the methods described, above, however, detection of DNA damage and oxidised DNA bases (COMET assay), excretion of genotoxic agents (COMET assay with colon cells [65, 90]) and GST induction are useful approaches.

18.7 Conclusions

In conclusion, the evaluation of dietary anticarcinogens will be based on a different type evaluation than isolated compounds to be used as chemopreventive agents. All evidence accumulated so far showed that high intake of plant foods will reduce risks associated with development of tumors in many tissues. When assuming an otherwise nutritionally well balanced diet, a high intake of a variety of vegetables and fruits will pose no toxicological risk and under normal circumstances should be no issue for cancer risk expectation. Therefore an evaluation of individual phytoprotectants in plant foods interests more on the basis of: Do they contribute to the cancer protective potential of plants? To which extent? Which compounds are most effective? And, will a high intake of these specific plant foods containing specific potentially anticarcinogenic factors actually ensure a risk reduction of tumors e.g. in breast, prostate. The future analysis of mechanisms by which individual compounds, in physiologically relevant concentrations, and in physiological prevailing combinations may alter cell metabolism and gene expression, within tumor target tissues will aid us in finding answers to these questions.

References

[1] Levi, F.; LaVecchia, C.; Lucchini, F.; Negri, E. (1995) Cancer mortality in Europe. *Eur. J. Cancer Prev.* **4**, 389–417.

[2] Doll, R.; Peto, R. (1981) The causes of cancer: Quantitative estimates of avoidable risks of cancer in the United States today. *J. Natl. Cancer Inst.* **66**, 1191–1308.

[3] Ames, B. N.; Gold, L. S. (1990) Too many rodent carcinogens: Mitogenesis increases mutagenesis. *Science* **249**, 970–971.

[4] Fearon, E. R. (1997) Human cancer syndromes: Clues to the origin and nature of cancer. *Science* **278**, 1043–1050.
[5] Hesketh, R. (1997) *The Oncogene and tumor suppressor gene facts book.* Academic Press, Harcourt and Brace Publishing Company, San Diego, London, Boston, New York, Sydney, Tokyo, Toronto.
[6] Doll, R. (1996) Nature and nurture: possibilities for cancer control. *Carcinogenesis* **17**, 177–184.
[7] Block, G.; Patterson, B.; Subar, A. (1992) Fruit, vegetables, and cancer prevention: A review of the epidemiological evidence. *Nutr. Cancer* **18**, 1–29.
[8] Steinmetz, K. A.; Potter, J. D. (1991) Vegetables, fruit, and cancer. I. Epidemiology. *Cancer Causes Control* **2**, 325–357.
[9] Hill, M. J. (1994) Fruit and vegetable consumption and cancer risk. *ECP-News* **26**, 3–5.
[10] Potter, J. D.; Steinmetz, K. A. (1996) Vegetables, fruit and phytoestrogens as preventive agents. In: Stewart, B. W.; McGregor, D. B.; Kleihues, P. (Eds.) *Principles of chemoprevention.* International Agency for research on Cancer, Lyon, p. 61–90.
[11] World Cancer Research Fund and American Institute for Cancer Research (1997) *Food, nutrition and the prevention of cancer: A global perspective.* American Institute for Cancer Research, Washington DC.
[12] Hong, W. K.; Sporn, M. B. (1997) Recent advances in chemoprevention of cancer. *Science* **278**, 1073–1078.
[13] Wattenberg, L. W. (1992) Inhibition of carcinogenesis by minor dietary constituents. *Cancer Res.* **52** (Suppl.), 2085s–2091s.
[14] Johnson, I. T.; Williamson, G.; Musk, S. R. R. (1994) Anti-carcinogenic factors in plant foods: A new class of nutrients? *Nutr. Res. Rev.* **7**, 175–204.
[15] Mirvish, S. S. (1983) The etiology of gastric cancer: Intragastric nitrosamide formation and other theories. *J. Natl. Cancer Inst.* **71**, 629–647.
[16] Tannenbaum, S. R.; Wishnok, J. S.; Leaf, C. D. (1991) Inhibition of nitrosamine formation by ascorbic acid. *Am. J. Clin. Nutr.* **53** (Suppl.), 247S–250 S.
[17] Yoshizawa, K.; Willett, W. C.; Morris, S. J.; Stampfer, M. J.; Spiegelman, D.; Rimm, E. B.; Giovannucci, E. (1998) Study of prediagnostic selenium level in toenails and the risk of advanced prostate cancer. *J. Natl. Cancer Inst.* **90**, 1219–1224.
[18] Bertram, B.; Lee, B. H.; Rajca, A.; Wießler, M.; Schmezer, P.; Pool, B. L. (1990) Mixed disulfides from the antimutagens/anti-carcinogens disulfiram and mesna: syntheses, biochemical and biological effects. 19th annual EEMS Meeting. *Mutat. Res.* **234**, 414–415 (abstract).
[19] DiMascio, P.; Kaiser, S.; Sies, H. (1989) Lycopene as the most efficient carotenoid singlet oxygen quencher. *Arch. Biochem. Biophys.* **274**, 532–538.
[20] Sies, H.; Stahl, W. (1995) Vitamins E and C, β-carotene, and other carotenoids as antioxidants. *Am. J. Clin. Nutr.* **62** (Suppl.), 1315S–1321 S.
[21] Omenn, G. S.; Goodman, G. E.; Thornquist, M.; Barnhart, S.; Balmes, J.; Chierniack, M.; Cullen, M. R.; Glass, A.; Keogh, J. P.; Liu, D.; Meyskens, F. L.; Perloff, M.; Valanis, B.; Williams, J. (1996) Chemoprevention of lung cancer: The β-Carotene and Retinol Efficacy Trial (CARET) in high-risk smokers and asbestos-exposed workers. In: Hakama, M.; Beral, V.; Buiatti, E.; Faivre, J.; Parkin, D. M. (Eds.) *Chemoprevention in cancer control.* International Agency for Research on Cancer, Lyon, p. 67–86.
[22] The α-Tocopherol, B. C. C. P. S. G. (1994) The effect of vitamin E and β-carotene on the incidence of lung cancer and other cancers in male smokers. *New Engl. J. Med.* **330**, 1029–1035.
[23] Huttunen, J. K. (1996) Why did antioxidants not protect against lung cancer in the α-Tocopherol, β-carotene cancer prevention study? In: Hakama, M.; Beral, V.; Buiatti, E.; Faivre, J.; Parkin, D.M. (Eds.) *Chemoprevention in cancer control.* International Agency for Research on Cancer, Lyon, p. 63–67.

[24] Pool-Zobel, B. L.; Bub, A.; Müller, H.; Wollowski, I.; Rechkemmer, G. (1997) Consumption of vegetables reduces genetic damage in humans: first results of an intervention trial with carotenoid-rich foods. *Carcinogenesis* **18**, 1847–1850.

[25] van Acker, S. A.; van den Berg, D. J.; Tromp, M. N.; Griffioen, D. H.; van Bennekom, W. P.; van der Vijgh, W. J.; Bast, A. (1996) Structural aspects of antioxidant activity of flavonoids. *Free Radical Biol. Med.* **20**, 331–342.

[26] Rice-Evans, C. A.; Miller, N. J.; Bolwell, P. G.; Bramley, P. M.; Pridham, J. B. (1995) The relative antioxidant activities of plant-derived polyphenolic flavonoids. *Free Radical Res.* **22**, 375–383.

[27] Haenen, G. R.; Paquay, J. B.; Korthouwer, R. E.; Bast, A. (1997) Peroxynitrite scavenging by flavonoids. *Biochem. Biophys. Res. Commun.* **236**, 591–593.

[28] Duthie, S. J.; Collins, A. R.; Duthie, G. G.; Dobson, V. L. (1997) Quercetin and myricetin protect against hydrogen peroxide-induced DNA damage (strand breaks and oxidised pyrimidines) in human lymphocytes. *Mutat. Res.* **393**, 223–231.

[29] Brown, J. E.; Khodr, H.; Hider, R. C.; Rice-Evans, C. A. (1998) Structural dependence of flavonoid interactions with Cu^{2+} ions: implications for their antioxidant properties. *Biochem. J.* **330**, 1173–1178.

[30] Hertog, M. G. L. (1994) *Flavonols and flavanes in foods and their relation with cancer and coronary heart disease risk. 1.* University Wegeningen, (GENERIC) Thesis/ Dissertation.

[31] Manach, C.; Texier, O.; Régérat, F.; Demigné, C.; Rémésy, C. (1998) Bioavailability of quercetin and its metabolites in rat and human plasma. In: COST 916 Bioactive plant cell wall components in nutrition and health: Polyphenols in food. (Ed. EU 181689) p. 85–90.

[32] Offord, E.; Mace, K.; Avanti, O.; Talalay, P.; Pfeiffer, A. (1998) Mechanisms involved in the chemoprotective effect of rosemary polyphenols. In: COST 916 Bioactive plant cell wall components in nutrition and health: Polyphenols in food. (Ed. EU 181689) p. 123–128.

[33] Uda, Y.; Price, K. R.; Williamson, G.; Rhodes, M. J. C. (1997) Induction of the anticarcinogenic marker enzyme, quinone reductase, in murine hepatoma cells *in vitro* by flavonoids. *Cancer Lett.* **120**, 213–216.

[34] Yannai, S.; Day, A.J.; Williamson, G.; Rhodes, M. J. C. (1998) Characterization of flavonoids as monofunctional or bifunctional inducers of quinone reductase in murine hepatoma cell lines. *Food Chem. Toxicol.* **36**, 623–630.

[35] Agullo, G.; Gamet-Payrastre, L.; Fernandez, Y.; Anciaux, N.; Demigné, C.; Rémésy, C. (1996) Comparative effects of flavonoids on the growth, viability and metabolism of a colonic adenocarcinoma cell line (HT29 cells). *Cancer Lett.* **105**, 61–70.

[36] Avila, M. A.; Cansado, J.; Harter, K. W.; Velasco, J. A.; Notario, V. (1996) Quercetin as a modulator of the cellular neoplastic phenotype: effects on the expression of mutated H-*ras* and p53 in rodent and human cells. In: American Institute for Cancer Research (Ed.) *Dietary Phytochemicals in cancer prevention and treatment.* Plenum Press, New York, London, p. 101–110.

[37] Fotsis, T.; Pepper, M. S.; Aktas, E.; Breit, S.; Rasku, S.; Adlercreutz, H.; Wähälä, K.; Montsano, R.; Schweigerer, L. (1997) Flavonoids, dietary-derived inhibitors of cell proliferation and *in vitro* angiogenesis. *Cancer Res.* **57**, 2916–2921.

[38] Kuo, S.-M. (1996) Antiproliferative potency of structurally distinct dietary flavonoids on human colon cancer cells. *Cancer Lett.* **110**, 41–48.

[39] Plaumann, B.; Fritsche, M.; Rimpler, H.; Brandner, G.; Hess, R. D. (1996) Flavonoids activate wild-type p53. *Oncogene* **13**, 1605–1614.

[40] Wei, H.; Bowen, R.; Cai, Q.; Barnes, S.; Wang, Y. (1995) Antioxidant and antipromotional effects of the soybean isoflavone genistein. *Proc. Soc. Exp. Biol. Med.* **208**, 124–130.

[41] Wei, H.; Wei, L.; Frenkel, K.; Bowen, R.; Barnes, S. (1993) Inhibition of tumor promoter-induced hydrogen peroxide formation *in vitro* and *in vivo* by genistein. *Nutr. Cancer* **20**, 1–12.

[42] Zava, D. T.; Duwe, G. (1997) Estrogenic and antiproliferative properties of genistein and other flavonoids in human breast cancer cells *in vitro*. *Nutr. Cancer* **27**, 31–40.

[43] Welshons, W. V.; Murphy, C. S.; Koch, R.; Calaf, G.; Jordan, V. C. (1987) Stimulation of breast cancer cells *in vitro* by the environmental estrogen enterolactone and the phytoestrogen equol. *Breast Cancer Res. Treat.* **10**, 169–175.

[44] Bedi, A.; Pasricha, P. J.; Akhtar, A. J.; Barber, J. P.; Bedi, G. C.; Giardiello, F. M.; Zehnbauer, B. A.; Hamilton, S. R.; Jones, R. J. (1995) Inhibition of apoptosis during development of colorectal cancer. *Cancer Res.* **55**, 1811–1816.

[45] Corcoran, G. B.; Fix, L.; Jones, D. P.; Moslen, M. T.; Nicotera, P.; Oberhammer, F. A.; Buttyan, R. (1994) Apoptosis: Molecular control point in toxicity. *Toxicol. Appl. Pharmacol.* **128**, 169–181.

[46] Baghurst, P. A.; Baghurst, K. I.; Record, S. J. (1996) Dietary Fibre, non-starch polysaccharides and resistant starch – a review. *Suppl. Food Aust.* **48**, S3–S35.

[47] Cummings, J. H.; Pomare, E. W.; Branch, W. J.; Naylor, C. P. E.; Macfarlane, G. T. (1987) Short chain fatty acids in human large intestine, portal, hepatic and venous blood. *Gut* **28**, 1221–1227.

[48] Hague, A.; Singh, B.; Paraskeva, C. (1997) Butyrate acts as a survival factor for colonic epithelial cells: further fuel for the *in vivo* versus *in vitro* debate. *Gastroenterology* **112**, 1036–1040.

[49] Hass, R.; Busche, R.; Luciano, L.; Reale, E.; v.Engelhardt, W. (1997) Lack of butyrate is associated with induction of Bax and subsequent apoptosis in the proximal colon of guinea pig. *Gastroenterology* **112**, 875–881.

[50] Hague, A.; Paraskeva, C. (1995) The short-chain fatty acid butyrate induces apoptosis in colorectal tumor cell lines. *Eur. J. Cancer Prev.* **4**, 359–364.

[51] Reddy, C. S.; Hayes, A. W. (1994) Food-borne toxicants. In: Hayes, A. W. (Ed.) *Principles and methods of toxicology.* Raven Press Ltd., New York, p. 317–360.

[52] Bingham, S. A.; Pignatelli, B.; Pollock, J. R. A.; Ellul, A.; Malaveille, C.; Gross, G.; Runswick, S.; Cummings, J. H.; O'Neill, I. K. O. (1996) Does increased endogenous formation of *N*-nitroso compounds in the human colon explain the association between red meat and colon cancer? *Carcinogenesis* **17**, 515–523.

[53] Rowland, I. R.; Granli, T.; Bockman, O. C.; Key, P. E.; Massey, R. C. (1991) Endogenous *N*-nitrosation in man assessed by measurement of apparent total *N*-nitroso compounds in feces. *Carcinogenesis* **12**, 1395–1401.

[54] Cerutti, P. A. (1985) Prooxidant states and tumor promotion. *Science* **227**, 375–381.

[55] Dreher, D.; Junod, A. F. (1996) Role of oxygen free radicals in cancer development. *Eur. J. Cancer* **32A**, 30–38.

[56] Liehr, J. G. (1997) Dual role of oestrogens as hormones and pro-carcinogens: tumor initiation by metabolic activation of oestrogens. *Eur. J. of Cancer Prevention* **6**, 3–10.

[57] Nebert, D. W. (1993) Elevated estrogen 16α-hydroxylase activity: Is this a genotoxic or nongenotoxic biomarker in human breast cancer risk. *J. Natl. Cancer Inst.* **85**, 1888–1891.

[58] Cheeseman, K. H. (1993) Lipid peroxidation and cancer. In: Halliwell, B.; Aruoma, O. I. (Eds.) *DNA and free radicals.* Ellis Horwood, New York, London, Toronto, Sydney, Tokyo, Singapore, p. 109–144.

[59] Rowland, I. R. (1993) Metabolism of toxic metals. In: *Role of the gut flora in toxicity and cancer,* Anonymous.

[60] Pool-Zobel, B. L.; Leucht, U. (1997) Induction of DNA damage in human colon cells derived from biopsies by suggested risk factors of colon cancer. *Mutat. Res.* **375**, 105–116.

[61] Pool-Zobel, B. L.; Abrahamse, S. L.; and Rechkemmer, G. (1998) Cellular effects of phytohormones: Modulation of oxidative DNA damage and signal transduction in human colon cancer cells. Proceedings of the 89th Annual Meeting of the American Association for *Cancer Res.* **39**, 641, conference proceeding.

[62] Brendler-Schwaab, S. Y.; Schmezer, P.; Liegibel, U. M.; Weber, S.; Michalek, K.; Tompa, A.; Pool-Zobel, B. L. (1994) Cells of different tissues for *in vitro* and *in vivo* studies in toxicology: Compilation of isolation methods. *Toxicol. in vitro* **8**, 1285–1302.

[63] McKelvey, V. J.; Green, M.; Schmezer, P.; Pool-Zobel, B. L.; De Meó, M. P.; Collins, A. R. (1993) The single cell gelelectrophoresis (SCGE) assay (COMET assay): A European review. *Mutat. Res.* **288**, 47–63.

[64] Martin, F. L.; Venitt, S.; Carmichael, P. L.; Crofton-Sleigh, C.; Stone, E. M.; Cole, K. J.; Gusterson, B.A.; Grover, P.L.; Phillips, D.H. (1997) DNA damage in breast epithelial cells: detection by the single-cell gel (COMET) assay and induction by human mammary lipid extracts. *Carcinogenesis* **18**, 2299–2305.

[65] Pool-Zobel, B. L.; Lotzmann, N.; Knoll, M.; Kuchenmeister, F.; Lambertz, R.; Leucht, U.; Schröder, H. G.; Schmezer, P. (1994) Detection of genotoxic effects in human gastric and nasal mucosa cells isolated from biopsy samples. *Environ. Mol. Mutagen.* **24**, 23–45.

[66] Pool-Zobel, B. L.; Bub, A.; Liegibel, U. M.; Treptow-van Lishaut, S.; Rechkemmer, G. (1998) Mechanisms by which Vegetable Consumption Reduces Genetic Damage in Humans. *Cancer Epidemiol. Biomarkers Prev.* **7**, 891–899.

[67] Rowland, I. R.; Bearne, C. A.; Fischer, R.; Pool-Zobel, B. L. (1996) The effect of lactulose on DNA damage induced by 1,2-dimethylhydrazine in the colon of human-flora-associated rats. *Nutr. Cancer* **26**, 38–47.

[68] Pool-Zobel, B. L.; Neudecker, C.; Domizlaff, I.; Ji, S.; Schillinger, U.; Rumney, C. J.; Moretti, M.; Villarini, M.; Scassellati-Sforzolini, G.; Rowland, I. R. (1996) *Lactobacillus-* and *Bifidobacterium*-mediated antigenotoxicity in colon cells of rats: Prevention of carcinogen-induced damage *in vivo* and elucidation of involved mechanisms. *Nutr. Cancer* **26**, 365–380.

[69] Wollowski, I.; Ji, S.; Bakalinsky, A. T.; Neudecker, C.; Pool-Zobel, B. L. (1998) Inactivation of carcinogens and prevention of DNA-damage in the colon of rats by bacteria used for the production of yoghurt. *submitted* (in press).

[70] Ames, B. N.; Shigenaga, M. K.; Hagen, T. M. (1993) Oxidants, antioxidants, and the degenerative diseases of aging. *Proc. Natl. Acad. Sci. USA* **90**, 7915–7922.

[71] Halliwell, B. (1993) Oxidative DNA damage: meaning and measurement. In: Halliwell, B.; Aruoma, O. I. (Eds.) *DNA and free radicals.* Ellis Horwood, New York, London,Toronto, Sydney, Tokyo, Singapore, p. 67–82.

[72] Crawford, D.; Zbinden, I.; Amstad, P.; Cerutti, P. A. (1988) Oxidant stress induces the proto-oncogenes c-*fos* and c-*myc* in mouse epidermal cells. *Oncogene* **3**, 27–32.

[73] Pool-Zobel, B. L.; Abrahamse, S. L.; Collins, A. R.; Kark, W.; Gugler, R.; Oberreuther, D.; Siegel, E. G.; Treptow-van Lishaut, S.; Rechkemmer, G. (1999) *Analysis of DNA strand breaks, oxidized bases and glutathione S-transferase P1 in human colon cells. Cancer Epidemiol. Biomarkers Prev.* **8**, 609–614.

[74] Zhang, Y.; Talalay, P.; Cho, C. G.; Posner, G. H. (1992) A major inducer of anti-carcinogenic protective enzymes from broccoli: Isolation and elucidation of structure. *Proc. Natl. Acad. Sci. USA* **89**, 2399–2403.

[75] Cimino, F.; Esposito, F.; Ammendola, R.; Russo, T. (1997) Gene regulation by reactive oxygen species. *Current Topics in Cellular Regulation* **35**, 123–148.

[76] Waleh, N. S.; Calaoagan, J.; Murphy, B. J.; Knapp, A. M.; Sutherland, R. M.; Laderoute, K. R. (1998) The redox-sensitive human antioxidant responsive element induces gene expresseion under low oxygen conditions. *Carcinogenesis* **19**, 1337 (abstract).

[77] Primiano, T.; Sutter, T. R.; Kensler, T. W. (1997) Antioxidant-inducible genes. *Adv. Pharmacol.* **38**, 293–328.

[78] Stillwell, W. G.; Kidd, L. C. R.; Wishnok, J. S.; Tannenbaum, S. R.; Sinha, R. (1997) Urinary excretion of unmetabolized and phase II conjugates of 2-amino-1-methyl-6-

phenylimidazo[4,5-*b*]pyridine and 2-amino-3,8-dimethylimidazo[4,5-*f*]quinoxaline in humans: relationship to cytochrome P450 1A2 and *N*-acetyltransferase activity. *Cancer Res.* **57**, 3457–3464.

[79] Bogaards, J. J. P.; Verhagen, H.; Willems, M. I.; van Poppel, G.; van Bladeren, P. J. (1994) Consumption of Brussels sprouts results in elevated α class glutathione S-transferase levels in human blood plasma. *Carcinogenesis* **15**, 1073–1075.

[80] Pool-Zobel, B. L.; Glei, M.; Spänkuch, B.; Rechkemmer, G. (2000) *Study on mechanisms by which carrot and tomato juice may reduce oxidative DNA-damages* in vivo*: In vitro effects of all*-trans-β-*carotene and lycopene*. This volume, Chapter IV, 19.

[81] Gibbs, J. B. (1992) Pharmacological probes of *ras* function. *Seminar Cancer Biol.* **3**, 383–390.

[82] Feig, L. A. (1993) Strategies for suppressing the function of oncogenic *ras* protein in tumors. *J. Natl. Cancer Inst.* **85**, 1266–1268.

[83] Kohl, N. E.; Mosser, S. D.; deSolms, J.; Giuliani, E. A.; Pompliano, D. L.; Graham, S. L.; Smith, R. L.; Scolnick, E. M.; Oliff, A.; Gibbs, J. B. (1993) Selective inhibition of *ras*-dependent transformation by a farnesyltransferase inhibitor. *Science* **260**, 1934–1937.

[84] Vigushin, D. M.; Poon, G. K.; Boddy, A.; English, J.; Halbert, G. W.; Pagonis, C.; Jarman, M.; Coombes, R. C. (1998) Phase I and pharmacokinetic study of D-limonene in patients with advanced cancer. Cancer Research Campaign Phase I/II Clinical Trials Committee. *Cancer Chemotherapy Pharmocol.* **42**, 111–117.

[85] McConnell, H. M.; Owicki, J. C.; Parce, J. W.; Miller, D. L.; Baxter, G. T.; Wada, H. G.; Pitchford, S. (1992) The cytosensor microphysiometer: biological applications of silicon technology. *Science* **257**, 1906–1912.

[86] Ashkenazi, A.; Dixit, V. M. (1998) Death receptors: Signaling and modulation. *Science* **281**, 1305–1308.

[87] Evan, G.; Littlewood, T. (1998) A matter of life and cell death. *Science* **281**, 1317–1322.

[88] Ahmad, N.; Feyes, D. K.; Nieminen, A.; Agarwal, R.; Mukhtar, H. (1997) Green tea constituent epigallocatechin-3-gallate and induction of apoptosis and cell cycle arrest in human carcinoma cells. *J. Natl. Cancer Inst.e* **89**, 1886.

[89] Liu, M.; Pelling, J. C.; Ju, J.; Chu, E.; Brash, D. E. (1998) Antioxidant action via p53-mediated apoptosis. *Cancer Res.* **58**, 1723–1729.

[90] Venturi, M.; Hambly, R. J.; Glinghammer, B.; Rafter, J. J.; Rowland, I. R. (1997) Genotoxic activity in human faecal water and the role of bile acids: a study using the alkaline COMET assay. *Carcinogenesis* **18**, 2353–2359.

[91] Rice-Evans, C. A.; Sampson, J.; Bramley, P. M.; Holloway, D. E. (1997) Why do we expect carotenoids to be antioxidants *in vivo*? *Free Radical Res.* **26**, 381–398.

[92] Barja, G.; López-Torres, M.; Pérez-Campo, R.; Rojas, C.; Cadenas, S.; Prat, J.; Pamplona, R. (1994) Dietary vitamin C decreases endogenous protein oxidative damage, malondialdehyde, and lipid peroxidation and maintains fatty acid unsaturation in the guinea pig liver. *Free Radical Biol. Med.* **17**, 105–115.

[93] Ip, C.; Lisk, D. J. (1996) The attributes of selenium-enriched garlic in cancer prevention. In: American Institute for Cancer Research (Ed.) *Dietary Phytochemicals in Cancer Prevention and Treatment*, Plenum Press, New York, London, p. 179–188.

[94] Barcelo, S.; Gardiner, J. M.; Gescher, A.; Chipman, J. K. (1998) CYP 2E1-mediated mechanism of anti-genotoxicity of the broccoli constituent sulforaphane. *Carcinogenesis* **17**, 277–282.

[95] Kensler, T. W. (1997) Chemoprevention by inducers of carcinogen detoxication enzymes. *Environ. Health Perspect.* **105**, 965–970.

[96] Ariazi, E. A.; Gould, M. N. (1996) Identifying differential gene expression in monoterpene-treated mammary carcinomas using subtractive display. *J. Biol. Chem.* **271**, 29 286–29 294.

[97] Crowell, P. L.; Chang, R. R.; Ren, Z.; Elson, C. E.; Gould, M. N. (1994) Selective inhibition of isoprenylation of 21–26-kDa proteins by the anti-carcinogen D-limonene and its metabolites. *J. Biol. Chem.* **266**, 17679–17685.

[98] Schulz, S.; Reinhold, D.; Schmidt, H.; Ansorge, S.; Höllt, V. (1997) Perillic acid inhibits *ras*/MAP kinase-driven IL-2 production in human T lymphocytes. *Biochem. Biophys. Res. Commun.* **241**, 720–725.

[99] Stayrook, K. R.; McKinzie, J. H.; Burke, Y. D.; Burke, Y. A.; Crowell, P. L. (1997) Induction of the apoptosis-promoting protein Bak by perillyl alcohol in pancreatic ductal adenocarcinoma relative to untransformed ductal epithelial cells. *Carcinogenesis* **18**, 1655–1658.

[100] Gradelet, S.; LeBon, A. M.; Bergès, R.; Suschetet, M.; Astorg, P. (1998) Dietary carotenoids inhibit aflatoxin B1-induced liver preneoplastic foci and DNA damage. *Carcinogenesis* **19**, 403–411.

[101] Haag, J. D.; Lindstrom, M. J.; Gould, M. N. (1992) Limonene-induced regression of mammary carcinomas. *Cancer Res.* **52**, 4021–4026.

[102] Levy, J.; Bosin, E.; Feldman, B.; Giat, Y.; Miinster, A.; Danilenko, M.; Sharoni, Y. (1995) Lycopene is a more potent inhibitor of human cancer cell proliferation than either α-carotene or β-carotene. *Nutr. Cancer* **24**, 257–266.

[103] Tsuda, T.; Watanabe, M.; Ohshima, K.; Norinobu, S.; Choi, S. W.; Kawakishi, S.; Osawa, T. (1994) Antioxidative activity of the anthocyanin pigments cyanidin 3-*O*-D-glucoside and cynidin. *J. Agric. Food Chem.* **1994**, 2407–2410.

[104] Tsuda, T.; Ohshima, K.; Kawakishi, S.; Osawa, T. (1996) Oxidation products of cyanidin 3-*O*-β-glucoside with a free radical initiator. *Lipids* **31**, 1259–1263.

[105] Tsuda, T.; Shiga, K.; Ohshima, K.; Kawakishi, S.; Osawa, T. (1996) Inhibition of lipid peroxidation and the active oxygen radical scavenging effect of anthocyanin pigments isolated from *Phaseolus vulgaris L. Biochem. Pharmacol.* **52**, 1033–1039.

[106] Dai, R.; Jacobson, K. A.; Robinson, R. C.; Friedman, F. K. (1997) Differential effects of flavonoids on testosterone metabolizing cytochrome P450s. *Life Sci.* **6**, 75–80.

[107] Rechkemmer, G.; Pool-Zobel, B. L. (1998) Antigenotoxic and physiological properties of anthocyanins/anthocyanidins in intestinal epithelial cells. In: COST 916 Bioactive plant cell wall components in nutrition and health: Polyphenols in food. (Ed. EU 181689) p. 131–136.

[108] Moon, J. Y.; Lee, D. W.; Park, K. H. (1998) Inhibition of 7-ethoxycoumarin O-deethylase activity in rat liver microsomes by naturally occurring flavonoids: stucture-activity relationships. *Xenobiotica* **28**, 117–126.

[109] Nanjo, F.; Goto, K.; Seto, R.; Suzuki, M.; Sakai, M.; Hara, Y. (1996) Scavenging effects of tea catechins and their derivatives on 1,1-diphenyl-2-picrylhydrazyl radical. *Free Radical Biol. Med.* **21**, 695–902.

[110] Yu, R.; Jiao, J. J.; Duh, J. L.; Gudehithlu, K.; Tan, T. H.; Kong, A. N. T. (1997) Activation of mitogen-activated protein kinases by green tea polyphenols: potential signaling pathways in the regulation of antioxidant-responsive element-mediated phase II enzyme gene expression. *Carcinogenesis* **18**, 451–456.

[111] Paschka, A. G.; Butler, R.; Young, C. Y. F. (1998) Induction of apoptosis in prostate cancer cell lines by the green tea component, (–)-epigallocatechin-3-gallate. *Cancer Lett.* **130**, 1–7.

[112] Wang, W.; Liu, L. Q.; Higuchi, C. M.; Chen, H. (1998) Induction of NADPH: Quinone reductase by dietary phytoestrogens in colonic Colo205 cells. *Biochem. Pharmacol.* **56**, 189–195.

[113] Helsby, N. A.; Williams, J.; Kerr, D.; Gescher, A.; Chipman, J. K. (1997) The isoflavones equol and geistein do not induce xenobiotic-metabolizing enzymes in mouse and in human cells. *Xenobiotica* **27**, 587–596.

[114] Brown, A.; Jolly, P.; Wei, H. (1998) Genistein modulates neuroblastoma cell proliferation and differentiaion through induction of apoptosis and regulation of tyrosine kinase activity and N-myc expression. *Carcinogenesis* **19**, 991–997.
[115] Wei, H.; Cai, Q.; Tian, L.; Lebwohl, M. (1998) Tamoxifen reduces endogenous and UV-light induced oxidative damage to DNA, lipid and protein *in vitro* and *in vivo*. *Carcinogenesis* **19**, 1013–1018.
[116] Montano, M. M.; Katzenellenbogen, B. S. (1997) A quinone reductase gene: a unique estrogen receptor-regulated gene that is activated by antiestrogens. *Proc. Natl. Acad. Sci. USA* **94**, 2581–2586.
[117] Barnes, S.; Sfakianos, J.; Coward, L.; Kirk, M. (1996) Soy isoflavonoids and cancer prevention: Underlying biochemical and pharmacological issues. In: American Institute for Cancer Research (Ed.) *Dietary phytochemicals in cancer prevention and treatment*. Plenum Press, New York, London, p. 87–100.
[118] Jones, P. A.; Baker, V. A.; Irwin, A. J. E.; Earl, L. K. (1998) Interpretation of the *in vitro* proliferation response of MCF-7 cells to potential oestrogens and non-oestrogenic substances. *Toxicol. in vitro* **12**, 373–382.
[119] Gehm, B. D.; McAndrews, J. M.; Chien, P.-Y.; Jaemeson, J. L. (1997) Resveratrol, a polyphenolic compound found in grapes and wine, is an agonist for the estrogen receptor. *Proc. Natl. Acad. Sci. USA* **94**, 14 138–14 143.
[120] Pool-Zobel, B. L.; Abrahamse, S. L.; Rechkemmer, G. (1996) Pretreatment of colon cells with sodium butyrate but not iso-butyrate protects them from DNA damage induced by hydrogen peroxide. In: American Institute for Cancer Research (Ed.) *Dietary phytochemicals in cancer prevention and treatment*. Plenum Press, New York, London p. 287.
[121] Abrahamse, S. L.; Pool-Zobel, B. L.; Rechkemmer, G. (1998) Potential of short chain fatty acids to modulate the induction of DNA damage and changes in the intracellular calcium concentration in isolated rat colon cells. *Carcinogenesis*, in press.
[122] Finnie, I. A.; Dwarakanath, A. D.; Taylor, B. A.; Rhodes, J. M. (1995) Colonic mucin synthesis is increased by sodium butyrate. *Gut* **36**, 93–99.
[123] Stein, J.; Schröder, O.; Bonk, M.; Oremek, G.; Lorenz, M.; Caspary, W. F. (1996) Induction of glutathione S-transferase-pi by short-chain fatty acids in the intestinal cell line caco-2. *Eur. J. Clin. Invest.* **26**, 84–87.
[124] Butzner, J. D.; Parmar, R.; Bell, C. J.; Dalal, V. (1996) Butyrate enema therapy stimulates mucosal repair in experimental colitis in the rat. *Gut* **38**, 568–573.
[125] Csordas, A. (1995) Toxicology of butyrate and short-chain fatty acids. In: Hill, M. J. (Ed.) *Role of gut bacteria in human toxicology and pharmacology*. Taylor and Francis, London, p. 105–127.
[126] Souleimani, A.; Asselin, C. (1993) Regulation of c-*myc* expression by sodium butyrate in the human colon carcinoma cell line Caco-2. *FEBS Lett.* **326**, 45–50.
[127] Vigushin, D. M.; Poon, G. K.; Boddy, A.; English, J.; Halbert, G. W.; Pagonis, C.; Jarman, M.; Coombes, R. C. (1998) Phase I and pharmacokinetic study of D-limonene in patients with advanced cancer. Cancer Research Campaign Phase I/II Clinical Trials Committee. *Cancer Chemotherapy Pharmocol.* **42**, 111–117.
[128] Rock, C. L.; Flatt, S. W.; Wright, F. A.; Faerber, S.; Newman, V.; Kealey, S.; Pierce, J. P. (1997) Responsiveness of carotenoids to a high vegetable diet intervention designed to prevent breast cancer recurrence. *Cancer Epidemiol. Biomarkers Prev.* **6**, 617–623.
[129] Müller, H.; Bub, A.; Rechkemmer, G. (1997) Trennung und Identifikation von Carotinoiden und ihren Oxidationsprodukten in Plasma und Lipoproteinen (LDL) des Menschen nach Tomatensaftverzehr. *Z. Ernaehrungswiss.* **36**, 62 (abstract).
[130] Oszmiansky, J.; Sapis, J. C. (198) Anthocyanidins in fruits and berries. *J. Food Sci.* **53**, 1241.
[131] Mazza, G. (1995) Anthocyanins in grapes and grape products. *Crit. Rev. Food Sci. Nutr.* **35**, 341–371.

[132] Goldbohm, R. A.; Hertog, M. G. L.; Brants, H. A. M.; van Poppel, G.; van den Brandt, P. A. (1998) Intake of flavonoids and cancer risk: a prospective cohort study. In: COST 916 Bioactive plant cell wall components in nutrition and health: Polyphenols in food. (Ed. EU 181689) p. 166.
[133] Manach, C.; Texier, O.; Régérat, F.; Demigné, C.; Rémésy, C. (1998) Bioavailability of quercetin and its metabolites in rat and human plasma. In: COST 916 Bioactive plant cell wall components in nutrition and health: Polyphenols in food. (Ed. EU 181689) p. 85–90.
[134] Yang, C. S.; Chen, L.; Lee, M. J.; Balentine, D.; Kuo, M. C.; Schantz, S. P. (1998) Blood and urine levels of tea catechins after ingestion of different amounts of green tea by human volunteers. *Cancer Epidemiol. Biomarkers Prev.* **7**, 351–354.
[135] Wähälä, K. (1998) Dietary isoflavonoids, their metabolites and their synthesis. In: p. 73–80. Anonymous
[136] Morton, M. S.; Wilcox, G.; Wahlqlvist, M. L.; Griffeth, K. (1994) Determination of lignans and isoflavonoids in human female plasma following dietary supplementation. *J. Endocrinol.* **142**, 251–259.
[137] Early Breast Cancer Trialists' Collaborative Group (EBCTCG) (1998) Tamoxifen for early breast cancer: an overview of the randomised trials. *Lancet* **351**, 1451–1467.
[138] Glinghammer, B.; Venturi, M.; Rowland, I. R.; Rafter, J. (1997) Shift from a dairy product-rich to a dairy product-free diet: influence on cytotoxicity and genotoxicity of fecal water – potential risk factors for colon cancer. *Am. J. Clin. Nutr.* **66**, 1277–1282.
[139] Gibson, G. R.; Roberfroid, M. B. (1995) Dietary modulation of the human colonic microbiota: Introducing the concept of prebiotics. *J. Nutr.* **125**, 1401–1412.
[140] Gibson, G. R.; Beatty, E. R.; Wang, X.; Cummings, J. (1995) Selective stimulation of bifidobacteria in the human colon by oligofructose and inulin. *Gastroenterology* **108**, 975–982.
[141] Mallett, A. K.; Rowland, I. R. (1990) Bacterial enzymes: Their role in the formation of mutagens and carcinogens in the intestine. *Drug Disp.* **8**, 71–79.

19 Carotenoids and Anthocyans

Gerhard Rechkemmer

19.1 Abstract

Carotenoids and anthocyanins are important pigments in plants. In human nutrition these phytochemicals are naturally present in various fruits and vegetables. Furthermore, they are used as colorants in food industry. The interest in these substances for human nutrition is related to the general observation that diets rich in fruits and vegetables are associated with a lower risk for cardiovascular disease and certain types of cancer [1, 2]. In general, vegetables contain higher amounts of carotenoids, whereas fruits, and particularly berries, have high concentrations of anthocyanins.

Epidemiological studies have shown that particularly diets rich in carotenoid–containing vegetables decrease certain tumor risks. Based on these observations and on *in vitro* data with *β*-carotene, respectively, two large prospective intervention trials (ATBC and CARET) have been conducted by randomly supplementing study participants with the major provitamin A carotenoid, *β*-carotene. However, both intervention studies did not show the expected decrease in lung cancer in high risk groups of heavy smokers and asbestos workers, but an increase in incidence and mortality.

A conclusion from these intervention studies could be that the association of elevated *β*-carotene plasma concentrations observed in groups consuming high amounts of fruits and vegetables and the decrease in tumor incidence is not directly related to *β*-carotene, although this substance may be a biomarker for a high consumption of fruits and vegetables. The protective substance(s) or factors and the mechanisms responsible for the decreased tumor risk, however, are not clearly identified.

A lower risk of prostate cancer by those men with a high consumption of tomato products was observed [3, 4]. In a recent meta-analysis a reduced risk was also found for a number of other tumor sites with a high consumption of tomatoes or tomato products. The overall relative risk for all cancer sites combined was 0.5 for the high versus the low tomato consumption [5].

In a dietary intervention study with healthy non-smoking men we observed a significant reduction in endogenous and oxidative DNA-damage in lymphocytes, an early biomarker, during periods with high lycopene or *β*-carotene plasma concentrations achieved by daily consumption of tomato or carrot juice, respectively [6].

Anthocyanins and their aglycones, the anthocyanidins, are much less studied in relation to carcinogenesis than the carotenoids. These polyphenolic compounds (flavonoids) are excellent antioxidants in *in vitro* systems and so far were particularly studied in relation to the pathogenesis of cardiovascular disease and especially the inhibition of low-density lipoprotein oxidation as a risk factor for the development of atherosclerosis. A protective effect of red wine consumption on the risk of coronary heart disease was attributed to the high flavonoid and especially anthocyanin content of red wine [7]. However, these potent antioxidant compounds may also affect cellular mechanisms involved in carcinogenesis.

19.2 Introduction

Many epidemiological studies have demonstrated that a high consumption of fruits and vegetables is associated with a lower incidence of certain cancers (for reviews see: [1, 8, 9]). The evidence for the cancer preventive effects of fruits and vegetables has recently been extensively reviewed by the World Cancer Research Fund and the American Institute for Cancer Research [10]. Fruits and vegetables contain complex mixtures of bioactive compounds such as vitamins, minerals, fibers and phytochemicals (for example carotenoids and/or anthocyanins). These components may have significant cancer preventive properties. Prospective as well as retrospective epidemiological studies have shown an inverse association between the consumption of carotene-rich fruits and vegetables or carotene plasma concentrations and risk of cancer, especially for lung and stomach cancer and much less consistent for breast and prostate cancer [9, 11–13]. But observational epidemiological studies of carotenoids and human health must be interpreted cautiously, as it is possible that observed effects may result from other dietary factors correlated with carotenoid intake rather than from carotenoids themselves [14]. Thus the evidence for a preventive activity of individual dietary components (e.g. β-carotene or lycopene) contained in fruits and vegetables is much less well supported by epidemiological studies [10, 15].

Carotenoids and anthocyans are phytochemicals present in many fruits and vegetables in considerable amounts and are thus an integral part of a balanced diet. Carotenoids and anthocyanins are also used by the food industry as efficient natural food colorants (in Europe E-numbers 160, 161 and 163, respectively). Only sparse information is available about the dietary consumption of carotenoids. It is estimated that about 1–2 mg of β-carotene are present in a typical Western diet [16]. No consumption data are available for anthocyans. However, in social groups consuming high amounts of vegetables and fruits, e.g. vegetarians or vegans, the uptake of these plant pigments may be as high as 50 mg/day on a regular basis. Particularly ripe berries and red wine contain high concentrations of anthocyans up to several 100 mg/100 g fresh weight or liter, respectively.

19.3 Carotenoids

More than 600 different carotenoids have been identified in plants, photosynthetic bacteria and animals, approximately 50 of which have vitamin A activity [14]. It is estimated that about 40 different carotenoids occur in the human diet. However, so far only approximately 14 carotenoids and metabolites of carotenoids have been detected in human plasma and tissue [11]. Carotenoids can be classified into two groups, those serving as precusors of vitamin A (for example: *trans*-*β*-carotene, 13-*cis*-*β*-carotene, *trans*-*α*-carotene, *trans*-*β*-cryptoxanthin) and others not having that function (e.g., *trans*-lycopene and the xanthophylls *trans*-lutein and *trans*-zeaxanthin). For the provitamin A carotenoids the discrimination between their vitamin A function and their carotenoid effects is often difficult. For the non-provitamin A compounds, e.g. lycopene, lutein or canthaxanthin their singlet-oxygen and radical quenching properties are held responsible for their protective effects [17–20].

Provitamin A carotenoids can be ultimately converted enzymatically in the intestinal mucosa to retinol (vitamin A), which is required for vision, maintenance of differentiated epithelia, mucus secretion and reproduction [14].

The occurrence of carotenoids and their concentrations varies greatly in fruits and vegetables. Most plants show a very distinct pattern of different carotenoids. In general, vegetables contain much higher concentrations of carotenoids compared to fruits. Prominent examples of carotenoid-containing vegetables are carrots containing high amounts of *β*- and *α*-carotene, tomatoes with high concentrations of lycopene, and green leafy vegetables like spinach or kale with high concentrations of lutein (a xanthophyll). Interestingly, carotenoid

Retinol
(all-trans-Retinol)

β-Carotene
(β,β-Carotene)

α-Carotene
(α-Carotene)

Lycopene
ψ,ψ-Carotene

Lutein
(β,ε-Carotene-3,3′-diol)

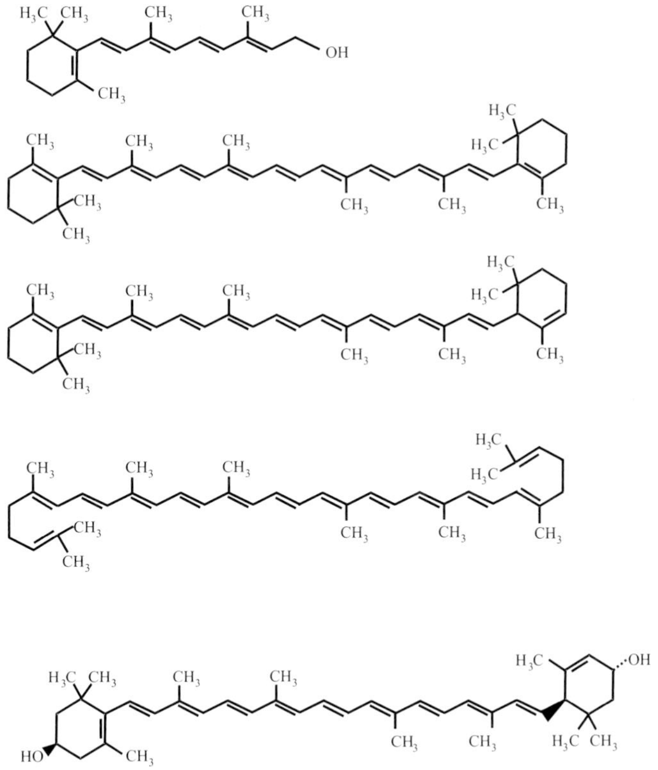

Figure 19.1: Chemical structure of different carotenoids.

patterns in different human organs also differ greatly. In the prostate for example lycopene is the major carotenoid [21].

Because of the epidemiological evidence for an inverse association between β-carotene plasma concentrations, the consumption of β-carotene rich fruits and vegetables and a decreased risk in lung cancer, two large intervention trials with β-carotene supplements have been carried out using the cancer incidence and mortality as primary endpoints.

Some of the functional effects of carotenoids on the processes of carcinogenesis might be associated with the provitamin A character. However, carotenoids can also act as "functional antioxidants", e.g. they are able to efficiently quench singlet oxygen by physical energy transfer reactions not leading to a chemical alteration of the carotenoid [17–19, 22–24]. The energy is transferred to the conjugated double bond structure of the carotenoids and is dissipated as heat. However, in chemical reactions with other reactive oxygen species (ROS), unlike the classical antioxidants vitamin E and C, respectively, carotenoids are destroyed and are not restored in subsequent redox reactions. Thus, in a purely chemical sense, they may not be regarded as "real antioxidants". Furthermore, carotenoids are able to induce cell communication through gap junctions [25–27]. Conversion of carotenoids to retinoids is not required for the antiproliferative effect of carotenoids on human colon cancer cells [28]. The modulation of genotoxic and related effects by carotenoids and vitamin A was recently reviewed extensively [29].

19.3.1 Evidence for anticarcinogenic activity of carotenoids

According to a recent survey of the World Cancer Research Fund [10] for carotenoids only an association between lung cancer and dietary carotenoid intake was probable, whereas for cancers of the gastrointestinal tract (esophagus, stomach, colon, rectum) and for breast and cervix a relation between carotenoids and the cancer incidence was possible. No such data are available so far for anthocyanins.

19.3.1.1 β-Carotene

Epidemiological studies clearly demonstrated that a high concentration of β-carotene in the plasma was a protective anticarcinogenic factor [13, 14, 30]. Several plausible mechanisms for a cancer preventive activity of β-carotene were described [13]:

- conversion to vitamin A and retinoids, possibly at the tissue level
- antioxidant potential to scavenge free radical species
- immune-enhancing effects
- stimulation of gap-junction-mediated cell-to-cell communication

Table 19.1: Evidence for anticarcinogenic actions of vegetables, fruits and carotenoids in different organs (adapted from [10]).

	Judgement of evidence		
	Convincing	**Probable**	**Possible**
Vegetables	Mouth and pharynx	Larynx	Liver
	Esophagus	Pancreas	Ovary
	Lung	Breast	Endometrium
	Stomach	Bladder	Cervix
	Colon		Prostate
	Rectum		Thyroid
			Kidney
Fruits	Mouth and pharynx	Larynx	Ovary
	Esophagus	Pancreas	Endometrium
	Lung	Breast	Cervix
	Stomach	Bladder	Thyroid
Carotenoids		Lung	Esophagus
			Stomach
			Colon
			Rectum
			Breast
			Cervix

In the Physicians Health Study (a randomized, double-blind, placebo-controlled trial) 22,071 healthy, male physicians (age 40–84 years), mostly non-smokers, participated and the randomly assigned intervention group (11,036 physicians) took 50 mg β-carotene every other day for 12 years. No significant benefit but also no increased risk of malignant neoplasms, cardiovascular disease, or death from all causes was observed with β-carotene supplementation [31]. In a recent follow-up study a marginally significant ($P = 0.07$) increased risk for prostate carcinoma was observed for those men in the lowest quartile for plasma β-carotene at baseline compared with those in the highest quartile. The risk of prostate carcinoma was significantly reduced in a randomly assigned subgroup starting with a low plasma β-carotene concentration but taking β-carotene as a supplement [32]. In a study carried out in New Zealand, however, it was concluded that vegetables rich in β-carotene are not protective against prostate cancer, whereas lycopene from tomato-based foods was found to be associated with a small reduction in risk [33].

Since in smokers the development of lung cancer may be associated with the generally observed low β-carotene plasma concentration [34], recent large prospective intervention trials with β-carotene investigated the possible association between lung cancer incidence and β-carotene status in smokers in Finland (ATBC: alpha-tocopherol and beta-carotene trial) [35, 36] and the USA (CARET: beta-carotene and retinol efficacy trial) [37, 38].

19.3.1.1.1 ATBC

The ATBC (α-tocopherol, β-carotene) study investigated Finnish male smokers taking 20 mg of β-carotene daily as supplement. This intervention trial was organized in a 2×2 factorial design, e.g. 29,133 men aged 50–69 years and all smoking 5 or more cigarettes daily were randomly assigned to four treatment groups receiving either α-tocopherol (50 mg), β-carotene (20 mg), a combination of α-tocopherol and β-carotene, or placebo, respectively [35]. The major result was an 18% higher incidence of lung cancer in the β-carotene group and an increased lung cancer risk (relative risk = 1.16). The increase in risk was more pronounced in heavy smokers (>20 cigarettes daily, RR = 1.25). A further risk increase was observed with alcohol consumption.

19.3.1.1.2 CARET

In the CARET (β-carotene, retinoid) study 18,314 male and female smokers, former smokers and asbestos workers (14,254 smokers, 4060 asbestos-exposed workers) were investigated taking 30 mg β-carotene and 25,000 IU of retinyl palmitate per day for 4 years in a randomized, double-blinded and placebo-controlled intervention trial. In this trial β-carotene supplementation was associated with increased lung cancer risk (RR = 1.36) and lung cancer mortality (RR = 1.59) [38].

19.3.1.1.3 Conclusions from ATBC and CARET

These studies clearly showed that β-carotene supplementation did not decrease lung cancer incidence or mortality in a high risk group, e.g. heavy smokers and asbestos workers. On the contrary, an increase in cancer incidence in the supplementation groups was found, causing the CARET study to be terminated earlier than originally planned.

19.3.1.2 Lycopene

Lycopene is a non–provitamin A carotenoid and is present in human blood and tissues [17, 39]. Lycopene added to cell culture medium is a more potent inhibitor of human cancer (endometrial, mammary and lung) cell proliferation *in vitro* than either α- or β-carotene [40].

In a recent health professionals follow-up study the risk of prostate cancer was associated with the consumption of tomato products [3]. In 1986 using a validated, semiquantitative food-frequency questionnaire the dietary intake for a 1 year period was assessed for a cohort of 47,894 eligible men initially free of diagnosed cancer. Follow-up questionnaires were sent to the entire cohort in 1988, 1990 and 1992. Between 1986 and 1992, 812 new cases of prostate cancer were documented. Intakes of the carotenoids β-carotene, α-carotene, lutein, and β-cryptoxanthin were not associated with risk of prostate cancer, only lycopene intake was related to a lower risk (age- and energy-adjusted RR = 0.79).

The combined intake of tomatoes, tomato sauce, tomato juice and pizza was inversely associated with risk of prostate cancer (multivariate RR = 0.65). The reduction of prostate cancer risk in groups with a high consumption of tomato-based foods was recently also confirmed by a study carried out in New Zealand [33].

19.3.1.3 Other carotenoids

Besides β-carotene and lycopene other carotenoids relevant in the human diet are lutein and zeaxanthin from green leafy vegetables and astaxanthin and canthaxanthin which are present in shell fish and fish but also in specific mushrooms. These other nutritionally relevant carotenoids have received much less attention compared to β-carotene or recently lycopene, particularly with respect to studies with patients, healthy volunteers or in human model systems. In mouse and rat model systems an anticarcinogenic effect of astaxanthin on bladder, oral and colon cancer was demonstrated [41–43]. A recent study showed an inverse association of lutein with colon cancer in man and woman, the greatest inverse association was observed among subjects in whom colon cancer was diagnosed when they were young and among those with tumors located in the proximal segment of the colon [44]. The major dietary sources of lutein in this study were spinach, broccoli, lettuce, tomatoes, oranges and orange juice, carrots, celery and greens. The association with other carotenoids were not remarkable in this study [44].

19.4 Anthocyanins

Anthocyans are a large group (>250 different chemical compounds) of water-soluble plant pigments [45]. They belong to the flavonoids. The basic chemical structure is a polyphenolic ring structure. In contrast to the other flavonoids, the

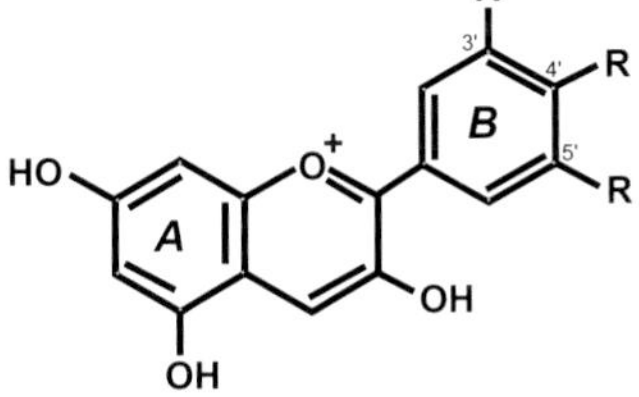

Figure 19.2: Basic chemical structure of the major nutritionally relevant anthocyanidins.

Table 19.2: Substitution patterns and colors of important anthocyanidins in fruits and vegetables (adapted from [47]).

	3′	4′	5′	Color
Cyanidin	OH	OH	H	Orange-red
Delphinidin	OH	OH	OH	Bluish-red
Malvidin	OCH_3	OCH_3	OCH_3	Bluish-red
Pelargonidin	H	OH	H	Orange
Peonidin	OCH_3	OH	H	Orange-red
Petunidin	OCH_3	OH	OH	Bluish-red

anthocyans carry a positive charge in the central ring structure and are thus cations. In plants they are present exclusively as glycosidic compounds. The number and nature of the different attached sugar moieties is responsible for the high number of anthocyanins. Anthocyanins show red-blue color and their color depends primarily on the pH and the binding of certain trace metal ions. Due to their polyphenolic nature the anthocyanins are efficient antioxidants in *in vitro* systems [46].

Besides these 6 most prominent anthocyanidins another 11 anthocyanidins have been detected in plants (apigeninidin, aurantinidin, capensinidin, europinidin, hirsutidin, 6-hydroxycyanidin, luteolinidin, 5-methylcyanidin, pulchellidin, rosinidin, tricetinidin).

19.4.1 Experimental evidence for anticarcinogenic activity of anthocyanins

Only very few studies have been performed to elucidate a potential anticarcinogenic activity of anthocyanins, despite their presence and importance in the human diet, particularly in fruits and fruit products (e.g. juices, red wine, jams, jellies). The high antioxidative potential of these compounds, at least under *in vitro* conditions, should also be a beneficial factor in carcinogenesis. However, the bioavailability of the anthocyanins is not well studied. It appears that it is very low compared to carotenoids. Thus, only the epithelial cells lining the intestinal tract are directly exposed to high concentrations of anthocyanins in the diet. The concentration in the blood and in other organs may be very low. Oxidative DNA damage induced by treatment with hydrogen peroxide in primary human colon cells obtained from biopsies and in a human colon tumor cell line (HT29) was significantly reduced, if the cells were preincubated with extracts of aronia, elderberry, macqui and tintorera fruits, all rich sources of anthocyanins [46, 48]. However, endogenous oxidative DNA damage was not affected by the plant extracts or the pure anthocyanins/anthocyanidins [46]. Preliminary data show that anthocyanidins efficiently inhibit the cellular effects of growth factors in human colonic tumor cells (HT29 clone 19A) and thus may have biological relevance in the later stages of carcinogenesis [48].

References

[1] Steinmetz, K. A.; Potter, J. D. (1996) Vegetables, fruit, and cancer prevention: a review. *J. Am. Diet. Assoc.* **96**, 1027–1039.

[2] Cummings, J. H.; Bingham, S. A. (1998) Diet and the prevention of cancer. *BMJ* **317**, 1636–1640.

[3] Giovannuci, E.; Ascherio, A.; Rimm, E. B.; Stampfer, M. J.; Colditz, G. A.; Willett, W. C. (1995) Intake of carotenoids and retinol in relation to risk of prostate cancer. *J. Natl. Cancer Inst.* **87**, 1767–1776.

[4] Giovannuci, E.; Clinton, S. K. (1998) Tomatoes, lycopene, and prostate cancer. *Proc. Soc. Exp. Biol. Med.* **218**, 129–139.

[5] Giovannuci, E. (1999) Tomatoes, tomato-based products, lycopene, and cancer: review of the epidemiologic literature. *J. Natl. Cancer Inst.* **91**, 317–331.

[6] Pool-Zobel, B. L.; Bub, A.; Müller, H.; Wollowski, I.; Rechkemmer, G. (1997) Consumption of vegetables reduces genetic damage in humans: first results of a human intervention trial with carotenoid-rich foods. *Carcinogenesis* **18**, 1847–1850.

[7] Renaud, S.; de Longeril, M. (1992) Wine alcohol, platelets, and the French paradox for coronary heart disease. *Lancet* **339**, 1523–1526.

[8] Steinmetz, K. A.; Potter, J. D. (1991) A review of vegetables, fruit, and cancer. I. Epidemiology. *Cancer Causes Control* **2**, 427–442.

[9] Block, G.; Patterson, B.; Subar, A. (1992) Fruit, vegetables, and cancer prevention: A review of epidemiological evidence. *Nutr. Cancer* **18**, 1–29.

[10] World Cancer Research Fund; American Institute for Cancer Research (1997) Food, nutrition and the prevention of cancer: a global perspective.

[11] Gerster, H. (1992) Anticarcinogenic effect of common carotenoids. *Internat. J. Vit. Nutr. Res.* **63**, 93–121.

[12] van Poppel, G. (1993) Carotenoids and cancer: an update with emphasis on human intervention studies. *Eur. J. Cancer* **29**, 1335–1344.

[13] van Poppel, G. (1996) Epidemiological evidence for β-carotene in prevention of cancer and cardiovascular disease. *Europ. J. Clin. Nutr.* **50**, S57–S61.

[14] Taylor Mayne, S. (1996) Beta-carotene, carotenoids, and disease prevention in humans. *FASEB J.* **10**, 690–701.

[15] Working Group on Diet and Cancer (1998) Nutritional aspects of the development of cancer. *Report on Health and Social Subjects* **48**, 1–274. The Stationery Office, Norwich, UK.

[16] Müller, H. (1995) Carotinoide in der Gesamtnahrung – Bestimmung der täglichen Zufuhrmenge an individuellen Derivaten. *Fat. Sci. Technol.* **97**, 397–402.

[17] Sies, H.; Stahl, W. (1998) Lycopene: Antioxidant and biological effects and its bioavailability in the human. *Proc. Soc. Exp. Biol. Med.* **218**, 121–124.

[18] Bast, A.; van der Plas, R. M.; van den Berg, H.; Haenen, G. R. M. M. (1996) Betacarotene as antioxidant. *Europ. J. Clin. Nutr.* **50**, S54–S56.

[19] Miller, N. J.; Sampson, J.; Candeias, L. P.; Bramley, P. M.; Rice–Evans, C. A. (1996) Antioxidant activities of carotenes and xanthophylls. *FEBS Letters* **384**, 240–242.
[20] Miller, N. J.; Sampson, J.; Candeias, L. P.; Bramley, P. M.; Rice–Evans, C. (1996) Antioxidant activities of carotenes and xanthophylls. *FEBS Letters* **384**, 240–242.
[21] Clinton, S. K.; Emenhiser, C.; Schwartz, S. J.; Bostwick, D. G.; Williams, A. W.; Moore, B. J.; Erdman, J. W. (1996) cis-trans lycopene isomers, carotenoids, and retinol in the human prostate. *Cancer Epidemiology, Biomarkers & Prevention* **5**, 823–833.
[22] Rice–Evans, C. A.; Sampson, J.; Bramley, P. M.; Holloway, D. E. (1997) Why do we expect carotenoids to be antioxidants in vivo? *Free Radical Research* **26**, 381–398.
[23] Palozza, P.; Luberto, C.; Calviello, G.; Ricci, P.; Bartoli, G. M. (1997) Antioxidant and prooxidant role of β-carotene in murine normal and tumor thymocytes: effects of oxygen partial pressure. *Free Radical Biology & Medicine* **22**, 1065–1073.
[24] Tsuchihashi, H.; Kigoshi, M.; Iwatsuki, M.; Niki, E. (1995) Action of β-carotene as an antioxidant against lipid peroxidation. *Arch. Biochem. Biophys.* **323**, 137–147.
[25] Zhang, L.-X.; Acevedo, P.; Guo, H.; Bertram, J. S. (1995) Upregulation of gap junctional communication and connexin43 expression by carotenoids in human dermal fibroblasts but not in human keratinocytes. *Molecular Carcinogenesis* **12**, 50–58.
[26] Zhang, L.-X.; Cooney, R. V.; Bertram, J. S. (1992) Carotenoids up-regulate connexin43 gene expression independent of their provitamin A or antioxidant properties. *Cancer Res.* **52**, 5702–5712.
[27] Stahl, W.; Nicolai, S.; Briviba, K.; Hanusch, M.; Broszeit, G.; Peters, M.; Martin, H.-D.; Sies, H. (1997) Biologcial activities of natural and synthetic carotenoids: induction of gap junctional communication and singlet oxygen quenching. *Carcinogenesis* **18**, 89–92.
[28] Onogi, N.; Okuno, M.; Matsushima–Nishiwaki, R.; Fukutomi, Y.; Moriwaki, H.; Muto, Y.; Kojima, S. (1998) Antiproliferative effect of carotenoids on human colon cancer cells without conversion to retinoic acid. *Nutr. Cancer* **32**, 20–24.
[29] DeFlora, S.; Bagnasco, M.; Vainio, H. (1999) Modulation of genotoxic and related effects by carotenoids and vitamin A in experimental models: mechanistic issues. *Mutagenesis* **14**, 153–172.
[30] Burri, B. J. (1997) Beta-carotene and human health: a review of current research. *Nutr. Res.* **17**, 547–580.
[31] Hennekens, C. H.; Buring, J. E.; Manson, J. E.; Stampfer, M.; Rosner, B.; Cook, N. R.; Belanger, C.; LaMotte, F.; Gaziano, J. M.; Ridker, P. M.; Willett, W. C.; Peto, R. (1996) Lack of effect of long-term supplementation with beta carotene on the incidence of malignant neoplasms and cardiovascular disease. *N. Engl. J. Med.* **334**, 1145–1149.
[32] Cook, N. R.; Stampfer, M. J.; Ma, J.; Manson, J. E.; Sacks, F. M.; Buring, J. E.; Hennekens, C. H. (1999) Beta-carotene supplementation for patients with low baseline levels and decreased risks of total and prostate carcinoma. *Cancer* **86**, 1783–1792.
[33] Norrish, A. E.; Jackson, R. T.; Sharpe, S. J.; Skeaff, C. M. (2000) Prostate cancer and dietary carotenoids. *Am. J. Epidemiol.* **151**, 119–123.
[34] Margetts, B. M.; Jackson, A. A. (1996) The determinants of plasma β-carotene: interaction between smoking and other lifestyle factors. *Europ. J. Clin. Nutr.* **50**, 236–238.
[35] Heinonen, O. P.; Albanes, D. (1994) The effect of vitamin E and beta carotene on the incidence of lung cancer and other cancers in male smokers. *N. Engl. J. Med.* **330**, 1029–1035.
[36] Albanes, D.; Heinonen, O. P.; Taylor, P. R.; Virtamo, J.; Edwards, B. K.; Rautalahti, M.; Hartmann, A. M.; Palmgren, J.; Freedman, L. S.; Haapakoski, J.; Barrett, M. J.; Pietinen, P.; Malila, N.; Tala, E.; Liippo, K.; Salomaa, E.-R.; Tangrea, J. A.; Teppo, L.; Askin, F. B.; Taskinen, E.; Erozan, Y.; Greenwald, P.; Huttunen, J. K. (1996) α-tocopherol and β-carotene supplements and lung cancer incidence in the alpha-toco-

pherol, beta-carotene cancer prevention study: effects of base-line characteristics and study compliance. *J. Natl. Cancer Inst.* **88**, 1560–1570.

[37] Omenn, G. S.; Goodman, G. E.; Thornquist, M. D.; Balmes, J.; Cullen, M. R.; Glass, A.; Keogh, J. P.; Meyskens, F. L.; Valanis, B.; Williams, J. H.; Barnhart, S.; Hammar, S. (1996) Effects of a combination of beta carotene and vitamin A on lung cancer and cardiovascular disease. *N. Engl. J. Med.* **334**, 1150–1155.

[38] Omenn, G. S.; Goodman, G. E.; Thornquist, M. D.; Balmes, J.; Cullen, M. R.; Glass, A.; Keogh, J. P.; Meyskens, F. L.; Valanis, B.; Williams, J. H.; Barnhart, S.; Cherniak, M. G.; Brodkin, C. A.; Hammar, S. (1996) Risk factors for lung cancer and for intervention effects in CARET, the beta-carotene and retinol efficacy trial. *J. Natl. Cancer Inst.* **88**, 1550–1559.

[39] Paetau, I.; Khachik, F.; Brown, E. D.; Beecher, G. R.; Kramer, T. R.; Chittams, J.; Clevidence, B. A. (1998) Chronic ingestion of lycopene-rich tomato juice or lycopene supplements significantly increases plasma concentrations of lycopene and related tomato carotenoids in humans. *American Journal of Clinical Nutrition* **68**, 1187–1195.

[40] Levy, J.; Bosin, E.; Feldman, B.; Giat, Y.; Miinster, A.; Danilenko, M.; Sharoni, Y. (1995) Lycopene is a more potent inhibitor of human cancer cell proliferation than either β-carotene or carotene. *Nutr. Cancer* **24**, 257–266.

[41] Tanaka, T.; Morishita, Y.; Suzui, M.; Kojima, T.; Okumura, A.; Mori, H. (1994) Chemoprevention of mouse urinary bladder carcinogenesis by the naturally occuring carotenoid astaxanthin. *Carcinogenesis* **15**, 15–19.

[42] Tanaka, T.; Makita, H.; Ohnishi, M.; Mori, H.; Satoh, K.; Hara, A. (1995) Chemoprevention of rat oral carcinogenesis by naturally occuring xanthophylls, astaxanthin and canthaxanthin. *Cancer Res.* **55**, 4059–4064.

[43] Tanaka, T.; Kawamori, T.; Ohnishi, M.; Makita, H.; Mori, H.; Satoh, K.; Hara, A. (1995) Suppression of azoxymethan-induced rat colon carcinogenesis by dietary administration of naturally occuring xanthophylls astaxanthin and canthaxanthin during the postinitiation phase. *Carcinogenesis* **16**, 2957–2963.

[44] Slattery, M. L.; Benson, J.; Curtin, K.; Ma, K.-N.; Schaeffer, D.; Potter, J. D. (2000) Carotenoids and colon cancer. *Am. J. Clin. Nutr.* **71**, 575–582.

[45] Mazza, G.; Miniati, E. (1993) Anthocyanins in Fruits, Vegetables, and Grains. 1–362.

[46] Pool-Zobel, B. L.; Bub, A.; Schröder, N.; Rechkemmer, G. (1999) Anthocyanins are potent antioxidants in model systems but do not reduce endogenous oxidative DNA damage in human colon cells. *Eur. J. Nutr.* **38**, 227–234.

[47] Mazza, G.; Miniati, E. (1993) Anthocyanins in Fruits, Vegetables, and Grains.

[48] Rechkemmer, G.; Pool-Zobel, B. L. (1998) Antigenotoxic and physiological properties of anthocyanin-containing fruit extracts in intestinal epithelial cells. **Polyphenols in food**, 131–137. COST 916. EU Commission Report.

20 Anticarcinogenesis by Isothiocyanates, Indole-3-carbinol, and *Allium* Thiols

Stephen S. Hecht

20.1 Abstract

This review discusses anticarcinogenesis by important constituents of family Cruciferae and genus *Allium* plants. There is convincing evidence that these plants contain anticarcinogens, otherwise known as cancer chemopreventive or chemoprotective agents. Cruciferae contain substantial amounts of glucosinolates which are hydrolyzed by the enzyme myrosinase to isothiocyanates and other products. Studies in laboratory animals demonstrate that Cruciferae, glucosinolates, and isothiocyanates inhibit carcinogenesis. Isothiocyanates have been extensively tested as chemopreventive agents. They are powerful inhibitors of lung and esophageal tumorigenesis in a variety of models and some also inhibit mammary tumor induction. The strongest effects occur when the isothiocyanates are administered before or concurrently with the carcinogen. Isothiocyanates inhibit cytochromes P450 that are involved in carcinogen activation and induce Phase 2 enzymes involved in carcinogen detoxification. There is convincing evidence that the major mechanism of inhibition of nitrosamine carcinogenesis by isothiocyanates is inhibition of cytochrome P450 mediated metabolic activation. Isothiocyanate induction of apoptosis may also play a role in chemoprevention. Indole-3-carbinol is produced upon hydrolysis of glucobrassican, a common constituent of broccoli and other Cruciferae. Indole-3-carbinol is an effective inhibitor of carcinogenesis when administered before or concurrently with the carcinogen, but appears to be a tumor promoter when given subsequent to carcinogen. Indole-3-carbinol induces Phase 1 and 2 enzymes, probably through the intermediacy of condensation products formed in the stomach. It also has favorable effects on estrogen metabolism and is being developed for chemoprevention of breast cancer. *Allium* thiols are formed when garlic or onions are cut or crushed. Many of these compounds inhibit carcinogenesis. Diallyl sulfide is the most extensively studied of the *Allium* thiols. It is an effective inhibitor of carcinogenesis by nitrosamines, hydrazines, polycyclic aromatic hydrocarbons, and other compounds. Diallyl sulfide inhibits nitrosamine carcinogenesis by inhibiting cytochrome P450 enzymes involved in their metabolic activation. It appears to inhibit polycyclic aromatic hydrocarbon carcinogenesis by inducing glutathione S-transferases. On balance, the available data are consistent with the hypothesis that individual constituents of vegetables are at least partially responsible for the protective effects of vegetables against cancer observed in epidemiologic studies.

There is no doubt that edible plants contain anticarcinogens, also known as cancer chemopreventive or chemoprotective agents. Among these, sulfur containing compounds have been studied extensively. Reviews by Wattenberg in 1978 already cited numerous examples of chemoprevention by these compounds [1, 2]. Recent reviews extensively document their chemopreventive activities and discuss relevant mechanisms [3–11]. Indole-3-carbinol has mixed activities. In some systems, it is a chemopreventive agent while in others it can promote tumorigenesis [12]. This paper will discuss isothiocyanates and indole-3-carbinol which are derived from glucosinolates that occur in vegetables of the family Cruciferae, and thiols present in genus *Allium* plants. The discussion will be limited to naturally occurring compounds and will attempt to provide representative examples of chemopreventive activities and mechanisms.

20.2 Isothiocyanates and glucosinolates

20.2.1 Occurrence and formation

Isothiocyanates occur in plants as thioglucoside conjugates called glucosinolates [13, 14]. Over one hundred glucosinolates have been identified, mainly in vegetables of the family Cruciferae [4, 13, 14]. Common vegetables of this family are summarized in Table 20.1 [4]. Hydrolysis of the glucosinolates is catalyzed by multiple forms of the enzyme myrosinase (thioglucoside glucohydrolase, EC

Table 20.1: Common vegetables of the family Cruciferae (from ref. [4]).

Genus	Species	Common Name
Armoracia	*Rusticana*	Horseradish
	Campestris	Turnip
	Chinensis	Pak choy
	Juncea	Brown mustard
	Napus	Rape, swede, rutabaga
	Nigra	Black mustard
	Oleracea	Cabbage, kale, Brussels sprouts, cauliflower, broccoli, kohlrabi
	Pekinensis	Chinese cabbage
Lepidium	*Sativum*	Garden cress
Nasturtium	*Officinale*	Watercress
Raphanus	*Sativus*	Radish
Sinapis	*Alba*	Mustard

3.2.3.1) which occur in the same plants, separated cellularly from the glucosinolates. When the plant is macerated or chewed, myrosinase mixes with the glucosinolate and effects the hydrolysis as illustrated in Figure 20.1. Myrosinase activity has also been found in some intestinal microflora, which is important with respect to intake of intact glucosinolates [4]. Myrosinase catalyzes hydrolysis of the glucosinolate *S*-sugar bond leading to an unstable thiohydroxamic acid which undergoes a Lossen rearrangement yielding the isothiocyanate. Depending on the nature of the R group and the conditions, other products such as nitriles and thiocyanates may also form.

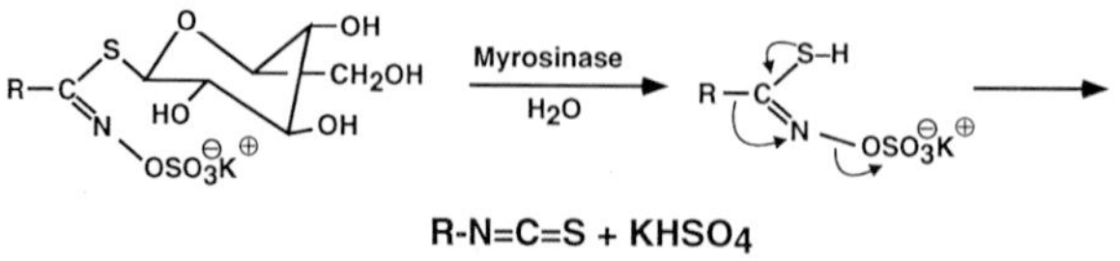

Figure 20.1: Formation of isothiocyanates in the myrosinase catalyzed hydrolysis of glucosinolates.

A large number of glucosinolates with many different R groups occur in substantial quantities in cruciferous plants and crops. This area has been extensively reviewed [13, 14]. Typical glucosinolate contents in agriculturally important plants such as cabbage, brussels sprouts, cauliflower, turnip, radish, and watercress range from approximately 0.5–3 mg/g [14].

20.2.2 Inhibition of carcinogenesis by isothiocyanates, glucosinolates, and cruciferous vegetables

Studies on inhibition of carcinogenesis by isothiocyanates are summarized in Table 20.2 [15–55]. A wide variety of isothiocyanates, both naturally occurring and synthetic, have been tested. Naturally occurring isothiocyanates with chemopreventive activity include benzyl ($R = PhCH_2$, BITC), 2-phenylethyl ($R = PhCH_2CH_2$, PEITC), 3-phenylpropyl ($R = PhCH_2CH_2CH_2$, PPITC), and sulforaphane ($R = CH_3S(O)(CH_2)_4$). Among these, BITC and PEITC are the most extensively studied. BITC is an effective inhibitor of rat mammary and mouse lung tumorigenesis by the polycyclic hydrocarbons DMBA and BaP. It is less effective in nitrosamine induced tumor models. In contrast, PEITC has broad inhibitory activity against tumors induced by nitrosamines. This includes inhibition of lung tumorigenesis in mice and rats by the tobacco-specific carcinogen NNK, inhibition of liver tumor induction by NDEA in the mouse, inhibition of esophageal tumor induction by NBMA in the rat, and inhibition of pancreas and lung tumorigenesis by BOP in the hamster. Inhibition of NNK-induced pulmonary carcinogenesis by PEITC has been demonstrated in multiple studies in mice

Table 20.2: Modification of carcinogenesis by isothiocyanates.

Isothiocyanate R–N=C=S; R=	Naturally Occurring?[a]	Carcinogen[b]	Species and Target Organ	Effect	Reference
α-Naphthyl–	No	3′-Me-DAB	Rat liver	inhibition	Sasaki 1963 [15]
		ethionine	Rat liver	inhibition	Sidransky *et al.* 1966 [16]
		AAF	Rat liver	inhibition	Sidransky *et al.* 1966 [16]
		DAB	Rat liver	inhibition	Lacassagne *et al.* 1970 [17]
		m-toluylenediamine	Rat liver	inhibition	Ito *et al.* 1969 [18]
		NDEA	Rat liver	no effect	Makiura *et al.* 1973 [19]
		BHBN	Rat bladder	inhibition	Ito *et al.* 1974 [20]
β-Naphthyl–	No	DAB	Rat liver	inhibition	Lacassagne *et al.* 1970 [17]
Ph–	Yes	DMBA	Rat mammary	inhibition	Wattenberg 1977 [21]
		NNK	Mouse lung	no effect	Morse *et al.* 1989 [22]
$PhCH_2$–	Yes	DMBA	Rat mammary	inhibition	Wattenberg 1977, 1981 [21, 23]
			Mouse forestomach	inhibition	Wattenberg 1977 [21]
			Mouse lung	inhibition	Wattenberg 1977 [21]
		BaP	Mouse lung	inhibition	Lin *et al.* 1993, Wattenberg 1987 [24, 25]
			Mouse forestomach	inhibition or no effect	Lin *et al.* 1993, Wattenberg 1987 [24, 25]
			Mouse skin	no effect	Lin *et al.* 1993 [24]
		NNK	Mouse lung	no effect	Morse *et al.* 1989, 1990 [22, 26]
		NDEA	Mouse forestomach	inhibition	Wattenberg 1987 [25]
			Mouse lung	no effect	Wattenberg 1987 [25]
			Rat liver	inhibition	Sugie *et al.* 1993 [27]
		MAM	Rat small intestine/ colon	inhibition	Sugie *et al.* 1994 [28]
		NBMA	Rat esophagus	no effect	Wilkinson *et al.* 1995 [29]
		NDEA + BHBN	Rat bladder	enhancement	Hirose *et al.* 1998 [30]
$Ph(CH_2)_2$–	Yes	DMBA	Rat mammary	inhibition or no effect	Wattenberg 1977, Lubet *et al.* 1997, Futakuchi *et al.* 1998 [21, 31, 32]
			Mouse forestomach	inhibition	Wattenberg 1977 [21]
		NNK	Mouse lung	inhibition	Wattenberg 1977 [21]
			Rat lung	inhibition	Chung *et al.* 1996, Hecht *et al.* 1996, Morse *et al.* 1989 [33–35]

Table 20.2 (continued)

Isothiocyanate R–N=C=S; R=	Naturally Occurring?[a]	Carcinogen[b]	Species and Target Organ	Effect	Reference
			Rat nasal cavity, liver	no effect	Morse *et al.* 1989 [35]
			Mouse lung	inhibition	Morse *et al.* 1989, 1989, 1991, 1992; Matzinger *et al.* 1995; El-Bayoumy *et al.* 1996; Jiao *et al.* 1997 [22, 36–41]
			Mouse lung	no effect	Morse *et al.* 1990 [26]
		NDEA	Mouse liver	inhibition	Pereira 1995 [42]
		NBMA	Rat esophagus	inhibition or no effect	Siglin *et al.* 1995; Stoner *et al.* 1991; Wilkinson *et al.* 1995 [29, 43, 44]
		BOP	Hamster pancreas and lung	inhibition	Nishikawa *et al.* 1996 [45]
		BaP	Mouse lung	no effect	Adam-Rodwell *et al.* 1993, Lin *et al.* 1993 [24, 46]
			Mouse skin	no effect	Lin *et al.* 1993 [24]
		NDEA + BHBN	Rat bladder	enhancement	Hirose *et al.* 1998 [30]
		NDEA + MNU + BHBN + DMH + DHPN	Rat lung, esophagus, kidney, liver, bladder	inhibition and enhancement	Ogawa *et al.* 1998 [47]
$Ph(CH_2)_3-$	Yes	NNK	Mouse lung	inhibition	Morse *et al.* 1989, 1991 [36, 37]
		NBMA	Rat esophagus	inhibition	Wilkinson *et al.* 1995 [29]
		BOP	Hamster lung	inhibition	Nishikawa *et al.* 1996 [48]
		NNN	Rat esophagus	inhibition	Stoner *et al.* 1998 [49]
$Ph(CH_2)_4-$	Yes	NNK	Mouse lung	inhibition	Morse *et al.* 1989, 1991 [36, 37]
		NBMA	Rat esophagus	inhibition	Wilkinson *et al.* 1995 [29]
$Ph(CH_2)_5-$	No	NNK	Mouse lung	inhibition	Morse *et al.* 1991 [37]
$Ph(CH_2)_6-$	No	NNK	Mouse lung	inhibition	Morse *et al.* 1991, 1992; Jiao *et al.* 1997 [37, 38, 41]
		NNK	Rat lung	inhibition	Chung *et al.* 1996, Hecht *et al.* 1996 [33, 50]
		BaP	Mouse skin	no effect	Lin *et al.* 1993 [24]
		NBMA	Rat esophagus	enhancement	Stoner *et al.* 1995 [51]
		AOM	Rat colon	enhancement	Rao *et al.* 1995 [52]
$Ph(CH_2)_8-$	No	NNK	Mouse lung	inhibition	Jiao *et al.* 1994 [53]

Table 20.2 (continued)

Isothiocyanate R–N=C=S; R=	Naturally Occurring?[a]	Carcinogen[b]	Species and Target Organ	Effect	Reference
$Ph(CH_2)_{10}$–	No	NNK	Mouse lung	inhibition	Jiao *et al.* 1994 [53]
$PhCH(Ph)CH_2$–	No	NNK	Mouse lung	inhibition	Jiao *et al.* 1994 [53]
$PhCH_2CH(Ph)$–	No	NNK	Mouse lung	inhibition	Jiao *et al.* 1994 [53]
CH_2=$CHCH_2$–	Yes	NNK	Mouse lung	no effect	Jiao *et al.* 1994 [53]
$CH_3(CH_2)_5$–	Yes	NNK	Mouse lung	inhibition	Jiao *et al.* 1994 [53]
$CH_3(CH_2)_3CH(CH_3)$–	?	NNK	Mouse lung	inhibition	Jiao *et al.* 1994 [53]
$CH_3(CH_2)_{11}$–	No	NNK	Mouse lung	inhibition	Jiao *et al.* 1994, 1996 [53, 54]
3-Pyr(CO)$(CH_2)_3$–	No	NNK	Mouse lung	no effect	Morse *et al.* 1989 [36]
9-Phenanthryl–	No	BaP	Mouse skin	no effect	Lin *et al.* 1993 [24]
9-Methylene-phenanthryl-	No	BaP	Mouse skin	no effect	Lin *et al.* 1993 [24]
6-Chrysenyl–	No	BaP	Mouse skin	no effect	Lin *et al.* 1993 [24]
6-Benzo[*a*]pyrenyl–	No	BaP	Mouse skin	no effect	Lin *et al.* 1993 [24]
4-methylsulphinyl-butyl–	Yes	DMBA	Rat mammary	inhibition	Zhang *et al.* 1994 [55]
exo-2-acetyl-*exo*-norborn-6-yl–	No	DMBA	Rat mammary	inhibition	Zhang *et al.* 1994 [55]
endo-2-acetyl-*exo*-norborn-6-yl–	No	DMBA	Rat mammary	inhibition	Zhang *et al.* 1994 [55]
exo-2-acetyl-*exo*-norborn-5-yl–	No	DMBA	Rat mammary	inhibition	Zhang *et al.* 1994 [55]

[a] Based on Fenwick *et al.* [13].
[b] Abbreviations: AOM, azoxymethane; BOP, *N*-nitroso-bis(2-oxopropyl)amine; 3′-Me-DAB, 3′-methyl-4-dimethylaminoazobenzene; DAB, 4-dimethylaminoazobenzene; AAF, 2-acetylaminofluorene; DHPN, 2,2′-dihydroxy-di-*n*-propylnitrosamine; DMBA, 7,12-dimethylbenz[*a*]anthracene; DMH, 1,2-dimethylhydrazine; MNU, *N*-methyl-*N*-nitrosourea; NNK, 4-(methylnitrosamino)-1-(3-pyridyl)-1-butanone; BaP, benzo[*a*]pyrene; NDEA, *N*-nitrosodiethylamine; MAM, methylazoxymethanol acetate; NBMA, *N*-nitrosobenzylmethylamine; BHBN, *N*-butyl-*N*-(4-hydroxybutyl)nitrosamine.

and rats; this compound is presently in Phase I clinical trials in healthy smokers [56]. Structure-activity studies demonstrate that increased isothiocyanate lipophilicity increases inhibitory potency [53]. Thus, single doses of 10-phenyldecyl isothiocyanate or 1-dodecyl isothiocyanate as low as 0.04–1 μmol are sufficient to inhibit mouse lung tumorigenesis induced by a single dose of 10 μmol NNK [53]. Further studies demonstrate that the isothiocyanate group, but not the phenyl ring, is necessary for inhibition and that lower reactivity with glutathione leads to better inhibitory potency [53, 54]. *N*-Acetylcysteine and glutathione conjugates of PEITC also show inhibitory activity against mouse lung tumorigenesis by NNK [41]. While PEITC is a superb inhibitor of nitrosamine induced carcinogenicity in multiple tumor models, it is less effective against polycyclic aromatic hydrocarbons (PAH). Bioassays carried out to date fail to demonstrate inhibition of BaP induced mouse lung or skin tumorigenesis by PEITC, possibly for pharmacokinetic reasons. Mixed results have been obtained in the DMBA rat mammary tumor model. Initial studies by Wattenberg, in which PEITC was given by gavage, showed inhibition of mammary tumorigenesis [21]. A recent study by Lubet *et al.* in which PEITC was given in the diet showed no effect or somewhat enhanced mammary tumorigenesis by DMBA [31]. However, another recent dietary study demonstrated that carcinoma volume, but not multiplicity or incidence, was decreased by PEITC [32]. The effects of PEITC on carcinogenesis by PAH require further study.

The studies summarized in Table 20.2 demonstrate inhibition of carcinogenesis mainly when isothiocyanates are given either before, or before and during carcinogen administration. Few studies demonstrate inhibition by isothiocyanates given after carcinogen treatment, although BITC does inhibit DMBA induced mammary carcinogenesis when administered in this way [23]. For reasons discussed below, it may be important to investigate further the ability of isothiocyanates to inhibit carcinogenesis when given in the post-initiation phase.

Enhancement of tumorigenesis has been observed in some studies with isothiocyanates. Both BITC and PEITC promote urinary bladder carcinogenesis in rats treated with NDEA and BHBN, although the dose employed was higher than that used for chemoprevention [30]. 6-Phenylhexyl isothiocyanate ($R = Ph(CH_2)_6$, PHITC), which is not known to be naturally occurring, enhances colon carcinogenesis and esophageal carcinogenesis in rat tumor models [51, 52].

Relatively few studies have been carried out on the effects of glucosinolates on carcinogenesis, probably because the compounds are generally less available in pure form. These studies have been reviewed recently [4]. Sinigrin, the glucosinolate with R=allyl, inhibits liver and tongue tumors in rat models, but has no effect on lung, liver, or nasal tumors induced by NNK [4, 57]. Sinigrin may enhance pancreatic tumorigenesis in NNK treated rats [57]. Sinigrin also inhibits DMH-induced aberrant crypt foci and induces apoptosis in rat colon [58]. Glucobrassican, the precursor to indole-3-carbinol, inhibits BaP induced lung and forestomach tumors in mice, while glucotropaeolin (R = benzyl), and glucosinalbin (R = 4-hydroxybenzyl) have little or no effect [59]. Glucobrassican and glucotropaeolin inhibit DMBA induced mammary tumors in the rat [59]. It should be noted that glucosinolates also have well documented toxic effects, particularly goitrogenicity [13, 14, 60].

A modest number of studies have investigated the effects of cruciferous vegetables on tumorigenesis; these have been reviewed [3, 4, 60]. Several studies show that cabbage or cauliflower decrease tumor formation in rat and mouse models; however, enhancement of pancreatic and skin tumorigenesis has been observed in cabbage-fed hamsters and mice [61]. A recent investigation demonstrates protective effects of cruciferous seed meals and hulls against colon cancer in mice [62]. The complexity of vegetables prevents direct assignment of their inhibitory properties to particular constituents. However, based on studies carried out to date, it is plausible that isothiocyanates and other hydrolysis products of glucosinolates in vegetables are at least partially responsible for inhibition of carcinogenesis by vegetables.

20.2.3 Mechanisms of chemoprevention by isothiocyanates

Isothiocyanates can profoundly affect carcinogen metabolism. Numerous studies demonstrate that isothiocyanates inhibit specific cytochrome P450 enzymes involved in the activation and detoxification of carcinogens. Other studies show that isothiocyanates induce Phase 2 enzymes such as glutathione S-transferases and quinone reductase. These studies have been reviewed [7, 63, 64]. While many studies have investigated the effects of isothiocyanates on these enzymes, fewer have looked at the effects of isothiocyanates on carcinogen metabolism in the specific models where inhibition of tumor development has been observed. For example, sulforaphane is known to be a potent inducer of glutathione S-transferases and quinone reductase, but there is no evidence that induction of these enzymes is specifically responsible for its inhibitory effects on DMBA induced mammary tumorigenesis [55, 65]. Moreover, sulforaphane also inhibits cytochrome P450 activity [66]. PEITC is the most extensively studied chemopreventive isothiocyanate with respect to mechanisms of inhibition of rat lung tumorigenesis by NNK and rat esophageal tumorigenesis by NBMA.

Studies on inhibition of NNK induced rat lung carcinogenesis by PEITC clearly show that its major effect is specific inhibition of cytochrome P450 enzymes in the rat lung which are responsible for the metabolic activation of NNK. In studies carried out under the conditions of the bioassay in which PEITC inhibited rat lung tumorigenesis by NNK, we demonstrated that PEITC had no effect on the distribution of NNK and its metabolites in different tissues of the rat, although levels of metabolites resulting from the metabolic activation of NNK were reduced in the lung of PEITC treated rats [67]. We also showed that while PEITC had no effect on hepatic microsomal metabolism of NNK and its major metabolite NNAL, it specifically inhibited the metabolic activation of both

NNK and NNAL in the rat lung [68]. In a third study, we examined the effects of PEITC on DNA adducts of NNK in the rat lung and in individual cell types of the lung [69]. The results demonstrated that PEITC significantly inhibited DNA pyridyloxobutylation by NNK, particularly in the Type II cells which are the targets of rat lung tumorigenesis by NNK.

Extensive studies also demonstrate that inhibition of cytochrome P450s by isothiocyanates is the major mechanism by which they inhibit NNK-induced lung tumorigenesis in the mouse [70–73]. This results in inhibition of O^6-methylguanine formation and tumorigenesis [22, 37]. When added to microsomal incubations, PEITC inhibits NNK oxidation by competitive and noncompetitive mechanisms [72, 73]. Longer chain arylalkyl isothiocyanates are stronger inhibitors of NNK metabolic activation than is PEITC, which correlates with the tumor inhibition data [71, 72]. For example, PHITC is a potent competitive inhibitor of NNK oxidation in mouse lung microsomes with an apparent K_i of 11–16 nM [71]. Dietary PEITC has significant effects on cytochrome P450 enzymes in the mouse, but little effect on Phase 2 enzymes [71, 72]. In the mouse, dietary PEITC and other isothiocyanates have differing effects on cytochrome P450 activities depending on the protocol employed, the dose, and the time after dosing [71–73]. In general, strong inhibitory effects are observed on pulmonary NNK metabolic activation but the inhibition does not correlate with the effects of the isothiocyanates on specific cytochrome P450 enzymes known to be involved in NNK activation. These results suggest that there are unknown cytochrome P450 enzymes in the mouse lung which metabolically activate NNK and are inhibited by isothiocyanates. Studies on the mechanisms by which PEITC inhibits BOP induced hamster lung tumorigenesis conclude that PEITC exerts its chemopreventive activity by decreasing cell turnover and DNA methylation in the target organs, and by influencing hepatic cytochrome P450 enzymes [74].

PEITC is a potent inhibitor of rat esophageal tumorigenesis induced by NBMA (Table 20.2). A comparative study demonstrates that PPITC is even more potent while BITC and PBITC have little effect on tumorigenesis [29]. PHITC enhances tumorigenesis in the same model [51]. Mechanistic studies clearly show that PEITC inhibits the metabolic activation of NBMA in the rat esophagus, probably through inhibition of a cytochrome P450 enzyme [75]. Concomitant with this inhibition, one observes inhibition of O^6-methylguanine formation in rat esophageal DNA [29]. The inhibitory effects: PPITC > PEITC > PBITC > BITC on tumorigenicity correlate with their inhibitory effects on O^6-methylguanine formation [29]. In contrast, effects on carcinogen activation could not explain the enhancing effect of PHITC on rat esophageal tumorigenesis [75]. Inhibition of NNN tumorigenicity in the rat esophagus also appears to be due to inhibition of its metabolic activation [49].

Studies in humans who consumed watercress, a rich source of PEITC, support the results obtained in laboratory animals. Watercress caused a decrease in the levels of oxidative metabolites of acetaminophen, probably due to inhibition of oxidative metabolism by P450 2E1 [76, 77]. Consumption of watercress by smokers altered the profile of NNK metabolism, based on measurements of urinary metabolites [78, 79]. The results indicated that watercress consumption inhibited oxidative metabolism of NNK by inhibition of P450 1A2 [78–80]. These results were consistent with observations in rats in which PEITC blocked NNK induced lung tumorigenesis. Watercress consumption had no effect on P450 2D6 activity [81].

While these studies clearly show that inhibition of cytochrome P450 enzymes involved in the metabolic activation of carcinogens is a major mechanism by which isothiocyanates inhibit tumorigenicity, a series of recent investigations demonstrates another potential avenue of inhibition. PEITC is a strong inducer of c-Jun *N*-terminal kinase 1 (JNK1); this may be involved in the induction of Phase 2 enzymes [82]. The sustained induction of JNK was associated with apoptosis induction in various cell types [83]. Induction of apoptosis by isothiocyanates may proceed through a caspase-3-dependent mechanism [84]. PEITC blocks tumor promoter induced cell transformation in mouse epidermal JB6 cells, and this inhibitory activity on cell transformation is correlated with induction of apoptosis. Moreover, apoptosis induction by PEITC occurs through a *p53*-dependent pathway [85]. These events may be involved in chemoprevention by isothiocyanates and suggest that these compounds may have beneficial properties beyond their favorable modification of carcinogen metabolism.

20.3 Indole-3-carbinol

20.3.1 Occurrence and formation

Glucobrassican, the glucosinolate precursor to indole-3-carbinol, is found in substantial quantities in a number of cruciferous vegetables. Typical levels in vegetables such as cabbage, cauliflower, brussels sprouts, and turnip range from 0.1–3.2 mmol/kg [60]. In the United Kingdom, mean daily intake of glucobrassican from cooked vegetables was estimated as 1.5–3.1 mg/person [86].

Myrosinase catalyzed hydrolysis of glucobrassican produces indole-3-carbinol and other products, as illustrated in Figure 20.2 [60]. At pH 7, the expected initial product is the corresponding isothiocyanate (**1**) but this has never been isolated or synthesized [60]. The isothiocyanate spontaneously hydrolyzes producing indole-3-carbinol (**3**). The latter self-condenses with the elimination of formaldehyde, producing 3,3′-diindolylmethane (**4**). When ascorbic acid is present, it reacts with indole-3-carbinol yielding ascorbigen (**5**). If the hydrolysis of glucobrassican occurs at pH 3–4, indole-3-acetonitrile (**2**) is produced [60].

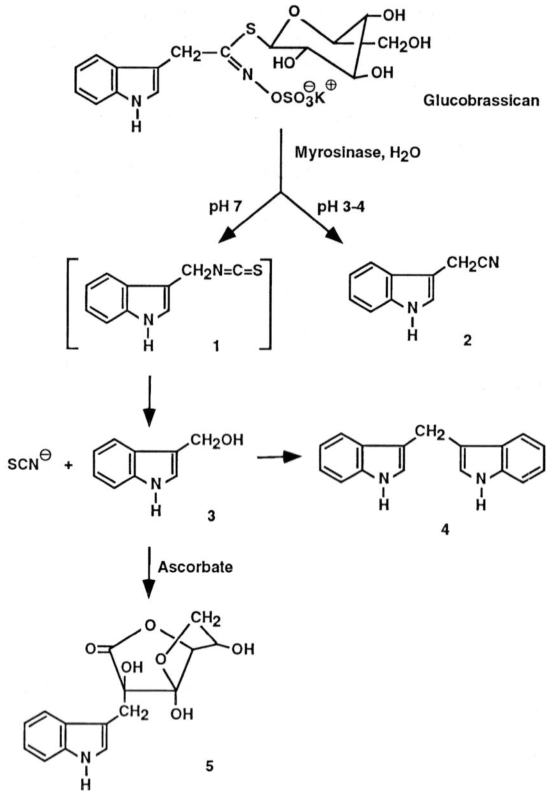

Figure 20.2: Hydrolysis of glucobrassican to indole-3-carbinol and related products.

20.3.2 Effects of indole-3-carbinol on carcinogenesis

Representative studies on the effects of indole-3-carbinol on carcinogenesis in animal models are summarized in Table 20.3 [12, 87–102]. In most cases, indole-3-carbinol demonstrates inhibitory activity when given before, or before and during carcinogen treatment. This is consistently observed in rat mammary tumor models using either directly or indirectly acting carcinogens, or in spontaneous tumor models [87–90]. Other targets of inhibition are rat endometrium, mouse forestomach, rat liver, rat tongue, mouse lung, and trout liver [87, 91–98]. Inhibition of carcinogenesis by a variety of different carcinogens including PAH, nitrosamines, nitro compounds, and aflatoxin B_1 is observed. An exception is DMH induced rat colon tumorigenesis which was enhanced.

In contrast, there is strong evidence that indole-3-carbinol is a tumor promoter when given in the post-initiation stage, e.g. after carcinogen administra-

Table 20.3: Modification of carcinogenesis by indole-3-carbinol.

Carcinogen[a]	Species and Target Organ	Effect	Reference
A. Initiation stage or throughout			
DMBA	Rat mammary	inhibition	Wattenberg and Loub 1978 [87] Grubbs *et al.* 1995 [88]
MNU	Rat mammary	inhibition	Grubbs *et al.* 1995 [88]
None	Mouse mammary	inhibition	Bradlow *et al.* 1991 [89] Malloy *et al.* 1997 [90]
None	Rat endometrial	inhibition	Kojima *et al.* 1994 [91]
BaP	Mouse forestomach	inhibition	Wattenberg and Loub 1978 [87]
4-NQO	Rat tongue	inhibition	Tanaka *et al.* 1992 [92]
NDEA	Rat liver	inhibition	Tanaka *et al.* 1990 [93] Kim *et al.* 1994 [94]
DMH	Rat colon	inhibition	Pence *et al.* 1986 [95]
AFB_1	Trout liver	inhibition	Nixon *et al.* 1984 [96] Dashwood *et al.* 1989 [97]
NNK	Mouse lung	inhibition	Morse *et al.* 1990 [98] El-Bayoumy *et al.* 1996 [40]
B. Post-initiation Stage			
4-NQO	Rat tongue	inhibition	Tanaka *et al.* 1992 [92]
NDEA + MNU + DHPN	Rat liver, thyroid	enhancement	Kim *et al.* 1997 [99]
NDEA	Rat liver	enhancement	Kim *et al.* 1994 [94]
NDEA	Mouse liver	inhibition	Oganesian *et al.* 1997 [100]
AFB_1	Trout liver	enhancement	Bailey *et al.* 1987 [101] Dashwood *et al.* 1991 [102]

[a] Abbreviations: see Table 20.2; additional abbreviations: 4-NQO, 4-nitroquinoline-1-oxide; AFB_1, aflatoxin B_1.

tion. This was first observed in the trout liver model and has been confirmed in rat liver and thyroid [94, 99, 101, 102]. Other studies, however, demonstrate inhibition of mouse liver and rat tongue carcinogenesis by indole-3-carbinol given in the post-initiation stage [92, 100].

Some unpublished studies support the results discussed above, and the U.S. National Cancer Institute is pursuing the clinical development of indole-3-carbinol. The target organ of highest clinical priority is the breast [103].

20.3.3 Mechanisms of chemoprevention by indole-3-carbinol

Mechanistic aspects of chemoprevention by indole-3-carbinol have been reviewed [60, 103–105]. The protective effects of indole-3-carbinol against carcinogenesis result partly from its ability to modify enzymes involved in carcinogen metabolic activation and detoxification. In rats, indole-3-carbinol induces cytochromes P450 1A1, 1A2, 2B1, and 3A as well as Phase 2 enzymes such as UDP-glucuronosyl transferase, epoxide hydrolase, and glutathione S-transferases [88, 106–108]. It also induces P450s and Phase 2 enzymes in mice [109, 110]. Treatment with indole-3-carbinol reduces carcinogen-DNA adducts indicating that the overall modification of enzyme activities favors detoxification [97]. Some of these effects are due not to the parent compound but rather to condensation products formed upon contact with gastric acid. Multiple products of this type are observed *in vivo* [111]. The acid condensation products are planar compounds which, unlike indole-3-carbinol itself, are agonists of the Ah receptor, resulting in the induction of P450 1A enzymes [104]. 3,3′-Diindolylmethane, one of the acid condensation products, is also an inhibitor of rat and human cytochrome P450 1A1, human P450 1A2, and rat P450 2B1 [112]. The condensation products were more effective than indole-3-carbinol as inhibitors of aflatoxin B_1 DNA binding and hepatocarcinogenesis in the trout, using an embryo microinjection model, indicating that their formation in the stomach is important in the expression of the biological activities of indole-3-carbinol [113].

While modification of carcinogen metabolic activation/detoxification ratios appears to be one mechanism of chemoprevention by indole-3-carbinol, changes in carcinogen distribution can also occur. In mice, indole-3-carbinol protects against NNK-induced pulmonary carcinogenesis by increasing the hepatic clearance of NNK and thereby decreasing bioavailability in the lung [98]. In these mice, urinary levels of two NNK metabolites – NNAL and NNAL-Gluc – decreased with a corresponding increase in levels of metabolites resulting from α-hydroxylation [98]. In smokers treated with indole-3-carbinol, decreased levels of NNAL and NNAL-Gluc in urine were also observed indicating that indole-3-carbinol has similar effects on hepatic NNK metabolism in humans and mice [114]. Enhanced metabolism of NNK in humans probably results from induction of P450 1A2 by indole-3-carbinol.

One of the indole-3-carbinol condensation products, indolo[3,2-*b*]carbazole, decreases estrogen receptor levels in cultured breast cancer cells [103, 115]. It is also a weak estrogen, but its action is mainly antiestrogenic in human breast cancer cells. Indole-3-carbinol also affects the metabolism of estradiol by increasing the ratio of 2-hydroxylation to 16α-hydroxylation. 2-Hydroxylation is catalyzed by P450 1A2 and 16α-hydroxylation by P450 3A4 [116]. Therefore, these results are consistent with induction of P450 1A2 by indole-3-carbinol. This ratio change is associated with inhibition of mammary and endometrial tumor development in rodents [117, 118]. Indole-3-carbinol also enhances estradiol 2-hydroxylation in humans and has been proposed as a chemopreventive agent for breast cancer [103, 118–120].

20.4 Thiols of *Allium* plants

20.4.1 Occurrence and formation

Plants of the genus *Allium*, particularly garlic and onion, have been thoroughly investigated with respect to the occurrence of sulfur-containing compounds which are responsible for their characteristic odors. This area has been reviewed [121–125]. When these plants are crushed, alliinases, which are C–S lyase enzymes, act on *S*-alkyl cysteine *S*-oxides to produce a wide variety of sulfur-containing compounds. Diallyl sulfide and related thiols have received the most attention with respect to chemoprevention.

20.4.2 Effects of *Allium* thiols on carcinogenesis

A number of studies demonstrate that onion and garlic oils inhibit tumorigenesis [126–131]. This has spurred interest in chemoprevention by their constituents. Representative studies are summarized in Table 20.4 [132–150]. Diallyl sulfide is an effective inhibitor of tumorigenesis by a variety of carcinogen types including hydrazines, nitrosamines, aromatic amines, vinyl carbamate, PAH, and others. A large number of different tissues are protected including mouse colon, lung, skin, and forestomach and rat esophagus, lung, and thyroid. In general, diallyl sulfide inhibits tumorigenesis when administered prior to or concurrently with the carcinogen. For example, when given prior to NBMA, it is a potent inhibitor of esophageal tumorigenesis in the rat but has no effect when given after carcinogen administration [134, 138]. Both enhancement and inhibition have been observed in other studies in which diallyl sulfide was administered after the carcinogen [136, 148, 150].

Diallyl disulfide is also an effective inhibitor of tumorigenesis in mouse and rat models. However, diallyl trisulfide shows marginal effects or enhancement. Mixed results have been obtained with allyl methyl trisulfide, while allyl methyl disulfide and allyl methyl sulfide both are inhibitory in experiments reported to date. Saturated analogues are generally less effective as chemopreventive agents than the corresponding allyl compounds.

Table 20.4: Modification of carcinogenesis by *Allium* thiols.

Thiol	Carcinogen[a]	Species and Target Organ	Effect	Reference
Diallyl sulfide	DMH	Mouse colon	inhibition	Wargovich 1987 [132]
	DMH	Rat liver	inhibition	Hayes *et al.* 1987 [133]
	BaP	Mouse lung and forestomach	inhibition	Sparnins *et al.* 1988 [146]
	NBMA	Rat esophagus	inhibition	Wargovich *et al.* 1988 [134]
	NDEA	Mouse forestomach and lung	inhibition or no effect	Wattenberg *et al.* 1989 [145]
	DMBA	Mouse skin	inhibition	Athar *et al.* 1990 [135]
	NDEA, MNU, DBN	Rat lung, thyroid	inhibition	Jang *et al.* 1991 [136]
	NNK	Mouse lung	inhibition	Hong *et al.* 1992 [137]
	NBMA	Rat esophagus	inhibition or no effect	Wargovich *et al.* 1992 [138]
	DMBA	Hamster cheek pouch	inhibition	Nagabhushan *et al.* 1992 [139]
	NDEA or NDEA, MNU, DMH, BHBN, DHPN	Rat liver	enhancement	Takahashi *et al.* 1992 [148]
	DMBA	Mouse skin	inhibition	Dwivedi *et al.* 1992 [147]
	AA	Rat forestomach	inhibition	Hadjiolov *et al.* 1993 [140]
	IQ	Rat liver	inhibition	Tsuda *et al.* 1994 [141]
	VC	Mouse skin	inhibition	Surh *et al.* 1995 [142]
	NDEA	Rat liver	enhancement	Takada *et al.* 1994 [150]
	NDEA	Mouse liver	inhibition	Pereira 1995 [42]
Diallyl trisulfide	BaP	Mouse lung and forestomach	inhibition [forestomach] no effect [lung]	Sparnins *et al.* 1988 [146]
	NDEA	Rat liver	enhancement	Takada *et al.* 1994 [150]
Diallyl disulfide	NDEA	Mouse lung and forestomach	inhibition	Wattenberg *et al.* 1989 [145]
	DMBA	Mouse skin	inhibition	Dwivedi *et al.* 1992 [147]
	NDEA, MNU, DMH, BHBN, DHPN	Rat kidney and colon	inhibition	Takahashi *et al.* 1992 [148]
	MNU	Rat mammary	inhibition	Schaffer *et al.* 1996 [149]
Allyl methyl trisulfide	BaP	Mouse lung and forestomach	inhibition (forestomach), no effect (lung)	Sparnins *et al.* 1986, 1988 [143, 146]
	NDEA	Rat liver	enhancement	Takada *et al.* 1994 [150]

Table 20.4 (continued)

Thiol	Carcinogen[a]	Species and Target Organ	Effect	Reference
Allyl methyl disulfide	BaP	Mouse lung and forestomach	inhibition	Sparnins *et al.* 1988 [146]
	NDEA	Mouse lung and forestomach	inhibition	Wattenberg *et al.* 1989 [145]
Allyl methyl sulfide	NDEA	Rat liver	inhibition	Takada *et al.* 1994 [150]
Allyl mercaptan	NDEA	Mouse lung and forestomach	inhibition	Wattenberg *et al.* 1989 [145]
Dipropyl trisulfide	BaP	Mouse lung and forestomach	inhibition (forestomach)	Sparnins *et al.* 1988 [146]
Dipropyl disulfide	NDEA	Mouse lung and forestomach	inhibition (forestomach)	Wattenberg *et al.* 1989 [145]
	DMBA	Mouse skin	no effect	Belman *et al.* 1989 [144]
Dipropyl sulfide	BaP	Mouse lung and forestomach	no effect	Sparnins *et al.* 1988[146]
	NDEA	Rat liver	enhancement	Takada *et al.* 1994 [150]
Propyl methyl trisulfide	BaP	Mouse lung and forestomach	no effect	Sparnins *et al.* 1988 [146]
Propyl methyl disulfide	BaP	Mouse lung and forestomach	no effect	Sparnins *et al.* 1988 [146]
	NDEA	Rat liver	inhibition	Takada *et al.* 1994 [150]
Di(1-propenyl)-sulfide	DMBA	Mouse skin	inhibition	Belman *et al.* 1989 [144]
Ajoene	DMBA	Mouse skin	inhibition	Belman *et al.* 1989 [144]
Propylene sulfide	NDEA	Rat liver	inhibition	Takada *et al.* 1994 [150]

[a] Abbreviations: see Tables 20.2 and 20.3; additional abbreviations: DBN, dibutylnitrosamine; AA, aristolochic acid; IQ, 2-amino-3-methylimidazo[4,5-*f*]quinoline; VC, vinyl carbamate.

20.4.3 Mechanisms of chemoprevention by *Allium* thiols

Modification of carcinogen metabolism is the major mechanism by which diallyl sulfide and other *Allium* thiols protect against tumorigenesis. These compounds affect both Phase 1 and Phase 2 enzymes. Mechanistic studies on diallyl sulfide demonstrate that it is converted to diallyl sulfoxide and diallyl sulfone metabolically. Diallyl sulfide and diallyl sulfone are strong competitive inhibitors of cytochrome P450 2E1, which is involved in the metabolic activation of DMH, NDEA, and VC, three carcinogens which are inhibited by diallyl sulfide [142, 151]. Consistent with this, diallyl sulfide and diallyl sulfone are effective inhibitors of carbon tetrachloride, *N*-nitrosodimethylamine, and acetaminophen induced hepatotoxicity [151]. Diallyl sulfide probably inhibits P450s involved in the metabolic activation of NBMA and NNK as well, although these enzymes have not been fully characterized.

Inhibition of P450s is probably not the major mechanism by which diallyl sulfide and related compounds inhibit PAH tumorigenesis [146, 152]. Comparative studies demonstrate that induction of glutathione S-transferase activity is the major protective mechanism operating in the mouse forestomach. Glutathione S-transferases are involved in the detoxification of the diol epoxide ultimate carcinogens of PAH such as BaP. A correlation has been observed between induction of glutathione S-transferase activity and chemopreventive activity of a number of allyl thiols in the mouse forestomach, but not the lung [146]. The allyl group is necessary for induction, which also parallels chemopreventive activity [146]. In contrast, there is little effect of diallyl sulfide and related compounds on ethoxyresorufin *O*-deethylase or epoxide hydrolase activity [152].

Less is known about the effects of onion and garlic components on the post-initiation phase of carcinogenesis. One study shows that some of these compounds are effective inhibitors of soybean lipoxygenase activity. The strongest inhibitor, di(1-propenyl)sulfide, inhibited both lipoxygenase activity and tumor promotion in mouse skin while the corresponding saturated compound, di(*n*-propyl)disulfide inhibited neither [144]. Several sulfides which enhanced hepatocarcinogenesis by NDEA when given after carcinogen treatment also enhanced ornithine decarboxylase activity, but did not affect levels of 8-oxodeoxyguanosine or lipid peroxidation. These results suggest that the promoting effect of the sulfides could be caused by increased cell proliferation and polyamine biosynthesis [150].

20.5 Conclusions

The results described here clearly demonstrate that Cruciferae and *Allium* plants as well as their constituents can inhibit carcinogenesis in a variety of animal models. The strongest evidence emanates from studies on the individual constituents because variables can be more readily controlled. Isothiocyanates derived from naturally occurring glucosinolates are generally potent inhibitors of chemical carcinogenesis, particularly when administered prior to, or concurrently with the carcinogen. The results suggest that isothiocyanates are stronger inhibitors of nitrosamine and PAH tumorigenesis than indole-3-carbinol or *Allium* thiols, but limited direct comparative data are available. For example, the lowest total gavage dose of PEITC required to significantly inhibit NNK-induced mouse lung tumorigenesis is 5 μmol, while the corresponding figures for indole-3-carbinol and diallyl sulfide are 100 and 105 μmol, respectively [70]. BITC also appears to be a stronger inhibitor of mouse lung tumorigenesis than diallyl sulfide [21, 24, 146]. There are few examples of enhancement of carcinogenesis in studies with naturally occurring isothiocyanates or *Allium* thiols. In contrast, there can be little doubt that indole-3-carbinol has tumor promoting activity. On balance, however, the available data are consistent with the hypothesis that specific chemopreventive agents in vegetables are at least partially responsible for the protective effects of vegetables against cancer that are seen in epidemiologic studies.

There are many complexities in attempting to evaluate the potential anticarcinogenic effects of vegetables. First, levels of specific chemopreventive agents vary widely depending on the particular species and cultivar. Cooking and eating conditions will also affect the uptake of specific agents. There are interindividual differences in metabolism of the chemopreventive agents. For example, a recent study suggests that people who are deficient in GSTM1 and who consume broccoli are protected against colon cancer because of less efficient metabolism of chemopreventive isothiocyanates [153, 154]. Another study demonstrates higher P450 1A2 activity in individuals who consume cruciferous vegetables and are GSTM1 null, presumably due to induction by isothiocyanates or related compounds [155]. Human exposure to carcinogens is also complex as is the metabolism of each carcinogen, where there are large inter-individual differences in multiple pathways of activation and detoxification. Measurement of human uptake and metabolism of chemopreventive agents in vegetables is necessary for evaluating the potential anticarcinogenic effects of vegetables. Few studies of this type have been carried out for the agents considered here, but new methods for assessing isothiocyanate uptake are becoming available [156–158]. It will be important to incorporate these into epidemiologic studies which also employ biomarkers of carcinogen metabolism.

Acknowledgements

Studies on chemoprevention in the author's laboratory are supported by Grant No. CA-46 535 from the U.S. National Cancer Institute.

References

[1] Wattenberg, L. W. (1978) Inhibition of chemical carcinogenesis. *J. Natl. Cancer Inst.* **60**, 11–18.

[2] Wattenberg, L. W. (1978) Inhibitors of chemical carcinogenesis. *Adv. Cancer Res.* **26**, 197–226.

[3] Stoewsand, G. S. (1995) Bioactive organosulfur phytochemicals in *brassica oleracea* vegetables – a review. *Food Chem. Toxicol.* **33**, 537–543.

[4] Verhoeven, D. T. H.; Verhagen, H.; Goldbohm, R. A.; Van den Brandt, P. A.; Van Poppel, G. (1997) A review of mechanisms underlying anti-carcinogenicity by brassica vegetables. *Chem.-Biol. Interact.* **103**, 79–129.

[5] Johnson, I. T.; Williamson, G.; Musk, S. R. R. (1994) Anti-carcinogenic factors in plant foods: a new class of nutrients. *Nutr. Res. Rev.* **7**, 175–204.

[6] Jongen, W. M. F. (1996) Glucosinolates in Brassica: occurrence and significance as cancer-modulating agents. *Proc. Nutr. Soc.* **55**, 433–446.

[7] Smith, T. J.; Yang, C. S. (1994) Effects of food phytochemicals on xenobiotic metabolism and tumorigenesis. In: Huang, M.-T.; Osawa, T.; Ho, C.-T.; Rosen, R. T. (Eds.) *Food phytochemicals for cancer prevention. I. Fruits and vegetables.* American Chemical Society, Washington, DC, p. 17–48.

[8] Hecht, S. S. (1995) Chemoprevention by isothiocyanates. *J. Cell. Biochem.* **22** (Suppl.), 195–209.

[9] Wattenberg, L. W. (1992) Chemoprevention of cancer by naturally occurring and synthetic compounds. In: Wattenberg, L. W.; Lipkin, M.; Boone, C. W.; Kelloff, G. J. (Eds.) *Cancer chemoprevention.* CRC Press, Boca Raton, FL, p. 19–39.

[10] Wargovich, M. J. (1992) Inhibition of gastrointestinal cancer by organosulfur compounds in garlic. In: Wattenberg, L. W.; Lipkin, M.; Boone, C. W.; Kelloff, G. J. (Eds.) *Cancer chemoprevention.* CRC Press, Boca Raton, FL, p. 195–203.

[11] Lea, M. A. (1996) Organosulfur compounds and cancer. In: American Institute for Cancer Research (Ed.) Dietary phytochemicals in cancer prevention and treatment. *Adv. Exp. Med. Biol.* **401**, Plenum Press, Inc., New York, p. 147–154.

[12] Dashwood, R. H. (1998) Indole-3-carbinol: anti-carcinogen or tumor promoter in brassica vegetables. *Chem.-Biol. Interact.* **110**, 1–5.

[13] Fenwick, G. R.; Heaney, R. K.; Mawson, R. (1989) Glucosinolates. In: Cheeke, P. R. (Ed.) *Toxicants of plant Origin, Volume II. Glycosides*, CRC Press, Inc., Boca Raton, FL, p. 2–41.

[14] Tookey, H. L.; Van Etten, C. H.; Daxenbichler, M. E. (1980) Glucosinolates. In: Liener, I. E. (Ed.) *Toxic Constituents of Plant Stuffs*, Academic Press, New York, p. 103–142.

[15] Sasaki, S. (1963) Inhibitory effects by _-naphthyl-isothiocyanate on liver tumorigenesis in rats treated with 3′-methyl-4-dimethyl-aminoazobenzene. *J. Nara Med. Assoc.* **14**, 101–115.

[16] Sidransky, H.; Ito, N.; Verney, E. (1966) Influence of α-naphthyl-isothiocyanate on liver tumorigenesis in rats ingesting ethionine and *N*-2-fluorenylacetamide. *J. Natl. Cancer Inst.* **37**, 677–686.

[17] Lacassagne, A.; Hurst, L.; Xuong, M. D. (1970) Inhibition, par deux naphthylisothiocyanates, de l'hepatocancérogenèse produit, chez le rat, par le *p*-diméthylaminoazobenzène (DAB). *C.R. Séances Soc. Biol. Fil.* **164**, 230–233.

[18] Ito, N.; Hiasa, Y.; Konishi, Y.; Marugami, M. (1969) The development of carcinoma in liver of rats treated with *m*-toluylenediamine and the synergistic and antagonistic effects with other chemicals. *Cancer Res.* **29**, 1137–1145.

[19] Makiura, S.; Kamamoto, Y.; Sugihara, S.; Hirao, K.; Hiasa, Y.; Arai, M.; Ito, N. (1973) Effect of 1-naphthyl isothiocyanate and 3-methylcholanthrene on hepatocarcinogenesis in rats treated with diethylnitrosoamine. *Jpn. J. Cancer Res.* **64**, 101–104.

[20] Ito, N.; Matayoshi, K.; Matsumura, K.; Denda, A.; Kani, T.; Arai, M.; Makiura, S. (1974) Effect of various carcinogenic and non-carcinogenic substances on development of bladder tumors in rats induced by *N*-butyl-*N*-(4-hydroxybutyl)nitrosoamine. *Jpn. J. Cancer Res.* **65**, 123–130.

[21] Wattenberg, L. W. (1977) Inhibition of carcinogenic effects of polycyclic hydrocarbons by benzyl isothiocyanate and related compounds. *J. Natl. Cancer Inst.* **58**, 395–398.

[22] Morse, M. A.; Amin, S. G.; Hecht, S. S.; Chung, F.-L. (1989) Effects of aromatic isothiocyanates on tumorigenicity, O^6-methylguanine formation, and metabolism of the tobacco-specific nitrosamine 4-(methylnitrosamino)-1-(3-pyridyl)-1-butanone in A/J mouse lung. *Cancer Res.* **49**, 2894–2897.

[23] Wattenberg, L. W. (1981) Inhibition of carcinogen-induced neoplasia by sodium cyanate, *tert*-butyl isocyanate, and benzyl isothiocyanate administered subsequent to carcinogen exposure. *Cancer Res.* **41**, 2991–2994.

[24] Lin, J.-M.; Amin, S.; Trushin, N.; Hecht, S. S. (1993) Effects of isothiocyanates on tumorigenesis by benzo[*a*]pyrene in murine tumor models. *Cancer Lett.* **74**, 151–159.

[25] Wattenberg, L. W. (1987) Inhibitory effects of benzyl isothiocyanate administered shortly before diethylnitrosamine or benzo[*a*]pyrene on pulmonary and forestomach neoplasia in A/J mice. *Carcinogenesis* **8**, 1971–1973.

[26] Morse, M. A.; Reinhardt, J. C.; Amin, S. G.; Hecht, S. S.; Stoner, G. D.; Chung, F.-L. (1990) Effect of dietary aromatic isothiocyanates fed subsequent to the administration of 4-(methylnitrosamino)-1-(3-pyridyl)-1-butanone on lung tumorigenicity in mice. *Cancer Lett.* **49**, 225–230.

[27] Sugie, S.; Okumura, A.; Tanaka, T.; Mori, H. (1993) Inhibitory effects of benzyl isothiocyanate and benzyl thiocyanate on diethylnitrosamine-induced hepatocarcinogenesis in rats. *Jpn. J. Cancer Res.* **84**, 865–870.

[28] Sugie, S.; Okamoto, K.; Okumura, A.; Tanaka, T.; Mori, H. (1994) Inhibitory effects of benzyl thiocyanate and benzyl isothiocyanate on methylazoxymethanol acetate-induced intestinal carcinogenesis in rats. *Carcinogenesis* **15**, 1555–1560.

[29] Wilkinson, J. T.; Morse, M. A.; Kresty, L. A.; Stoner, G. D. (1995) Effect of alkyl chain length on inhibition of *N*-nitrosomethylbenzylamine-induced esophageal tumorigenesis and DNA methylation by isothiocyanates. *Carcinogenesis* **16**, 1011–1015.

[30] Hirose, M.; Yamaguchi, T.; Kimoto, N.; Ogawa, K.; Futakuchi, M.; Sano, M.; Shirai, T. (1998) Strong promoting activity of phenylethyl isothiocyanate and benzyl isothiocyanate on urinary bladder carcinogenesis in F344 male rats. *Int. J. Cancer* **77**, 773–777.

[31] Lubet, R. A.; Steele, V. E.; Eto, I.; Juliana, M. M.; Kelloff, G. J.; Grubbs, C. J. (1997) Chemopreventive efficacy of anethole thithione, *N*-acetyl-L-cysteine, miconazole and phenethylisothiocyanate in the DMBA-induced rat mammary cancer model. *Int. J. Cancer* **72**, 95–101.

[32] Futakuchi, M.; Hirose, M.; Miki, T.; Tanaka, H.; Ozaki, M.; Shirai, T. (1998) Inhibition of DMBA-initiated rat mammary tumor development by 1-*O*-hexyl-2,3,5-trimethylhydroquinone, phenylethyl isothiocyanate, and novel synthetic ascorbic acid derivatives. *Eur. J. Cancer Prev.* **7**, *153–159.*

[33] Chung, F.-L.; Kelloff, G.; Steele, V.; Pittman, B.; Zang, E.; Jiao, D.; Rigotty, J.; Choi, C.-I.; Rivenson, A. (1996) Chemopreventive efficacy of arylalkyl isothiocyanates and *N*-acetylcysteine for lung tumorigenesis in Fischer rats. *Cancer Res.* **56**, 772–778.

[34] Hecht, S. S.; Trushin, N.; Rigotty, J.; Carmella, S. G.; Borukhova, A.; Akerkar, S. A.; Rivenson, A. (1996) Complete inhibition of 4-(methylnitrosamino)-1-(3-pyridyl)-1-bu-

tanone induced rat lung tumorigenesis and favorable modification of biomarkers by phenethyl isothiocyanate. *Cancer Epidemiol. Biomarkers Prev.* **5**, 645–652.

[35] Morse, M. A.; Wang, C.-X.; Stoner, G. D.; Mandal, S.; Conran, P. B.; Amin, S. G.; Hecht, S. S.; Chung, F.-L. (1989) Inhibition of 4-(methylnitrosamino)-1-(3-pyridyl)-1-butanone-induced DNA adduct formation and tumorigenicity in lung of F344 rats by dietary phenethyl isothiocyanate. *Cancer Res.* **49**, 549–553.

[36] Morse, M. A.; Eklind, K. I.; Amin, S. G.; Hecht, S. S.; Chung, F.-L. (1989) Effects of alkyl chain length on the inhibition of NNK-induced lung neoplasia in A/J mice by arylalkyl isothiocyanates. *Carcinogenesis* **10**, 1757–1759.

[37] Morse, M. A.; Eklind, K. I.; Hecht, S. S.; Jordan, K. G.; Choi, C.-I.; Desai, D. H.; Amin, S. G.; Chung, F.-L. (1991) Structure-activity relationships for inhibition of 4-(methylnitrosamino)-1-(3-pyridyl)-1-butanone lung tumorigenesis by arylalkyl isothiocyanates in A/J mice. *Cancer Res.* **51**, 1846–1850.

[38] Morse, M. A.; Eklind, K. I.; Amin, S. G.; Chung, F. L. (1992) Effect of frequency of isothiocyanate administration on inhibition of 4-(methylnitrosamino)-1-(3-pyridyl)-1-butanone-induced pulmonary adenoma formation in A/J mice. *Cancer Lett.* **62**, 77–81.

[39] Matzinger, S. A.; Crist, K. A.; Stoner, G. D.; Anderson, M. W.; Pereira, M. A.; Steele, V. E.; Kelloff, G. J.; Lubet, R. A.; You, M. (1995) K-*ras* mutations in lung tumors from A/J and A/JxTSG-*p53* F_1 mice treated with 4-(methylnitrosamino)-1-(3-pyridyl)-1-butanone and phenethyl isothiocyanate. *Carcinogenesis* **16**, 2487–2492.

[40] El-Bayoumy, Upadhyaya, P.; Desai, D. H.; Amin, S.; Hoffmann, D.; Wynder, E. L. (1996) Effects of 1,4-phenylenebis(methylene)selenocyanate, phenethyl isothiocyanate, indole-3-carbinol, and D-limonene individually and in combination on the tumorigenicity of the tobacco-specific nitrosamine 4-(methylnitrosamino)-1-(3-pyridyl)-1-butanone in A/J mouse lung. *Anticancer Res.* **16**, 2709–2712.

[41] Jiao, D.; Smith, T. J.; Yang, C. S.; Pittman, B.; Desai, D.; Amin, S.; Chung, F.-L. (1997) Chemopreventive activity of thiol conjugates of isothiocyanates for lung tumorigenesis. *Carcinogenesis* **18**, 2143–2147.

[42] Pereira, M. A. (1995) Chemoprevention of diethylnitrosamine-induced liver foci and heptacellular adenomas in C3H mice. *Anticancer Res.* **15**, 1953–1956.

[43] Siglin, J. C.; Barch, D. H.; Stoner, G. D. (1995) Effects of dietary phenethyl isothiocyanate, ellagic acid, sulindac and calcium on the induction and progression of *N*-nitrosomethylbenzylamine-induced esophageal carcinogenesis in rats. *Carcinogenesis* **16**, 1101–1106.

[44] Stoner, G. D.; Morrissey, D.; Heur, Y.-H.; Daniel, E.; Galati, A.; Wagner, S. A. (1991) Inhibitory effects of phenethyl isothiocyanate on *N*-nitrosobenzylmethylamine carcinogenesis in the rat esophagus. *Cancer Res.* **51**, 2063–2068.

[45] Nishikawa, A.; Furukawa, F.; Uneyama, C.; Ikeyaki, S.; Tanakamaru, Z.; Chung, F.-L.; Takahashi, M.; Hayashi, Y. (1996) Chemopreventive effects of phenethyl isothiocyanate in lung and pancreatic tumorigenesis in *N*-nitrosobis(2-oxopropyl)amine-treated hamsters. *Carcinogenesis* **17**, 1381–1384.

[46] Adam-Rodwell, G.; Morse, M. A.; Stoner, G. D. (1993) The effects of phenethyl isothiocyanate on benzo[*a*]pyrene-induced tumors and DNA adducts in A/J mouse lung. *Cancer Lett.* **71**, 35–42.

[47] Ogawa, K.; Futakuchi, M.; Hirose, M.; Boonyaphiphat, P.; Mizoguchi, Y.; Miki, T.; Shirai, T. (1998) Stage and organ dependent effects of 1-*O*-hexyl-2,3,5-trimethylhydroquinone, ascorbic acid derivatives, *N*-heptadecane-8,10-dione and pheylethyl isothiocyanate in a rat multiorgan carcinogenesis model. *Int. J. Cancer* **76**, 851–856.

[48] Nishikawa, A.; Furukawa, F.; Ikezaki, S.; Tanakamaru, Z.-Y.; Chung, F.-L.; Takahashi, M.; Hayashi, Y. (1996) Chemopreventive effects of 3-phenylpropyl isothiocyanate on hamster lung tumorigenesis initiated with *N*-nitrosobis(2-oxopropyl)amine. *Jpn. J. Cancer Res.* **87**, 122–126.

[49] Stoner, G. D.; Adams, C.; Kresty, L. A.; Hecht, S. S.; Murphy, S. E.; Morse, M. A. (1998) Inhibition of *N'*-nitrosonornicotine-induced esophageal tumorigenesis by 3-phenylpropyl isothiocyanate. *Carcinogenesis,* in press.

[50] Hecht, S. S.; Trushin, N.; Rigotty, J.; Carmella, S. G.; Borukhova, A.; Akerkar, S. A.; Desai, D.; Amin, S.; Rivenson, A. (1996) Inhibitory effects of 6-phenylhexyl isothiocyanate on 4-(methylnitrosamino)-1-(3-pyridyl)-1-butanone metabolic activation and lung tumorigenesis in rats. *Carcinogenesis* **17**, 2061–2067.

[51] Stoner, G. D.; Siglin, J. C.; Morse, M. A.; Desai, D. H.; Amin, S. G.; Kresty, L. A.; Toburen, A. L.; Heffner, E. M.; Francis, D. J. (1995) Enhancement of esophageal carcinogenesis in male F344 rats by dietary phenylhexyl isothiocyanate. *Carcinogenesis* **16**, 2473–2476.

[52] Rao, C. V.; Rivenson, A.; Simi, B.; Hamid, R.; Kelloff, G. J.; Steele, V.; Reddy, B. S. (1995) Enhancement of experimental colon carcinogenesis by dietary 6-phenylhexyl isothiocyanate. *Cancer Res.* **55**, 4311–4318.

[53] Jiao, D.; Eklind, K. I.; Choi, C. I.; Desai, D. H.; Amin, S. G.; Chung, F. L. (1994) Structure-activity relationships of isothiocyanates as mechanism-based inhibitors of 4-(methylnitrosamino)-1-(3-pyridyl)-1-butanone-induced lung tumorigenesis in A/J mice. *Cancer Res.* **54**, 4327–4333.

[54] Jiao, D.; Smith, T. J.; Kim, S.; Yang, C. S.; Desai, D.; Amin, S.; Chung, F. L. (1996) The essential role of the functional group in alkyl isothiocyanates for inhibition of tobacco-nitrosamine-induced lung tumorigenesis. *Carcinogenesis* **17**, 755–759.

[55] Zhang, Y.; Kensler, T. W.; Cho, C.-G.; Posner, G. H.; Talalay, P. (1994) Anti-carcinogenic activities of sulforaphane and structurally related synthetic norbornyl isothiocyanates. *Proc. Natl. Acad. Sci. USA* **91**, 3147–3150.

[56] National Cancer Institute. Chemoprevention Branch and Agent Development Committee. (1996) Clinical development plan: phenethyl isothiocyanate. *J. Cell. Biochem.* **265**, 149–157.

[57] Morse, M. A.; Wang, C.-X.; Amin, S. G.; Hecht, S. S.; Chung, F.-L. (1988) Effects of dietary sinigrin or indole-3-carbinol on O^6-methylguanine-DNA-transmethylase activity and 4-(methylnitrosamino)-1-(3-pyridyl)-1-butanone-induced DNA methylation and tumorigenicity in F344 rats. *Carcinogenesis* **9**, 1891–1895.

[58] Smith, T. K.; Lund, E. K.; Johnson, I. T. (1998) Inhibition of dimethylhydrazine-induced aberrant crypt foci and induction of apoptosis in rat colon following oral administration of the glucosinolate sinigrin. *Carcinogenesis* **19**, 267–273.

[59] Wattenberg, L. W.; Hanley, A. B.; Barany, G.; Sparnins, V. L.; Lam, L. K. T.; Fenwick, G. R. (1986) Inhibition of carcinogenesis by some minor dietary constituents. In: Hayashi, Y. (Ed.) *Diet, nutrition and cancer.* Japan Sci. Soc. Press, Tokyo, p. 193–203.

[60] McDannel, R.; McLean, A. E. M.; Hanley, A. B.; Heaney, R. K.; Fenwick, G. R. (1988) Chemical and biological properties of indole glucosinolates (glucobrassicans): a review. *Food Chem. Toxicol.* **26**, 59–70.

[61] Birt, D. F.; Pelling, J. C.; Pour, P. M.; Tibbels, M. G.; Schweickart, L.; Bresnick, E. (1987) Enhanced pancreatic and skin tumorigenesis in cabbage-fed hamsters and mice. *Carcinogenesis* **8**, 913–917.

[62] Barrett, J. E.; Klopfenstein, C. F.; Leipold, H. W. (1998) Protective effects of cruciferous, seed meals and hulls against colon cancer in mice. *Cancer Lett.* **127**, 83–88.

[63] Zhang, Y.; Talalay, P. (1994) Anti-carcinogenic activities of organic isothiocyanates: chemistry and mechanism. *Cancer Res.* **54** (Suppl.), 1976s–1981s.

[64] Yang, C. S.; Smith, T. J.; Hong, J.-Y. (1994) Cytochrome P450 enzymes as targets for chemoprevention against chemical carcinogenesis and toxicity: opportunities and limitations. *Cancer Res.* **54**, 1982s–1986s.

[65] Zhang, Y.; Talalay, P.; Cho, C. G.; Posner, G. H. (1992) A major inducer of anti-carcinogenic protective enzymes from broccoli: Isolation and elucidation of structure. *Proc. Natl. Acad. Sci.* USA **89**, 2399–2403.

[66] Maheo, K.; Morel, F.; Lanouet, S.; Kramer, H.; Le Ferrec, E.; Ketterer, B.; Guillouzo, A. (1997) Inhibition of cytochromes P450 and induction of glutathione S-transferases by sulforaphane in primary human and rat hepatocytes. *Cancer Res.* **57**, 3649–3652.

[67] Staretz, M. E.; Hecht, S. S. (1995) Effects of phenethyl isothiocyanate on the tissue distribution of 4-(methylnitrosamino)-1-(3-pyridyl)-1-butanone and metabolites in F344 rats. *Cancer Res.* **55**, 5580–5588.

[68] Staretz, M. E.; Koenig, L.; Hecht, S. S. (1997) Effects of long term phenethyl isothiocyanate treatment on microsomal metabolism of 4-(methylnitrosamino)-1-(3-pyridyl)-1-butanone and 4-(methylnitrosamino)-1-(3-pyridyl)-1-butanol in F344 rats. *Carcinogenesis* **18**, 1715–1722.

[69] Staretz, M. E.; Foiles, P. G.; Miglietta, L. M.; Hecht, S. S. (1997) Evidence for an important role of DNA pyridyloxobutylation in rat lung carcinogenesis by 4-(methyl-nitrosamino)-1-(3-pyridyl)-1-butanone: effects of dose and phenethyl isothiocyanate. *Cancer Res.* **57**, 259–266.

[70] Hecht, S. S. (1998) Biochemistry, biology, and carcinogenicity of tobacco-specific *N*-nitrosamines. *Chem. Res. Toxicol.* **11**, 559–603.

[71] Guo, Z.; Smith, T. J.; Wang, E.; Eklind, K. I.; Chung, F.-L.; Yang, C. S. (1993) Structure-activity relationships of arylalkyl isothiocyanates for the inhibition of 4-(methylnitrosamino)-1-(3-pyridyl)-1-butanone metabolism and the modulation of xenobiotic-metabolizing enzymes in rats and mice. *Carcinogenesis* **14**, 1167–1173.

[72] Smith, T. J.; Guo, Z.; Li, C.; Ning, S. M.; Thomas, P. E.; Yang, C. S. (1993) Mechanisms of inhibition of 4-(methylnitrosamino)-1-(3-pyridyl)-1-butanone bioactivation in mouse by dietary phenethyl isothiocyanate. *Cancer Res.* **53**, 3276–3282.

[73] Smith, T. J.; Guo, Z.; Thomas, P. E.; Chung, F.-L.; Morse, M. A.; Eklind, K.; Yang, C. S. (1990) Metabolism of 4-(*N*-methyl-*N*-nitrosamino)-1-(3-pyridyl)-1-butanone in mouse lung microsomes and its inhibition by isothiocyanates. *Cancer Res.* **50**, 6817–6822.

[74] Nishikawa, A.; Lee, I.-S.; Uneyama, C.; Furukawa, F.; Kim, H.-C.; Kasahara, K.-i, Huh, N.; Takahashi, M. (1997) Mechanistic insights into chemopreventive effects of phenethyl isothiocyanate in *N*-nitrosobis(2-oxopropyl)amine-treated hamsters. *Jpn. J. Cancer Res.* **88**, 1137–1142.

[75] Morse, M. A.; Lu, J.; Gopalakrishnan, R.; Peterson, L. A.; Wani, G.; Stoner, G. D. (1997) Mechanism of enhancement of esophageal tumorigenesis by 6-phenylhexyl isothiocyanate. *Cancer Lett.* **112**, 119–125.

[76] Chen, L.; Mohr, S. N.; Yang, C. S. (1996) Decrease of plasma and urinary oxidative metabolites of acetaminophen after consumption of watercress by human volunteers. *Clin. Pharmacol. Ther.* **60**, 651–660.

[77] Li, Y.; Wang, E.; Chen, L.; Stein, A.; Reuhl, K.; Yang, C. (1997) Effects of phenethyl isothiocyanate on acetaminophen metabolism and hepatotoxicity in mice. *Toxicol. Appl. Pharmacol.* **144**, 306–314.

[78] Hecht, S. S.; Chung, F.-L.; Richie Jr, J. P.; Akerkar, S. A.; Borukhova, A.; Skowronski, L.; Carmella, S. G. (1995) Effects of watercress consumption on metabolism of a tobacco-specific lung carcinogen in smokers. *Cancer Epidemiol. Biomarkers Prev.* **4**, 877–884.

[79] Carmella, S. G.; Borukhova, A.; Akerkar, S. A.; Hecht, S. S. (1997) Analysis of human urine for pyridine-*N*-oxide metabolites of 4-(methylnitrosamino)-1-(3-pyridyl)-1-butanone, a tobacco-specific lung carcinogen. *Cancer Epidemiol. Biomarkers Prev.* **6**, 113–120.

[80] Smith, T. J.; Guo, Z.; Guengerich, F. P.; Yang, C. S. (1996) Metabolism of 4-(methylnitrosamino)-1-(3-pyridyl)-1-butanone (NNK) by human cytochrome P450 1A2 and its inhibition by phenethyl isothiocyanate. *Carcinogenesis* **17**, 809–813.

[81] Caporaso, N.; Whitehouse, J.; Monkman, S.; Boustead, C.; Issaq, H.; Fox, S.; Morse, M. A.; Idle, J. R.; Chung, F-L. (1994): *In vitro* but not *in vivo* inhibition of CYP 2D6

by phenethyl isothiocyanate (PEITC), a constituent of watercress. *Pharmacogenetics* **4**, 275–280.

[82] Yu, R.; Jiao, J.-J.; Duh, J.-L.; Tan, T.-H.; Kong, A.-N. T. (1996) Phenethyl isothiocyanate, a natural chemopreventive agent, activates c-Jun *N*-terminal kinase 1. *Cancer Res.* **56**, 2954–2959.

[83] Chen, Y.-R.; Wang, W.; Kong, A.-N. T.; Tan, T.-H. (1998) Molecular mechanisms of c-Jun *N*-terminal kinase-mediated apoptosis induced by anti-carcinogenic isothiocyanates. *J. Biol. Chem.* **273**, 1769–1775.

[84] Yu, R.; Mandlekar, S.; Harvey, K. J.; Ucker, D. S.; Kong, A.-N. T. (1998) Chemopreventive isothiocyanates induce apoptosis and caspase-3-like protease activity. *Cancer Res.* **58**, 402–408.

[85] Huang, C.; Ma, W.; Li, J.; Hecht, S. S.; Dong, Z. (1998) Essential role of *p53* in phenethyl isothiocyanate (PEITC)-induced apoptosis. *Cancer Res.*, in press.

[86] Sones, K.; Heaney, R. K.; Fenwick, G. R. (1984): An estimate of the mean daily intake of glucosinolates from cruciferous vegetables in the UK. *J. Sci. Food Agric.* **35**, 712–720.

[87] Wattenberg, L. W.; Loub, W. D. (1978) Inhibition of polycyclic aromatic hydrocarbon-induced neoplasia by naturally occurring indoles. *Cancer Res.* **38**, 1410–1413.

[88] Grubbs, C. J.; Steele, V. E.; Casebolt, T.; Juliana, M. M.; Eto, I.; Whitaker, L. M.; Dragnev, K. H.; Kelloff, G. J.; Lubet, R. L. (1995) Chemoprevention of chemically-induced mammary carcinogenesis by indole-3-carbinol. *Anticancer Res.* **5**, 709–716.

[89] Bradlow, H. L.; Michnovicz, J. J.; Telang, N. T.; Osborne, M. P. (1991) Effects of dietary indole-3-carbinol on estradiol metabolism and spontaneous mammary tumors in mice. *Carcinogenesis* **12**, 1571–1574.

[90] Malloy, V. L.; Bradlow, H. L.; Oreutreich, N. (1997) Interaction between a semisynthetic diet and indole-3-carbinol on mammary tumor incidence in Balb/cfC3H mice. *Anticancer Res.* **17**, 4333–4338.

[91] Kojima, T.; Tanaka, T.; Mori, H. (1994) Chemoprevention of spontaneous endometrial cancer in female Donryu rats by dietary indole-3-carbinol. *Cancer Res.* **54**, 1446–1449.

[92] Tanaka, T.; Toshihiro, K.; Morishita, Y.; Mori, H. (1992) Inhibitory effects of the natural products indole-3-carbinol and sinigrin during initiation and promotion phases of 4-nitroquinoline-1-oxide-induced rat tongue carcinogenesis. *Cancer Res.* **83**, 835–842.

[93] Tanaka, T.; Mori, Y.; Morishita, Y.; Hara, A.; Ohno, T.; Kojima, T.; Mori, H. (1990) Inhibitory effect of sinigrin and indole-3-carbinol on diethylnitrosamine-induced hepatocarcinogenesis in male ACI/N rats. *Carcinogenesis* **18**, 1403–1406.

[94] Kim, D. J.; Lee, K. K.; Han, B. S.; Ahn, B.; Bae, J. H.; Jang, J. J. (1994) Biphasic modifying effect of indole-3-carbinol on diethylnitrosamine-induced preneoplastic glutathione S-transferase placental form-positive liver cell foci in Sprague-Dawley rats. *Jpn. J. Cancer Res.* **85**, 578–583.

[95] Pence, B. C.; Buddingh, F.; Yang, S. P. (1986) Multiple dietary factors in the enhancement of dimethylhydrazine carcinogenesis: main effect of indole-3-carbinol. *J. Natl. Cancer Inst.* **77**, 269–276.

[96] Nixon, J. E.; Hendricks, J. D.; Pawlowski, N. E.; Pereira, C. B.; Sinhuber, R. O.; Bailey, G. S. (1984) Inhibition of aflotoxin B_1 carcinogens in rainbow trout by flavone and indole compounds. *Carcinogenesis* **5**, 615–619.

[97] Dashwood, R. H.; Arbogast, D. N.; Fong, A. T.; Pereira, C.; Hendricks, J. D.; Bailey, G. S. (1989) Quantitative inter-relationships between aflatoxin B_1 carcinogen dose, indole-3-carbinol anti-carcinogen dose, target organ DNA adduction and final tumor response. *Carcinogenesis* **10**, 175–181.

[98] Morse, M. A.; LaGreca, S. D.; Amin, S. G.; Chung, F.-L. (1990) Effects of indole-3-carbinol on lung tumorigenesis and DNA methylation induced by 4-(methylnitros-

amino)-1-(3-pyridyl)-1-butanone (NNK) and on the metabolism and disposition of NNK in A/J mice. *Cancer Res.* **50**, 2613–2617.

[99] Kim, D. J.; Han, B. S.; Ahn, B.; Hasegawa, R.; Shirai, T.; Ito, N.; Tsuda, H. (1997) Enhancement by indole-3-carbinol of liver and thyroid gland neoplastic development in a rat medium-term multiorgan carcinogenesis model. *Carcinogenesis* **18**, 377–381.

[100] Oganesian, A.; Hendricks, J. D.; Williams, D. E. (1997) Long term dietary indole-3-carbinol inhibits diethylnitrosamine-initiated hepatocarcinogenesis in the infant mouse model. *Cancer Lett.* **118**, 87–94.

[101] Bailey, G. S.; Hendricks, J. D.; Shelton, D. W.; Nixon, J. E.; Pawlowski, N. E. (1987) Enhancement of carcinogenesis by the natural anti-carcinogen indole-3-carbinol. *J. Natl. Cancer Inst.* **78**, 931–934.

[102] Dashwood, R.; Fong, A. T.; Williams, D. E.; Hendricks, J. D.; Bailey, G. S. (1991) Promotion of aflatoxin B_1 carcinogenesis by the natural tumor modulator indole-3-carbinol: influence of dose, duration, and intermittent exposure on indole-3-carbinol promotional potency. *Cancer Res.* **51**, 2362–2365.

[103] National Cancer Institute, Chemoprevention Branch and Agent Development Committee (1996) Clinical development plan: indole-3-carbinol. *J. Cell. Biochem.* **265**, 127–136.

[104] Bradfield, C. A.; Bjeldanes, L. F. (1991) Modification of carcinogen metabolism by indolylic autolysis products of *Brassica Oleraceae*. In: Friedman, M., (Ed.) *Nutritional and toxicological consequences of food processing*. Plenum Press, Inc., New York, p. 153–163.

[105] Williams, D. E.; Lech, J. J.; Buhler, D. R. (1998) Xenobiotics and xenoestrogens in fish: modulation of cytochrome P450 and carcinogenesis. *Mutat. Res.* **399**, 179–192.

[106] Schertzer, H. G.; Sainsbury, M. (1991) Chemoprotective and hepatic enzyme induction properties of indole and indenoindole antioxidants in rats. *Food Chem. Toxicol.* **29**, 391–400.

[107] Wortelboer, H. M.; Van Der Linden, E. C. M.; De Kruif, C. A.; Noordhoek, J.; Blaarboer, B. J.; Van Bladeren, P. J.; Falk, H. E. (1992) Effects of indole-3-carbinol on biotransformation enzymes in the rat: *in vivo* changes in liver and small intestinal mucosa in comparison with primary hepatocyte cultures. *Fd. Chem. Toxicol.* **30**, 589–599.

[108] Stresser, D. M.; Bailey, G. S.; Williams, D. E. (1994) Indole-3-carbinol and β-naphthoflavone induction of aflatoxin B_1 metabolism and cytochromes P450 associated with bioactivation and detoxification of aflatoxin B_1 in the rat. *Drug Metab. Dispos.* **22**, 383–391.

[109] Schertzer, H. G.; Sainsbury, M. (1991) Intrinsic acute toxicity and hepatic enzyme inducing properties of the chemoprotectants indole-3-carbinol and 5,10-dihydroindeno-[1,2-*b*]indole in mice. *Fd. Chem. Toxicol.* **29**, 237–242.

[110] Baldwin, W. S.; Leblanc, G. A. (1992) The anti-carcinogenic plant compound indole-3-carbinol differentially modulates P450-mediated steroid hydroxylase activities in mice. *Chem.-Biol. Interactions* **83**, 155–169.

[111] Stresser, D. M.; Williams, D. E.; Griffin, D. A.; Bailey, G. S. (1995) Mechanisms of tumor modulation by indole-3-carbinol. Disposition and excretion in male Fischer 344 rats. *Drug Metab. Dispos.* **23**, 965–975.

[112] Stresser, D. M.; Bjeldanes, L. F.; Bailey, G. S.; Williams, D. E. (1995): The anti-carcinogen 3,3′-diindolylmethane is an inhibitor of cytochrome P450. *J. Biochem. Toxicol.* **10**, 191–201.

[113] Dashwood, R. H.; Fong, A. T.; Arbogast, D. N.; Bjeldanes, L. F.; Hendricks, J. D.; Bailey, G. S. (1994) Anti-carcinogenic activity of indole-3-carbinol and products: ultrasensitive bioassay by trout embryo microinjection. *Cancer Res.* **54**, 3617–3619.

[114] Taioli, E.; Garbers, S.; Bradlow, H. L.; Carmella, S. G.; Akekar, S.; Hecht, S. S. (1997) Effects of indole-3-carbinol on the metabolism of 4-(methylnitrosamino)-1-(3-pyridyl)-1-butanone (NNK) in smokers. *Cancer Epidemiol. Biomarkers Prev.* **6**, 517–522.
[115] Liu, H.; Wormke, M.; Safe, S. H.; Bjeldanes, L. F. (1994) Indolo[3,2-*b*]carbazole: a dietary-derived factor that exhibits both antiestrogenic and estrogenic activity. *J. Natl. Cancer Inst.* **86**, 1758–1765.
[116] Yamazaki, H.; Shaw, P. M.; Guengerich, F. P.; Shimada, J. (1998) Roles of cytochrome P450 1A2 and 3A4 in the oxidation of estradiol and estrone in human liver microsomes. *Chem. Res. Toxicol.* **11**, 659–665.
[117] Bradlow, H. L.; Hershcopf, R. J.; Martucci, C. P.; Fishman, J. (1985) Estradiol 16α-hydroxylation in the mouse correlates with mammary tumor incidence and presence of murine mammary tumor virus: a possible model for the hormonal etiology of breast cancer in humans. *Proc. Natl. Acad. Sci. USA* **82**, 6295–6299.
[118] Michnovicz, J. J.; Bradlow, H. L. (1990) Induction of estradiol metabolism by dietary indole-3-carbinol in humans. *J. Natl. Cancer Inst.* **11**, 947–949.
[119] Bradlow, H. L.; Michnovicz, J. J.; Halper, M.; Miller, D. G.; Wong, G. Y. C.; Osborne, M. P. (1994) Long-term responses of women to indole-3-carbinol or a high fiber diet. *Cancer Epidemiol. Biomarkers Prev.* **3**, 591–595.
[120] Michnovicz, J. J.; Adlercreutz, H.; Bradlow, H. L. (1997) Changes in levels of urinary estrogen metabolites after oral indole-3-carbinol treatment in humans. *J. Natl. Cancer Inst.* **89**, 718–723.
[121] Fenwick, G. R.; Hanley, A. B. (1985) The genus *Allium*, Part 1. *CRC Crit. Rev. Food Sci. Nutr.* **22**, 199–271.
[122] Fenwick, G. R.; Hanley, A. B. (1985) The genus *Allium*, Part 2. *CRC Crit. Rev. Food Sci. Nutr.* **22**, 273–377.
[123] Fenwick, G. R.; Hanley, A. B. (1985) The genus *Allium*, Part 3. *CRC Crit. Rev. Food Sci. Nutr.* **23**, 1–73.
[124] Block, E. (1992) The organosulfur chemistry of the genus *Allium* – implications for the organic chemistry of sulfur. *Angew. Chem. Int. Ed. Engl.* **31**, 1135–1178.
[125] Block, E. (1996) Recent results in the organosulfur and organoselenium chemistry of genus *Allium* and *Brassica* plants. *Adv. Exp. Med. Biol.* **401**, 155–169.
[126] Belman, S. (1983) Onion and garlic oils inhibit tumor promotion. *Carcinogenesis* **8**, 1063–1065.
[127] Sadhana, A. S.; Rao, A. R.; Kucheria, K.; Bijani, V. (1988) Inhibitory action of garlic oil on the initiation of benzo[*a*]pyrene-induced skin carcinogenesis in mice. *Cancer Lett.* **40**, 193–197.
[128] Nishino, H.; Iwashima, A.; Itahura, Y.; Matsuura, H.; Fuwa, T. (1989) Antitumor-promoting activity of garlic extracts. *Oncology* **46**, 277–280.
[129] Perchellet, J.-P.; Perchellet, E. M.; Belman, S. (1990) Inhibition of DMBA-induced mouse skin tumorigenesis by garlic oil and inhibition of two tumor-promotion stages by garlic and onion oils. *Nutr. Cancer* **14**, 183–193.
[130] Liu, J.; Lin, R. I.; Milner, J. A. (1992) Inhibition of 7,12-dimethylbenz[*a*]anthracene-induced mammary tumors and DNA adducts by garlic powder. *Carcinogenesis* **13**, 1847–1851.
[131] El-Mofty, M. M.; Sakr, S. A.; Essawy, A.; Gawad, H. S. A. (1994) Preventive action of garlic on aflatoxin B_1-induced carcinogenesis in the toad Bufo regularis. *Nutr. Cancer* **21**, 95–100.
[132] Wargovich, M. J. (1987) Diallyl sulfide, a flavor component of garlic (*Allium sativum*) inhibits dimethylhydrazine-induced colon cancer. *Carcinogenesis* **8**, 487–489.
[133] Hayes, M. A.; Rushmore, T. H.; Goldberg, M. T. (1987) Inhibition of hepatocarcinogenic responses to 1,2-dimethylhydrazine by diallyl sulfide, a component of garlic oil. *Carcinogenesis* **8**, 1155–1157.

[134] Wargovich, M. J.; Woods, C.; Eng, V. W. S.; Stephens, L. C.; Gray, K. (1988) Chemoprevention of *N*-nitrosomethylbenzylamine-induced esophageal cancer in rats by the naturally occurring thioether, diallyl sulfide. *Cancer Res.* **48**, 6872–6875.

[135) Athar, M.; Raza, H.; Bickers, D.M.; Mukhtar, H. (1990) Inhibition of benzoyl peroxide-mediated tumor promotion in 7,12-dimethylbenz[*a*]anthracene-initiated skin of Sencar mice by antioxidants nordihydroguaiaracetic acid and diallyl sulfide. *J. Invest. Dermatol.* **94**, 162–165

[136] Jang, J. J.; Cho, K. J.; Lee, Y. S.; Bae, J. H. (1991) Modifying responses of allyl sulfide, indole-3-carbinol and geranium in a rat multi-organ carcinogenesis model. *Carcinogenesis* **12**, 691–695.

[137] Hong, J.-Y.; Wang, Z. Y.; Smith, T. J.; Zhou, S.; Shi, S.; Pan, J.; Yang, C. S. (1992) Inhibitory effects of diallyl sulfide on the metabolism and tumorigenicity of the tobacco-specific carcinogen 4-(methylnitrosamino)-1-(3-pyridyl)-1-butanone (NNK) in A/J mouse lung. *Carcinogenesis* **13**, 901–904.

[138] Wargovich, M. J.; Imada, O.; Stephens, L. C. (1992) Initiation and post-initiation chemopreventive effects of diallyl sulfide in esophageal carcinogenesis. *Cancer Lett.* **64**, 39–42.

[139] Nagabhusan, M.; Line, D.; Polverini, P. J.; Solt, D. B. (1992) Anti-carcinogenic action of diallyl sulfide in hamster buccal pouch and forestomach. *Cancer Lett.* **66**, 207–216.

[140] Hadjiolov, D.; Fernando, R. C.; Schmieser, H. H.; Wiessler, M.; Hadjiolov, N.; Pirajnov, G. (1993) Effect of diallyl sulfide on aristolochic acid-induced forestomach carcinogenesis in rats. *Carcinogenesis* **14**, 407–410.

[141] Tsuda, H.; Vehara, N.; Iwahori, Y.; Asamoto, M.; Iigo, M.; Nagao, M.; Matsumoto, K.; Ito, M.; Hirono, I. (1994) Chemopreventive effects of β-carotene, α-tocopherol and five naturally occurring antioxidants on initiation of hepatocarcinogenesis by 2-amino-3-methylmidazo[4,5-*f*]quinoline in the rat. *Jpn. J. Cancer Res.* **85**, 1214–1219.

[142] Surh, Y.-J.; Lee, R. C.-J.; Park, K.-K.; Mayne, S. T.; Liem, A.; Miller, J. A. (1995) Chemoprotective effects of capsaicin and diallyl sulfide against mutagenesis or tumorigenesis by vinyl carbamate and *N*-nitrosodimethylamine. *Carcinogenesis* **16**, 2467–2471.

[143] Sparnins, V. L.; Mott, A. W.; Barany, G.; Wattenberg, L. W. (1986) Effects of allyl methyl trisulfide on glutathione S-transferase activity and BP-induced neoplasia in the mouse. *Nutr. Cancer* **8**, 211–215.

[144] Belman, S.; Solomon, J.; Segal, A.; Block, E.; Barany, G. (1989) Inhibition of soybean lipoxygenase and mouse skin tumor promotion by onion and garlic components. *J. Biochem. Toxicol.* **4**, 151–160.

[145] Wattenberg, L. W.; Sparnins, V. L.; Barany, G. (1989) Inhibition of *N*-nitroso-diethylamine carcinogenesis in mice by naturally occurring organosulfur compounds and monoterpenes. *Cancer Res.* **49**, 2689–2692.

[146] Sparnins, V. L.; Barany, G.; Wattenberg, L. W. (1988) Effects of organosulfur compounds from garlic and onions on benzo[*a*]pyrene induced neoplasia and glutathione S-transferase activity in the mouse. *Carcinogenesis* **9**, 131–134.

[147] Dwivedi, C.; Rohlfs, S.; Jarvis, D.; Engineer, F. N. (1992) Chemoprevention of chemically induced skin tumor development by diallyl sulfide and diallyl disulfide. *Pharm. Res.* **9**, 1668–1670.

[148] Takahashi, S.; Hakoi, K.; Yada, H.; Hirose, M.; Ito, N.; Fukushima, S. (1992) Enhancing effects of diallyl sulfide on hepatocarcinogenesis and inhibitory actions of the related diallyl disulfide on colon and renal carcinogenesis in rats. *Carcinogenesis* **13**, 1513–1518.

[149] Schaffer, E. M.; Liu, J.-Z.; Green, J.; Dangler, C. A.; Milner, J. A. (1996) Garlic and associated allyl sulfur components inhibit *N*-methyl-*N*-nitrosourea induced rat mammary carcinogenesis. *Cancer Lett.* **102**, 199–204.

[150] Takada, N.; Matsuda, T.; Otoshi, T.; Yano, Y.; Otani, S.; Hasegawa, T.; Nakae, D.; Konishi, Y.; Fukushima, S. (1994) Enhancement by organosulfur compounds from garlic and onions of diethylnitrosamine-induced glutathione S-transferase positive foci in the rat liver. *Cancer Res.* **54**, 2895–2899.
[151] Hong, J.-Y.; Lin, M. C.; Wang, Z. Y.; Wang, E.-J.; Yang, C. S. (1994) Inhibition of chemical toxicity and carcinogenesis by diallyl sulfide and diallyl sulfone. In: Huang, M.-T.; Osawa, T.; Ho, C.-T.; and Rosen, R. J. (Eds.) *Food phytochemicals for cancer prevention I: Fruits and vegetables.* ACS Symposium Series 546, American Chemical Society, Washington, DC, p. 97–101.
[152] Srivastava, S. K.; Hu, X.; Xia, H.; Zaren, H. A.; Chatterjee, M. L.; Agarwal, R.; Singh, S. V. (1997) Mechanism of differential efficacy of garlic organosulfides in preventing benzo[*a*]pyrene-induced cancer in mice. *Cancer Lett.* **118**, 61–67.
[153] Lin, H. J.; Probst-Hensch, N. M.; Louie, A. D.; Kan, I. H.; Witte, J. S.; Ingles, S. A.; Frankl, H. D.; Lee, E. R.; Haile, R. W. (1998) Glutathione transferase null genotype, broccoli, and lower prevalence of colorectal adenomas. *Cancer Epidemiol. Biomarkers Prev.* **7**, 647–652.
[154] Ketterer, B. (1998) Dietary isothiocyanates as confounding factors in the molecular epidemiology of colon cancer. *Cancer Epidemiol. Biomarkers Prev.* **7**, 645–646.
[155] Probst-Hensch, N. M.; Tannenbaum, S. R.; Chan, K. K.; Coetzee, G. A.; Ross, R. K.; Yu, M. C. (1998) Absence of glutathione S-transferase M1 gene increases cytochrome P450 1A2 activity among frequent consumers of cruciferous vegetables in a Caucasian population. *Cancer Epidemiol. Biomarkers Prev.* **7**, 635–638.
[156] Chung, F.-L.; Morse, M. A.; Eklind, K. I.; Lewis, J. (1992) Quantitation of human uptake of the anti-carcinogen phenethyl isothiocyanate after a watercress meal. *Cancer Epidemiol. Biomarkers Prev.* **1**, 383–388.
[157] Zhang, Y.; Wade, K. L.; Prestera, T.; Talalay, P. (1996) Quantitative determination of isothiocyanates, dithiocarbamates, carbon disulfide, and related thiocarbamates, carbon disulfide, and related thiocarbonyl compounds by cyclocondensation with 1,2-benzenedithiol. *Anal. Biochem.* **239**, 160–167.
[158] Chung, F.-L.; Jiao, D.; Getahun, S. M.; Yu, M. C. (1998) A urinary biomarker for uptake of dietary isothiocyanates in humans. *Cancer Epidemiol. Biomarkers Prev.* **7**, 103–108.

21 Tea and Cancer: What Do We Know and What Do We Need to Know?

Chung S. Yang, Guang-yu Yang, Sungbin Kim, Mao-Jung Lee, Jie Liao and Jee Y. Chung

21.1 Abstract

Many studies have demonstrated the inhibition of tumorigenesis by green tea, black tea, and tea polyphenols in animal models. Yet, such inhibitory activities have not been convincingly shown in human populations. Whereas some epidemiological studies have observed an inverse correlation between tea consumption and cancer incidence, other studies have not shown such a correlation or have indicated a positive association between tea consumption and incidence of certain cancers. In order to understand the effects of tea consumption on human carcinogenesis, it is important for us to know the tea components which effect the anticarcinogenic actions, the tissue concentration of these compounds after tea consumption, and the key mechanisms by which these compounds inhibit the carcinogenic processes.

Many studies have suggested that tea catechins: (–)-epigallocatechin-3-gallate (EGCG), (–)-epigallocatechin (EGC), (–)-epicatechin-3-gallate (ECG), and (–)-epicatechin (EC) are important for the inhibition of tumor formation in animals and of cancer cell growth in culture. Recent studies also indicated that theaflavins also possess such activities. Theaflavins, however, only account for 1–2% of the dry weight of the water-extractable materials in black tea. The bioavailability and biological activities of thearubigens, the major constituents of black tea, however, are not known.

The absorption and tissue distributions of tea catechins are just beginning to be understood. Peak values of EGCG (~300 ng/ml), EGC (~560 ng/ml), and EC (~200 ng/ml) were observed in the blood 1 to 2 hours after drinking a green tea preparation (equivalent to approximately 5 cups of tea) by human volunteers [38]. Substantial amounts of these catechins were detected in colon mucosa and prostate tissues in surgical samples from patients who consumed tea 12 hours before surgery. The absorption, pharmacokinetics, and tissue distribution of tea catechins have been studied in rats [60]. Their levels in mouse blood and some tissues have also been measured. The bioavailability of EGCG was low (much lower than EGC) in rats, moderate in humans, and higher in mice.

Tea polyphenols are well recognized as antioxidants and this property has been considered to be closely related to their anti-cancer activities. Prooxidant activities, however, may be important for the induction of apoptosis by tea polyphenols. Tea is one of the few agents which can inhibit carcinogenesis at the promotion and progression stages in animal models. Antiproliferative and growth inhibitory activities of tea polyphenols have been demonstrated in mice and human cancer cell lines. A common mechanism for this action may be the inhibition of the signal transduction pathways which are key to cell growth. The inhibition of AP-1 and NF-κB by tea polyphenols have been demonstrated. These and other activities which inhibit abnormal growth, if they can be put forth by physiological levels of tea polyphenols in human tissues, could be very useful for the prevention of cancer (supported by NIH grant CA56673).

21.2 Introduction

Tea, made from the leaves of the plant *Camellia sinensis*, is the most popular beverage worldwide besides water. Many studies have demonstrated the inhibition of tumorigenesis by green tea, black tea, and tea polyphenols in animal models. Yet, such inhibitory activities have not been convincingly shown in human populations. Whereas some epidemiological studies have observed an inverse correlation between tea consumption and cancer incidence, other studies have not shown such a correlation or have indicated a positive association between tea consumption and incidence of certain cancers.

Green tea is manufactured by drying fresh tea leaves; therefore, its composition resembles that of fresh tea leaves. Green tea contains characteristic polyphenolic compounds, commonly known as catechins, which include (–)-epigallocatechin-3-gallate (EGCG), (–)-epigallocatechin (EGC), (–)-epicatechin-3-gallate (ECG) and (–)-epicatechin (EC). These catechins usually account for about 10 % of the dry weight of the tea leaves and 30 % of the dry weight of the solids in brewed tea or tea water extracts [1]. Caffeine, theobromine, and theophylline are the principal alkaloids, and they usually account for about 4 % of the dry weight of tea leaves. In the manufacturing of black tea, the fresh tea leaves are crushed, and the polyphenol oxidase is released to catalyze the oxidative polymerization of catechins to form theaflavins, thearubigens, and other oligomers in a process known as fermentation. Theaflavin, theaflavin-3-gallate, theaflavin-3′-gallate, and theaflavin-3,3′-digallate, which give the characteristic color and taste of black tea, account for 1–2 % of the dry weight of the water-extractable material. Thearubigens, which are poorly characterized chemically, may account for 6 % of the dry weight of black tea leaves and 17 % of the solid weight of water extract of black tea [1].

In this chapter, we plan to briefly review the results of studies with animal models and human population, and discuss several issues that are important in understanding the effects of tea consumption on human carcinogenesis. These include the tea components which effect the anticarcinogenic actions, the tissue concentrations of these compounds after tea consumption, and the key mechanisms by which these compounds inhibit the carcinogenic processes. Because of space limitations, review articles instead of the original articles will be cited on many occasions.

21.3 Studies with animal models

The cancer preventive activity of tea has been demonstrated in many animal models including those with cancers of the skin, lung, esophagus, stomach, liver, duodenum and small intestine, pancreas, colon, prostate, and mammary glands [2–5]. Some examples are used in this section to illustrate the nature of the inhibitory activities against tumorigenesis by tea.

21.3.1 Inhibition of lung tumorigenesis

The tobacco carcinogen 4-(methylnitrosamino)1-(3-pyridyl)-1-butanone (NNK) effectively induces adenomas and adenocarcinomas in the lungs of A/J mice. In typical experiments, decaffeinated green or black tea (DGT or DBT) was given to mice in the drinking fluid as 0.3 or 0.6 % solution of tea solids. The tea solids were dehydrated water extracts of tea leaves. Tea was administered to mice during the NNK treatment period (for 3 weeks beginning 2 weeks before a single dose of NNK, 103 mg/kg, i. p.) or after the NNK treatment period (beginning 1 week after the NNK treatment until the termination of the experiment 15 weeks later). The tea treatment markedly inhibited tumorigenesis, especially tumor multiplicity [6]. In these experiments, the inhibitory activities of decaffeinated green and black tea were comparable, although green tea might be slightly more effective in some experiments. Administration of DGT for a shorter period of time during (or after) the carcinogen treatment period also inhibited tumorigenesis, although it was less effective than the above described protocol [7]. When the mice were given black tea 16 weeks after the NNK injection, at which time almost all of the mice developed adenomas, the progression of these tumors to adenocarcinomas was significantly inhibited. These parameters included cancer incidence, cancer multiplicity, and the volume of the adenocarcinoma analyzed at 52 weeks [8]. Brewed black tea and green tea infusions (brewed in a Bunn tea brewing machine, similar to those consumed by humans), when given to A/J mice as the sole source of drinking fluid, also inhibited spontaneously developed lung tumors [9]. In this experiment, rhabdomyosarcomas were also found in the mice, and the tumor incidence and size were inhibited by black and green tea. These experiments indicate that tea has broad inhibitory activity against both spontaneous and chemically-induced tumorigenesis, and can inhibit carcinogenesis when administered during the initiation, promotion, or progression stages of carcinogenesis. Such broad inhibitory activity of tea has also been demonstrated in the skin carcinogenesis models.

21.3.2 Inhibition of skin tumorigenesis

The inhibitory action of tea and tea constituents have been studied extensively in UV and chemically-induced skin carcinogenesis models. In one study, mice were irradiated with UVB light for 10 days, and then promoted with 12-O-tetradecanoylphorbol-13-acetate (TPA); the formation of skin tumors was inhibited by oral tea administration during the initiation period [6]. When 7,12-dimethylbenz[a]anthracene (DMBA) was used as an initiating agent and either UVB or TPA was used as the tumor promoter, tumorigenesis was inhibited by the administration of green tea in drinking fluid during the promotion period. Using a similar DMBA-UVB induced skin carcinogenesis model, administration of brewed black tea, green tea, decaffeinated black tea, or decaffeinated green tea as the drinking fluid during the UVB promotion period (31 weeks) significantly inhibited skin carcinogenesis [10]. Using a UVB light-induced complete carcinogenesis model with SKH-mice, green tea (administered in drinking fluid) was a more effective inhibitor than black tea. Regular green tea and black tea were more effective than the decaffeinated preparations, and oral administration of caffeine had substantial inhibitory effects on UVB-induced carcinogenesis [11]. In another set of experiments, skin papillomas were initiated in mice by a dose of DMBA or UVB and then promoted by TPA for 25–30 weeks. Administration of green tea or EGCG in drinking fluid inhibited the growth of papillomas and caused partial tumor regression [12].

21.3.3 Studies on colon and breast cancers

Different results have been reported concerning the effects of tea on colon carcinogenesis. Colon cancer formation induced by azoxymethane in rats and by 1,2-dimethylhydrazine in mice was inhibited by low concentrations of green tea polyphenols (e.g. 0.01 or 0.1% solution as drinking fluid) and EGCG [13–15]. On the other hand, treatment of rats with black tea (0.6 or 2.5% solution) in drinking fluid did not inhibit azoxymethane-induced colon carcinogenesis. Treatment of rats with 1.25% black tea after azoxymethane dosing may have increased the multiplicity of exophytic carcinomas [16]. One possible interpretation of these two different types of results is that green tea polyphenols inhibit colon carcinogenesis whereas some black tea constituents (or their metabolites) are much less effective and may cause irritation and enhance carcinogenesis. This speculation needs careful examination. Black tea (1.25 or 2.5% solution) was found not to inhibit DMBA-induced mammary gland tumorigenesis in rats fed an AIN-76A diet, but to reduce mammary tumorigenesis in rats on a high fat diet [17]. Two earlier studies [18, 19] also indicated that EGCG and green tea extracts given in drinking water did not inhibit virus-induced mammary carcinogenesis.

21.4 Epidemiological studies

Many ecological, case-control, and cohort studies have been conducted to investigate the effects of tea consumption on human cancer incidence, but no clear cut conclusion can be drawn [2, 3, 20]. For example, studies in northern Italy have suggested the protective effect of tea against oral, pharyngeal, and laryngeal cancer [21, 22]. In a case-control study in Shanghai, frequent consumption of green tea has been shown to be associated with a lower incidence of esophageal cancer, especially among nonsmokers and non-alcohol-drinkers [23]. Yet many earlier studies have suggested that tea drinking is a risk factor for esophageal cancer. Although the hot temperature of tea may account for most of the risk, there is still a concern on the possible deleterious effects of tea [2, 4]. Similar studies in Shanghai have suggested that green tea consumption is associated with lower risk of pancreatic and colorectal cancer [24]. The protective effect against gastric cancer by tea has also been suggested from studies in Kyushu (Japan), northern Turkey, and central Sweden, but not from many other studies in different geographic areas [4, 25]. In studies in Saitama, Japan, women consuming more than 10 cups of tea daily have been shown to have lower risk for cancer (all sites combined) and increased tea consumption to be associated with lower risk for breast cancer metastasis and recurrence [26, 27]. In a prospective cohort study of postmenopausal women in Iowa, tea (mostly black tea) drinking has been shown to be associated with a lower risk for digestive tract cancers and urinary tract cancers [28]. On the other hand, in the Netherlands Cohort Study on Diet and Cancer, consumption of black tea has not been found to affect the risk for stomach, colorectal, lung, and breast cancers [29]. A recent study on middle aged Finnish men has indicated a positive association between increased tea consumption and colon cancer risk [30].

A gross analysis indicates that most reports on the cancer prevention effects were from studies on Asians who drink predominantly green tea, whereas studies on black-tea drinking Europeans rarely observed any protective effect. One suggestion is that the cancer prevention activity can be effected by green tea but not by black tea. Yet studies in animal models indicated that black tea is also effective in inhibiting tumorigenesis. Milk is usually added to black tea before drinking, whereas green tea is usually drunk without milk. The possibility that milk may affect the absorbability of tea polyphenols needs further investigation. It is also possible that the different results on tea and cancer are due to the different etiological factors involved in different geographic areas. For example, the gastric cancer in Japan may be closely related to nitrosamines and nitrosamides whose endogenous formation could be blocked by tea polyphenols, whereas in Europe, gastric cancer may involve different etiological factors which are not affected by tea consumption. Additional research, especially well designed cohort and intervention studies, is needed to elucidate the relationship between tea consumption and human cancer risk.

21.5 Possible active components and their bioavailability

Many authors have considered EGCG as the active component of green tea because EGCG is the most abundant catechin, and cancer inhibitory activity of EGCG has been demonstrated [2–4, 31, 32]. However, other components should also be considered based on their biological activities and bioavailabilities. For example, EGC, which is also abundant in green tea, is more bioavailable and has similar anti-proliferative and inhibitory activities, and could also be important.

21.5.1 Possible active components

From our studies with three human lung cancer cell lines (H661, H1299, and H441) and one colon cancer cell line (HT-29), we observed that the potency of inhibiting cell growth had the rank order of EGCG = EGC > ECG > EC [33]. The potency of theaflavin-3,3′-digallate was similar to EGCG and higher than that of theaflavin-3,3′-gallate which was still higher than theaflavin. The growth inhibitory activity of green tea extracts appeared to be due to the summation of activities of EGCG, EGC, ECG, and EC. These catechins, together with the theaflavins, may account for most of the growth inhibitory activity of black tea extracts. The growth inhibitory activity of thearubigens, the major components of black tea, is not known.

It has been demonstrated that EGCG inhibits the carcinogenesis of skin, stomach, colon and lung [31, 32], as well as the growth of human prostate and breast tumors in athymic mice [34]. Theaflavins have been shown to inhibit lung and esophageal carcinogenesis [35, 36]. The activity of EGC has not been tested in carcinogenesis models, but short term studies showed that it has about the same anti-proliferative activity as EGCG (unpublished results). In this experiment, EGC and EGCG were administered 24 h after a dose of NNK injection, and NNK-induced hyperproliferation of bronchiolar epithelial cells was assayed on day 5. In many studies, black tea has comparable or slightly lower inhibitory activities. The results suggest that the remaining catechin in black tea, the theaflavins, and other components in black tea all contribute to the cancer inhibitory activity. The bioactivities of thearubigens are not known. The inhibitory activity of caffeine has been demonstrated in different animal models [11, 32]. In the complete UV-carcinogenesis model, caffeine seems to play an important role and decaffeinated black tea was not effective [11]. In a recent study on the inhibition of NNK-induced lung and liver tumorigenesis in rats, caffeine, at a concentration corresponding to that in 2% black tea extract, displays inhibitory activity comparable to, or stronger than, that of 2% black tea [37].

21.5.2 Absorption and tissue distribution in animals

The absorption and tissue distribution of tea catechins are just beginning to be understood. Following *i.v.* injection of decaffeinated green tea to rats, the plasma concentration-time curves of EGCG, EGC, and EC could be fitted into a two-compartment model. The β elimination half-lives ($t_{1/2(\beta)}$) were 212, 45, and 41 min for EGCG, EGC, and EC, respectively. When pure EGCG was given *i.v.*, however, a shorter $t_{1/2(\beta)}$ (135 min) for EGCG was observed, suggesting that other components in DGT could affect the plasma concentration and elimination of EGCG. After *i.g.* administration of DGT, about 14% of EGC and 31% of EC appeared in the plasma, but less than 1% of EGCG was bioavailable. Conversion of *i.g.* administered EGCG to EGC was not observed. After *i.v.* administration of DGT, the level of EGCG was found to be the highest in the intestines and the highest levels of EGC and EC were observed in the kidney.

When tea (0.6% green tea polyphenols) was administered through the drinking fluid, the blood levels of EGC and EC were much higher than that of EGCG. A large amount of EGCG was found in the feces. The blood EGC and EC levels increased in the first two weeks (peaked at 800–1000 ng/ml), and then the levels were markedly decreased to 300–350 ng/ml on day 28. A similar decrease in urinary excretion of EGC and EC was also observed. When a second cycle of tea administration was initiated after a 10 day wash-out period, the EGC and EC levels were much lower than the high levels found on days 8 and 14 in the first cycle. Substantial amounts of EGC and EC were found in the rat esophagus (185–195 ng/g tissue), large intestine (300–930), kidney (400–500), bladder (800–810), lung (190–230), and prostate (240–250) but were low in the liver, spleen, heart, and thyroid. The amount of EGCG were higher in the esophagus and large intestine because of direct contact, but lower in other organs because of poor systemic absorption of EGCG by the rat. A similar pattern of increase and then decrease in blood catechin levels was also seen in mice except that the decrease took place at 4 days after treatment with tea. In mice, the bioavailability of EGCG was much higher than that in rats. The highest levels of these catechins were in the low micromolar range.

21.5.3 Blood and saliva levels in humans

Administration of 1.5, 3.0, and 4.5 g of DGT (in 500 ml of water) resulted in maximal plasma concentrations (C_{max}) of 326 ng, 550 ng, and 190 ng/ml for EGCG, EGC, and EC in human volunteers, respectively [38]. These C_{max} values were observed at 1.4 to 2.4 h after the ingestion of the tea preparation. A good dose-response relationship was not observed between the doses of 3 g and 4.5 g. The half-life of EGCG (5.0–5.5 h) appeared to be higher than those of EGC and EC (2.5–3.4 h). EGC and EC, but not EGCG, were excreted through the urine. Over 90 % of the total urinary EGC and EC was excreted within 8 h. Substantial amounts of the catechins were detected in colon mucosa and prostate tissues in surgical samples from patients who consumed tea 12 h before surgery.

Because of the possible application of tea in the prevention of oral and esophageal cancers, the salivary levels of tea catechins were determined in 6 human volunteers after drinking tea [39]. Saliva samples were collected after thoroughly rinsing the mouth with water. After drinking green tea preparations (equivalent to 2 to 3 cups of tea), peak saliva levels of EGC (11.7–43.9 μg/ml), EGCG (4.8–22 μg/ml), and EC (1.8–7.5 μg/ml) were observed after a few minutes. These levels were two orders of magnitude higher than those in the plasma. The $t_{1/2}$ of the salivary catechins were 10–20 min, much shorter than that of the plasma. Holding a tea solution in the mouth for a few minutes (without swallowing) produced even higher salivary catechin levels, but taking tea solids in capsules resulted in no detectable salivary catechin level. Holding EGCG solution in the mouth resulted in EGCG and EGC in the saliva and subsequently EGC in the urine. The results suggest that EGCG was converted to EGC in the oral cavity, and both catechins were absorbed through the oral mucosa. A catechin esterase activity which converts EGCG to EGC was found in the saliva. The results suggest that slowly drinking tea is a very effective way of delivering rather high concentrations of catechins to the oral cavity and then the esophagus.

21.6 Possible mechanisms for the inhibitory actions of tea on tumorigenesis

Many mechanisms have been proposed concerning the inhibitory action of tea against tumorigenesis [2, 4]. As was pointed out previously [4, 40], in the search for relevant mechanisms, we need to consider the tissue concentrations of the effective components. Inhibitory activities elicited by low micromolar or lower concentrations are likely to be more relevant than activities demonstrated with higher concentrations of tea components. Many studies have demonstrated the inhibition of carcinogen activation by tea and tea polyphenols *in vitro*, but such activities have been observed only in some studies but not others *in vivo*. Moderate enhancement of glutathione peroxidase, catalase, glutathions *S*-transferase, NADPH-quinone oxidoreductase, UDP-glucuronolyl transferase, and methoxyresorufin *O*-dealkylase activities by tea administration has also been reported [41–43]; the effects of these enzyme inductions on carcinogenesis are not clear.

Reactive oxygen species may play important roles in carcinogenesis through damaging DNA, altering gene expression, or affecting cell growth and differentiation [44, 45]. The anti-carcinogenic activities of tea polyphenols are believed to be related, but not entirely due to their antioxidative properties. The findings that green tea polyphenol fractions inhibited TPA-induced hydrogen peroxide formation in mouse epidermis [46] and 8-hydroxydeoxyguanosine (8-OHdG) formation in different systems [32, 47] are consistent with this concept. On the other hand, the prooxidant activity of tea polyphenols may play a role in inducing apoptosis. In the induction of apoptosis with human lung cancer H661 cells by EGCG, the activity could be inhibited by catalase [33]. Inhibition of tumor promotion-related enzymes, such as TPA-induced epidermal ornithine decarboxylase [46, 48], protein kinase C [49], lipoxygenase and cyclooxygenase [50, 51], by tea preparations has been demonstrated. Other mechanisms, relating to the quenching of activated carcinogens [52], antiviral activity [53], and enhancing of immune functions [54] have also been suggested, but their relevance to carcinogenesis is unknown. Inhibition of nitrosation by tea preparations has been demonstrated *in vitro* and in humans [55, 56], and this may be an important factor in the prevention of certain cancers, e.g., gastric cancer, if the endogenously formed *N*-nitroso compounds are causative factors.

The inhibitory activities of tea and tea polyphenols may be due to their ability to inhibit growth related signal transduction pathways in many cases. For example, EGCG and theaflavins inhibit epidermal growth factor (EGF)- or TPA-induced transformation of JB6 cells, and this inhibition was correlated to the inhibition of AP-1-dependent transcriptional activity [57]. The inhibition of AP-1 activation occurs through the inhibition of a c-jun NH2-terminal kinase-dependent pathway. In H-ras transformed JB6 (the 30.7b Ras 12 cells), H-ras activated AP-1 pathway is a major growth stimulant. In these cells, the AP-1 activation was inhibited by EGCG, EGC, theaflavin-3,3′-digallate and other polyphenols, and the phosphorylation of both c-jun and extracellular signal-regulated protein kinase (Erk) was inhibited [58]. Since the ras genes are activated in many animal carcinogenesis models and in human cancers, the inhibition of the phosphorylation of c-jun and Erk could be an important mechanism for the inhibition of cancer formation and growth. Inhibition of tumor necrosis factor α (TNF-α) has also been proposed as a possible mechanism for the cancer preventive activity of EGCG [59]. The inhibition of nuclear factor-κB (NF-κB) by EGCG which has been demonstrated recently [60], may be related to the decreased TNF-α expression and contribute to the anti-carcinogenic effect of tea polyphenols in many situations.

21.7 Concluding remarks

Judging from the diverse inhibitory activities observed in different animal carcinogenesis systems and different cancer cell lines as well as the different mechanisms that have been proposed, it is likely that there are multiple mechanisms by which tea constituents elicit their inhibitory effects against carcinogenesis. It is a challenge for us to determine which mechanisms are relevant to human cancer prevention. In this context, only activities displayed by tea constituents at concentrations which are achievable in human tissues are important. Therefore, information on the bioavailability and tissue levels of EGCG, EGC, theaflavins, and other tea constituents should be a vital part of future mechanistic studies. The effects of different doses of tea on tissue levels of tea polyphenols are rather intriguing. Further studies are needed to determine the best dosage and route to deliver certain tea constituents to a certain organ. Large doses of pure compounds in capsules may not be the best way to effectively deliver the active components to many tissues.

The effects of tea on human carcinogenesis are still not clear. The possible difference between green and black tea should be considered. The recently gained knowledge on the tissue levels and biological activities of tea polyphenols as well as possible biomarkers of tea consumption would be useful in the planning of future epidemiological studies and human cancer prevention trials.

Abbreviations

EGCG	(–)-epigallocatechin-3-gallate
EGC	(–)-epigallocatechin
ECG	(–)-epicatechin-3-gallate
EC	(–)-epicatechin
NNK	4-(methylnitrosamino)-1-(3-pyridyl)-1-butanone
DGT	decaffeinated green tea
DBT	decaffeinated black tea
TPA	12-*O*-tetradecanoylphorbol-13-acetate
DMBA	7,12-dimethylbenz[a]anthracene
$(t_{1/2(\beta)})$	β elimination half-lives
TNF-α	tumor necrosis factor α

References

[1] Balentine, D. A. (1992) Manufacturing and chemistry of tea. In: Ho, C.-T.; Huang, M.-T.; Lee, C. Y. (Eds.) Phenolic Compounds in Food and Their Effects on Health I: Analysis, Occurrence, and Chemistry, American Chemical Society, Washington, DC, p. 102–117.

[2] Yang, C. S.; Wang, Z.-Y. (1993) Tea and Cancer: A review. *J. Natl. Cancer Inst.* (Invited Review) **58**, 1038–1049.

[3] Katiyar, S. K.; Mukhtar, H. (1996) Tea in chemoprevention of cancer: Epidemiological and experimental studies (Review). *Int. J. Oncology* **8**, 221–238.

[4] Yang, C. S.; Chen, L.; Lee, M.-J.; Landau, J. M. (1996) Effects of tea on carcinogenesis in animal models and humans. Edited under the auspices of the American Institute for Cancer Research. Dietary Phytochemicals in Cancer Prevention and Treatment. Plenum Press, New York, 51–61.

[5] Dreosti, I. E.; Wargovich, M. J.; Yang, C. S. (1997) Inhibition of carcinogenesis by tea: The evidence from experimental studies. *Crit. Rev. Food Sci. Nutr.* **37**, 761–770.

[6] Wang, Z. Y.; Hong, J.-Y.; Huang, M. T.; Teuhl, K.; Conney, A. H.; Yang, C. S. (1992) Inhibition of *N*-nitrosodiethylamine and 4-(methylnitrosamino)-1-(3-pyridyl)-1-butanone-induced tumorigenesis in A/J mice by green tea and black tea. *Cancer Res.* **52**, 1943–1947.

[7] Shi, S. T.; Wang, Z.-Y.; Smith, T. J.; Hong, J.-Y.; Chen, W.-F.; Ho, C.-T.; Yang, C. S. (1994) Effects of green tea and black tea on 4-(methylnitrosamino)-1-(3-pyridyl)-1-butanone bioactivation, DNA methylation, and lung tumorigenesis in A/J mice. *Cancer Res.* **54**, 4641–4647.

[8] Yang, G.-Y.; Wang, Z.-Y.; Kim, S.; Liao, J.; Seril, D.; Chen, X.; Smith, T. J.; Yang, C. S. (1997) Characterization of early pulmonary hyperproliferation, tumor progression and their inhibition by black tea in a 4-(methylnitrosamino)-1-(3-pyridyl)-1-butanone (NNK) induced lung tumorigenesis model with A/J mice. *Cancer Res.* **57**, 1889–1894.

[9] Landau, J. M.; Wang, Z.-Y.; Yang, G.-Y.; Ding, W.; Yang, C. S. (1998) Inhibition of spontaneous formation of lung tumors and rhabdomyosarcomas in A/J mice by black and green tea. *Carcinogenesis* **19**, 501–507.
[10] Wang, Z.-Y.; Huang, M.-T.; Lou, Y.-R.; Xie, J.-G.; Reuhl, K. R.; Newmark, H. L.; Ho, C. T.; Yang, C. S.; Conney, A. H. (1994) Inhibitory effects of black tea, green tea, decaffeinated black tea, and decaffeinated green tea on ultraviolet B light-induced skin carcinogenesis in 7,12-dimethylbenz[a]anthracene-initiated SKH-1 mice. *Cancer Res.* **54**, 3428–3435.
[11] Huang, M.-T.; Xie, J.-G.; Wang, Z.-Y.; Ho, C.-T.; Lou, Y.-R.; Wang, C.-X.; Hard, G. C.; Conney, A. H. (1997) Effects of tea, decaffeinated tea, and caffeine on UVB light induced complete carcinogenesis in SKH-1 mice: demonstration of caffeine as a biologically important constituent of tea. *Cancer Res.* **57**, 2623–2629.
[12] Wang, Z. Y.; Huang, M.-T.; Ho, C.-T.; Chang, R.; Ma, W.; Ferraro, T.; Reuhl, K. R.; Yang, C. S.; Conney, A. H. (1992) Inhibitory effect of green tea on the growth of established skin papillomas in mice. *Cancer Res.* **52**, 6657–6665.
[13] Yamane, T.; Hagiwara, N.; Tateishi, M.; Akachi, S.; Kim, M.; Okuzumi, J.; Kitao, Y.; Inagake, M.; Kuwata, K.; Takahashi, T. (1991) Inhibition of azoxymethane-induced colon carcinogenesis in rat by green tea polyphenol fraction. *Jpn. J. Cancer Res.* **82**, 1336–1339.
[14] Yin, P.; Zhao, J.; Cheng, S.; Zhu, Q.; Liu, Z.; Zhengguo, L. (1994) Experimental studies of the inhibitory effects of green tea catechin on mice large intestinal cancers induced by 1,2-dimethylhydrazine. *Cancer Lett.* **79**, 33–38.
[15] Kim, M.; Hagiwara, N.; Smith, S. J.; Yamamoto, T.; Yamane, T.; Takahashi, T. (1994) Preventive effect of green tea polyphenols on colon carcinogenesis. In: Hunag, M.-T.; Osaswa, T.; Ho, C.-T.; Rosen, R. T. (Eds.) *Food phytochemicals in cancer prevention.* ACS Symposium Series **546**, 51–55.
[16] Weisburger, J. H.; Rivenson, A.; Reinhardt, J.; Aliaga, C.; Braley, J.; Pittman, B.; Zang, E. (1998) Effect of black tea on azoxymethane-induced colon cancer. *Carcinogenesis* **19**, 229–232.
[17] Rogers, A. E.; Hafer, L. J.; Iskander, Y. S.; Yang, S. (1998) Black tea and mammary gland carcinogenesis by 7,12-dimethylbenz[*a*]anthracene in rats fed control or high fat diets. *Carcinogenesis* **19**, 1269–1273.
[18] Fujiki, J.; Suganuma, M.; Suguri, H.; Takagi, K.; Yoshizawa, S.; Ootsuyama, A.; Tanooka, H.; Okuda, T.; Kobayashi, M.; Sugimura, T. (1990) New antitumor promoters: (–)-epigallocatechin gallate and sarcophytol A and B. In: Kuroda, Y. U.; Shankel, D. M.; Walters, M. D. (Eds.) *Antimutagenesis and anticarcinogenesis mechanisms II*, Plenum Press, New York, NY, p. 205–212.
[19] Sakata, K.; Ikeda, T.; Imoto, S.; Wada, N.; Takeshima, K.; Yoneyama, K.; Enomoto, K.; Kitajima, M. (1995) Effect of green tea extract and tamoxifen on mammary carcinogenesis and estrogen metabolism of mouse. Pro. San Antonio Breast Cancer Symposium 101.
[20] Blot, W. J.; McLaughlin, J. K.; Chow, W.-H. (1997) Cancer rates among drinkers of black tea. *Crit. Rev. Food Sci. Nutr.* **37**(8), 739–760.
[21] Franceschi, S.; Serraino, D. (1992) Risk factors for adult soft tissue sarcoma in northern Italy. *Ann. Oncol.* **3**(2), 85–88.
[22] La Vecchia, C.; Negri, E.; Franceschi, S.; D'Avanzo, B.; Boyle, P. (1992) Tea consumption and cancer risk. *Nutr. Cancer* **17**, 27–31.
[23] Gao, Y. T.; McLaughlin, J. K.; Blot, W. J.; Ji, B. T.; Dai, W.; Fraumeni, J. J. (1994) Reduced risk of esophageal cancer associated with green tea consumption. *J. Natl. Cancer Inst.* **86**, 855–858.
[24] Ji, B.-T.; Chow, W.-H.; Yang, G.; McLaughlin, J. K.; Gao, R.-N.; Zheng, W.; Shu, X.-O.; Jin, F.; Fraumeni, J. F. J. (1996) The influence of cigarette smoking, alcohol, and green tea consumption on the risk of carcinoma of the cardia and distal stomach in Shanghai, China. *Cancer* **77**, 2449–2457.

[25] Wang, Z.-Y.; Chen, L.; Lee, M.-J.; Yang, C. S. (1996) Tea and cancer prevention. In: Finley, J. W.; Armstrong, D.; Robinson, S. F.; Nagy, S. (Eds.) *Hypernutritious food*, American Chemical Society Symposium, p. 239–260.
[26] Nakachi, K.; Suemasu, K.; Suga, K.; Takeo, T.; Imai, K.; Higashi, Y. (1998) Influence of drinking green tea on breast cancer malignancy among Japanese patients. *Jpn. J. Cancer Res.* **89**, 254–261.
[27] Imai, K.; Suga, K.; Nakachi, K. (1997) Lead Article: Cancer-preventive effects of drinking green tea among a Japanese population. *Prev. Med.* **26**, 769–775.
[28] Zheng, W.; Doyle, T. J.; Kushi, L. H.; Sellers, T. A.; Hong, C.-P.; Folsom, A. R. (1996) Tea consumption and cancer incidence in a prospective cohort study of postmenopausal women. *Am. J. Epidemiol.* **144**, 175–182.
[29] Goldbohm, A. R.; Hertog, M. G. L.; Brants, H. A. M.; Poppel, G. V.; van den Brandt, P. A. (1996) Consumption of black tea and cancer risk: a prospective cohort study. *J. Natl. Cancer Inst.* **88**, 93–100.
[30] Hartman, T. J.; Tangrea, J. A.; Pietinen, P.; Malila, N.; Virtanen, M.; Taylor, P. R.; Albanes, D. (1998) Tea and coffee consumption and risk of colon and rectal cancer in middle-aged Finnish men. *Nutr. Cancer* **31**, 41–48.
[31] Yamane, T.; Takahashi, T.; Kuwata, K.; Oya, K.; Inagake, M.; Kitao, Y.; Suganuma, M.; Fujiki, H. (1995) Inhibiton of *N*-methyl-*N'*-nitro-*N*-nitrosoguanidine-induced carcinogenesis by (–)-epigallocatechin gallate in the rat glandular stomach. *Cancer Res.* **55**, 2081–2084.
[32] Xu, Y.; Ho, C.-T.; Amin, S. G.; Han, C.; Chung, F.-L. (1992) Inhibition of tobacco-specific nitrosamine-induced lung tumorigenesis in A/J mice by green tea and its major polyphenol as antioxidants. *Cancer Res.* **52**, 3875–3879.
[33] Yang, G.-Y.; Liao, J.; Kim, K.; Yurkow, E. J.; Yang, C. S. (1998) Inhibition of growth and induction of apoptosis in human cancer cell lines by tea polyphenols. *Carcinogenesis* **19**, 611–616.
[34] Liao, S.; Umekita, Y.; Guo, J.; Kokontis, J. M.; Hiipakka, R. A. (1995) Growth inhibition and regression of human prostate and breast tumors in athymic mice by tea epigallocatechin gallate. *Cancer Lett.* **96**, 239–243.
[35] Yang, G.-Y.; Liu, Z.; Seril, D. N.; Liao, J.; Ding, W.; Kim, S.; Bondoc, F.; Yang, C. S. (1997) Black tea constituents, theaflavins, inhibit 4-(methylnitrosamino)-1-(3-pyridyl)-1-butanone (NNK)-induced lung tumorigenesis in A/J mice. *Carcinogenesis* **18**, 2361–2365.
[36] Morse, M. A.; Kresty, L. A.; Steele, V. E.; Kelloff, G. J.; Boone, C. W.; Balentine, D. A.; Harbowy, M. E.; Stoner, G. D. (1997) Effects of theaflavins on *N*-nitrosomethylbenzylamine-induced esophageal tumorigenesis. *Nutr. Cancer* **29**, 7–12.
[37] Chung, F.-L.; Wang, M.; Rivenson, A.; Iatropoulos, M. J.; Reinhardt, J. C.; Pittman, B.; Ho, C.-T.; Amin, S. G. (1998) Inhibition of lung carcinogenesis by black tea in Fischer rats treated with a tobacco-specific carcinogen: caffeine as an important constituent. *Cancer Res.* **58**, 4096–4101.
[38] Yang, C. S.; Chen, L.; Lee, M.-J.; Balentine, D.; Kuo, M. C.; Schantz, S. P. (1998) Blood and urine levels of tea catechins after ingestion of different amounts of green tea by human volunteers. *Cancer Epidemiol. Biomarkers Prev.* **7**, 351–354.
[39] Yang, C. S.; Lee, M.-J.; Chen, L. (1998) Human salivary tea catechin levels and catechin esterase activities: implication in human cancer prevention studies. (paper submitted for publication).
[40] Yang, C. S. (1997) Inhibition of carcinogenesis by tea. *Nature* **389**, 134–135.
[41] Khan, S. G.; Katiyar, S. K.; Agarwal, R. (1992) Enhancement of antioxidant and phase II enzymes by oral feeding of green tea polyphenols in drinking water to SKH-1 hairless mice: possible role in cancer chemoprevention. *Cancer Res.* **52**, 4050–4052.
[42] Sohn, O. S.; Surace, A.; Fiala, E. S.; Richie, J. P. J.; Colosimo, S.; Zang, E.; Weisburger, J. H. (1994) Effects of green and black tea on hepatic xenobiotic metabolizing systems in the male F344 rat. *Xenobiotica* **24**, 119–127.

[43] Bu-Abbas, A.; Clifford, M. N.; Walker, R.; Ioannides, C. (1994) Selective induction of rat hepatic CYP1 and CYP4 proteins and of peroxisomal proliferation by green tea. *Carcinogenesis* (Lond.) **15**, 2575–2579.

[44] Cerutti, P. A. (1989) Mechanisms of action of oxidant carcinogens. *Cancer Det. Prev.* **14**, 281–284.

[45] Feig, D. I.; Reid, T. M.; Loeb, L. A. (1994) Reactive oxygen species in tumorigenesis. *Cancer Res.* **54** (Suppl.), 1890s–1894s.

[46] Huang, M.-T.; Ho, C.-T.; Wang, Z. Y.; Ferraro, T.; Finnegan-Oliver, T.; Lou, Y.-R.; Mitchell, J. M.; Laskin, J. D.; Newmark, H.; Yang, C. S.; Conney, A. H. (1992) Inhibitory effect of topical application of a green tea polyphenol fraction on tumor initiation and promotion in mouse skin. *Carcinogenesis* **13**, 947–954.

[47] Bhimani, R.; Troll, W.; Grunberger, D.; Frenkel, K. (1993) Inhibition of oxidative stress in HeLa cells by chemopreventive agents. *Cancer Res.* **53**, 4528–4533.

[48] Wang, Z.-Y.; Agarwal, R.; Bickers, D. R.; Mukhtar, H. (1991) Protection against ultraviolet B radiation-induced photocarcinogenesis in hairless mice by green tea polyphenols. *Carcinogenesis* (Lond.) **12**, 1527–1530.

[49] Yoshizawa, S.; Horiuchi, T.; Sugimura, M. (1992) Penta-*O*-galloyl-β-D-glucose and (–)-epigallocatechin gallate: cancer prevention agent. In: Huang, M.-T.; Ho, C.-T.; Lee, C. Y. (Eds.) *Phenolic compounds in foods and health II: Antioxidant & cancer prevention*, American Chemical Society, Washington, DC, 316–325.

[50] Katiyar, S.; Agarwal, R.; Wood, G. S. (1992) Inhibition of 12-*O*-tetradecanoylphorbol-13-acetate-caused tumor promotion in 7,12-dimethylbenz[*a*]anthracene-initiated SENCAR mouse skin by a polyphenolic fraction isolated from green tea. *Cancer Res.* **52**, 6890–6897.

[51] Lou, F. Q. (1987) A study of prevention of atherosclerosis with tea-pigment. The Int. Tea-Quality-Human Health Sympo. (China), Abstract 141–143.

[52] Khan, W. A.; Wang, Z. Y.; Athar, M.; Bickers, D. R.; Mukhtar, H. (1988) Inhibition of the skin tumorigenicity of ($\pm$)7β,8α-epoxy-7,8,9,10-tetrahydrobeno[*a*]pyrene by tannic acid, green tea polyphenols and quercetin in Sencar mice. *Cancer Lett.* **42**, 7–12.

[53] Shimamura, T. (1994) Inhibition of influenza virus infection by tea polyphenols. In: Huang, M.-T.; Osawa, T.; Ho, C.-T.; Rosen, R. T. (Eds.) *Food phytochemicals for cancer prevention I*, ACS Symposium Series, Washington, DC, 101–104.

[54] Yan, Y. S. (1992) Effect of Chinese tea extract on the immune function of mice bearing tumors and their antitumor activity. *Chung Hua Yu Fang I Hsueh Tsa Chih* **26**, 5–7.

[55] Nakamamura, M.; Kawabata, T. (1981) Effect of Japanese green tea on nitrosamine formation *in vitro*. *J. Food Sci.* **46**, 306–307.

[56] Stich, H. F. (1992) Teas and tea components as inhibitors of carcinogen formation in model system and man. *Prev. Med.* **21**, 377–384.

[57] Dong, Z.; Ma, W.-Y.; Huang, C.; Yang, C. S. (1997) Inhibition of tumor promoter-induced AP-1 activation and cell transformation by tea polyphenols, (–)-epigallocatechin gallate and theaflavins. *Cancer Res.* **57**, 4414–4419.

[58] Chung, J. Y.; Huang, C.; Meng, X.; Dong, Z.; Yang, C. S. (1999) Inhibition of activator protein I activity and cell growth by purified green tea and black tea polyphenols in H-*ras*-transformed cells: Structure-activity relationship and mechanisms involved. *Cancer Res.* **59**, 4610–4617.

[59] Suganuma, M.; Okabe, S.; Sueoka, E.; Iida, N.; Komori, A.; Kim, S.-J.; Fujiki, H. (1996) A new process of cancer prevention mediated through inhibition of tumor necrosis factor α expression. *Cancer Res.* **56**, 3711–3715.

[60] Lin, Y.-L.; Lin, J. (1997) (–)-Epigallocatechin-3-gallate blocks the induction of nitric oxide synthase by down-regulating lipopolysaccharide-induced activity of transcription factor nuclear factor-κB. *Mol. Pharmacol.* **52**, 465–472.

[61] Chen, L.; Lee, M.-J.; Li, H.; Yang, C. S. (1997) Absorption, distribution, and elimination of tea polyphenols in rats. *Drug Metal Dispos.* **25**, 1045–1050.

22 Lignans and Isoflavones

Lilian U. Thompson

22.1 Abstract

Diets that are rich in fruits, vegetables, whole grains and soybeans have been suggested to be protective against cancer and epidemiological studies suggest that their phytoestrogen content may in part be responsible. The main phytoestrogens are the lignans, which are found widely in foods but are present in the highest concentration in flaxseed, and the isoflavones which are the richest in soybeans. Flaxseed and soybean as well as their purified lignans and isoflavones, respectively, have been shown to reduce tumor development in carcinogen-treated animals but this has yet to be confirmed in long term prospective studies in humans. *In vitro* studies suggest that both hormone and non-hormone related mechanisms are involved in their anticancer effects. Although isoflavones and lignans have similarity in chemical structures, there are differences in their metabolism, bioavailability and biological activities.

22.2 Introduction

Phytoestrogens are plant components with estrogen-like activities. The three main classes are the isoflavones, lignans and coumestans. The main isoflavones are daidzein, genistein and glycitein and their glycosides (daidzin, genistin and glycitin) and methylated precursors (formononetin and biochanin A) (Figure 22.1) with glycitein being relatively minor compared to the other two [1, 2]. The isoflavones are more commonly found in soybeans, lentils, chickpeas, beans and red clover, with soybeans and soybean-derived foods being the richest sources (values >100-times other foods) [1, 2]. The major plant lignans are secoisolariciresinol diglycoside (SDG) and matairesinol (Figure 22.2) [1]. They are more widely distributed in foods including cereals, oilseeds, legumes, fruits, vegetables and seaweeds but are present in the highest concentration in flaxseed (value >100-times other foods), also known as linseed [1, 3, 4]. The main coumestan is coumestrol, which is found in bean sprouts and fodder crops such as alfalfa.

Dietary phytoestrogens in the glycoside form are hydrolyzed by bacterial β-glycosidases in the colon to their respective aglucones and then are absorbed,

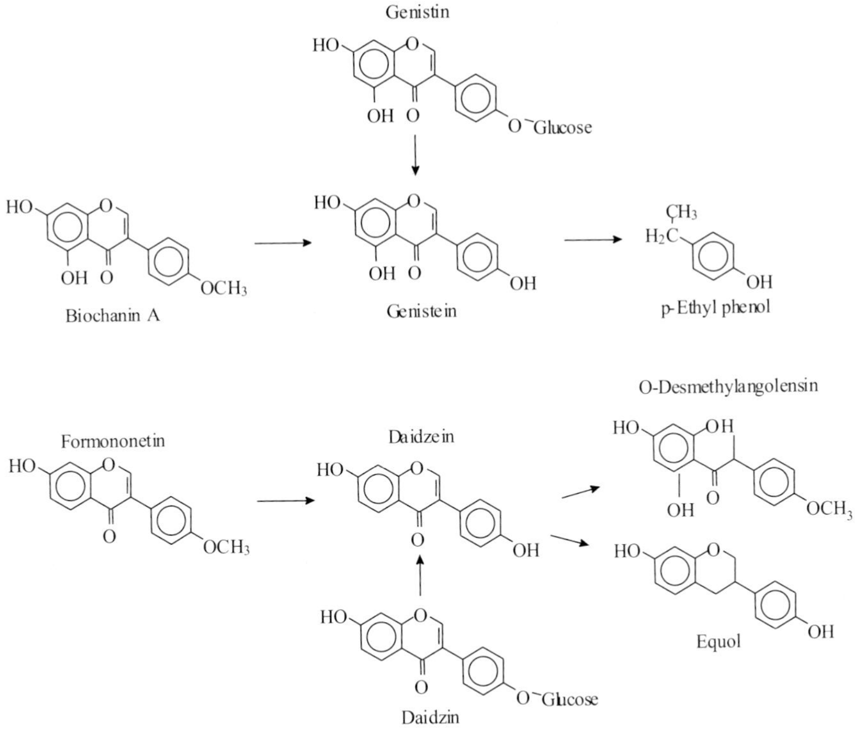

Figure 22.1: Major isoflavones and their metabolism.

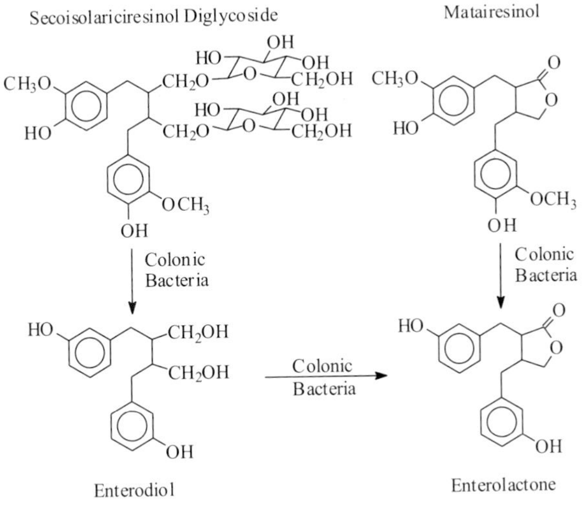

Figure 22.2: Major plant lignans and their metabolism to mammalian lignans.

or further metabolized before absorption. The SDG and matairesinol are metabolized to the mammalian lignans enterodiol (ED) and enterolactone (EL), respectively (Figure 22.2) [3, 5], while formononetin and biochanin A may be metabolized to daidzein and genistein, respectively (Figure 22.1) [1, 2]. Daidzein may be further transformed to equol or *O*-dimethylangolensin and genistein to *p*-ethyl phenol.

Like estrogens, the phytoestrogens undergo enterohepatic circulation and some eventually are excreted in the urine [6, 7]. Thus urinary level has been used as an index of phytoestrogen intake. There is a large individual variability in phytoestrogen metabolism due to the individual variability in the colonic bacterial flora. This in turn may in part be responsible for the different responses to dietary phytoestrogens.

This paper will describe some of the epidemiological, *in vitro*, animal and clinical studies suggesting the potential of the isoflavones and lignans or their major source, soybean and flaxseed, respectively, to influence carcinogenesis particularly that of the breast (mammary), colon and prostate. Comments are also made on the safety of lignans and isoflavones as it is relevant to the potential application of these phytoestrogens as anti-carcinogenic agents. Whenever possible, the effect of soybean or isoflavones are compared to that of flaxseed or lignans. Comments on future research issues are also provided.

22.3 Anticancer effects

22.3.1 Epidemiological studies

Cross sectional studies have shown that, on traditional diets, Finnish and Americans with a high risk of breast cancer have lower urinary excretion of total phytoestrogens than the Japanese with low cancer risk [8–10]. Urinary lignan excretion was also lower in breast cancer patients and omnivores (with high cancer risk) than in vegetarians (with low cancer risk) [9, 10]. This was supported by a recent case control study of breast cancer patients which showed that an increasing urinary excretion of daidzein, equol and enterolactone was associated with a significant reduction in breast cancer development [11]. The odds ratio estimates suggest a four-fold reduction in risk with equol and a three-fold risk reduction with enterolactone. The isoflavone levels in the plasma and prostatic fluid of men in Hongkong and China, whose diets are rich in soybean, were also much higher than those in Portugal and Great Britain [12]. On the other hand, the Portuguese men whose diets may be more "vegetarian" and richer in whole grain products than soybeans have higher lignan levels in their prostatic fluid than the Chinese and British men. Of the three locations, the incidence of prostate cancer is the lowest in Hongkong and China, the highest in Great Britain and intermediate in Portugal.

Although the above observations suggest a potential role for both lignans and isoflavones in reducing cancer risk, epidemiological data on the relationship between soybean intake and cancer have not been consistent [13]. For example, case control studies showed an inverse association between soy protein intake and breast cancer risk in Chinese women in Singapore [14], while in China, no difference was observed between cases and controls despite similar soy protein intake as those in Singapore [14]. In Japan, no differences were observed between cases and control in soybean consumption [16]. On the other hand, prospective studies in Japan have shown an inverse association between the consumption of soy products (miso, soybean paste and miso soup) and breast cancer risk [17, 18]. Similarly, the intake of tofu was found to be more than twice in Asian immigrants than the US born Asians but it decreased with years of residence in the US [19]. The risk of breast cancer decreased with increasing intake of tofu in these subjects. Unlike soybean, no epidemiological data are available on the potential effect of flaxseed on carcinogenesis because it has not traditionally been used as food.

22.3.2 *In vitro* studies

The chemical structural similarity of the isoflavones and lignans to estradiol and the antiestrogen tamoxifen used for breast cancer treatment has led to the suggestion that they may have weak estrogenic/antiestrogenic properties; this, in turn, have been suggested to be responsible for their anticancer effects. Indeed, several *in vitro* studies have shown that they have estrogen agonist/antagonist activities and can stimulate/prevent the growth of estrogen receptor positive breast cancer cells (e.g. MCF-7 cells) depending on their concentrations and the presence of estrogen [1]. For example, at low levels (e.g. 0.5–10 μM), isoflavones and lignans have been reported to be estrogenic in the absence of estrogen and to increase cell proliferation in estrogen receptor positive mammary tumor cells [20, 21], while at high levels (e.g. >50 μM), they are antiestrogenic and antiproliferative [22, 23]. They have been shown to bind to rat nuclear type II binding site (bioflavonoid receptor) [24], bind to α-fetoprotein competing with estradiol and estrone [25], inhibit steroid binding by sex steroid binding protein [26], stimulate sex hormone binding synthesis by HepG2 liver cells [24], and induce pS2 expression in MCF-7 breast cancer cells [27]. They also have been shown to inhibit enzymes involved in estrogen biosynthesis such as aromatase [28, 29], 5α-reductase, 17β-hydroxysteroid dehydrogenase [30] as well as those involved in cell proliferation and transformation (e.g. tyrosine kinase, topoisomerase II) [31, 32]. They can also inhibit platelet activating factor and EGF-induced expression of c-fos [33], EGF and erB2/neu receptor [34], lipid peroxidation [35, 36], tumor promoter-induced H_2O_2 formation [37] and angiogenesis [38]. They may also induce cell differentiation, inhibit cell multiplication and/or induce apoptosis as has been observed in several cell lines (e. g. human myeloid leukemia cells, human melanoma cells, mouse leukemia cells, mouse embryonal carcinoma cells, rat pheochromocytoma cell) [39–44]. Other activities reported to be inhibited as well by genistein include insulin-stimulated glucose transport, alcohol dehydrogenase, ribosomal S6 kinase, phosphatidylinositol turnover and β-galactosidase [13].

Although the isoflavones and lignans have similar biological properties, they differ in effectiveness. In general, genistein is more effective than daidzein and enterolactone is more effective than enterodiol but it cannot be generalized that the isoflavones are more effective than the lignans. For example, enterolactone is more effective than genistein in inhibiting the binding of estradiol and testosterone to sex steroid binding protein [26] and in inhibiting aromatase enzyme [28]. On the other hand, genistein is more effective than enterolactone in decreasing endothelial cell proliferation [38] and in inhibiting 5α-reductase in genital skin fibroblast [30].

The *in vitro* studies suggest several potential mechanisms, both hormone and non-hormone related, whereby isoflavones and lignans may influence carcinogenesis. However, in many cases, the levels found effective *in vitro* are much higher (1–100 μM) compared with plasma levels of these compounds (<1 μM) in animals and humans after the intake of soybean, flaxseed or purified isoflavones or lignans [45]. It is nevertheless conceivable that multiple mechanisms may take place at the same time. Low concentrations of several phytoestrogens together may also additively or synergistically prevent tumor development as a high dose of a single compound. Indeed, it has been shown that a cocktail of seven phytoestrogens (0.5 μM) can inhibit the 5α-reductase and 17β-hydroxysteroid dehydrogenase activities at the same level as 100 μM of enterolactone [30].

22.3.3 Animal studies

22.3.3.1 Soybean and isoflavones

22.3.3.1.1 Breast cancer

Seven out of nine studies previously reviewed reported a lower rate of tumor development in animals fed soybean or soybean products [13]. Since soybean is rich in other substances that may also influence carcinogenesis such as saponins, protease inhibitors and phytic acid, its effect cannot be solely attributed to its isoflavone content. However, support comes from a study which demonstrated that soybean devoid of the isoflavones do not have an influence on tumor growth [46]. Also in rats, daily injection of genistein (0.8 mg) for 180 days significantly reduced the mammary tumor multiplicity although it reduced the incidence only marginally [47]. Similar treatment with daidzein was not effective at reducing the incidence of mammary tumors. The isoflavones have been shown to reduce protein tyrosine kinase and topoisomerase activities *in vitro* [31, 32], but these effects were not observed in this study suggesting that the effect of genistein was independent of inhibition of these enzymes. However, in agreement with the *in vitro* studies, genistein acted as an estrogen *in vivo* in the absence of estrogen [48]. In ovarirectomized athymic mice injected with MCF breast cancer cells, tumors did not grow in the absence of estrogen while genistein promoted the growth although at a slower rate than with the estrogen treatment.

There are indications that early exposure to the isoflavones may protect against breast cancer later in life. Rats injected three-times with 5 mg genistein neonatally on postnatal days 2, 4 and 6, or with 500 μg genistein/g body weight prepubertally on postnatal days 16, 18 and 20, resulted in longer tumor latency and lower tumor incidence and multiplicity [49]. Exposure to genistein neonatally reduced the number of terminal end buds with no significant effect on the lobules II while prepubertal treatment reduced the terminal end buds and increased the number of lobules II. Since the terminal end buds are the least mature of the terminal duct structures of the mammary gland and also the most susceptible to carcinogenesis, the change in differentiation state caused by the isoflavones are thought to be responsible for the reduced tumor development in the rats exposed to genistein neonatally or prepubertally [49].

22.3.3.1.2 Other cancers

Soy has not been tested for effects on breast cancer metastasis but 10 %, 15 % or 20 % soy protein isolate fed in place of casein has been shown to reduce the incidence (by 35 %, 43 % and 43 %, respectively), number (by 74 %, 82 % and 83 %, respectively) and size (by 53 %, 53 % and 53 %, respectively) of lung metastasis of melanoma cell line B16BL6 injected directly to the blood stream in mice [50]. This effect is likely due to genistein since genistein (200 umol/kg body weight) given orally to mice injected with B16F10 melanoma cells have been shown to inhibit lung tumor nodule formation by 54 % and increase the lifespan of tumor bearing animals by 48 % [51]. The lung collagen hydroxyproline content and serum sialic acid levels, markers of metastases, were also reduced. Daidzein had no significant effects. The effects seen may be related to the ability of genistein to decrease the invasion of the basolateral membrane, a major step in cancer metastasis [52].

The effect of soy or its isoflavones on colon tumorigenesis is equivocal. After a 5-week feeding with genistein (75 or 150 mg/kg), significant reductions in colon aberrant crypt foci (preneoplastic marker) were observed in azoxymethane-treated rats [53]. Similarly, in comparison with soy protein concentrate devoid of isoflavones as control, the feeding of soy flakes, soy flour or 150 mg/kg genistein resulted in significant reductions in the number of aberrant crypt foci [54]. In contrast, the feeding of soy protein isolate or genistein (0.3 or 1 mg/kg) at the initiation or promotion stage of colon carcinogenesis did not result in significant reductions in aberrant crypts in comparison with the casein control [55]. At the same genistein level, no significant effect was observed in tumor development of human colon tumor cells transplanted to nude mice [55]. No significant differences were also observed in incidence, multiplicity, size and distribution of intestinal tumors between the ApcMin mouse fed the low isoflavone (16 mg/kg) or high isoflavone (475 mg/kg) diets [56]. However, genistein (250 mg/kg) significantly increased the noninvasive and total adenocarcinoma multiplicity but had no effect on the colon adenocarcinoma incidence nor on the multiplicity of invasive adenocarcinoma in carcinogen-treated rats [57].

A few studies suggest that the isoflavones or lignans may influence prostatic cancer development in rats. Using a neonatally estrogenized mice model, soy was found to reduce prostatic dysplasia [58]. The incidence of induced prostate related cancer was reduced and the disease-free period was prolonged by the high isoflavone diet compared with the low isoflavone diet when fed before carcinogen induction [59]. A study in rats transplanted with Dunning R3327 PAP prostate tumor and fed diets containing 33% soy flour for 24 weeks showed that soy delayed but did not stop the development of the prostate tumor [60, 61]. In contrast, genistein injection reduced the growth of tumor implanted in the dorsolateral prostate of Lobund-Wistar rats [62, 63]. This in part was attributed to the ability of genistein to downregulate the EGF and erB2/neu receptors in the rat dorsolateral prostate [63].

22.3.3.2 Flaxseed and lignans

22.3.3.2.1 Breast cancer

Short term feeding studies in rats showed that 5% and 10% flaxseed or defatted flaxseed can reduce the mammary epithelial cell proliferation and nuclear aberration [64]. The effect was significantly negatively related to the urinary lignan excretion. When 5% flaxseed was fed to rats for 4 weeks prior to injection of carcinogen (initiation stage), or continuously until sacrifice, the tumor incidence and number were reduced but not the tumor size when compared to those fed the control high fat diet [65]. On the other hand, when rats were provided the high fat diet prior to carcinogen treatment and then given the 5% flaxseed, their tumor size was significantly reduced by over 50% while the tumor incidence and mutiplicity were not. Flaxseed at the 2.5% or 5% level fed to rats for 7 weeks starting when their tumors were already about 1 cm diameter, also reduced the established tumor size by over 50% [66]. A similar effect was observed with purified SDG or flaxseed oil fed at levels equivalent to that taken in the 5% flaxseed diet suggesting that the effect seen with flaxseed was in part due to its lignan. This was supported by a significant negative relationship between the established tumor size and the urinary lignan excretion in this experiment. While both the oil and lignan component of flaxseed reduced the tumor size, only the lignans significantly reduced the number and size of new tumors that appeared during the feeding period. This is in agreement with observations that feeding rats SDG at levels (1.5 mg/day) equivalent to that in 5% flaxseed at the promotion stage of carcinogenesis can reduce the tumor multiplicity [67].

As with genistein, early exposure to flaxseed or SDG may also reduce the risk of breast cancer. We observed a significant reduction in the number of terminal end bud structure of the mammary gland of the offspring of dams fed 5% flaxseed during pregnancy and lactation [unpublished data]. Since a similar effect was observed upon feeding SDG equivalent to that in 5% flaxseed but not with the equivalent amount of flaxseed oil, the effect of flaxseed was likely due to the lignans.

All these results suggest that flaxseed has a protective effect against mammary cancer and this is in part due to the mammalian lignans derived from its major precursor, SDG. However, the nature of the effect depends upon the experimental design (i. e. the amount and time of lignan exposure).

22.3.3.2.2 Other cancers

Flaxseed has been tested for its potential to reduce colon carcinogenesis by using the formation of colon aberrant crypt as the endpoint. In a short term feeding study (4 weeks), 5% and 10% flaxseed or defatted flaxseed reduced the number of aberrant crypt and aberrant crypt foci by about 50% in carcinogen-treated rats [68]. In a longer term study (100 days), a similar reduction in aberrant crypt and aberrant crypt foci as well as a reduction in the number of microadenoma and polyps were observed upon feeding 2.5% or 5% flaxseed or defatted flaxseed [69]. More importantly, the SDG at the level found in 5% flaxseed produced the same effect as the 5% flaxseed, suggesting that the effect seen with flaxseed was due to its SDG. The fact that the defatted flaxseed, which contained the same amount of lignans as the full fat flaxseed, produced the same effect as the full fat flaxseed also suggested that the effect seen was not due to the flaxseed oil component. A significant negative relationship was observed between urinary lignan excretion and the number of aberrant crypts per focus, providing further support for the role of lignans on colon cancer development [69].

Like soybean, flaxseed was tested for its effect on the lung metastasis of the melanoma cell line B16BL6 injected directly to the blood stream [70]. Feeding 2.5%, 5% or 10% flaxseed two weeks before and two weeks after injection resulted in a 32%, 54% and 63% reduction in the number of lung tumors, respectively. The tumor size was also reduced in a dose-dependent manner. The SDG at levels found in the above flaxseed levels produced similar reductions in the incidence and number of lung metastases indicating that the lignans are in part responsible for the effect seen with flaxseed [unpublished data].

To determine whether lignans may influence prostate tumor development, rats transplanted with Dunning R3327 PAP prostate tumor were fed diets containing either 33% rye bran, heat treated rye bran or rye endosperm for 24 weeks [61]. Like the effect seen with the isoflavone-containing soybean diet, the rye bran diets which contain lignans did not stop but did delay the development of prostate tumors. The rye endosperm diet with no lignans had no significant effect. Interestingly, the 33 % rye bran diets had a similar effect as the 33 % soy diet despite the fact that total urinary phytoestrogen excretion was 10-times lower with the rye bran diet than the soy diet.

22.3.4 Clinical studies

Many prospective clinical trials involving the feeding of flaxseed, soybean products or isoflavones and measurement of some cancer risk markers are still in progress. Because the proposed mechanisms whereby the isoflavones and lignans may influence carcinogenesis relate to their weak estrogenic/antiestrogenic effect, past clinical studies have concentrated on demonstrating the hormonal effects of soybean or flaxseed. It has been suggested that an antiestrogenic response may be cancer protective.

In premenopausal women, 10 g raw flaxseed per day fed for three months did not change the menstrual cycle length but lengthened the luteal phase length [71]. No significant changes in sex hormones were observed except for a decreased plasma estradiol to progesterone ratio. No significant changes in sex hormones were also observed in men fed 13.5 g flaxseed in bread for 6 weeks [72]. In contrast, feeding 60 g/day soybean protein containing 45 mg isoflavones to premenopausal women for one month lowered the plasma gonadotrophins although follicle stimulating hormone was lowered instead of the lutenizing hormone [73]. Also, the follicular phase length of the menstrual cycle was lengthened instead of the luteal phase. Lu *et al.* [74] observed a reduction in plasma estrogen in premenopausal women after a daily consumption of soy protein at about half the level used in the above study [73]; the low plasma level persisted 2–3 menstrual cycles after the soy intake was stopped. Overall, these results suggest that soybean and flaxseed exert an antiestrogenic effect in premenopausal women. However, it is unclear why flaxseed and soybean and their respective lignans and isoflavones lengthen the different phases of the menstrual cycle and what the long term implications of the finding might be.

In postmenopausal women, an increase in vaginal cell maturation index with an intake of soybean and/or flaxseed have been reported [75]. Incidence of hot flushes was decreased although the effect was about equivalent to the placebo in some studies [76]. There is inconsistency in the correlation of vaginal cytology with hot flushes, however, this is probably due to differences among study designs and subjects. Although these findings suggest an estrogenic effect of soybean and flaxseed when estrogen levels are low, as is usually the case in postmenopausal women, other studies showed no effect of feeding soybean on the sex hormones of postmenopausal women [77].

Contrary to the above results, an increased proliferation rate of breast lobular epithelium in premenopausal women was observed after a 14-day intake of 60 g of soy containing 45 mg isoflavones, suggesting an estrogenic effect [78]. A pilot study involving 24 pre- and postmenopausal women fed 37.4 g soy protein isolate containing 37.4 mg genistein for 6 months also observed a similar estrogenic effect in premenopausal women [79]. Plasma estradiol levels and nipple aspirate fluid volumes were increased in premenopausal but not in postmenopausal women. Hyperplastic cells in breast nipple aspirates of 4 of 14 (28.6%) premenopausal women and 3 of 10 (30%) postmenopausal women were also detected. Increased cell proliferation has been associated with increased cancer risk. Evidently, the study did not support the authors' hypothesis that nipple aspirate fluid characteristics of these women will be modified by the soy intake so that they more closely resemble those of Asian women who are known to have lower risk of breast cancer. Studies in a larger number of human subjects should probably be conducted to further establish the protective role of soybean in carcinogenesis. No similar study has been conducted using flaxseed and lignans; this should be addressed in future investigations.

22.4 Safety

The recent report of hyperproliferation in human breast epithelial cells upon soy intake [78, 79] is a concern as it suggests an increased risk of breast cancer. An additional concern is the finding that infants fed soy milk consume phytoestrogens several-fold higher than those taking mother's or cow's milk [80, 81]. Their plasma isoflavone levels are 13,000–22,000-times higher than plasma estradiol concentrations in early life and their daily exposure to isoflavones was estimated to be 6–11-fold higher than the level (based on body weight) that can cause hormonal effects in adults consuming soy foods. Although animal studies suggest that early exposure to phytoestrogens may reduce the risk of breast cancer at a later age [49, unpublished data], there is currently no human study which has demonstrated that adverse effects (e.g. in reproductive function) are not associated with such early exposure.

Feeding flaxseed at 5% or 10% levels or SDG at levels equivalent to that in 5% flaxseed during pregnancy and lactation did not affect pregnancy and pregnancy outcome except for a lower birth weight in offspring of dams fed 10% flaxseed [82]. With 10% flaxseed, the female offspring also had shortened anogenital distance, higher uterine and ovarian weights, earlier age and lighter weight at puberty, lengthened estrous cycle and persistent estrus while the male offspring had lower postnatal growth and increased prostate cell proliferation indicating hormonal effects with implications for cancer and reproductive capacity [82].

At high concentrations, genistein has been shown to be genotoxic in *in vitro* studies [83] but enterolactone, enterodiol, matairesinol and secoisolariciresinol appear to be devoid of aneuploidogenic and clastogenic potential [84]. Overall, all these studies suggest that caution should be taken in making recommendations for the use of phytoestrogen-containing foods for prevention of cancer until the safety at the various stages of the human life cycle has been determined.

22.5 Conclusion and future research

Epidemiological studies involving the measurement of urinary excretion of phytoestrogens or intake of soybeans are suggestive but not conclusive regarding the cancer protective effect of isoflavones and lignans. *In vitro* studies showed biological properties of isoflavones and lignans from which anticancer mechanisms may be deduced. However, many of the biological properties determined *in vitro* have yet to be demonstrated in *in vivo* situations. In studies where they have been measured *in vivo*, the data sometimes did not relate to tumor data or did not support the *in vitro* observations. Also, levels found effective *in vitro* are often much higher than the plasma levels observed in animal and human studies. Animal studies are more consistent in demonstrating a cancer protective effect of soybean and flaxseed or their purified isoflavones and SDG, respectively, in the mammary and prostate but not in the colon in the case of soybean and genistein. This is perhaps due to differences in experimental design and the dose fed.

Whether a cancer protective effect of isoflavones and lignans will be demonstrated in long term prospective studies in humans remains in question. The few clinical studies done in premenopausal women suggest an antiestrogenic and possibly cancer protective effect of soybean or flaxseed based on the menstrual cycle and/or sex hormone changes. However, the increased breast epithelial cell proliferation observed in soy-fed premenopausal women suggests an estrogenic response and, possibly, a cancer promoting effect of soybean. While the estrogenic response reported for postmenopausal women may be desirable as far as reduction in menopausal symptoms and risk of osteoporesis and cardiovascular disease are concerned, its implication on hormone related cancer such as the breast and the endometrium remains to be determined. The reproduction and cancer risk implication of early human exposure to isoflavones and lignans should also be addressed in the future.

There is certainly a need to understand how *in vitro* data would relate to animal and ultimately human conditions. Bioavailability studies are helpful in the understanding of the mechanisms of lignan action, in relating *in vitro* anticancer effects with *in vivo* situations, and in establishing the timing and level of intake of phytoestrogen in foods needed to maintain the blood and tissue levels of phytoestrogens considered cancer protective. Although many such studies have been conducted [85–98], the results have not yet provided all the answers.

The recommended level and frequency of intake which would result in more health benefits than adverse effects remain to be established.

While the isoflavones and lignans have been compared in several *in vitro* studies, there are only very few studies which compared them or their major sources (i.e. soybean and flaxseed), in animal or human studies. Comparative studies should be conducted since the lignans are more widely found in vegetarian diets than the isoflavones.

Undoubtedly significant advances have been made, particularly in the last decade, towards the understanding of the potential anticancer effects of the isoflavones and lignans. While some of the results are encouraging, more questions need to be answered before a definitive conclusion can be made regarding the anticancer effects of these compounds.

References

[1] Adlercreutz, H.; Mazur, W. (1997) Phytoestrogens and western diseases. *Ann. Med.* **29**, 95–120.

[2] Bingham, S. A.; Atkinson, C.; Liggins, J.; Bluck, L.; Coward, A. (1998) Phyto-oestrogens: where are we now? *Br. J. Nutr.* **79**, 393–406.

[3] Axelson, M.; Sjovall, J.; Gustafsson, B. E.; Setchell, K. D. (1982) Origin of lignans in mammals and identification of a precursor from plants. *Nature* **298**, 659–660.

[4] Thompson, L. U.; Robb, P.; Serraino, M.; Cheung, F. (1991) Mammalian lignan production from various foods. *Nutr. Cancer* **16**, 43–52.

[5] Borriello, S. P.; Setchell, K. D.; Axelson, M.; Lawson, A. M. (1985) Production and metabolism of lignans by the human faecal flora. *J. Appl. Bacteriol.* **58**, 37–43.

[6] Axelson, M.; Setchell, K. D. (1980) Conjugation of lignans in human urine. *FEBS Lett.* **122**, 49–53.

[7] Setchell, K. D..; Adlercreutz, H. (1988) Mammalian lignans and phytoestrogens: recent studies on their formation, metabolism and biological role in health and disease. In: Rowland, I. R. (Ed.) *Role of the gut flora, toxicity and cancer.* Academic Press, London, p. 315–345.

[8] Adlercreutz, H.; Honjo, H.; Higashi, A.; Fotsis, T.; Hamalainen, E.; Hasegawa, T.; Okada, H. (1991) Urinary excretion of lignans and isoflavonoid phytoestrogens in Japanese men and women consuming a traditional Japanese diet. *Am. J. Clin. Nutr.* **54**, 1093–1100.

[9] Adlercreutz, H. T.; Goldin, B. R.; Gorbach, S. L.; Hockerstedt, K. A. V.; Watanabe, S.; Hamalainen, E. K.; Markkanen, M. H.; Makela, T. H.; Wahala, K. T.; Hase, T. A.; Fotsis, T. (1995) Soybean phytoestrogen intake and cancer risk. *J. Nutr.* **125**, 757S–770 S.

[10]. Adlercreutz, H.; Fotsis, T.; Heikkinen, R.; Dwyer, J. T.; Woods, M.; Goldin, B. R.; Gorbach, S.L. (1982) Excretion of the lignans enterolactone and enterodiol and of equol in omnivorous and vegetarian postmenopausal women and women with breast cancer. *Lancet* **2**, 1295–1299.

[11] Ingram, D.; Sanders, K.; Kolybaba, M.; Lopez, D. (1997) Case-control study of phytooestrogens and breast cancer. *Lancet* **350**, 990–994.

[12] Morton, M. S.; Chan, P. S. F.; Cheng, C.; Blacklock, N.; Matos-Ferreira, A.; Abranches-Monteiro, L.; Correia, R.; Lloyd, S.; Griffiths, K. (1997) Lignans and isoflavonoids in plasma and prostatic fluid in men: samples from Portugal, Hong Kong, and the United Kingdom. *Prostate* **32**, 122–128.
[13] Messina, M. J.; Persky, V.; Setchell, K. D.; Barnes, S.; (1994) Soy intake and cancer risk: A review of the *in vitro* and *in vivo* data. *Nutr. Cancer* **21**, 113–131.
[14] Lee, H. P.; Gourley, L.; Duffy, S. W.; Esteve, J.; Lee, J.; Day, N. E.; (1991) Dietary effects on breast cancer risk in Singapore. *Lancet* **337**, 1197–1200.
[15] Yuan, J.; Wang, Q.; Ross, R. K.; Henderson, B. E.; Yu, M. C. (1995) Diet and breast cancer in Shanghai and Tianjin China. *Br. J. Cancer* **71**, 1353–1358.
[16] Hirohata, T.; Shigematsu, T.; Nomura, A. M.; Nomura, Y.; Horie, A.; Hirohayta, I. (1985) Occurrence of breast cancer in relation to diet and reproductive history. A case control study in Fukuoka Japan. *Natl. Cancer Inst. Monograph.* **69**, 187–190.
[17] Nomura, A. M. Y.; Henderson, B. E.; Lee, J. (1978) Breast cancer and diet among the Japanese in Hawaii. *Am. J. Clin. Nutr.* **31**, 2020–2025.
[18] Hirayama, T. (1986) A large scale cohort study on cancer risks by diet with special reference to the risk reducing effects of green, yellow vegetable consumption. In: Hayashi, Y; Nagao, M.; Sugimura, T.; Takayama, S.; Tomatis, L.; Wattenberg, L. W.; Wogan, G. N. (Eds.) *Diet, nutrition and cancer.* Japan Scientific Society Press, Tokyo, p. 41–53.
[19] Wu, A. H.; Ziegler, R. G.; Horn-Ross, P. L.; Nomura, A. M.; West, D. W.; Kolonel, L. N.; Rosenthal, J. F.; Hoover, R. N.; Pike, M. C. (1996) Tofu and risk of breast cancer in Asian-Americans. *Cancer Epidemiol. Biomarkers Prev.* **5**, 901–906.
[20] Welshons, W. V.; Murphy, C. S.; Koch, R.; Calaf, G.; Jordan, V. C. (1987) Stimulation of breast cancer cells *in vitro* by the environmental estrogen enterolactone and the phytoestrogen equol. *Breast Cancer Res. Treat.* **10**, 169–175.
[21] Mousavi, Y.; Adlercreutz, H. (1992) Enterolactone and estradiol inhibit each other's proliferative effect on MCF-7 breast cancer cells in culture. *J. Steroid Biochem. Mol. Biol.* **41**, 615–619.
[22] Hirano, T.; Fukuoka, K.; Oka, K.; Naito, T.; Hosaka, K.; Mitsuhashi, H.; Matsumoto, Y. (1990) Antiproliferative activity of mammalian lignan derivatives against the human breast carcinoma cell line, ZR-75-1. *Cancer Invest.* **8**, 595–602.
[23] Wang, C.; Kurzer, M. S. (1997) Phytoestrogen concentration determines effects on DNA synthesis in human breast cancer cells. *Nutr. Cancer* **28**, 236–247.
[24] Adlercreutz, H.; Mousavi, Y.; Clark, J.; Hockerstedt, K.; Hamalainen, E.; Wahala, K.; Makela, T.; Hase, T. (1992) Dietary phytoestrogens and cancer: *in vitro* and *in vivo* studies. *J. Steroid Biochem. Mol. Biol.* **41**, 331–337.
[25] Garreau, B.; Vallette, G.; Adlercreutz, H.; Wahala, K.; Makela, T.; Benassayag, C.; Nunez, E. A. (1991) Phytoestrogens: new ligands for rat and human α-fetoprotein. *Biochim. Biophys. Acta* **1094**, 339–345.
[26] Martin, M. E.; Haourigui, M.; Pelissero, C.; Benassayag, C.; Nunez, E. A. (1996) Interactions between phytoestrogens and human sex steroid binding protein. *Life Sci.* **58**, 429–436.
[27] Sathyamoorthy, N.; Wang, T. T. Y.; Phang, T. M. (1994) Stimulation of pS2 expression by diet-derived compounds. *Cancer Res.* **54**, 957–961.
[28] Adlercreutz, H.; Bannwart, C.; Wahala, K.; Makela, T.; Brunow, G.; Hase, T.; Arosemena, P. J.; Kellis, J. T., Jr.; Vickery, L. E. (1993) Inhibition of human aromatase by mammalian lignans and isoflavonoid phytoestrogens. *J. Steroid Biochem. Mol. Biol.* **44**, 147–153.
[29] Wang, C.; Makela, T.; Hase, T.; Adlercreutz, H.; Kurzer, M. S. (1994) Lignans and flavonoids inhibit aromatase enzyme in human preadipocytes. *J. Steroid Biochem. Mol. Biol.* **50**, 205–212.
[30] Evans, B. A. J.; Griffiths, K.; Morton, M. S. (1995) Inhibition of 5-reductase in genital

skin fibroblasts and prostate tissue by dietary lignans and isoflavonoids. *J. Endocrinol.* **147**, 295–302.
[31] Akiyama, T.; Ishida, J.; Nakagawa, S.; Ogawara, A.; Watanabe, S.; Itoh, N.; Shibuya, M.; Fuani, Y. (1987) Genistein, a specific inhibitor of tyrosine specific protein kinases. *J. Biol. Chem.* **262**, 5592–5595.
[32] Markovits, J.; Linassier, C.; Fosses, P.; Couprie, J.; Pierre, J.; Jacquemin-Sablon, A.; Saucier, J. M.; Lee Pecq, J. B.; Larsen, A. K. (1989) Inhibitory effects of the tyrosine kinase inhibitor genistein on mammalian DNA topoisomerase II. *Cancer Res.* **49**, 5111–5117.
[33] Tripathi, Y. B.; Lim, R. W.; Fernandez-Gallardo, S.; Kandala, J. C.; Guntaka, R. V.; Shukla, S.D. (1992) Involvement of tyrosine kinase and protein kinase C in platelet-activating factor-induced c-fos gene expression in A-431 cells. *Biochem. J.* **286**, 527–533.
[34] Dalu, A.; Haskell, J. F.; Coward, L.; Lamartiniere, C. A. (1998) Genistein, a component of soy, inhibits the expression of the EGF and ErbB2/neu receptor in the rat dorsolateral prostate. *Prostate* **37**, 36–43.
[35] Jha, H. C.; von Recklinghausen, G.; Ziliiken, F.(1985) Inhibition of *in vitro* microsomal lipid peroxidation by isoflavonoids. *Biochem. Pharmacol.* **34**, 1367–1369.
[36] Prasad, K. (1997) Hydroxyl radical-scavenging property of secoisolariciresinol diglucoside (SDG) isolated from flax-seed. *Mol. Cell Biochem.* **168**, 117–123.
[37] Wei, H.; Wei, L.; Frankel, K.; Bowen, R.; Barnes, S. (1993) Inhibition of tumor-promoter-induced hydrogen peroxide formation *in vitro* and *in vivo* by genistein. *Nutr. Cancer* **20**, 1–12.
[38] Fotsis, T.; Pepper, M.; Adlercreutz, H.; Fleischmann, G.; Hase, T.; Montesano, R.; Schweigerer, L. (1993) Genistein, a dietary-derived inhibitor of *in vitro* angiogenesis. *Proc. Natl. Acad. Sci. USA* **90**, 2690–2694.
[39] Constantinou, A.; Figuchi, F.; Huberman, E. (1990) Induction of differentiation and DNA strand breakage in human HL-60 and K-562 leukemia cells by genistein. *Cancer Res.* **50**, 2618–2624.
[40] Kiguchi, K.; Constantinou, A. I.; Huberman, E. (1990). Genistein-induced cell differentiation and protein-linked DNA strand breakage in human melanoma cells. *Cancer Commun.* **2**, 271–277.
[41] Kondo, K.; Tsuneizumi, K.; Watanabe, T.; Oishi, M. (1991) Induction of *in vitro* differentiation of mouse embryonal carcinoma (F9) cells by inhibition of topoisomerases. *Cancer Res.* **51**, 5398–5404.
[42] Miller, D. R.; Lee, G. M.; Maness, P. F. (1993) Increased neurite outgrowth induced by inhibition of protein tyrosine kinase activity in PC 12 pheochromocytoma cells. *J. Neurochem.* **60**, 2134–2144.
[43] Watanabe, T.; Kondo, F.; Oishi, M. (1991) Induction of *in vitro* differentiation of mouse erythroleukemia cells by genistein, an inhibitor of tyrosine protein kinases. *Cancer Res.* **51**, 764–768.
[44] Yanagihara, K.; Ito, A.; Toge, T.; Numoto, M. (1993) Antiproliferative effects of isoflavones on human cancer cell lines established from the gastrointestinal tract. *Cancer Res.* **53**, 5815–5821.
[45] Thompson, L. U. (1998) Experimental studies on lignans and cancer. In: Adlercreutz, H. (Ed.) *Phytoestrogens*. Bailliere, London, p. 691–705.
[46] Barnes, S.; Grubbs, C.; Setchell, K. D.; Carlson, J. (1990) Soybean inhibits mammary tumors in models of breast cancer. In: Pariza, M. (Ed.) *Mutagens and carcinogens in the diet*. Liss, New York, p. 239–253.
[47] Constantinou, A. I.; Mehta, R. G.; Vaughan, A. (1996) Inhibition of *N*-methyl-*N*-nitrosourea-induced mammary tumors in rats by the soybean isoflavones. *Anticancer Res.* **16**, 3293–3298.
[48] Hsieh, C. Y.; Santell, R. C.; Haslam, S. Z.; Helferich, W. G. (1998) Estrogenic effects

of genistein on the growth of estrogen receptor-positive human breast cancer (MCF-7) cells *in vitro* and *in vivo*. *Cancer Res.* **58**, 3833–3838.
[49] Lamartiniere, C. A.; Murrill, W. B.; Manzolillo, P. A.; Zhang, J. X.; Barnes, S.; Zhang, X.; Wei, H.; Brown, N. M. (1998) Genistein alters the ontogeny of mammary gland development and protects against chemically-induced mammary cancer in rats. *Proc. Soc. Exp. Biol. Med.* **217**, 358–364.
[50] Yan, L.; Yee, J. L.; McGuire, M. H.; Graef, G. L. (1997) Effect of dietary supplementation of soybeans on experimental metastasis of melanoma cells in mice. *Nutr. Cancer* **29**, 1–6.
[51] Menon, L. G.; Kuttan, R.; Nair, M. G.; Chang, Y. C.; Kuttam, G. (1998) Effect of isoflavones genistein and daidzein in the inhibition of lung metastasis in mice induced by B16F-10 melanoma cells. *Nutr. Cancer* **30**, 74–77.
[52] Scholar, E. M.; Toews, M. L. (1994) Inhibition of invasion of murine mammary carcinoma cells by the tyrosine kinase inhibitor genistein. *Cancer Lett.* **87**, 159–162.
[53] Steele, V. E.; Pereira, M. A.; Sigman, C. C.; Kelloff, G. I. (1995) Cancer chemoprevention agent development strategies for genistein. *J. Nutr.* **125**, 713S–716S.
[54] Bennink, M. K.; Thiagarajan, L. D.; Bourquin, L. D.; Kavar, A. (1996) Prevention of precancerous colonic lesions in rats by soy flakes, soy flour, genistein and calcium. In: *Proceedings of the 2nd International Symposium on the Role of Soy in Preventing and Treating Chronic Disease*, St. Louis, MO, p. 29.
[55] Gallaher, D. D.; Galllaher, C. M.; An, Z.; Hoffman, R. M. (1996) Soy protein isolates and genistein: Effects on initiation, promotion and progression of colon cancer. In: *Proceedings of the 2nd International Symposium on the Role of Soy in Preventing and Treating Chronic Disease*, St. Louis, MO, p. 29–30.
[56] Sorensen, I. K.; Kristiansen, E.; Mortensen, A.; Nicolaisen, G. M.; Wijnands, J. A.; van Kranen, H. J.; van Kaejl, C. F. *Cancer Lett.* **130**, 217–225.
[57] Rao, C. V.; Wang, C. X.; Sim, B.; Lubet, R.; Kelloff, G.; Steele, V.; Reddy, B. S. (1997) Enhancemnet of experimental colon cancer by genistein. *Cancer Res.* **57**, 3717–3722.
[58] Makela, S. I.; Pylkkanen, L. H.; Santii, R. S.; Adlercreutz, H. (1995) Dietary soybean may be antiestrogenic in male mice. *J. Nutr.* **125**, 437–445.
[59] Pollard, M.; Luckert, P. H. (1997) Influence of isoflavones in soy protein isolate on development of induced prostate-related cancer in L-W rats. *Nutr. Cancer* **28**, 41–45.
[60] Zhang, J. X.; Hallmans, G.; Landstrom, M.; Bergh, A.; Damber, J. E.; Aman, P.; Adlercreutz, H. (1997) Soy and rye diets inhibit the development of Dunning R3327 prostatic adenocarcinoma in rats. *Cancer Lett.* **114**, 313–314.
[61] Landstrom, M.; Zhang, J. X.; Hallmans, G.; Aman, P.; Bergh, A.; Damber, J. E.; Mazur, W.; Wahala, K.; Adlercreutz, H. (1998) Inhibitory effects of soy and rye diets on the development of Dunning R3327 prostate adenocarcinoma in rats. *Prostate* **36**, 151–161.
[62] Schleicher, R.; Zheng M.; Zhang, M.; Lamartinier, C.A. (1996) Genistein inhibition of prostate cancer cell growth and metastasis *in vivo*. In: *Proceedings of the 2nd International Symposium on the Role of Soy in Preventing and Treating Chronic Disease*, St. Louis, MO, p. 47.
[63] Dalu, A.; Haskell, J. F.; Coulard, L.; Lamartiniere, C. A. (1998) Genistein, a component of soy, inhibits the expression of the EGF and ErbB2/Neu receptors in the rat dorsolateral prostate. *Prostate* **37**, 36–43.
[64] Serraino, M.; Thompson, L. U. (1991) The effect of flaxseed supplementation on early risk markers for mammary carcinogenesis. *Cancer Lett.* **60**, 135–142.
[65] Serraino, M.; Thompson, L. U. (1992) The effect of flaxseed supplementation on the initiation and promotional stages of mammary tumorigenesis. *Nutr. Cancer* **17**, 153–159.
[66] Thompson, L. U.; Rickard, S. E.; Orcheson, L. J.; Seidl, M. M. (1996) Flaxseed and its lignan and oil components reduce mammary tumor growth at a late stage of carcinogenesis. *Carcinogenesis* **17**, 1373–1376.

[67] Thompson, L. U.; Seidl, M.; Rickard, S.; Orcheson, L.; Fong, H. (1996) Antitumorigenic effect of a mammalian lignan precursor from flaxseed. *Nutr. Cancer* **26**, 159–165.
[68] Serraino, M.; Thompson, L. U. (1992) Flaxseed supplementation and early markers of colon carcinogenesis. *Cancer Lett.* **63**, 159–165.
[69] Jenab, M.; Thompson, L. U. (1996) The influence of flaxseed and lignans on colon carcinogenesis and β-glucuronidase activity. *Carcinogenesis* **17**, 1343–1348.
[70] Yan, L.; Yee, J. A.; Li, D.; McGuire, M. H.; Thompson, L. U. (1998) Dietary flaxseed supplementation and experimental metastasis of melanoma cells in mice. *Cancer Lett.* **124**, 181–186.
[71] Phipps, W. R.; Martini, M. C.; Lampe, J. W.; Slavin, J. L.; Kurzer, M. S. (1993) Effect of flax seed ingestion on the menstrual cycle. *J. Clin. Endocrinol. Metab.* **77**, 1215–1219.
[72] Shultz, T. D.; Bonorden, W. R.; Seaman, W. R. (1991) Effect of short-term flaxseed consumption on lignan and sex hormone metabolism. *Nutr. Res.* (NY) **11**, 1089–1100.
[73] Cassidy, A.; Bingham, S.; Setchell, K. D. (1994) Biological effects of a diet of soy protein rich in isoflavones on the menstrual cycle of premenopausal women. *Am. J. Clin. Nutr.* **60**, 333–340.
[74] Lu, L. J.; Anderson, K. E.; Grady, J. J.; Nagamani, M. (1996) Effects of soya consumption for one month on steroid hormones in premenopausal women: implications for breast cancer risk reduction. *Cancer Epidemiol. Biomarkers Prev.* **5**, 63–70.
[75] Wilcox, G.; Wahlqvist, M. L.; Burger, H. G.; Medley, G. (1990) Oestrogenic effects of plant foods in postmenopausal women. *Br. Med. J.* **301**, 905–906.
[76] Murkies, A. L.; Wilcox, G.; Davis, S. R. (1998) Clinical review 92: Phytoestrogens. *J. Clin. Endocrinol. Metab.* **83**, 297–303.
[77] Baird, D. D.; Umbach, D. M.; Lansdell, L.; Hughes, C. L.; Setchell, K. D.; Weinberg, C. R.; Haney, A. F.; Wilcox, A. J.; McLachlan, J. A. (1995) Dietary intervention study to assess estrogenicity of dietary soy among postmenopausal women. *J. Clin. Endocrinol. Metab.* **80**, 1685–1690.
[78] McMichael-Phillips, D. F.; Harding, C.; Morton, M.; Potten, C. S.; Bundred, N. J. (1996) The effects of soy supplementation on epithelial proliferation in the normal breast. In: *Proceedings of the 2nd International Symposium on the Role of Soy in Preventing and Treating Chronic Disease*, St. Louis, MO, p. 35.
[79] Petrakis, N. L.; Barnes, S.; King, E. B.; Lowenstein, J.; Wiencke, J.; Lee, M. M.; Miike, R.; Kirk, M.; Coward, L. (1996) Stimulatory influence of soy protein isolate on breast secretion in pre- and postmenopausal women. *Cancer Epidemiol. Biomarkers Prev.* **5**, 785–794.
[80] Setchell, K. D.; Zimmer-Nechemias, L.; Cai, J.; Heubi, J. E. (1997) Exposure of infants to phyto-oestrogens from soy-based infant formula. *Lancet* **350**, 23–27.
[81] Klein, K. O. (1998) Isoflavones, soy-based infant formulas, and relevance to endocrine function. *Nutr. Rev.* **56**, 193–204.
[82] Tou, J.; Chen, J.; Thompson, L. (1998) Flaxseed and its lignan precursor, secoisolariciresinol diglycoside, affect pregnancy outcome and reproductive development in rats. *J. Nutr.* **128**, 1861–1868.
[83] Kulling, S. E.; Metzler, M. (1997) Induction of micronuclei, DNA strand breaks and HPRT mutations in cultured Chinese hamster V79 cells by the phytoestrogen coumestrol. *Food Chem. Toxicol.* **35**, 605–613.
[84] Kulling, S. E.; Jacobs, E.; Pfeiffer, E.; Metzler, M. (1998) Studies on the genotoxicity of the mammalian lignans enterolactone and enterodiol and their metabolic precursors at various endpoints *in vitro*. *Mutat. Res.* **416**, 115–124.
[85] Kirkman, L. M.; Lampe, J. W.; Campbell, D. R.; Martini, M. C.; Slavin, J. L. (1995) Urinary lignan and isoflavonoid excretion in men and women consuming vegetable and soy diets. *Nutr. Cancer* **24**, 1–12.

[86] Kurzer, M. S.; Lampe, J. W.; Martini, M. C.; Adlercreutz, H. (1995) Fecal lignan and isoflavonoid excretion in premenopausal women consuming flaxseed powder. *Cancer Epidemiol. Biomarkers Prev.* **4**, 353–358.
[87] Lampe, J. W.; Martini, M. C.; Kurzer, M. S.; Adlercreutz, H.; Slavin, J. L. (1994) Urinary lignan and isoflavonoid excretion in premenopausal women consuming flaxseed powder. *Am. J. Clin. Nutr.* **60**, 122–128.
[88] Lu, L. J.; Grady, J. J.; Marshall, M. V.; Ramanujam, V. M.; Anderson, K. E. (1995) Altered time course of urinary daidzein and genistein excretion during chronic soya diet in healthy male subjects. *Nutr. Cancer* **24**, 311–323.
[89] Nesbitt, P. D.; Lam, Y.; Thompson, L. U. (1998) Mammalian lignans from flaxseed in humans. *Am. J. Clin. Nutr.*, in press.
[90] Nesbitt, P. D.; Thompson, L. U. (1997) Lignans in homemade and commercial products containing flaxseed. *Nutr. Cancer* **29**, 222–227.
[91] Morton, M. S.; Wilcox, G.; Wahlqvist, M. L.; Griffiths, K. (1994) Determination of lignans and isoflavonoids in human female plasma following dietary supplementation. *J. Endocrinol.* **142**, 251–259.
[92] Morton, M. S.; Matos-Ferreira, A.; Abranches-Monteiro, L.; Correia, R.; Blacklock, N.; Chan, P. S.; Cheng, C.; Lloyd, S.; Chieh-Ping, W.; Griffiths, K. (1997) Measurement and metabolism of isoflavonoids and lignans in the human male. *Cancer Lett.* **114**, 145–151.
[93] Rickard, S. E.; Orcheson, L. J.; Seidl, M. M.; Luyengi, L.; Fong, H. H.; Thompson, L. U. (1996) Dose-dependent production of mammalian lignans in rats and *in vitro* from the purified precursor secoisolariciresinol diglycoside in flaxseed. *J. Nutr.* **126**, 2012–2019.
[94] Rickard, S. E.; Thompson, L. U. (1998) Chronic exposure to secoisolariciresinol diglycoside alters lignan disposition in rats. *J. Nutr.* **128**, 615–623.
[95] Thompson, L. U.; Rickard, S. E.; Cheung, F.; Kenaschuk, E. O.; Obermeyer, W. R. (1997) Variability in the anticancer lignan levels in flaxseed. *Nutr. Cancer* **27**, 26–30.
[96] Xu, X.; Harris, K. S.; Wang, H. J.; Murphy, P. A.; Hendrich, S. (1995) Bioavailability of soybean isoflavones depends upon gut microflora in women. *J. Nutr.* **125**, 2307–2315.
[97] Xu, X.; Wang, W. J.; Murphy, P. A.; Cook, L.; Hendrich, S. (1994) Daidzein is a more bioavailable soymilk isoflavone than is genistein in adult women. *J. Nutr.* **124**, 825–832.
[98] Hutchins, A. M.; Lampe, J. W.; Martini, M. C.; Campbell, D. R.; Slavin, J. L. (1995) Vegetables, fruits, and legumes: effect on urinary isoflavonoid phytoestrogen and lignan excretion. *J. Am. Diet. Assoc.* **95**, 769–774.

23 Chemoprevention of Colon Carcinogenesis by Non-Nutritive Compounds in Foods

Takuji Tanaka

23.1 Abstract

In addition to mutagens and/or carcinogens a number of modulators in carcinogenesis are present in our environment. Some of them are contained in our regular foods and therefore dietary factors play a role in the development of some types of cancers including colon cancer. Epidemiological studies have suggested that a diet rich in fruits and vegetables is associated with reduced risk for a number of common cancers. There are still many unknown constituents and/or factors in foods that could either enhance or reduce the possibilities for developing cancer. Animal studies in experimental chemical carcinogenesis have indicated that several non-nutritive components, belonging to different chemical groups, in foods protect against certain type of cancer including colonic neoplasm. These chemicals are known as "chemopreventive agents". Many of them are antioxidants and might suppress carcinogenesis through (i) inhibition of Phase I enzymes, (ii) induction of Phase II enzymes, (iii) scavenging DNA reactive agents, (iv) suppression of hyper-cell proliferation induced by carcinogen, and/or (v) inhibition of certain properties of neoplastic cells. With increasing the incidence of colon cancer rising certainly, there is an ever increasing need to determine the most effective means to prevent colon cancer and to understand the mechanism(s) underlying successful prevention. Previous studies in our laboratory demonstrated protective effects of several naturally occurring products against rat colon tumorigenesis. This article will introduce our recent studies in search for the chemopreventive effects of flavonoids (diosmin and hesperidin) and other phytochemicals (1′-acetoxychavicol acetate and auraptene) in edible plants on rat colon carcinogenesis.

23.2 Introduction

Many compounds are known to modify the development of tumors in response to various carcinogenic agents in experimental animals [1]. Epidemiological studies have suggested that specific, pharmacologically active agents present in the diet may reduce the relative risk of exposure of carcinogens. As for colon cancer, marked variations in dietary habits among populations of different cultures and life-styles have been associated with a risk of this malignancy [2, 3]. Also, there is an inverse correlation between the intake of vegetables/fruits and human colon cancer [4–6]. Thus, a relationship between the risk of the development of colon cancer and dietary habits is important [7, 8], although etiology of colon cancer is multifactorial and complex.

An important consideration in cancer research today is that exposure to pharmacologically active chemicals may play an important role in reducing the relative risks resulting from exposure to carcinogenic chemicals. Chemoprevention of cancer might be defined as the deliberate introduction of these selected non-toxic substances into the diet for the purpose of reducing cancer development. Numerous epidemiological studies on the relationship between diet and carcinogenesis have demonstrated a protective effect of the consumption of fruits and vegetables against various forms of cancers [6, 9, 10]. Potential chemopreventive agents are to be found both among nutrients and non-nutrients. Epidemiological and experimental studies have revealed that a number of micronutrients may have cancer preventive properties in several organs including large bowel [11]. Examples are vitamins A, C, and β-carotene, selenium, and calcium. We have demonstrated cancer chemopreventive ability of two xanthophylls without provitamin A activity in the rat colon and oral cavity [12, 13]. Most of the above-mentioned compounds are antioxidants which could serve as an explanation for their mode of action. The well-known non-nutritive chemopreventives in colon tumorigenesis is dietary fiber, a variety of ingestible carbohydrates [14]. Since the modifying effects of the major dietary factors on rodents colon carcinogenesis resulted in heterogeneous [15], we focus on non-nutritive inhibitors derived from vegetables and fruits in experimental colon carcinogenesis. Wattenberg suggested protective effects of some minor non-nutrients in the diet on colon tumorigenesis [16]. In 1985, he roughly classified chemopreventive agents into blocking and suppressing agents based on the time period that agents appear to have activity in animal models of carcinogenesis [17]. Since then, several naturally occurring compounds and synthetic chemicals have been intensively investigated for their chemopreventive ability on chemically induced colon carcinoma. These include the inorganic and organic selenium salts, phenolic antioxidants, non-steroidal anti-inflammatory drugs (NSAIDs), ornithine decarboxylase (ODC) inhibitors, etc. (Table 23.1). Our group also found several natural or synthetic chemopreventive agents against colon carcinogenesis (Table 23.2). Indeed food chemists and natural product scientists have identified hundreds of "phytochemicals" that are being evaluated for the prevention of cancer [18, 19]. Among the non-nutrients dietary components believed to exert a chemopreventive effect are flavonoids, polyphenolic derivatives of benzo(γ)pyrone that are widely distributed in edible plants [20]. There are several major classes of flavonoids, which may occur as glycosides or aglycones. Total dietary intake of flavonoids has been estimated as high as 1 g/day, equivalent to 50,000 ppm in diet [21], although more recent studies have indicated that intake varies widely [22].

Table 23.1: Possible chemopreventive agents against colon carcinogenesis.

Chemopreventive agents	Mechanisms of action
Dietary supplements	
Fiber; wheat bran	Decrease faecal diacylglycerol: decrease PKC activation
Calcium	Bind bile salts; direct antiproliferative effect on cryptal cells
Vitamin D	Normalize differentiation in crypt epithelium
Folic acid	Correct DNA methylation imbalance
Selenium	Antioxidant activity
Allyl sulfides, isothiocyanates, indoles	Induce GST and other detoxifying enzymes
Vitamin C, vitamin E, β-carotene, flavonoids	Scavenging oxygen radicals, preventing DNA damage
Inositol, phytic acid	Modulate transmembrane signaling
Caffeic acid, other plant phenolics	Inhibit nitrosation to form carcinogenic nitrosamines, antioxidants *in vivo*; reduce AA metabolism
ω-3 fatty acids	Reduce AA metabolism, thereby reducing PG activity
Conjugated linoleic acid	Alter membrane phospholipids; decrease cryptal cells proliferation; induce epithelial differentiation
Drugs	
Sulindac and related NSAIDs	Block PG activity; induce apoptosis
Aspirin	Block PG activity
Specific COX-2 inhibitors	Block COX-2 activity
DFMO	Inhibit ODC
N-Acetyl-L-cysteine	Increase DNA repair capability
Oltipraz	Induce GST and other detoxifying enzymes

PKC, protein kinase C; GST, glutathione *S*-transferase; AA, arachidonic acid; NSAIDs, non steroidal anti-inflammatory drugs; PG, prostaglandin; COX-2, cyclooxygenase-2; DFMO, DL-α-difluoromethylornithine.

Table 23.2: Natural and synthetic chemopreventive agents against colon carcinogenesis.

	Agents	Carcinogens	Year
Natural compounds	Chlorogenic acid	MAM acetate	1986
	Magnesium hydroxide	MAM acetate, DMH	1989
	Flavoglaucin	AOM	1991
	Shikonin	AOM	1992
	Gingerol	AOM	1992
	Protocatechuic acid	AOM	1992
	Benzyl isothiocyanate	AOM	1994
	Benzyl thiocyanate	AOM	1994
	Astaxanthin	AOM	1994
	Canthaxanthin	AOM	1995
	Hesperidin	AOM	1995
	Diosmin	AOM	1995
	Costunolide	AOM	1994
	S-Methyl methane	AOM	1995
	Thiosulfonate	AOM	1995
	1′-Acetoxychavicol acetate	AOM	1995

Table 23.2 (continued)

	Agents	Carcinogens	Year
Synthetic compounds	*p*-Methoxybenzeneselenol	AOM	1985
	Disulfiram	AOM	1995
	Indomethacin	1-Hydroxyanthraquinone	1991
	KYN-54	AOM	1992
	DFMO	AOM	1995
	Mofarotene	AOM	1995

MAM acetate, methylazoxymethanol acetate; DMH, dimethylhydrazine; AOM, azoxymethane; KYN-54, 5-hydroxy-4-(2-phenyl-(*E*)-ethenyl)-2(5H)-furanone; DFMO, DL-α-difluoromethylornithine.

Recently, it is proposed that aberrant crypt foci (ACF, Figure 23.1) being present in carcinogen-treated colons of rodent and in the colons of humans with a high risk for colon cancer have been employed to study modulators of colon carcinogenesis (Table 23.3) [23–25], since ACF are possible precursor lesions for colon cancer in rodents [24] and humans [26]. ACF possess several biological aberrations including gene mutations and amplification [24]. ACF also have increased cell proliferation activity compared to surrounding normal crypts [27, 28]. We recently have confirmed their results (Table 23.4). Certain chemopreventive compounds are reported to reduce such hyper-cell proliferation in ACF [29, 30] and to inhibit c-*myc* expression induced by methylazoxymethanol (MAM) acetate [31]. For demonstrating the inhibitory action of compounds in colon carcinogenesis, we have used two experimental animal bio-

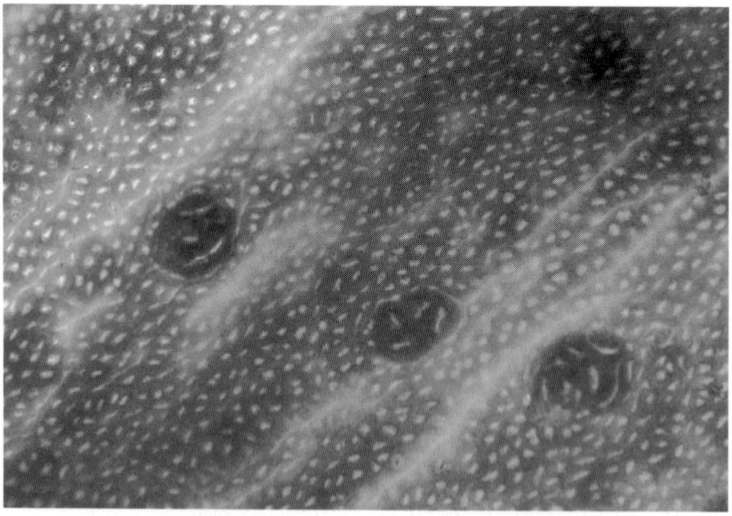

Figure 23.1: Aberrant crypt foci (ACF) induced by azoxymethane (AOM).

Table 23.3: Compounds tested for inhibition of ACF.

Anti-inflammatories/ Analgesics including NSAIDs	Piroxicam, Sulfasalazine, Ibuprofen, Ketoprofen, Indomethacin, COX-2 inhibitors, etc.
Anti-helminthics	Levamisole, Oltipraz, etc.
Organosulfur compounds	Diallyl sulfide, Sodium thiosulfate, Mesna, etc.
Minerals	Sodium selenite, Sodium molybdate, Calcium, etc.
β-Glucosidase inhibitors	Potassium glucarate, Calcium glucarate, β-Sitosterol, etc.
Phenolic antioxidants	Ellagic acid, Rutin, Propyl gallate, Butyl hydroxyanisole, Curcumin, Quercetin, Nordihydroguaiaretic acid, etc.
Indoles/Isothio-compounds	Benzyl isothiocyanate, Indole-3-carbinol, Phenylethyl-isothiocyanate, etc.
Vitamins	Ascorbyl palmitate, Follic acid, Vitamin D_3, etc.
Epicatechins	Catechin, etc.
Differentiation agents	Dehydroepiandrosterone, Sodium butyrate, 18β-Glycyrrhetinic acid, Fluocinolone acetonide, Inositol hexaphosphate, etc.
Others	Silymarin, Arginin, Purpurin, D-Mannitol, Sodium cromolyn, Rebaudioside A, Liquiritin, Phyllodulcin, Hydrangenol, Oleanolic acid, Costunolide, Soyasaponin A_2, etc.

NSAIDs, non-steroidal anti-inflammatory drugs.

Table 23.4: Proliferative activity of colonic pathological lesions induced by AOM in rats.

Lesions	BrdU-labeling index (%)	PCNA-positive nuclei (%)	AgNORs number (/nucleus)
Normal crypts (without AOM)	5.9	18	1.18
Normal appearing crypts (with AOM)	7.9	20	1.68
ACF	18.6	31	2.97
Adenoma	21.1	33	3.07
Adenocarcinoma	28.3	58	3.78

AOM, azoxymethane; BrdU, 5′-bromodeoxyuridine; PCNA, proliferative nuclear antigen; AgNORs, silver-stained nucleolar organizer regions.

assays: (1) a 5-week short-term bioassay of ACF for screening natural compounds, which are present in vegetables and fruits, with possible chemopreventive ability (Figure 23.2) and (2) a long-term rat colon carcinogenesis model for evaluating their inhibitory effects against colon carcinoma (Figure 23.3). In these bioassays, several biochemical and morphologic biomarkers are used (Table 23.5). Cell proliferation plays an important role in multistage carcinogenesis [32–35]. ODC and polyamines are intimately involved in normal cellular proliferation and are likely to play a role in carcinogenesis including colon tumorigenesis [36, 37]. 5′-Bromodeoxyuridine (BrdU)-labelling index, proliferating cell

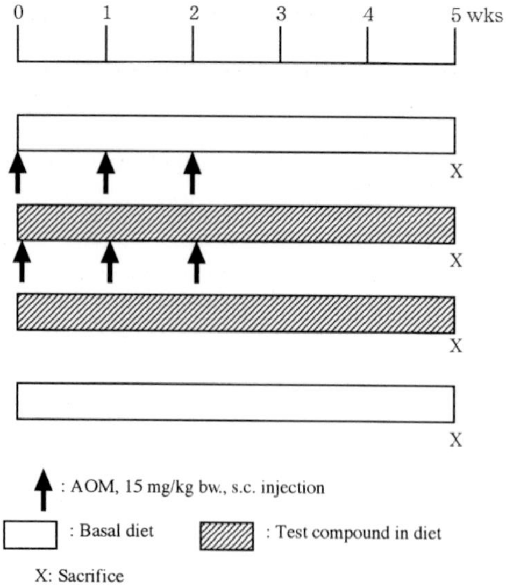

Figure 23.2: Experimental design for screening chemopreventive agents against colon tumorigenesis. At sacrifice, ACF and cell proliferation biomarkers are measured.

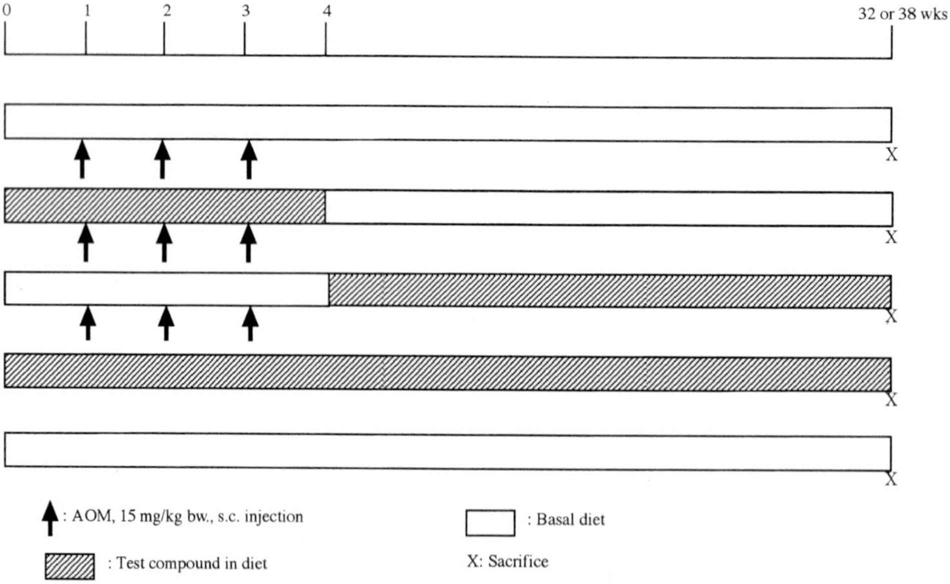

Figure 23.3: Experimental design for detecting chemopreventive agents against colon carcinogenesis. At sacrifice, colonic neoplasms and various cell proliferation biomarkers' expression are measured.

Table 23.5: Biomarkers used for detecting chemopreventive compounds against colon carcinogenesis.

Proliferation biomarkers	BrdU-labelling index, PCNA-labelling index, AgNORs number, etc.
Biochemical biomarkers	ODC activity, Polyamine levels, GST activity, QR activity, MDA, 4-HNE
Histological biomarkers	ACF, Adenoma, Adenocarcinoma

BrdU, 5′-bromodeoxyuridine; PCNA, proliferative nuclear antigen; AgNORs, silver-stained nucleolar regions; ODC, ornithine decarboxylase; GST, glutathione *S*-transferase; QR, quinone reductase; MDA, malondialdehyde; 4-HNE, 4-hydroxy-2(*E*)-nonenal; ACF, aberrant crypt foci.

nuclear antigen (PCNA)-labelling index, and silver-stained nucleolar regions (AgNORs) number are also known to be proliferation biomarkers [38]. Current data suggest that the balance between the Phase I carcinogen-activating enzymes and the Phase II detoxifying enzymes is critical to determining an individual's risk for cancer [39]. Human deficiencies in Phase II enzyme activity, specifically glutathione *S*-transferase (GST), have been identified and associated with increased risk for colon cancer [40]. Therefore, Phase II detoxifying enzymes, such as GST and quinone reductase (QR), might be useful as a biomarker for chemopreventive studies.

The present report will introduce our recent data demonstrating chemopreventive properties of two flavonoids diosmin (DIO, Figure 23.4) and hesperidin (HPD, Figure 23.4) [41] and other antioxidative natural products 1′-acetoxychavicol acetate (ACA, Figure 23.5) [42, 43] and auraptene (AUR, Figure 23.6) [44, 45], which are present in vegetables and fruits, in colon carcinogenesis.

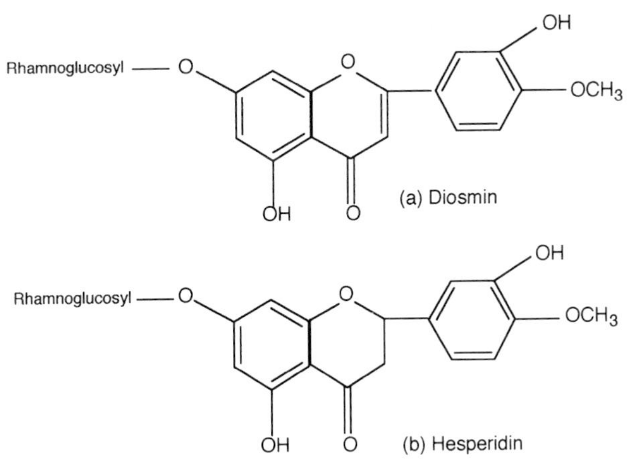

Figure 23.4: Chemical structures of (a) diosmin and (b) hesperidin.

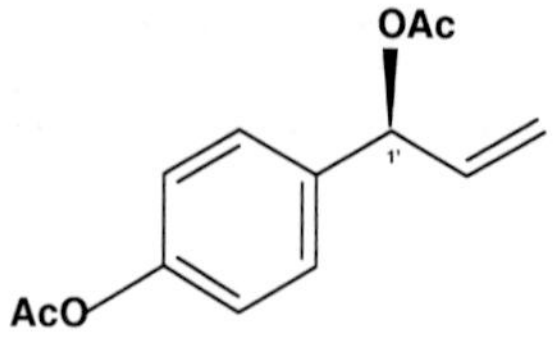

Figure 23.5: Chemical structure of 1′-acetoxychavicol acetate.

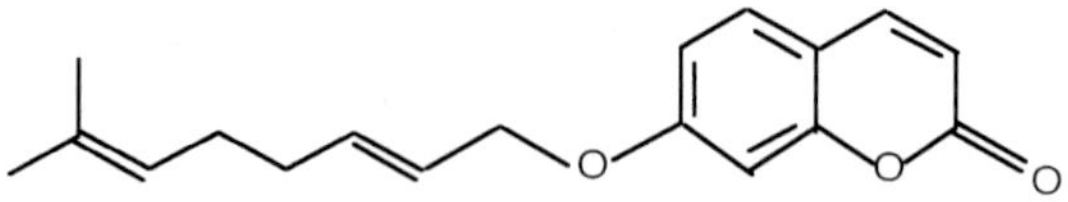

Figure 23.6: Chemical structure of auraptene.

23.3 Screening of possible chemopreventive agents against rat colon tumorigenesis ability using a 5-week short-term bioassay of ACF

As the first bioassay for pilot-studies, we investigated the modifying effects on test compounds DIO, HPD, ACA, and AUR on the development of ACF. ACF could be induced by subcutaneous injections of azoxymethane (AOM, 15 mg/kg body weight, 3-times) and test chemicals in the basal diet at various dose levels were administered to male F344 rat for 5 weeks, starting 1 week before AOM dosing (Figure 23.2). At the end of the study, ACF were counted and expression of several biomarkers was examined. The biomarkers included ornithine decarboxylase (ODC) activity and polyamine level in the colonic mucosa, number of silver-stained nucleolar organizer regions (AgNORs) protein/nucleus in the colonic crypts, and/or activities of GST and QR in the colonic mucosa (Table 23.5).

23.4 Evaluation of chemopreventive ability of selected compounds using a long-term rat colon carcinogenesis model

Based on the results in the pilot studies, the second bioassay for evaluating the chemopreventive effects of compounds, which have been screened by a short-term pilot study, on colon carcinogenesis were conducted. Male F344 rats were given subcutaneously injections of AOM (15 mg/kg body weight, weekly, 3-times) to induce colon cancer (Figure 23.3). For "initiation" feeding, oral administration of these compounds in the diets was begun 1 week before the AOM exposure and continued for 4 weeks, and for "post-initiation" feeding, experimental diets containing test compounds, beginning 1 week after the last dosing of AOM, were given for 28 weeks (DIO and HPD) or for 32 weeks (ACA and AUR). Biomarkers used were as follows: polyamine level activities of ODC, GST, and QR and in the colonic mucosa, number of silver-stained nucleolar organizer regions (AgNORs) protein/nucleus in the colonic crypts, and/or level of aldehydic lipid peroxidation products, malondialdehyde (MDA) and 4-hydroxy-2(*E*)-nonenal (4-HNE), in the colonic mucosa (Table 23.5).

23.5 Chemoprevention of rat colon carcinogenesis by DIO and HPD

Two flanonoids DIO and HES are present in citrus fruits: 0.036 mg DIO and 34.707 mg HPD/g fresh mass of *Citrus unshu* Marc [46]. These have several biological activities including antioxidant property, anti-inflammatory effect, and inhibition of prostaglandin (PG) synthesis [41]. Since alteration of PGs biosynthesis could modulate colon carcinogenesis [47], these compounds were suspected to affect colon tumorigenesis.

In the pilot study using ACF enumeration, male F344 rats received s.c. injections of AOM (15 mg/kg bw) once a week for 3 weeks. They were also fed the diets contained 0.1% DIO, 0.1% HPD, and 0.09% DIO plus 0.01% HPD, respectively, for 5 weeks, starting 1 week before the first injection of AOM. Two compounds, both alone (0.1% DIO or 0.09% HPD in diet) or in combination (0.1% DIO plus 0.09% HPD in diet), significantly inhibited the occurrence of ACF induced by AOM (44%–73% reduction, Figure 23.7). The combination regimen inhibited all sizes of ACF. As shown in Figure 23.8, dietary administration of these compounds suppressed colonic mucosal ODC activity. In the subsequent long-

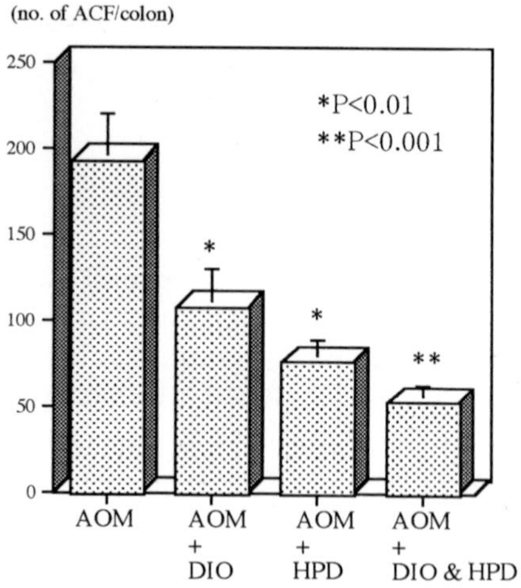

Figure 23.7: Incidence of ACF in a pilot study in rats treated with AOM plus DIO (0.1%), HPD (0.1%), or DIO (0.09%) + HPD (0.01%).

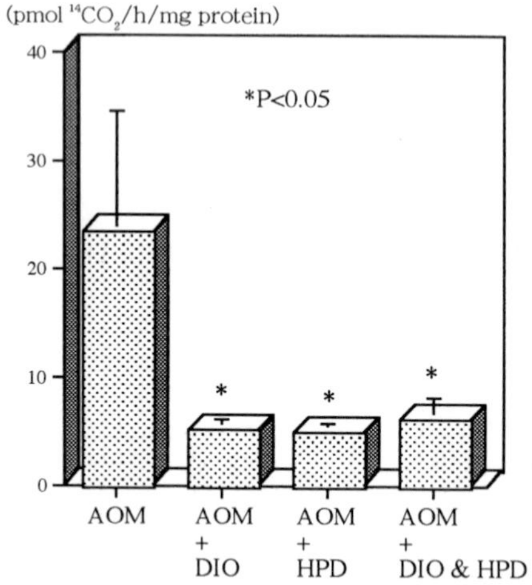

Figure 23.8: ODC activity of colonic mucosa in a pilot study. Rats were treated with AOM plus DIO (0.1%), HPD (0.1%), or DIO (0.09%) + HPD (0.01%).

term bioassay, "initiation" (4 weeks) or "post-initiation" feeding (28 weeks) of two test compounds (both alone and in combination) effectively reduced the incidence and multiplicity of colonic adenocarcinoma induced by AOM (Figures 23.9 and 23.10). Expression of biomarkers for cell proliferation, such as BrdU-labelling index and ODC activity in the colonic mucosa was also suppressed by these treatment (Figure 23.11). These data indicate that dietary administration of two flavonoids DIO and HPD, both alone and in combination, during either the initiation

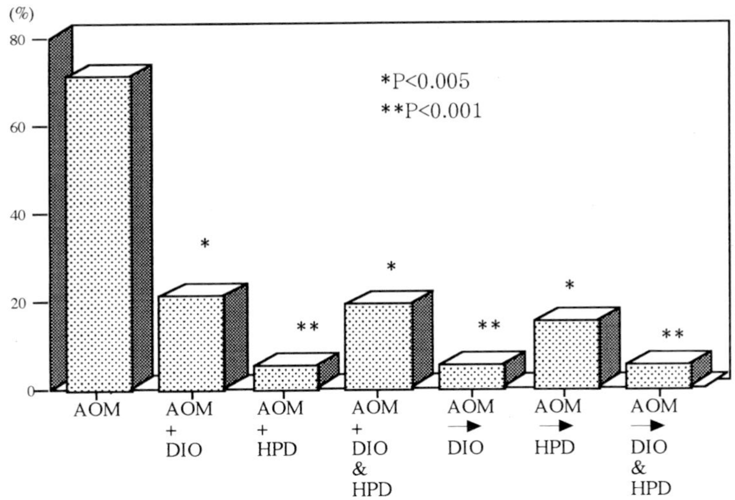

Figure 23.9: Incidence of colonic adenocarcinoma in rats. Rats were treated with AOM plus DIO (0.1%), HPD (0.1%), or DIO (0.09%) + HPD (0.01%).

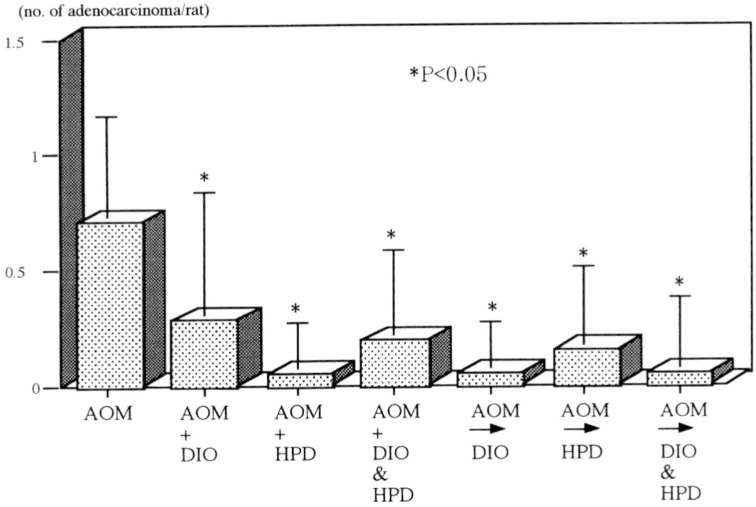

Figure 23.10: Multiplicity of colonic adenocarcinoma in rats treated with AOM plus DIO (0.1%), HPD (0.1%), or DIO (0.09%) + HPD (0.01%).

or post-initiation phase, significantly inhibited AOM-induced colon carcinogenesis. Such effects may be partly due to suppression of cell proliferation in the colonic crypts. Other mechanisms of action, such as inhibition of PGs biosynthesis, induction of Phase II enzymes [48], and antioxidant property, are also considered. Recently, Ciolino *et al.* [49] have suggested that chemopreventive effect of dios-

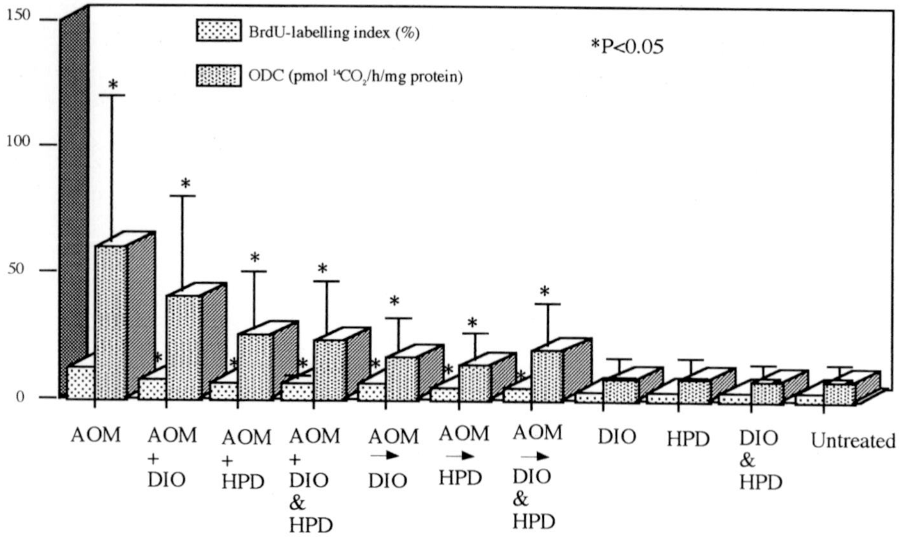

Figure 23.11: BrdU-labelling index and ODC activity in the colonic mucosa of rats treated with AOM and/or DIO (0.1% in diet), HPD (0.1% in diet), DIO (0.09% in diet) + HPD (0.01% in diet).

min may be due to the potent inhibitory activity of diosmetin on CYP 1A1 enzyme activity. The combined regimen (0.09% DIO plus 0.01% HPD) used in the study is based on the constituents of the drug "Daflon", which is used for the treatment of venous insufficiency in Europe [50]. Given the considerable interest in this drug as a possible chemopreventive agent, it would be intriguing to know whether chronic use of "Daflon" could reduce the risk for colon cancer in patients with venous insufficiency. The tumor data in the study, however, did not reflect any beneficial effect from DIO and HPD administered together as opposed to when these compounds were given individually.

23.6 Chemoprevention of rat colon carcinogenesis by ACA

ACA is present in seeds or a rhizome of *Languas galanga* (Zingiberaceae), used as a ginger substitute and a stomachic medicine in Thailand. The compound has been reported to suppress tumor promoter-induced Epstein-Barr virus activation *in vitro* [51]. ACA is known to reduce superoxide anion production by inhibiting the xanthine oxidase and NADPH oxidase system [52] and this activity has been suggested to be partly responsible for its cancer chemopreventive effects [53].

In the pilot study, male F344 rats were given three weekly s. c. injections of AOM (15 mg/kg bw) and fed the diet containing 0.01% or 0.02% ACA for 5 weeks, starting 1 week before the first dosing of AOM. Dietary feeding of ACA at the both dose levels caused significant reduction in the frequency of ACF (41% inhibition by 0.1% ACA feeding and 37% inhibition by 0.02% ACA feeding) (Figure 23.12). Feeding of ACA also suppressed expression of cell proliferation biomarkers, such as colonic mucosal ODC activity (Figure 23.13),

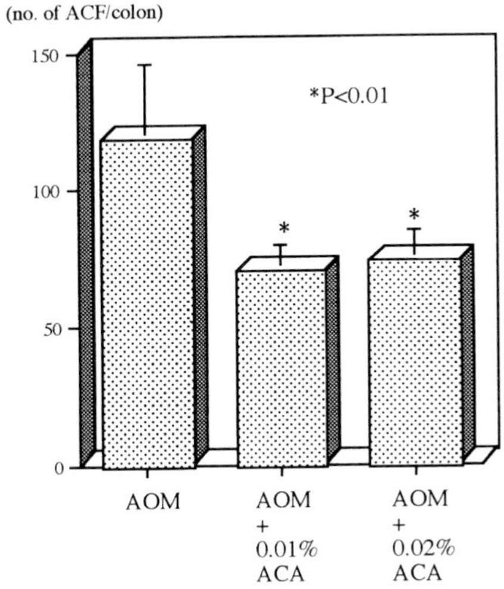

Figure 23.12: Incidence of ACF in a pilot study in rats treated with AOM plus ACA (0.01% or 0.02% in diet).

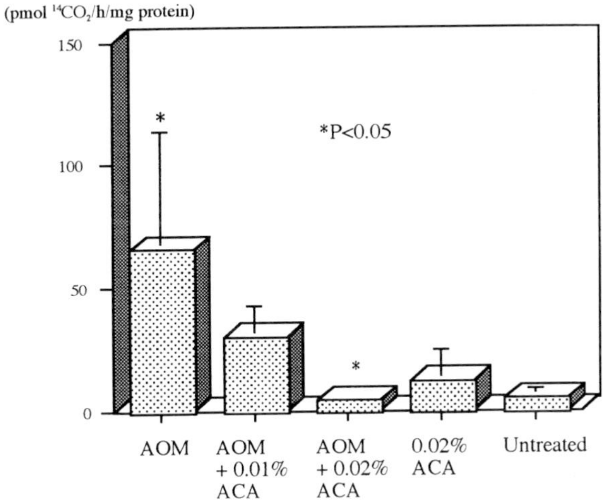

Figure 23.13: Colonic mucosal ODC activity in a pilot study. Rats were treated with AOM plus ACA (0.01% or 0.02% in diet), ACA (0.02% in diet), or nothing.

AgNORs number, and blood polyamine level. Subsequent long-term study for evaluating the chemopreventive ability of ACA when fed at dose levels of 0.01% and 0.05% during the initiation (4 weeks) or post-initiation phase (34 weeks) demonstrated dose-dependent inhibition in the incidence (Figure 23.14) and multiplicity (Figure 23.15) of colonic adenocarcinoma induced by AOM. ACA feeding resulted in low activity of ODC (Figure 23.16) and poly-

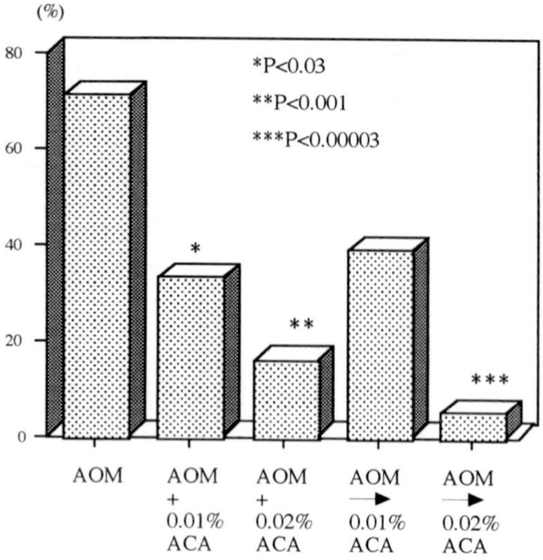

Figure 23.14: Incidence of colonic adenocarcinoma in rats treated with AOM plus ACA (0.01% or 0.02% in diet).

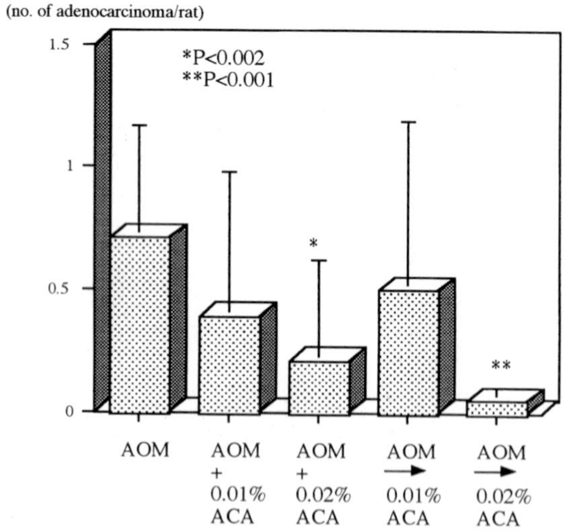

Figure 23.15: Multiplicity of colonic adenocarcinoma in rats treated with AOM plus ACA (0.01% or 0.02% in diet).

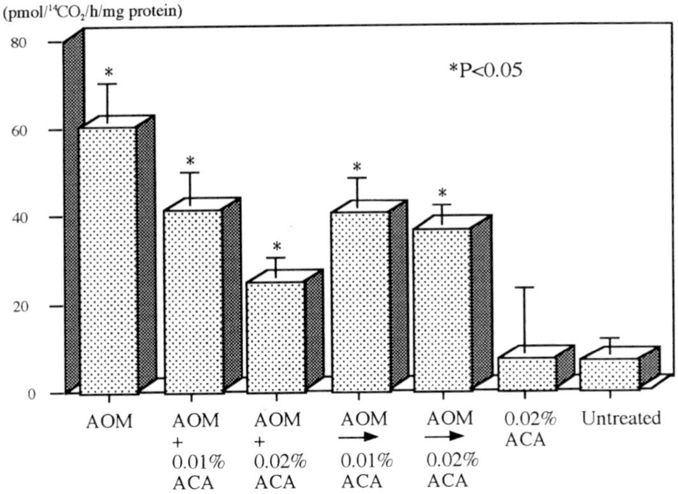

Figure 23.16: Colonic mucosal ODC activity. Rats were treated with AOM and/or ACA (0.01 % or 0.02 % in diet).

amine content in the colonic mucosa. In addition, GST (Figure 23.17 a) and QR activities (Figure 23.17 b) in the liver and colon were significantly elevated in rats gavaged with ACA. These findings suggest possible chemopreventive ability of ACA against colon tumorigenesis and the effect may be due to its suppression of cell proliferation in the colonic mucosa and its induction of detoxifying enzymes GST and QR. More recently, ACA has been reported to inhibit ni-

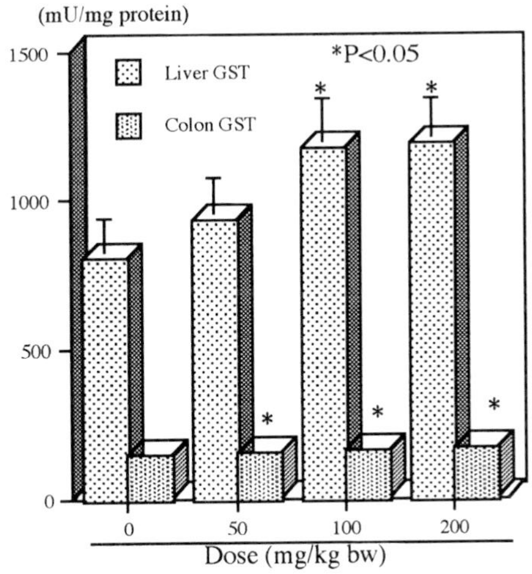

Figure 23.17 a: GST activities in the liver and colon of rats treated with ACA (0, 50, 100 or 200 mg/kg bw). Rats were gavaged with ACA for 5 days and GST activities were measured 30 min after the last administration.

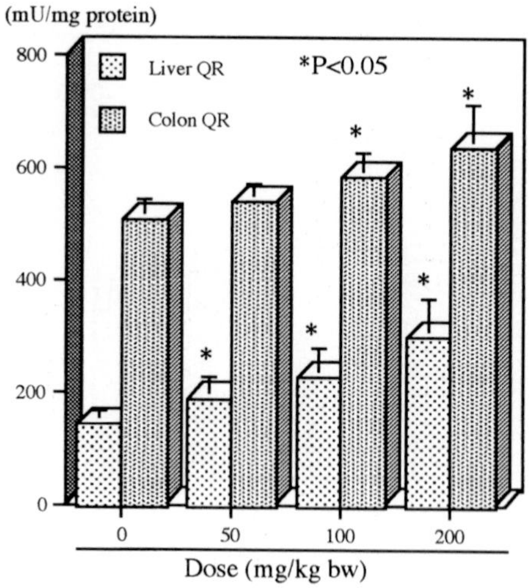

Figure 23.17b: QR activities in the liver and colon of rats treated with ACA (0, 50, 100 or 200 mg/kg bw). Rats were gavaged with ACA for 5 days and QR activities were measured 30 min after the last administration.

tric oxide (NO) production, apparently mediated by modulation of several transcription factors [54]. Since excessive production of NO at inflammatory sites is causatively involved in the process of multistage carcinogenesis [55], such activity also contributes to the anti-carcinogenic properties of ACA.

23.7 Chemoprevention of rat colon carcinogenesis by AUR

A known coumarin, AUR, is present in certain orange peels: 0.04 % in *Citrus natsudaidai* Hayata, 0.01–0.02 % in grapefruit and 100 μg/100 ml in grapefruit juice. Antiplatelet action of this compound has been reported [56], but the other biological properties are not known. Recently, antitumor promoting effect of AUR on mouse skin carcinogenesis has been found and AUR could suppress superoxide generation induced by 12-*O*-tetradecanoylphorbol-13-acetate [57]. Therefore, possible chemopreventive effect of AUR on colon carcinogenesis was examined in rats.

In the pilot study, male F344 rats were given three weekly s. c. injections of AOM (15 mg/kg bw) and fed the diet containing 0.01% or 0.05% AUR for 5 weeks, starting 1 week before the first exposure of AOM. Dietary feeding of AUR caused a significant reduction in the frequency of ACF in a dose-dependent manner (Figure 23.18). AUR at the both dose levels also suppressed expression of cell proliferation biomarkers, such as colonic mucosal ODC activity (Fig-

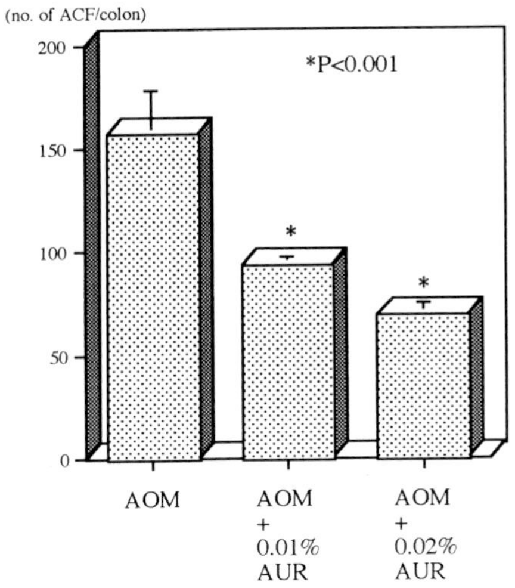

Figure 23.18: Incidence of ACF in a pilot study in rats treated wit AOM plus AUR (0.01% or 0.02% in diet).

ure 23.19), AgNORs number (Figure 23.20), BrdU-labelling index, and polyamine content. Subsequent long-term experiment for evaluating the chemopreventive efficacy of AUR was conducted using male F344 rats. They received AOM and were fed at dose levels of 0.01% and 0.05% during the initiation (4 weeks) or post-initiation phase (34 weeks). At the termination of the study, dietary AUR caused dose-dependent inhibition in the incidence (Figure 23.21) and multiplicity

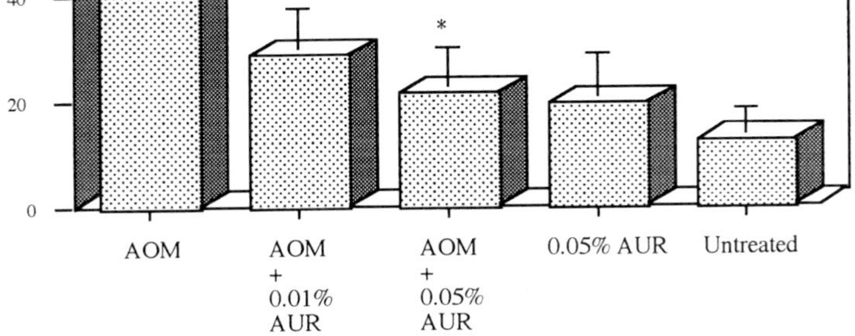

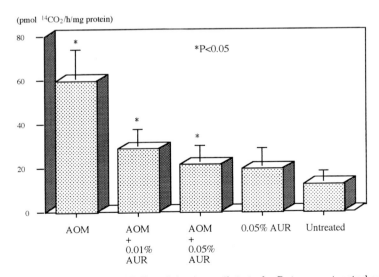

Figure 23.19: Colonic mucosal ODC activity in a pilot study. Rats were treated with AOM plus AUR (0.01% or 0.05% in diet), AUR (0.05% in diet), or nothing.

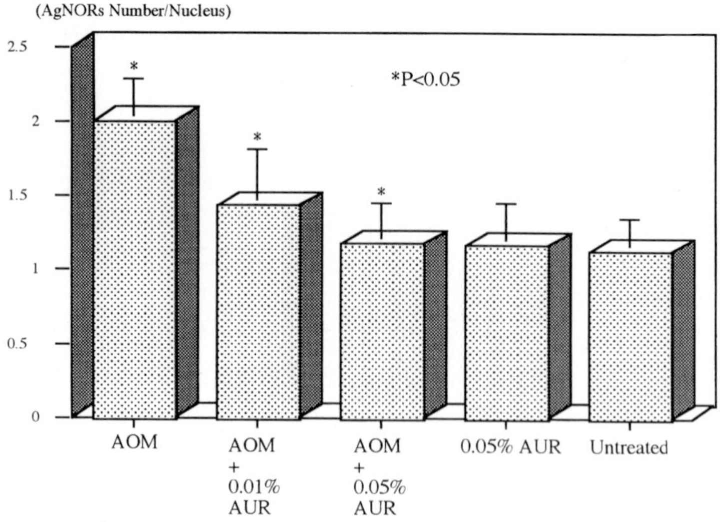

Figure 23.20: Number of AgNORs in colonic cryptal cells of rats treated with AOM plus AUR (0.01% or 0.05% in diet), AUR (0.05% in diet), or nothing.

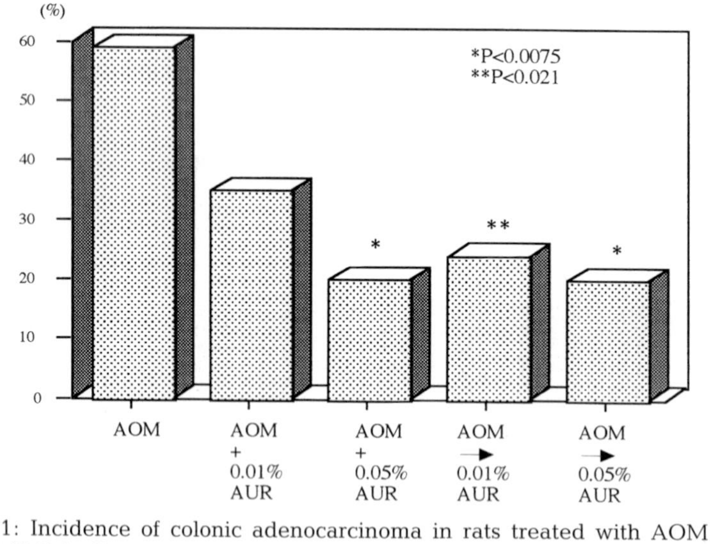

Figure 23.21: Incidence of colonic adenocarcinoma in rats treated with AOM plus AUR (0.01% or 0.05% in diet).

(Figure 23.22) of AOM-induced colonic adenocarcinoma. Dietary administration of AUR also suppressed the expression of cell proliferation biomarkers, such as ODC (Figure 23.23) and polyamine content in the colonic mucosa. In addition, gavage of AUR increased GST (Figure 23.24 a) and QR activities (Figure 23.24 b) in the liver and colon. AUR feeding could reduce the amounts of MDA and 4-HNE (Figure 23.25). Increased levels of the products of lipid peroxidation, including

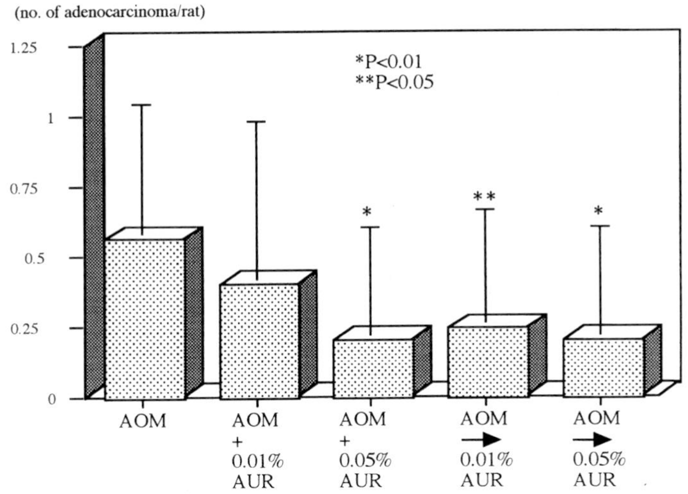

Figure 23.22: Multiplicity of colonic adenocarcinoma in rats treated with AOM plus AUR (0.01% or 0.05% in diet).

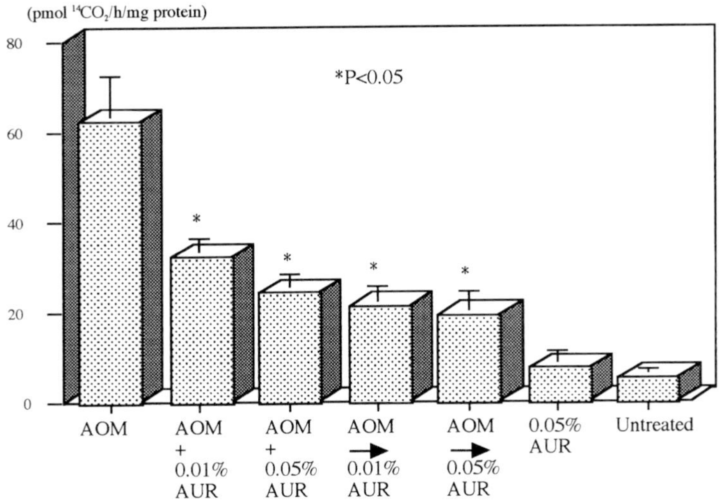

Figure 23.23: Colonic mucosal ODC activity. Rats were treated with AOM and/or AUR (0.01% or 0.05% in diet).

MDA and 4-HNA, were found in colon carcinogenesis [58, 59]. These findings suggested that the chemopreventive effects of AUR on AOM-induced colon tumorigenesis at the initiation level might be associated, in part, with increased activity of Phase II enzymes, and those at the post-initiation stage might be related to suppression of cell proliferation and lipid peroxidation in the colonic mucosa.

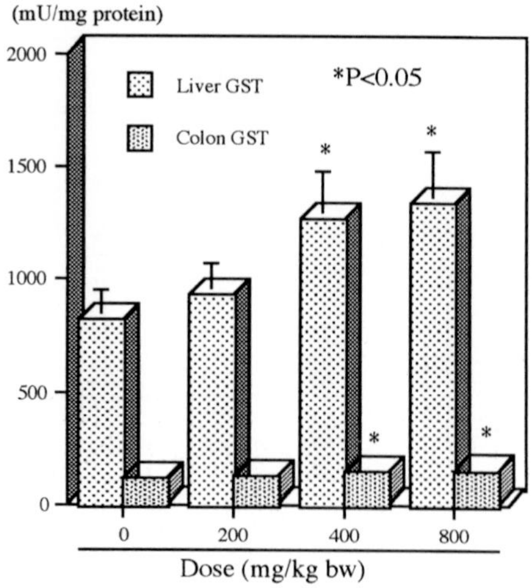

Figure 23.24 a: GST activities in the liver and colon of rats treated with AUR (0, 200, 400 or 800 mg/kg bw). Rats were gavaged with ACA for 5 days and GST activities were measured 30 min after the last administration.

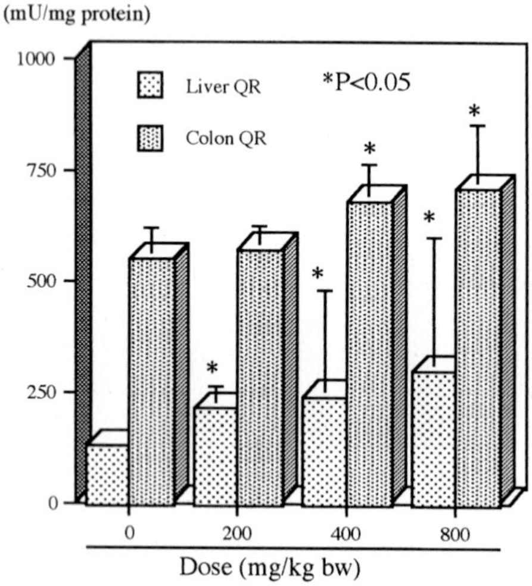

Figure 23.24 b: QR activities in the liver and colon of rats treated with AUR (0, 200, 400 or 800 mg/kg bw). Rats were gavaged with ACA for 5 days and QR activities were measured 30 min after the last administration.

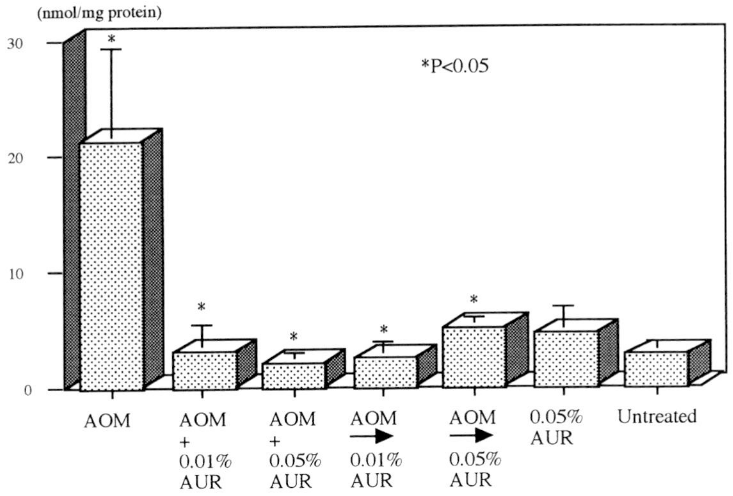

Figure 23.25: Amounts of MDA plus 4-HNE in the colonic mucosa of rats treated with AOM and/or AUR (0.01% or 0.05% in diet).

23.8 Discussion

Our recent data on the chemopreventive effects of naturally occurring compounds, DIO, HPD, ACA, and AUR, present in certain vegetables and fruits against AOM-induced colon tumorigenesis are described. All these compounds are antioxidants. In general, plants are complicated mixtures of numerous chemicals, and interactions with their components may affect the effectiveness of the antioxidant. However, the metabolic pathway and action of naturally occurring antioxidative compounds is not clear. Flavonoids compounds, which are widely distributed in the plant kingdom and occur in considerable quantities, show a wide range of pharmacological activities other than their antioxidative properties. These compounds have been used to treat various pathological conditions including allergies, inflammation, and diabetes. Experimental data including this report showing their antitumor activities is accumulating; their chemopreventive potential, however, has not been fully proven clinically. Their behavior and fate should be investigated *in vivo*.

As reported, commonly consumed foods contain non-nutritive phytochemicals capable to inhibit colon cancer in an animal model. The diet provides a rich abundance of these compounds which have the ability to intervene in all phases of carcinogenesis. Mechanisms of action include (i) inhibition of Phase I enzymes, (ii) induction of Phase II enzymes, (iii) scavenging DNA reactive agents, (iv) suppression of hyper-cell proliferation induced by carcinogen, and/ or (v) inhibition of certain properties of neoplastic cells. Each of these mechanisms have been studied in isolation. For explanation of reduced risk for cancer in populations with a greater reliance on fruits and vegetables in the daily diet, future research should focus on potential combinations of foods and the protective components within them.

The association of certain malignancies with chronic inflammation has been reconized for many years [60]. The link between inflammation and subsequent malignancy in visceral sites is known. Examples include large bowel cancer after ulcerative colitis or Crohn's disease [60, 61]. Evidence is now accumulating that excessive production of NO is a causative factor for carcinogenesis [55]. Therefore, it is suggested that inhibition of excessive NO production by inflammatory cells could be beneficial in prevention of carcinogenesis. ACA is able to suppress NO production through inhibition of induction of inducible NO synthase (iNOS) gene transcription. From the evidence mentioned above, our search for chemopreventives against colon cancer focuses on several flavonoids and some other compounds possessing certain biological activities including anti-inflammatory and/or antioxidative properties present in foods. Approximately 2,000 individual members of the flavonoid class have been described and the flavonoids are consumed in rather large amounts through dietary vegetables and fruits.

Recently, up-regulation of COX-2, but not COX-1, gene expression was reported in human colorectal neoplasms [62]. New drugs, specific for inhibition of COX-2, may provide effective tumor prevention with reduced side effects [63–65]. The elevation of COX-2 expression can protect intestinal epithelial cells from apoptosis [66]. Certain COX-2 inhibitors can induce apoptosis [67] and inhibit tumor angiogenesis [68]. Recently, synthetic antioxidants have reported to reduce COX-2 expression, PG production, and cell proliferation of colorectal cancer cells [69]. This may suggest that COX-2 may provide a new chemopreventive target in colorectal malignancies [70], if there are the natural products being a specific inhibitor of COX-2 expression in edible plants.

In conclusion, certain flavonoids and other substances with biological activity including antioxidative and/or anti-inflammatory properties, which are present in vegetables and fruits, could exert chemopreventive action in rat colon carcinogenesis as described. However, more work needs to be done to better understand the underlying mechanism(s) of action and to confirm their safety for use in humans. Plants are complex mixtures of chemicals. The potential for finding new chemopreventive agents in plants is high. Studies are underway to identify new agents in edible plants with chemopreventive potential. For screening chemopreventive agents based on different mechanisms, new *in vitro* co-culture model might be useful [71]. The effects of these agents on colon carcinogenesis should be carefully studied to assist the discovery and development of new chemopreventive agents, and to understand carcinogenesis mechanisms.

Acknowledgments

Research in our laboratory is supported by a Grant-in-Aid for the Second Term Comprehensive 10-year Strategy for Cancer Control from the Ministry of Health and Welfare in Japan; a Grant-in-Aid for Cancer Research from the Ministry of Health and Welfare in Japan; a Grant-in-Aid from the Ministry of Education, Science, Sports, and Culture of Japan; a grant from the Japanese Research and Development Association for New Food Creation; a grant from the Program for Promotion of Basic Research Activities for Innovative Biosciences; and Grant C98-1 for Collaborative Research from Kanazawa Medical University.

References

[1] Slaga, T. J. (Ed.) (1989) *Modifiers of chemical carcinogenesis: An approach to the biochemical mechanism and cancer prevention – A comprehensive survey.* New York, Raven Press, p. 1–275.
[2] Weisburger, J. H. (1991) Causes, relevant mechanisms, and prevention of large bowel cancer. *Semin. Oncol.* **18**, 316–336.
[3] Reddy, B. S. (1986) Diet and colon cancer: evidence from human and animal model studies. In: Reddy, B. S.; Cohen, L. A. (Eds.) *Diet, nutrition and cancer: A critical evaluation*, CRC Press, Boca Raton, FL, p. 47–65.
[4] Steinmetz, K. A.; Potter, J. D. (1991) Vegetable, fruits, and cancer. I. Epidemiology. *Cancer Causes Control* **2**, 325–357.
[5] Steinmetz, K. A.; Potter, J. D. (1991) Vegetable, fruits, and cancer. II. Mechanisms. *Cancer Causes Control* **2**, 427–442.
[6] Block, G.; Patterson, B.; Subar, A. (1992) Fruit, vegetables, and cancer prevention: a review of epidemiological evidence. *Nutr. Cancer* **18**, 1–29.
[7] Reddy, B. S. (1993) Nutritional factors as modulators of colon carcinogenesis. In: Park, D. V.; Ioannides, C.; Walkers, R. (Eds.) *Food, nutrition and chemical toxicity*, Smith-Gordon/Nishimura, London/Niigata, p. 325–336.
[8] Tanaka, T. (1997) Effect of diet on human carcinogenesis. *Crit. Rev. Oncol./Hematol.* **25**, 73–95.
[9] Steinmetz, K. A.; Potter, J. D. (1996) Vegetables, fruit, and cancer prevention: a review. *J. Am. Diet. Assoc.* **96**, 1027–1039.
[10] Hebert, J. R.; Landon, J.; Miller, D. R. (1993) Consumption of meat and fruit in relation to oral and esophageal cancer: a cross-national study. *Nutr. Cancer* **19**, 169–179.
[11] Micozzi, M. S. (1989) Foods, micronutrients, and reduction of human cancer. In: Moon, T. E.; Micozzi, M. S. (Eds.) *Nutrition and cancer prevention: Investigating the role of micronutrients*, Marcel Dekker, Inc., New York, p. 213–242.
[12] Tanaka, T.; Makita, H.; Ohnishi, M.; Mori, H.; Satoh, K.; Hara, A. (1995) Chemoprevention of rat oral carcinogenesis by naturally occuring xanthophylls, astaxanthin and canthaxanthin. *Cancer Res.* **55**, 4059–4064.

[13] Tanaka, T.; Kawamori, T.; Ohnishi, M.; Makita, H.; Mori, H.; Satoh, K.; Hara, A. (1995) Suppression of azoxymethane-induced rat colon carcinogenesis by dietary administration of naturally occurring xanthophylls astaxanthin and canthaxanthin during the post-initiation phase. *Carcinogenesis* **16**, 2957–2963.
[14] Weisburger, J. H.; Reddy, B. S.; Rose, D. P.; Cohen, L. A.; Kendall, M. E., Wynder, E. L. (1993) Protective mechanisms of dietary fibers in nutritional carcinogenesis. *Basic life Sci.* **61**, 45–64.
[15] Angres, G.; Beth, M. (1991) Effects of dietary constituents on carcinogenesis in different tumor models. An overview from 1975 to 1988. In: Alfin-Slater, R. B.; Kritchesvskys, D. (Eds.) *Cancer and nutrition.* Plenum Press, New York, p. 337–485.
[16] Wattenberg, L. W. (1983) Inhibition of neoplasia by minor dietary constituents. *Cancer Res.* **43**, 2448s–2453s.
[17] Wattenberg, L. W. (1985) Chemoprevention of cancer. *Cancer Res.* **45**, 1–8.
[18] Am. Inst. Cancer Res. (Ed.) (1996) *Dietary phytochemicals in cancer prevention and treatment*, Plenum Press, New York, p. 1–340.
[19] Huang, M.-T.; Osawa, T.; Ho, C.-T.; Rosen, R. T. (Ed.) (1994) *Food phytochemicals for cancer prevention I and II. Fruits and vegetables.* Am. Chem. Soc., Washington, DC.
[20] Formica, J. V.; Regelson, W. (1995) Review of the biology of quercetin and related bioflavonoids. *Food Chem. Toxicol.* **33**, 1061–1080.
[21] Pierpoint, W. S. (1986) Flavonoids in the human diet. *Prog. Clin. Biol. Res.* **213**, 125–140.
[22] Hertog, M. G.; Kromhout, D.; Aravanis, C.; Blackburn, H.; Buzina, R.; Fidanza, F.; Giampaoli, S.; Jansen, A.; Menotti, A.; Nedeljkovic, S.; Pekkarienen, M.; Simic, B. S.; Toshima, H.; Feskens, E. J. M.; Hollman, P. C. H.; Katan, M. B. (1995) Flavonoid intake and long-term risk of coronary heart disease and cancer in the seven countries study. *Arch. Int. Med.* **155**, 381–386.
[23] Pereira, M. A., Barnes, L. H.; Rassman, V. L.; Kelloff, G. V.; Steele, V. E. (1994) Use of azoxymethane-induced foci of aberrant crypts in rat colon to identify potential cancer chemopreventive agents. *Carcinogenesis* **15**, 1049–1054.
[24] Bird, R. P. (1995) Role of aberrant crypt foci in understanding the pathogenesis of colon cancer. *Cancer Lett.* **93**, 55–71.
[25] Kawamori, T.; Tanaka, T.; Hara, A.; Yamahara, J.; Mori, H. (1995) Modifying effects of naturally occurring products on the development of colonic aberrant crypt foci induced by azoxymethane in F344 rats. *Cancer Res.* **55**, 1277–1282.
[26] Pretlow, T. P.; Barrow, B. J.; Ashton, W. S.; O'Riordan, M. A.; Pretlow, T. G.; Jurcisek, J. A.; Stellato, T. A. (1991) Aberrant crypts: putative preneoplastic foci in human colonic mucosa. *Cancer Res.* **51**, 1564–1567.
[27] Yamashita, N.; Minamoto, T.; Onda, M.; Esumi, H. (1994) Increased cell proliferation of azoxymethane-induced aberrant crypt foci of rat colon. *Jpn. J. Cancer Res.* **85**, 692–698.
[28] Pretlow, T. P.; Cheter, C.; O'Riordan, M. A. (1994) Aberrant crypt foci and colon tumors in F344 rats have similar increases in proliferative activity. *Int. J. Cancer* **56**, 599–602.
[29] Zheng, Y.; Kramer, P. M.; Olson, G.; Lubet, R. A.; Steele, V.; Kelloff, G. J.; Pereira, M. A. (1997) Prevention by retinoids of azoxymethane-induced tumors and aberrant crypt foci and their modulation of cell proliferation in the colon of rats. *Carcinogenesis* **18**, 2119–2125.
[30] Li, H.; Kramer, P. M.; Lubet, R. A.; Steele, V. E.; Kelloff, G. J.; Pereira, M. A. (1998) Effect of calcium on azoxymethane-induced aberrant crypt foci and cell proliferation in the colon of rats. *Cancer Lett.* **124**, 39–46.
[31] Wang, A.; Yoshimi, N.; Tanaka, T.; Mori, H. (1993) Inhibitory effects of magnesium hydroxide on c-*myc* expression and cell proliferation induced by methylazoxymethanol acetate in rat colon. *Cancer Lett.* **75**, 73–78.
[32] Tanaka, T. (1992) Cancer chemoprevention. *Cancer J.* **5**, 11–16.

[33] Cohen, S. M.; Ellwein, L. B. (1990) Cell proliferation and carcinogenesis. *Science* **249**, 1007–1011.

[34] Pegg, a. E. (1988) Polyamine metabolism and its importance in neoplastic growth as a target for chemotherapy. *Cancer Res.* **48**, 759–774.

[35] Lipkin, M. (1991) Intermediate biomarkers and studies of cancer prevention in the gastrointestinal tract. *Prog. Clin. Biol. Res.* **369**, 397–406.

[36] Luk, G. D.; Hamilton, S. R.; Yang, P.; Smith, J. A.; O'Ceallaigh, D.; McAvinchey, D.; Hyland, J. (1986) Kinetic changes in mucosal ornithine decarboxylase activity during azoxymethane-induced colonic carcinogenesis in the rat. *Cancer Res.* **46**, 4449–4452.

[37] LaMuraglia, G. M.; Lacaine, F.; Malt, R. A. (1986) High ornithine decarboxylase activity and polyamine levels in human colorectal neoplasia. *Ann. Surg.* **204**, 89–93.

[38] Tanaka, T. (1997) Chemoprevention of human cancer: biology and therapy. *Crit. Rev. Oncol./Hematol.* **25**, 139–174.

[39] Wilkinson, J. I.; Clapper, M. L. (1997) Detoxication enzymes and chemoprevention. *Proc. Soc. Exp. Biol. Med.* **216**, 192–200.

[40] Szarka, C. E.; Pfeiffer, G. R.; Hum, S. T.; Everley, L. C.; Balshem, A. M.; Moore, D. F.; Litwin, S.; Goosenberg, E. B.; Frucht, H.; Engstrom, P. F.; Clapper, M. L. (1995) Glutathione *S*-transferase activity and glutathione *S*-transferase μ expression in subjects with risk for colorectal cancer. *Cancer Res.* **55**, 2789–2793.

[41] Tanaka, T.; Makita, H.; Kawabata, K.; Mori, H.; Kakumoto, M.; Satoh, K.; Hara, A.; Sumida, T.; Tanaka, T.; Ogawa, H. (1997) Chemoprevention of azoxymethane-induced rat colon carcinogenesis by the naturally occurring flavonoids, diosmin and hesperidin. *Carcinogenesis* **18**, 857–956.

[42] Tanaka, T.; Makita, H.; Kawamori, T.; Kawabata, K.; Mori, H.; Murakami, A.; Satoh, K.; Hara, A.; Ohigashi, H.; Koshimizu, K. (1997) A xanthine oxidase inhibitor 1'-acetoxychavicol acetate inhibits azoxymethane-induced colonic aberrant crypt foci. *Carcinogenesis* **18**, 1113–1118.

[43] Tanaka, T.; Kawabata, K.; Kakumoto, M.; Makita, H.; Matsunaga, K.; Mori, H.; Satoh, K.; Hara, A.; Murakami, A.; Koshimizu, K.; Ohigashi, H. (1997) Chemoprevention of azoxymethane-induced rat colon carcinogenesis by a xanthine oxidase inhibitor, 1'-acetoxychavicol acetate. *Jpn. J. Cancer Res.* **88**, 821–830.

[44] Tanaka, T.; Kawabata, K.; Kakumoto, M.; Makita, H.; Hara, A.; Mori, H.; Satoh, K.; Hara, A.; Murakami, A.; Kuki, W.; Takahashi, Y.; Yonei, H.; Koshimizu, K.; Ohigashi, H. (1997) *Citrus* auraptene inhibits chemically induced colonic aberrant crypt foci in male F344 rats. *Carcinogenesis* **18**, 2155–2161.

[45] Tanaka, T.; Kawabata, K.; Kakumoto, M.; Hara, A.; Murakami, A.; Kuki, W.; Takahashi, Y.; Yonei, H.; Maeda, M.; Ota, T.; Odashima, S.; Yamane, T.; Koshimizu, K.; Ohigashi, H. (1998) *Citrus* auraptene exerts dose-dependent chemopreventive activity in rat large bowel tumorigenesis: the inhibition correlates with suppression of cell proliferation and lipid peroxidation and with induction of phase II drug-metabolizing enzymes. *Cancer Res.* **58**, 2550–2556.

[46] Nogata, Y.; Ohta, H.; Yoza, K.; Berhow, M.; Hasegawa, S. (1994) High-performance liquid chromatographic determination of naturally occurring flavonoids in *Citrus* with a photodiode-ray detector. *J. Chromatogr.* **667**, 59–66.

[47] Reddy, B. S. (1992) Inhibitors of the arachidonic acid cascade and their chemoprevention of colon carcinogenesis. In: Wattenberg, L.; Lipkin, M.; Boone, C. W.; Kelloff, G. J. (Eds.) *Cancer chemoprevention*, CRC Press, Boca Raton, FL, p. 153–164.

[48] Boutin, J. A.; Meunier, F.; Lambert, P.-H.; Hennig, P.; Bertin, D.; Serkiz, B.; Volland, J.-P. (1993) *In vivo* and *in vitro* glucuronidation of the flavonoid diosmetin in rats. *Drug Metab. Dispos.* **21**, 1157–1166.

[49] Ciolino, H. P.; Wang, T. T. Y.; Yeh, G. C. (1998) Diomin and diosmetin are agonists of the aryl hydrocarbon receptor that differentially affect cytochrome P450 1A1 activity. *Cancer Res.* **58**, 2754–2760.

[50] Labrid, C. (1994) Pharmacologic properties of Daflon 500 mg. *Angiology* **45**, 524–530.
[51] Kondo, A.; Ohigashi, H.; Murakami, A.; Suratwadee, J.; Koshimizu, K. (1993) 1′-Acetoxychavicol acetate as a potent inhibitor of tumor promoter-induced Epstein-Barr virus activation from *Languas galanga*, a traditional Thai condiment. *Biosci. Biotechnol. Biochem.* **57**, 1344–1345.
[52] Noro, T.; Sekiya, T.; Katoh (nee Abe), M.; Oda, Y.; Miyase, T.; Kuroyanagi, M.; Ueno, A.; Fukushima, S. (1988) Inhibitors of xanthine oxidase from *Alpinia galanga*. *Chem. Pharm. Bull.* **36**, 244–248.
[53] Pence, C. B.; Reiners, J. J. J. (1987) Murine epidermal xanthine oxidase activity: correlation with degree of hyperplasia induced by tumor promoters. *Cancer Res.* **47**, 6388–6392.
[54] Ohta, T.; Fukuda, K.; Murakami, A.; Ohigashi, H.; Sugimura, T.; Wakabayashi, K. (1998) Inhibition by 1′-acetoxychavicol acetate of lipopolysaccharide- and interferon-γ-induced nitric oxide production through suppression of inducible nitric oxide synthase gene expression in RAW264 cells. *Carcinogenesis* **19**, 1007–1012.
[55] Ohshima, H.; Bartsch, H. (1994) Chronic infections and inflammatory processes as cancer risk factors: possible role of nitric oxide in carcinogenesis. *Mutat. Res.* **305**, 253–264.
[56] Teng, C. M.; Li, H. L.; Wu, T. S.; Huang, S. C.; Huang, T. F. (1992) Antiplatelet actions of some coumarin compounds isolated from plant sources. *Thrombosis Res.* **66**, 549–557.
[57] Murakami, A.; Kuki, W.; Takahashi, Y.; Yonei, H.; Nakamura, Y.; Ohto, Y.; Ohigashi, H.; Koshimizu, K. (1997) Auraptene, a citrus coumarin, inhibits 12-*O*-tetradecanoylphorbol-13-acetate-induced tumor promotion in ICR mouse skin possibly through suppression of superoxide generation in leukocytes. *Jpn. J. Cancer Res.* **88**, 443–452.
[58] Kang, J. O.; Slater, G.; Aufses, A. H. J.; Cohen, G. (1988) Production of ethane by rats treated with the colon carcinogen, 1,2-dimethylhydrazine. *Biochem. Pharmacol.* **37**, 2967–2971.
[59] Deschner, E. E.; Zedeck, M. S. (1986) Lipid peroxidation in liver and colon of methylazoxymethanol treated rats. *Cancer Biochem. Biophys.* **9**, 25–29.
[60] Gordon, L. I.; Weitzman, S. A. (1993) Inflammation and cancer. *Cancer J.* **6**, 2257–2261.
[61] Collins, R. H. J.; Feldman, M.; Fordtran, J. S. (1987) Colon cancer, dysplasia, and surveillance in patients with ulcerative colitis. *New Engl. J. Med.* **316**, 1654–1658.
[62] Eberhart, C. E.; Coffey, R. J.; Radhika, A.; Giardiello, F. M.; Ferrenbach, S.; Dubois, R. N. (1994) Up-regulation of cyclooxygenase 2 gene expression in human colorectal adenomas and adenocarcinomas. *Gastroenterology* **107**, 1183–1188.
[63] Oshima, M.; Dinchuk, J. E.; Kargman, S. L.; Oshima, H.; Hancock, B.; Kwong, E.; Trzaskos, J. M.; Evans, J. F.; Taketo, M. M. (1996) Suppression of intestinal polyposis in $APC^{\Delta 716}$ knockout mice by inhibition of prostaglandin endoperoxide synthase-2 (COX-2). *Cell* **87**, 803–809.
[64] Reddy, B. S.; Rao, C. V.; Seibert, K. (1996) Evaluation of cyclooxygenase-2 inhibitor for potential chemopreventive properties in colon carcinogenesis. *Cancer Res.* **56**, 4566–4569.
[65] Sheng, H.; Shao, J.; Kirkland, S. C.; Isakson, P.; Coffey, R.; Morrow, J.; Beauchamp, R. D.; DuBois, R. N. (1997) Inhibition of human colon cancer cell growth by selective inhibition of cyclooxygenase-2. *J. Clin. Invest.* **99**, 2254–2259.
[66] Tsuji, M.; DuBois, R. N. (1995) Alterations in cellular adhesion and apoptosis in epithelial cells overexpressing prostaglandin endoperoxide synthase-2. *Cell* **83**, 493–501.
[67] Hara, A.; Yoshimi, N.; Niwa, M.; Ino, N.; Mori, H. (1997) Apoptosis induced by NS-398, a selective cyclooxygenase-2 inhibitor, in human colorectal cancer cell lines. *Jpn. J. Cancer Res.* **88**, 600–604.

[68] Tsuji, M.; Kawano, S.; Tsuji, S.; Sawaoka, H.; Hori, M.; DuBois, R. N. (1998) Cyclooxygenase regulates angiogenesis induced by colon cancer cells. *Cell* **93**, 705–716.
[69] Chinery, R.; Beauchamp, D.; Shyr, Y.; Kirkland, S. C.; Coffey, R. J.; Morrow, J. D. (1998) Antioxidants reduce cyclooxygenase-2 expression, prostaglandin production, and proliferation in colorectal cancer cells. *Cancer Res.* **58**, 2323–2327.
[70] Rustgi, A. K. (1998) Cyclooxygenase-2: The future is now. *Nature Med.* **4**, 773–774.
[71] Mace, K.; Offord, E. A.; Harris, C. C.; Pfeifer, A. M. A. (1998) Development of *in vitro* models for cellular and molecular studies in toxicology and chemoprevention. *Arch. Toxicol.* **20** Suppl., 227–236.

24 Chemoprevention of Intestinal Neoplasia

Henk J. van Kranen and C. F. van Kreijl

24.1 Abstract

Protection by vegetables and fruits for several diseases, including intestinal cancer, is widely claimed from various epidemiological studies. However, causality for specific constituents and even for a complete vegetables-fruit mixture remains largely to be experimentally proven. In classical chemically-induced rodent model systems for intestinal neoplasia, the effects observed, if any, are strongly influenced by the rodent strain and initiating carcinogen applied. Recent knowledge of genetic alterations in the process of colorectal cancer (CRC) development, identified APC as the goalkeeper gene of CRC. Various mice with different (truncating) mutations in their Apc gene demonstrate associated different phenotypes, i. e. differences in the number of polyps they develop. We have used two Apc deficient strains the Apc^{1638N} and the Apc^{Min} mice to test several polyphenolic compounds (crysin, isoflavonoids, lignans) as well as a complete vegetable and fruit mixture for their anti-carcinogenic potencies on a high risk western style diet.

So far, no protection could be demonstrated for intestinal neoplasia in these mice. Moreover, for a vegetables-fruit mixture and rye bran even a slight increase in polyp number was observed at terminal sacrifice. For flaxseed preliminary data indicate a two-fold increase in polyp number at day 55, wheras at terminal sacrifice (day 107) no differences with the control diet were observed.

24.2 Manuscript

For several diseases, including intestinal cancer, protection by vegetables and fruits is widely claimed in various epidemiological studies. However, a causal relationship for specific constituents and even for a complete vegetables-fruit mixture still remains largely to be proven experimentally. In classical chemically-induced rodent models for intestinal neoplasia, the effects observed, if any, appear to be strongly influenced by the rodent strain and initiating carcinogen. This has been nicely demonstrated, for example, by the work of Alink and coworkers [1–4]. To further dissect the complex dietary and genetic interactions into their primary causes, model systems are required where each variable is introduced into a homogeneous genetic background and a controlled environment. At present, this is best provided by inbred mouse models, and for intestinal cancer more specifically, by mice with an (genetically) altered *Apc* gene.

The *adenomatous polyposis coli* (*Apc*) gene on chromosome 5q is considered to be a gatekeeper of intestinal tumorigenesis. Its inactivation is involved in the onset of CRC, i.e. the formation of a benign adenomas (polyps) [5, 6]. In the majority of colorectal neoplasia, both inherited and somatic, the *Apc* tumor suppressor gene is either mutationally inactivated by the introduction of premature stop codons and/or deleted by loss of heterozygosity (LOH) [7, 8]. In recent years, several mouse lineage's heterozygous for specific mutations at the endogenous *Apc* gene have been developed and characterized with respect to their intestinal tumor multiplicity (reviewed in [9, 10]). Two of these strains of mutant *Apc* mice, Apc^{Min} [11] and $Apc^{\Delta 716}$ [12] are characterized by a relatively high tumor multiplicity in an identical inbred C57BL6 genetic background (approx. 20–90 and 200–500 small intestinal tumors per mice in *Min* and $Apc^{\Delta 716}$ respectively), whereas the Apc^{1638N} [13] only develops 5 to 6 intestinal tumors on average. In our experiments we have restricted ourselves to the Apc^{1638N} and the Apc^{Min} mice. Several polyphenolic compounds (crysine, isoflavonoids, lignans, green and black tea) as well as a complete vegetable and fruit mixture were investigated for their anti-carcinogenic potencies in a high risk western style diet. These experiments were carried out in close collaboration with Dr. Alink and coworkers at the Wageningen Agricultural University and with Dr. Sorensen and coworkers at the Danish Veterinary and Food Administration in Copenhagen and were in part sponsored by a grant from the EU-FAIR programme (CT95-0894). Some of these results will be briefly discussed in the following section, starting with our first experiments in the Apc^{1638N} mice.

Using a western style high risk diet, i.e. high in fat (20%wt/wt, 40 energy%) and low in fiber and calcium, we investigated whether addition of chrysine or a complete vegetable-fruit mixture (VFM) to this diet were able to modulate the development of small and large intestinal polyps in the Apc^{1638N} mice. No (down) modulation at all was observed with either chrysine or VFM. Because the mild phenotype of the Apc^{1638N} mouse results in only 5–6 intestinal polyps we switched to the Apc^{Min} mouse (20–90 polyps/mouse) to increase our range of sensitivity and also extended the VFM experiment with dietary fat as a variable. The results have been recently published [14], and can be summarized as follows. Four different diets (A–D) were compared, which were either low in fat (20 energy%, diets A/B), or high in fat (40 energy%, diets C/D). Further, 19.5% (wt/wt) of the carbohydrates in diets B and D was replaced by a freeze-dried vegetables-fruit mixture (VFM). The diets were balanced as such that they only differed among each other in fat/carbohydrate content and the presence of specific plant-constituents. Because the initiation of intestinal tumors in Apc^{Min} mice occurs relatively early in life, exposure to the diets was started *in utero*. The following results were obtained: Without the addition of VFM, mice maintained at a high-fat diet, did not develop significantly higher numbers of small or large intestinal adenomas than mice maintained at a low-fat diet. A vegetables-fruit mixture added to a low-fat diet significantly lowered multiplicity of small intestinal polyps, but not of colon tumors, in male Apc^{Min} mice only. Strikingly, addition of a vegetables-fruit mixture to female mice maintained on a low-fat diet and to both sexes maintained on a high-fat diet significantly enhanced intestinal polyp multiplicity. In conclusion, these results indicate that neither a lower fat intake nor consumption of a vegetables-fruit mixture included in a high-fat diet decreases the development of polyps in mice genetically predisposed to intestinal tumor development.

The results of the next experiment with soy isoflavones have also been published this year [15] and are summarized below. Two different isolates of soy protein (applied as protein source in the diet) were tested in a "western type high risk diet". For the control and test group this resulted in administration of about 16 and 475 mg of total isoflavones per kg diet, respectively. As positive control, the low isoflavones diet supplemented with 300 ppm sulindac was administered to a third group of mice. No significant differences in the incidence, multiplicity and distribution of intestinal tumors were observed between Apc^{Min} mice that were fed the low and high isoflavones containing diets. However, a clear reduction in the number of small intestinal tumors was observed for the sulindac diet. Thus, in contrast to epidemiological studies, these results demonstrate that high amounts of soy isoflavones present in a "western type high risk" diet, do not protect against intestinal tumor development in Apc^{Min} mice.

To test whether plant lignans (i.e. matairesinol, secoisolariciresinol) can modulate polyp development in Apc^{Min} mice, an experiment with rye bran as a source of these lignans was conducted. Based on the same (isoflavone-free) diet as descibed for the soy experiment, a diet supplemented with 300 g/kg rye bran was prepared and tested together with the isoflavone-free diet as a control. As a positive control in this experiment, sulindac was administered in the drinking water. Our preliminary data indicate, in contrast to our expectations, an approximately 50% increase in the number of intestinal polyps, both in male and female mice on the rye bran diet. To be sure that the concentrations of the lignans were not too low we switched to flaxseed, the richest source of lignans (75–800-times more than other plant foods). Starting again with the isoflavone-free diet as mentioned previously as the control diet, 5% of flaxseed was added to the test diet. In this experiment, a interim section at day 55 was included, in order to follow polyp development more closely in time, i. e. not only at day 105 at sacrifice. Preliminary data indicate no significant differences in the number of polyps at day 105 between mice fed the control- or the flaxseed diet. However, at the intermediate stage (day 55), mice maintained on the flaxseed diet developed twice as much polyps compared to the mice on the control diet. Finally, mixtures of green and black tea are currently being studied for their ability to modulate the development of polyps in the Apc^{Min} mice and a study with different mixtures of n-3 and n-6 polyunsaturated fatty acids will start before the end of this year (i.c.w. Dr. Steerenberg of the Laboratory of Pathology and Immunobiology).

In conclusion, no protection could be demonstrated for intestinal neoplasia in these Apc deficient mice with the substances tested so far. Moreover, for a vegetables-fruit mixture and rye bran (rich in lignans) even a slight increase in polyp numbers was observed at terminal sacrifice.

References

[1] Alink, G. M.; Kuiper, H. A.; Beems, R. B.; Koeman, J. H. (1989) A study on the carcinogenicity of human diets in rats: the influence of heating and the addition of vegetables and fruit. *Food Chem. Toxicol.* **27**, 427–436.

[2] Alink, G. M.; Kuiper, H. A.; Hollanders, V. M.; Koeman, J. H. (1993) Effect of heat processing and of vegetables and fruit in human diets on 1,2-dimethylhydrazine-induced colon carcinogenesis in rats. *Carcinogenesis* **14**, 519–524.

[3] Rijnkels, J. M.; Hollanders, V. M.; Woutersen, R. A.; Koeman, J. H.; Alink, G. M. (1997) Modulation of dietary fat-enhanced colorectal carcinogenesis in *N*-methyl-*N'*-nitro-*N*-nitrosoguanidine-treated rats by a vegetables-fruit mixture. *Nutr. Cancer* **29**, 90–95.

[4] Rijnkels, J. M.; Hollanders, V. M.; Woutersen, R. A.; Koeman, J. H.; Alink, G. M. (1997) Interaction of dietary fat with a vegetables-fruit mixture on 1,2-dimethylhydrazine-induced colorectal cancer in rats. *Nutr. Cancer* **27**, 261–266.

[5] Powell, S. M.; Zilz, N.; Beazerbarclay, Y.; Bryan, T. M.; Hamilton, S. R.; Thibodeau, S. N.; Vogelstein, B.; Kinzler, K. W. (1992) APC mutations occur early during colorectal tumorigenesis. *Nature* **359**, 235–237.

[6] Kinzler, K. W.; Vogelstein, B. (1997) Cancer-susceptibility genes. Gatekeepers and caretakers. *Nature* **386**, 761.

[7] Fodde, R.; Smits, R.; Breukel, C.; Hofland, N.; Edelmann, W.; Kucherlapati, R.; Meera Khan, P. (1995) Genotype-phenotype correlations in intestinal carcinogenesis: the lessons from the mouse models. In: Muller, H.; Scott, R. J.; Weber, W. (Eds.) *Hereditary cancer.* Karger, Basel, p. 35–45.

[8] Dove, W. F.; Gould, K. A.; Luongo, C.; Moser, A.; Shoemaker, A. R. (1995) Emergent issues in the genetics of intestinal neoplasia. *Cancer Surv.* **25**, 335–355.

[9] Bilger, A.; Shoemaker, A. R.; Gould, K. A.; Dove, W. F. (1996) Manipulation of the mouse germline in the study of Min-induced neoplasia. *Semin. Cancer Biol.* **7**, 249–260.

[10] Shoemaker, A. R.; Gould, K. A.; Luongo, C.; Moser, A. R.; Dove, W. F. (1997) Studies of neoplasia in the Min mouse. *Biochim. Biophys. Acta* **1332**, F25–F48.
[11] Moser, A. R.; Pitot, H. C.; Dove, W. F. (1989) A dominant mutation that predisposes to multiple intestinal neoplasia in the mouse. *Science* **247**, 322–324.
[12] Oshima, M.; Oshima, H.; Kitagawa, K.; Kobayashi, M.; Itakura, C.; Taketo, M. (1995) Loss of Apc heterozygosity and abnormal tissue building in nascent intestinal polyps in mice carrying a truncated Apc gene. *Proc. Natl. Acad. Sci. USA* **92**, 4482–4486.
[13] Fodde, R.; Edelmann, W.; Yang, K.; van Leeuwen, .; Carlson, C.; Renault, B.; Breukel, C.; Alt, E.; Lipkin, M.; Meera Khan, P.; *et al.* (1994) A targeted chain-termination mutation in the mouse Apc gene results in multiple intestinal tumors. *Proc. Natl. Acad. Sci. USA* **91**, 8969–8973.
[14] van Kranen, H. J.; van Iersel, P. W.; Rijnkels, J. M.; Beems, D. B.; Alink, G. M.; van Kreijl, C. F. (1998) Effects of dietary fat and a vegetable-fruit mixture on the development of intestinal neoplasia in the ApcMin mouse. *Carcinogenesis* **19**, 1597–1601.
[15] Sorensen, I. K.; Kristiansen, E.; Mortensen, A.; Nicolaisen, G. M.; Wijnands, J. A.; van Kranen, H. J.; van Kreijl, C. F. (1998) The effect of soy isoflavones on the development of intestinal neoplasia in ApcMin mouse. *Cancer Lett.* **130**, 217–225.

E Biomarkers

25 Exocyclic DNA Adducts as Biomarkers for Oxidative Stress, Dietary Risk Factors and Cancer Chemopreventive Approaches

Helmut Bartsch and J. Nair

25.1 Abstract

Among exocyclic DNA adducts, promutagenic etheno (ε) bases are generated by reactions of DNA bases with lipid peroxidation (LPO) products derived from endogenous sources or from exposure to rodent and human carcinogens such as urethane or vinyl chloride. The availability of sensitive methods based on GC-MS, HPLC/RIA or on immunoaffinity/^{32}P-postlabelling has made it possible to detect three ε-adducts (εdA, εdC, N^2,3-εdG) *in vivo* and to study their formation and repair in experimental animals and humans. Highly variable background levels of ε-adducts were detected in tissues from unexposed humans and rodents, suggesting an endogenous pathway of formation, probably arising from reactions of enals generated by LPO with nucleic acid bases. Elevated levels of ε-adducts were found in the liver of patients and rats with genetic metal storage disorders and in spleen of mice in association with overproduction of NO (*via* iNOS), the latter is a consequence of oxidative stress during inflammatory processes that generate reactive oxygen/nitrogen intermediates. High dietary intake of ω-6 polyunsaturated fatty acids (PUFA) increased ε-DNA adduct levels in white blood cells (WBC) *in vivo*, particularly in female subjects (about 40-fold). This review examines the usefulness of exocyclic, in particular of etheno, DNA adducts, (i) to evaluate the etiological contributions of dietary fat intake, oxidative stress, chronic inflammatory/infectious processes, (ii) to verify the efficacy of chemopreventive agents on endogenous DNA damage and cancer risk, and (iii) to gain mechanistic insights into the role of oxidative stress/LPO-derived damage in the initiation and progression of human cancer.

25.2 Introduction

Dietary factors and chronic inflammation/infections are risk factors associated with about one third to one fourth of all cancer cases worldwide [1]. However, the type and sequence of molecular events that lead to human carcinogenesis are still poorly understood. The underlying mechanisms that lead to human carcinogenesis as a result of DNA damage produced endogenously by oxidative stress from dietary factors or by chronic inflammation/infections have so far escaped detection by conventional biomonitoring and epidemiological techniques. Many current detection methods are too insensitive for human biomonitoring, and the most widely used marker 8-oxodG gives results that vary over several orders of magnitude, likely due to its artefactual formation [2]. Oxidative stress leads to the attack of lipids, notably the polyunsaturated fatty acids, resulting in reactive α,β-unsaturated dialdehydes, keto compounds such as croton aldehyde, malondialdehyde and 4-hydroxyalkenals which can react with DNA bases (Figure 25.1) to form exocyclic DNA adducts such as propano

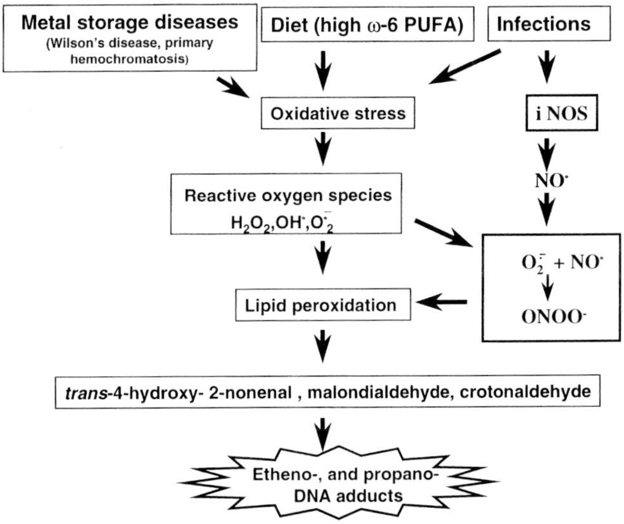

Figure 25.1: A hypothetical scheme for carcinogenic factors leading to endogenous reactive species and exocyclic DNA-base damage.

and etheno (ε) DNA base adducts [3]. These adducts are more stable secondary oxidation products and appear to be more useful markers of oxidative stress-derived DNA damage.

DNA, proteins and lipids are damaged by these reactive species, resulting in mutations and altered functions. Therefore, identification of endogenous sources for DNA damage and the resulting oxidative modification of cellular components will give new insights into the carcinogenesis process as well as provide a better basis for cancer chemoprevention. Recent development of sensitive detection methods made it possible to detect these exocyclic adducts *in vivo* and to study their formation and repair in experimental animals and humans. Highly variable background levels of these propano and ε-DNA adducts (for structures see Figure 25.2) were detected in tissues from unexposed humans and rodents [4–6]. These findings have attracted great attention to find out more about the sources, toxicological significance and genetic consequences of these background DNA lesions, the kind of pathological conditions that can lead to their increase and whether these adducts, when generated through oxidative stress-mediated reactions, could play a causative role in human carcinogenesis.

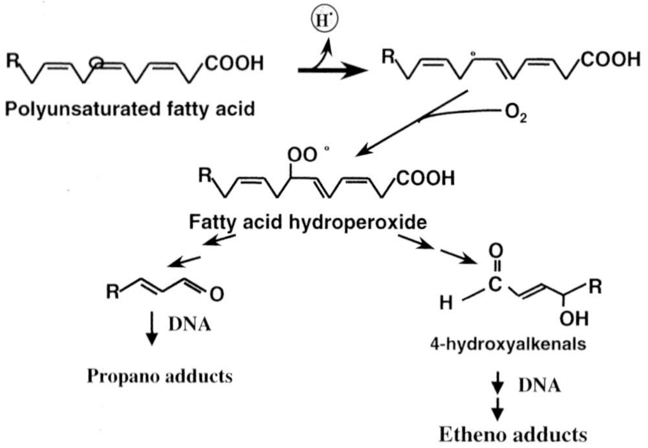

Figure 25.2: Suggested mechanism for the formation of exocyclic adducts from DNA nucleosides and LPO products of polyunsaturated fatty acids (PUFA).

Concerning the sources and origins of the background levels, evidence that exocyclic adducts are derived from DNA-reactive lipid peroxidation products is now supported by the following findings: (i) etheno and propano adducts are formed *in vitro* in the presence of DNA bases and α,β-unsaturated aldehydes derived from LPO products such as hydroxynonenal, its epoxide, fatty acid hydroperoxides (Figure 25.3) [3], and (ii) conditions that increase LPO in microsomal membranes and polyunsaturated fatty acids *in vitro* by bivalent iron, cumene hydroperoxide increased the exocyclic adduct levels [7, 8].

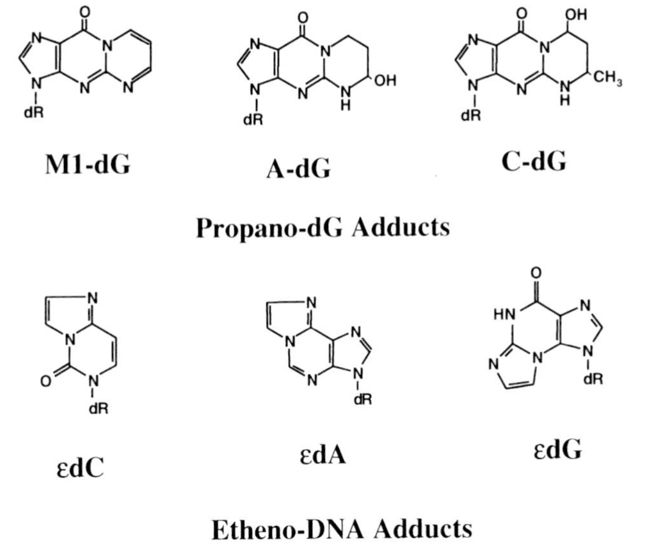

Figure 25.3: Structures of exocyclic DNA adducts described in the text; M1-dG: malondialdehyde-derived deoxyguanosine adduct, A-dG: acrolein-derived 1,N^2-propanodeoxyguanosine, C-dG: crotonaldehyde-derived 1,N^2-propanodeoxyguanosine, εdA: 1,N^6-ethenodeoxyadenosine, εdC: 3,N^4-ethenodeoxycytidine, εdG: N^2,3-ethenodeoxyguanosine.

25.3 Results and discussion

25.3.1 Increased exocyclic adduct formation as a consequence of LPO

25.3.1.1 Etheno adducts

Formation of 1,N^6-ethenodeoxyadenosine (εdA) and 3,N^4-ethenodeoxycytidine (εdC) has extensively been studied in LPO experiments *in vivo*, using an ultrasensitive immunoaffinity-^{32}P-postlabelling method [6] in rodents and in humans where LPO was increased as a result of hereditary metal storage disorders or high intake of dietary linoleic acid (ω-6 polyunsaturated fatty acid, PUFA). Increased formation of etheno-DNA adducts was detected in the liver of Long Evans Cinnamon (LEC) rats, a Long Evans strain with hereditary abnormal copper metabolism, that develop spontaneous hepatitis and later hepatocellular carcinoma. The results demonstrated an age- and copper-dependent formation of etheno adducts [9]. Treatment of Sprague-Dawley rats with iron fumarate and CCl_4 or ethanol led to increased etheno adduct levels [10]. Elevated εdA and εdC were found in liver DNA of patients suffering from the genetic metal storage disorders Wilson's disease (WD) and primary hemochromatosis (PH), whereby the etheno adduct levels were highly correlated with the hepatic copper content in normal and WD samples [11]. LPO in metal storage diseases arises from excess $OH^{\bullet}$ radical production by the transition metal ions Cu and Fe through Haber-Weiss- and Fenton-type reactions. Analysis of total white blood cell DNA from volunteers in a dietary study revealed that high intake of linoleic acid resulted in 40-fold higher εdA and εdC levels in females, as compared to the volunteers on an oleic acid-rich diet [12]. In order to investigate specific DNA damage caused by nitric oxide (NO) induced lipid peroxidation, the formation of the promutagenic etheno adducts εdA and εdC was measured in spleen DNA of SJL mice that had received an injection of RcsX (B cell lymphoma) cells, leading to an overproduction of NO [13]. RcsX-injected mice showed about 6-fold elevated levels of these adducts, as compared to control mice; administration of N^{G}-methyl-L-arginine (an inhibitor of inducible NO synthase) significantly reduced the adduct levels compared to control values [14]. Preliminary data obtained so far in experimental animals showed that the ε-adduct levels in liver are dependent on the type of diet given to mice and rats. For example, AIN diet exhibited highest ε-adduct levels in mice, as compared to NIH-07 diet. Interestingly, the former has been associated also with a higher spontaneous incidence of liver neoplasia in mice. The difference could be related to cancer-protective antioxidants, present in NIH-07 which could also inhibit lipid peroxidation and thus protect against endogenous damage [10].

25.3.1.2 Malondialdehyde-dG and propano-dG adducts

Using GC-MS methods M1-dG adducts were detected in human liver as a background. Adduct levels were increased in rat liver treated with CCl_4 [4]. M1-dG adducts were detected in human white blood cells by ^{32}P-postlabelling [15], and high dietary intake of linoleic acid resulted in a ~2.5-fold and ~4-fold increased adduct level in male and female volunteers, respectively [16]. Acrolein and crotonaldehyde were shown to produce several propano-dG adducts which are present in human and rodent organs as background DNA adducts [5]. Increased propano-dG adducts were also detected in LEC rats [17].

25.3.1.3 Perspectives

It is now well accepted that LPO plays a major role in tumor promotion and progression. However, only recently the availability of ultrasensitive methods has allowed to establish that exocyclic DNA adducts are present as background levels in human and rodent tissue and that the adduct levels could be increased upon triggering LPO. Studies in humans and in experimental models reported by us and others have unequivocally demonstrated that exocyclic DNA adducts are produced as a result of (i) NO overproduction that occurs under conditions of chronic infections and inflammations, (ii) genetic metal storage diseases, where transitional metals like copper and iron accumulate in tissues, and (iii) high intake of polyunsaturated fatty acids, in particular ω-6 PUFA. The formation of ε-adducts as a consequence of *in vivo* NO overproduction suggests that this type of DNA damage may play a role in the development of human cancers associated with chronic viral, bacterial and parasitic infections and cancers that have an inflammatory component in their etiology. Conventional oxidative stress markers such as 8-oxodG have not been increased under similar conditions, probably due to its further degradation by peroxynitrite [18]. Patients with Wilson's disease, if not treated by chelating therapy, accumulate copper in the liver as a consequence of a defective gene that codes for a copper-binding P-type protein leading to elevated oxidative stress and severe liver injury by $OH^{\bullet}$ radicals due to a Haber-Weiss reaction. Genetic primary hemochromatosis, another inherited disease associated with iron accumulation in the liver, leads to highly increased liver cancer risk. Here again, it is assumed that iron-catalyzed Fenton reactions lead to persistent oxidative stress and liver injury. Supportive evidence was provided by the LEC rat model, in which both etheno and propano adducts increased as a function of age of the animals; similarly an iron-overload led to higher ε-adduct levels in rats. High dietary intake of ω-6 PUFA is associated with increased breast and colon cancer risk in humans and experimental animals. The results available so far indicated that exocyclic DNA adducts, M1-dG and εdA, εdC are increased, in particular in females, as a result of linoleic acid-rich diet.

LPO-derived exocyclic adducts (together with other oxidative damage) are likely to increase mutation rates and genomic instability that may contribute to drive cells to full malignancy. If true, the scheme (Figure 25.1) could apply to a number of human tumors that are caused by infectious agents or persistent inflammatory processes. Not only could ε-adducts serve as markers to measure the accumulation of oxidative damage in disease progression, but they are already known to be procarcinogenic lesions with the following characteristics: (i) they are promutagenic DNA lesions in mammalian cells [19, 20]; (ii) they are repaired by specific DNA-glycosylases, confirming that ε-adducts are recognized as hazardous by cellular defense mechanisms [21, 22]; (iii) they are considered as the initiating lesions in vinyl chloride- and urethane-induced carcinogenesis; in both cases the adducts accumulate in the target organ upon exposure [23, 24]; (iv) there is now strong evidence that the background of exocyclic DNA adducts detected in tissues from unexposed humans and rodents arises from endogenous LPO products. The adduct levels were found to be highly variable and in part seemed to be associated with different dietary intake of antioxidants, ω-6 PUFA and induction of persistent oxidative stress. Therefore, exocyclic adducts are useful markers in molecular epidemiological studies on human cancers with ill-defined etiology and mechanisms and, most importantly, in chemopreventive studies aimed at reducing oxidative stress and LPO-derived DNA damage.

Acknowledgements

The authors' research in this area was in part supported by EU contracts EV5V-CT940409 and STEP-CT910145. The authors wish to acknowledge the work contributed by A. Barbin (IARC, Lyon, France) and the collaborations provided by M. Nagao (NCI, Tokyo, Japan), D. Phillips (Haddow Laboratories, Sutton, U.K.), C. Vaca (Karolinska Institute, Stockholm, Sweden), M. Mutanen (University of Helsinki, Helsinki, Finland) and S. Tannenbaum, G. Wogan (MIT, Boston, USA). We thank Mrs. G. Bielefeldt for skilled secretarial assistance.

References

[1] Ames, B. N.; Gold, L. S.; Willett, W. C. (1995) The causes and prevention of cancer. *Proc. Natl. Acad. Sci. USA* **92**, 5258–5265.

[2] Collins, A.; Cadet, J.; Epe, B.; and Gedik, C. (1997) Problems in the measurement of 8-oxoguanine in human DNA. Report of a workshop, DNA oxidation, held in Aberdeen, UK, 19–21 January, 1997. *Carcinogenesis* **18**, 1833–1836

[3] Chung, F.-L.; Chen, H.-J. C.; Nath, R. G. (1996) Lipid peroxidation as a potential endogenous source for the formation of exocyclic DNA adducts. *Carcinogenesis* **17**, 2105–2111.

[4] Chaudhary, A. K.; Nokubo, M.; Reddy, G. R.; Yeola, S. N.; Morrow, J. D.; Blair, I. A.; Marnett, L. J. (1994) Detection of endogenous malondialdehyde-deoxyguanosine adducts in human liver. *Science* **265**, 1580–1582.

[5] Nath, R. G.; Ocando, J. E.; Chung, F. L. (1996) Detection of 1,N^2-propanodeoxyguanosine adducts as potential endogenous DNA lesions in rodent and human tissues. *Cancer Res.* **56**, 452–456.

[6] Nair, J.; Barbin, A.; Guichard, Y.; Bartsch, H. (1995) 1,N^6-ethenodeoxyadenosine and 3,N^4-ethenodeoxycytine in liver DNA from humans and untreated rodents detected by immunoaffinity/^{32}P-postlabelling. *Carcinogenesis* **16**, 613–617.

[7] El Ghissassi, F.; Barbin, A.; Nair, J.; Bartsch, H. (1995) Formation of 1,N^6-ethenoadenine and 3,N^4-ethenocytosine by lipid peroxidation products and nucleic acid bases. *Chem. Res. Toxicol.* **8**, 278–283.

[8] Chen, H. J.; Chung, F. L. (1994) Formation of etheno adducts in reactions of enals *via* autoxidation. *Chem. Res. Toxicol.* **7**, 857–860.

[9] Nair, J.; Sone, H.; Nagao, M.; Barbin, A.; Bartsch, H. (1996) Copper-dependent formation of miscoding etheno-DNA adducts in the liver of Long Evans Cinnamon (LEC) rats developing hereditary hepatitis and hepatocellular carcinoma. *Cancer Res.* **56**, 1267–1271.

[10] Nair, J.; Barbin, A.; Velic, I.; Bartsch, H. (1999) Etheno DNA-base adducts from endogenous reactive species. *Mutat. Res.* **424**, 59–69.

[11] Nair, J.; Carmichael, P. L.; Fernando, R. C.; Phillips, D. H.; Bartsch, H. (1998) Elevated 1,N^6-ethenodeoxyadenosine (εdA) and 3,N^4-ethenodeoxycytidine (εdC) DNA

adducts in liver of patients with Wilson's disease and primary hemochromatosis. *Cancer Epidemiol. Biomarkers Prev.* **7**, 435–440.

[12] Nair, J.; Vaca, C. E.; Velic, I.; Mutanen, M.; Valsta, L. M.; Bartsch, H. (1997) High dietary _-6 polyunsaturated fatty acids drastically increase the formation of etheno-DNA base adducts in white blood cells of female subjects. *Cancer Epidemiol. Biomarkers Prev.* **6**, 597–601.

[13] Gal, A.; Tamir, S.; Tannenbaum, S. R.; Wogan, G. N. (1996) Nitric oxide production in SJL mice bearing the RcsX lymphoma: a model for *in vivo* toxicological evaluation of NO. *Proc. Natl. Acad. Sci. USA* **93**, 11499–11503.

[14] Nair, J.; Gal, A.; Tamir, S.; Tannenbaum, S.; Wogan, G.; Bartsch, H. (1998) Etheno adducts in spleen DNA of SJL mice stimulated to overproduce nitric oxide. *Carcinogenesis* **19**, 2081–2084.

[15] Vaca, C. E.; Fang, J. L.; Mutanen, M.; Valsta, L. (1995) ^{32}P-postlabelling determination of DNA adducts of malonaldehyde in humans: total white blood cells and breast tissue. *Carcinogenesis* **16**, 1847–1851.

[16] Fang, J. L.; Vaca, C. E.; Valsta, L. M.; Mutanen, M. (1996) Determination of DNA adducts of malonaldehyde in humans: effects of dietary fatty acid composition. *Carcinogenesis* **17**, 1035–1040.

[17] Chung, F. L.; Nath, R. G.; Nagao, M.; Nishikawa, A.; Zhou, G.-D.; Randerath, K. (1999) Endogenous formation and significance of 1,N^2-propanodeoxyguanosine adducts. *Mutat. Res.* **424**, 71–81.

[18] Uppu, R. M.; Cueto, R.; Squadrito, G. L.; Salgo, M. G.; Pryor, W. A. (1996) Competitive reactions of peroxynitrite with 2′-deoxyguanosine and 7,8-dihydro-8-oxo-2′-deoxyguanosine (8-oxodG): relevance to the formation of 8-oxodG in DNA exposed to peroxynitrite. *Free Radical Biol. Med.* **21**, 407–411.

[19] Bartsch, H.; Barbin, A.; Marion, M. J.; Nair, J.; Guichard, Y. (1994) Formation, detection, and role in carcinogenesis of ethenobases in DNA. *Drug Metab. Rev.* **26**, 349–371.

[20] Pandya, G. A.; Moriya, M. (1996) 1,N^6-Ethenodeoxyadenosine, a DNA adduct highly mutagenic in mammalian cells. *Biochemistry* **35**, 11487–11492.

[21] Saparbaev, M.; Kleibl, K.; Laval, J. (1995) *Escherichia coli, Saccharomyces cerevisiae*, rat and human 3-methyladenine DNA glycosylases repair 1,N^6-ethenoadenine when present in DNA. *Nucleic Acids Res.* **23**, 3750–3755.

[22] Saparbaev, M.; Laval, J. (1998) 3,N^4-ethenocytosine, a highly mutagenic adduct, is a primary substrate for *Escherichia coli* double-stranded uracil-DNA glycosylase and human mismatch-specific thymine-DNA glycosylase. *Proc. Natl. Acad. Sci. USA* **95**, 8508–8513.

[23] Guichard, Y.; El Ghissassi, F.; Nair, J.; Bartsch, H.; Barbin, A. (1996) Formation and accumulation of DNA ethenobases in adult Sprague-Dawley rats exposed to vinyl chloride. *Carcinogenesis* **17**, 1553–1559.

[24] Fernando, R. C.; Nair, J.; Barbin, A.; Miller, J. A.; Bartsch, H. (1996) Detection of 1,N^6-ethenodeoxyadenosine and 3,N^4-ethenodeoxycytidine by immunoaffinity/^{32}P-postlabelling in liver and lung DNA of mice treated with ethyl carbamate (urethane) or its metabolites. *Carcinogenesis* **17**, 1711–1718.

26 Biological Markers in Chemoprevention Research

LaVerne A. Mooney and Frederica P. Perera

26.1 Abstract

Epidemiologic studies have found that a diet rich in fruits and vegetables is protective against a variety of cancers (including lung), and that high plasma levels of antioxidants in untreated subjects have been associated with decreased risk. Hence, there is great interest in testing the effect of antioxidant vitamins in chemoprevention studies. Biological markers or biomarkers can augment chemoprevention research through improved understanding of disease etiology, confirmation of mechanisms, and identification of high-risk populations. In addition, select markers may function as surrogate endpoints, or as endpoints to monitor intervention trials.

Our group has carried out a series of studies in untreated subjects at high risk of lung cancer to select and validate markers. We have shown that genetic polymorphisms, antioxidant vitamin levels, and gender play a role in mitigating genetic damage from smoking. Although the accumulated evidence is supportive of a protective role for antioxidants, intervention studies using biomarker endpoints are needed to test whether DNA damage can be modulated by supplementation with antioxidants. We have recently initiated a Phase II b (biomarker modulation) trial of antioxidant vitamins C and E in smokers, which is measuring antioxidant status, biomarkers of smoking-related and oxidative DNA damage, and genetic polymorphisms. We are also collaborating on a nested case-control study of biomarkers in blood and lung tissue from the Physicians' Health Study, a Phase III trial of aspirin and β-carotene.

Here we summarize the approach taken to select and validate indicators in lung cancer research, the criteria for the use of biomarkers in chemoprevention trials, and the role of antioxidant vitamins (C, E and carotenoids) in lung carcinogenesis. This research illustrates the advantages and limitations of applying biomarkers to chemoprevention of lung cancer.

26.2 Introduction

Biological markers have been used routinely in etiologic studies to identify populations at high risk of cancer due to exposure or susceptibility factors. Measurement of nutritional status and genetic polymorphisms allowed researchers to postulate mechanisms for inherent or acquired susceptibility to disease. Markers have been incorporated into chemoprevention research either to identify populations at high risk, as surrogate endpoints for cancer, or to assess the effectiveness of interventions.

We are carrying out a Phase II b (biomarker modulation) [1, 2] randomized clinical trial in smokers using antioxidant vitamins C and E. The laboratory and transitional studies providing the rationale for assessing antioxidant vitamins and other biomarkers (DNA damage, genotype etc.) in healthy smokers enrolled in vitamin intervention trials are summarized. This paper provides examples of results from these transitional studies that guided the selection of biomarkers for the trial.

26.3 Validation studies

The criteria for employing biomarkers in environmental epidemiology and chemoprevention research have been outlined [2, 3]. Markers should be measurable, reliable, reflect exposure, disease or susceptibility, and be on the causal pathway. High positive predictive value (PPV) and attributable proportion (AP) are also criteria to evaluate surrogate endpoints [4].

Adequate validation studies are the cornerstone to using markers in chemoprevention research. In a series of transitional studies in high-risk populations we have applied a battery of markers that were putatively involved in lung cancer. The markers (Figure 26.1) include early markers such as carcinogen-DNA damage (PAH-DNA), carcinogen-hemoglobin adducts, autoantibodies to oxidative damage; as well as genetic susceptibility markers glutathione S-transferase M1 (GSTM1) and cytochrome P450 (CYP 1A1), and antioxidant vitamin status (vitamins A, C, E, and carotenoids). In addition, pre-clinical effect markers (p53, cyclin, *ras* and erb-B2) are being validated in a nested case-control study of lung cancer (Phase III trial). See Table 26.1 for a list of criteria for selection of biomarkers for chemoprevention trials. Some highlights illustrating the criteria follow.

In order to determine whether specific markers were reflecting exposure to cigarette smoke, we assessed whether there was a dose-response with exposure and whether the markers persisted after cessation of the exposure. Blood sam-

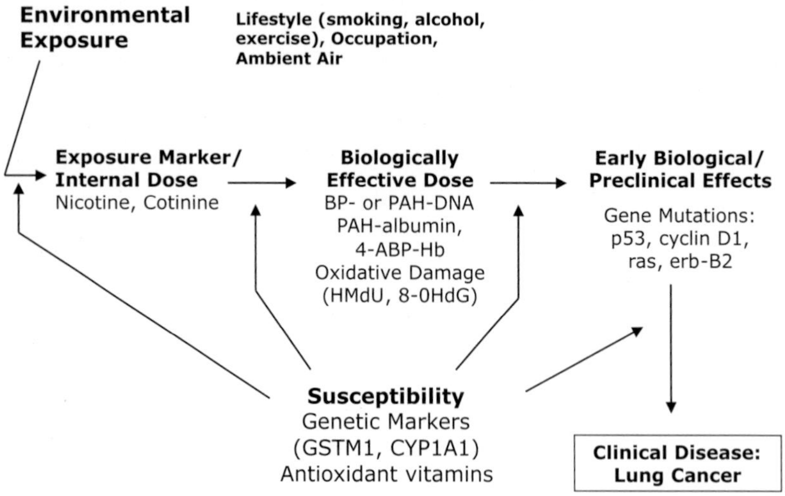

Figure 26.1: Biomarker continuum for lung carcinogenesis.

Table 26.1: Criteria for selecting biomarkers for chemoprevention studies.

1. Measurable and quantitative
2. Validated
   - Sensitivity/Specificity
   - Reliability
   - Biological Half-life
   - Inter- and Intra-Individual Variation
   - Identify and Measure Confounders
   - Feasibility
3. Associated with disease (AP, PPV, RR, OR)
4. Modulated by chemoprevention agent

ples from a longitudinal study of smokers, each of whom gave multiple samples while smoking, then 10 weeks, 8 and 14 months after quitting, were analyzed for several markers in an effort to determine whether or not they were smoking-related.

There was a significant reduction in mean PAH-DNA and 4-aminobi-phenyl-hemoglobin (ABP-Hb) adducts following cessation in all persons who were cotinine-verified quitters ($\geq$25 ng/ml) for $\geq$8 months ($p<0.05$). (Cotinine was used to assess compliance with the cessation protocol.) The substantial reduction (50–75%) in PAH-DNA and 4-ABP-Hb adduct levels after quitting verified that these carcinogen adducts reflect exposure to cigarette smoke [5].

The estimated half-life of the PAH-DNA adducts in leukocytes was 9 to 13 weeks by inspection of the means, and 23 (95% C.I.: 10–36 weeks) using a linear regression model adjusting for background. Women had higher levels of 4-ABP-Hb adducts at baseline and after smoking cessation (with adjustment for amount of smoking) suggesting that women may be more susceptible than men to this carcinogenic exposure. For 4-ABP-hemoglobin adducts, the estimated half-life was 7 to 9 weeks from inspection of the means, which is consistent with the lifetime of hemoglobin. This can be compared with an estimate of 12 weeks (95% C.I., 10–14 weeks) using a regression model that adjusted for background. The variability in the confidence intervals for the regression half-lives illustrates that some individuals may be less efficient in eliminating genetic damage than others.

In another validation study in smokers, we have investigated the relationship between IgM autoantibody (aAb) to oxidative DNA damage and vitamin levels. *Anti*-5-hydroxymethyl-2′-deoxyuridine (HMdU) aAb titers were measured in 140 heavy smokers by ELISA. *Anti*-HMdU aAb levels were significantly elevated in women compared to men after adjustment for smoking dose [6]. Men smoked more cigarettes per day than women and had higher plasma cotinine levels. Although women had significantly higher aAb than men while smoking and at the 8 and 14 month time points, the difference was not significant in post-cessation samples. These findings suggest a differential response to cigarette smoking exposure by gender that may have implications for disease risk.

26.3.1 Inter- and intra-individual variation

In order to measure the relative contributions of intra- and inter-individual variability to observed marker variance in DNA adducts, 20 smokers and 21 non-smokers were serially sampled every three weeks for several months. The sample results were analyzed using a random effects model for longitudinal data [7]. Inter-individual variance accounted for 70% of the total variance, with intra- and laboratory variability accounting for 30% [8]. The variability was 52-fold for GSTM1 null individuals, while for those with the gene it was 14-fold, supporting the role of DNA damage and GSTM1 in lung carcinogenesis.

26.3.2 Transitional studies: Nutritional and genetic factors and DNA damage

We have analyzed the relationship between levels of micronutrients (vitamins C, E, β-carotene, retinol, and the carotenoids) in peripheral blood, and biomarkers of genetic damage, in two populations of smokers who were not supplemented with vitamins.

Plasma concentrations of α-tocopherol and vitamin C were significantly inversely correlated with PAH-DNA adducts in mononuclear leukocytes measured in 63 healthy current smokers [9]. The inverse association between adducts and

α-tocopherol was stronger in those with the GSTM1 null genotype than in those with the gene present. Plasma vitamin C levels were also inversely correlated with DNA adducts; the results were of marginal statistical significance among subjects with the GSTM1 null genotype and not significant in GSTM1(+/+ or +/–) group.

In a second cross-sectional study of 159 heavy smokers entering a smoking cessation program, PAH-DNA adducts in total white blood cells (drawn while subjects were still smoking) were significantly higher (2-fold) in subjects with the CYP 1A1 exon 7 (Ile-Val) variant genotype compared with those without, but were only slightly elevated in subjects who had the MspI RFLP polymorphism but without the exon 7 variant [10]. These results, in conjunction with the *in vitro* and lung-cancer studies mentioned above support the hypothesis that the exon 7 polymorphism is the functional mutation [11, 12]. In the same subjects, plasma levels of retinol, α-tocopherol, and β-carotene were inversely associated with PAH-DNA adduct levels in total white blood cells. In serial samples from a subset of 40 subjects who were able to quit smoking, a strong inverse relationship between adducts and vitamins (retinol, α-tocopherol, and zeaxanthin) was observed – but only in the subjects with the GSTM1 null genotype [13]. Since there was no effect of GSTM1 genotype alone, these results suggest that some individuals may be at increased risk of DNA damage due to a combination of low plasma antioxidant/micronutrient levels and susceptible genotypes.

26.3.3 PAH-DNA as a marker of lung cancer risk

In a lung cancer case-control study, a significant association was found between lung cancer risk and PAH-DNA adduct levels in white blood cells (WBC) [14]. The odds of having elevated PAH-DNA adducts in WBCs was about 7-times higher in cases than in controls [14], indicating the potential of adducts as a risk biomarker. This lung cancer case-control study also confirmed that the GSTM1 deletion is a risk factor for lung carcinogensis. The combined effect of adducts and GSTM1 genotype on lung cancer risk was multiplicative [15]. Of course these findings support the biological relevance of adducts, but suffer from the limitation of all case-control studies that the disease may in fact alter the marker. Although extremely valuable, case/control studies cannot delineate the sequence of pro-carcinogenic events and the biomarkers may be affected by the disease itself.

26.3.4 Nutritional status and lung cancer risk

More than 80 % of lung cancers are attributed to smoking, however, only 10 % of smokers develop the disease in their lifetime. Inter-individual differences in lung cancer risk resulting from a combination of factors including nutritional status and genetic predisposition are thought to explain some of this variation.

Epidemiologic studies are convincing of a protective effect of fruits and vegetables rich in vitamin C and β-carotene for epithelial cancers in many organs including the lung ([16, 17] for review). Twenty-nine of 31 studies showed a significant protective effect for vitamin C and β-carotene and lung cancer [16]. Overall, for the major epithelial cancers, 120/130 studies showed a statistically significant reduction in risk by vitamins C, E, β-carotene or food sources rich in these micronutrients.

Recent reports from three large clinical trials of β-carotene in high risk smokers have found that supplementation with β-carotene is not protective against lung cancer and may even be harmful [18–20]. The elevated incidence of lung cancer in the treatment groups from the β-Carotene and Retinol Efficacy Trial (CARET) and the α-Tocopherol and β-Carotene Cancer Prevention Trial (ATBC) have engendered concern, resulting in the termination of CARET trial. However, in the ATBC trial, individuals with the highest levels of β-carotene at baseline (from dietary sources) had a lower lung cancer incidence than those with lowest levels of β-carotene. The finding of increased incidence of lung cancer in the high dose β-carotene treatment group, but not with baseline levels of β-carotene, suggests that there may be a deleterious interaction between cigarette smoke constituents and β-carotene at high doses. There are also indications that the increased risk may be associated with heavier smoking and alcohol use [19].

The experimental evidence for a protective effect of micronutrients in lung cancer is reviewed elsewhere [3, 16]. The experimental data largely concern the antioxidants and free radical scavengers (vitamin C, β-carotene and the carotenoids) and retinol. Micronutrients, including the antioxidants mentioned above and retinol, are involved in a variety of functions and mechanisms that affect metabolism of endogenous and exogenous chemicals. They may act directly to quench oxidants, including polycyclic aromatic hydrocarbons (PAH), and thereby reduce DNA damage. In addition, they may operate indirectly by modulating immune function or gene expression. There is also evidence that retinol and the retinoids enhance cell differentiation, suppress malignant transformation and counteract the effect of tumor promoters. Experimentally, they reduce the induction of DNA adducts, DNA damage, mutation and/or SCE by diverse carcinogens (including benzo[*a*]pyrene (BP), aflatoxin B_1 (AFB_1), and *N*-nitrosodimethylamine (DMN)). In human lung cancer cells, retinoids increased the expression of the tumor suppressor gene, p53. Both antioxidants and retinoids have been reported to impact the expression of metabolic enzyme activity. For example, antioxidants increase P450 levels in certain tissues. Activity of the cytochrome P450-dependent mixed-function oxidases, the major enzyme system involved in BP metabolism and activation, has been found to decrease with vitamin A deficiency. Antioxidants induce glutathione S-transferase (GST) *in vitro*; and GST activity is reportedly diminished in various cells and tissues with vitamin A deficiency.

In humans there appear to be multiple mechanisms to evade DNA damage. A study of smokers suggests that α-tocopherol may be the primary antioxidant defense in the lungs in response to cigarette smoke exposure [21]. In plasma, vitamin C acts directly as an antioxidant in cigarette smoke-exposed subjects [22].

26.3.5 Genetic susceptibility to carcinogen exposure

Cigarette smoke, diet, and ambient air are plentiful sources of benzo[*a*]pyrene (BP). Like other PAH, BP is metabolically activated by the P450 enzymes and detoxified by GSTM1. Epidemiologic evidence suggests that cancer risk from PAH is mediated in part by the cytochrome P450 (CYP 1A1) aryl hydrocarbon hydroxylase (AHH), which can transform these xenobiotics into reactive DNA-binding intermediates. Elevated risk of lung cancer has been associated with induction of AHH and/or polymorphisms in CYP 1A1 [23–26]. Heightened AHH enzyme activity *in vitro* has been associated with a CYP 1A1 allele that contains an isoleucine (Ile) to valine (Val) substitution in exon 7 [12, 27–29]. This exon 7 Ile/Val polymorphism is linked with a MspI RFLP [29]. Both of these CYP 1A1 polymorphisms have been associated with increased risk of lung cancer in Japanese populations [26, 30]. Both polymorphisms were associated with increased risk in Mexican and African Americans [31], while the combined effect of these two polymorphisms was significant in a study of African Americans [32]. A Brazilian study found that increased risk was associated only with the exon 7 polymorphism [11]. In several Caucasian populations, researchers have not detected significant associations between CYP 1A1 polymorphisms and risk [33–35]. However, the MspI polymorphism was associated with 2-fold increased risk in an American Caucasian population and the exon 7 polymorphism was associated with a similar risk in a German population [36, 37].

Another genetic susceptibility factor, the inherited absence of the glutathione S-transferase μ gene (the GSTM1 null genotype), has been associated with higher risk of lung cancer [26, 38–40]. GSTM1 codes for a phase II detoxification enzyme that is deleted in 30–50% of Caucasian populations [41]. The GSTM1 enzyme can catalyze the detoxification of PAH, such as BPDE, that have been activated by CYP 1A1 [42]. A meta-analysis of the GSTM1 genotype and lung cancer estimates that persons with the GSTM1 null genotype have a 40% increased risk of lung cancer compared to those with the gene [43]. Studies in Japanese subjects found that combined CYP 1A1 and GSTM1 "at risk" genotypes result in a 5 to 9-fold increased risk for lung cancer [26].

BP intermediate metabolites can bind covalently to DNA, sometimes resulting in characteristic mutations. Studies of BP-induced rodent lung tumors and *in vitro* mammalian cell assays have demonstrated the ability of benzo[*a*]-pyrene diol epoxide to predominantly cause G→T transversions [44]. Transversions (G→T) in the p53 gene are the most common base substitution observed in lung cancer cases [45]. This distinct mutation was found in approximately half of all non-small cell lung cancers, and was found to be associated with lifetime cigarette consumption [46]. Cigarette-smoking associated p53 mutations were 4-fold higher in smokers with either of the rare CYP 1A1 polymorphisms than in those with wild type alleles [47]. The G→T mutations are also found in the K-*ras* proto-oncogene of human lung adenocarcinomas. The ability of BP to bind to DNA and preferentially cause this type of mutation in critical genes is supportive of the hypothesis that smoking-related mutations could be instrumental in lung carcinogenesis by activating oncogenes and inactivating tumor suppressor genes.

26.4 Early *vs* intermediate markers

DNA and protein-carcinogen adducts are molecular markers that appear early in carcinogenesis. Adducts can be useful as they reflect not only exposure, but host genetic and nutritional susceptibility and metabolic capacity as well. They are frequent events that occur early enough in the disease process to provide the opportunity for intervention. Oncogene activation and tumor suppressor gene mutations tend to occur later in the disease process and hence are rare in subjects without apparent clinical disease. Susceptibility factors such as antioxidants and genetic polymorphisms may act early or late in the process of cancer development.

Some researchers have suggested that intermediate markers must have a high positive predictive value in terms of disease. This may be problematic for several reasons. PPV is dependent on the prevalence of the disease, so will be strongly influenced by the cohort selected for the validation study. Due to their placement in the scheme of carcinogenesis, early markers such as adducts may not have a high positive predictive value (PPV), while still reflecting individual response to exposure. If they are also clearly modulated by antioxidants/vitamins they have potential in chemoprevention studies, preferably coupled with "intermediate" preclinical effect markers such as altered oncogenes and tumor suppressor genes or cell proliferation and differentiation. Markers that are clearly on the causal pathway, such as intraepithelial neoplasia, are already being used as surrogate endpoints [48].

26.5 Modulation of biomarkers in an antioxidant vitamin trial in smokers

These observational studies suggest that chemoprevention strategies using antioxidant vitamins C and E found in varying degrees in the human diet can protect individuals from environmental exposures and possibly cancer. The evidence that high intake of fruits and vegetables is so convincing that there is a campaign to encourage Americans to eat 5 servings a day. However, before we can make Public Health recommendations for any specific constituents of the diet, well-designed randomized clinical trials are needed.

The next step was to test whether or not treatment with antioxidant vitamins C and E could modulate DNA damage in smokers, and whether or not the effect is mainly in those with the GSTM1 deleted genotype. We are currently enrolling patients into a Phase II b chemoprevention trial of lung cancer to achieve this goal (Figure 26.2). The study is a double blinded placebo-controlled two-arm randomized clinical trial of 500 mg of vitamin C and 400 I.U. of vitamin E or placebo in a population of 300 male and female smokers. The design for the study is shown in Figure 26.2.

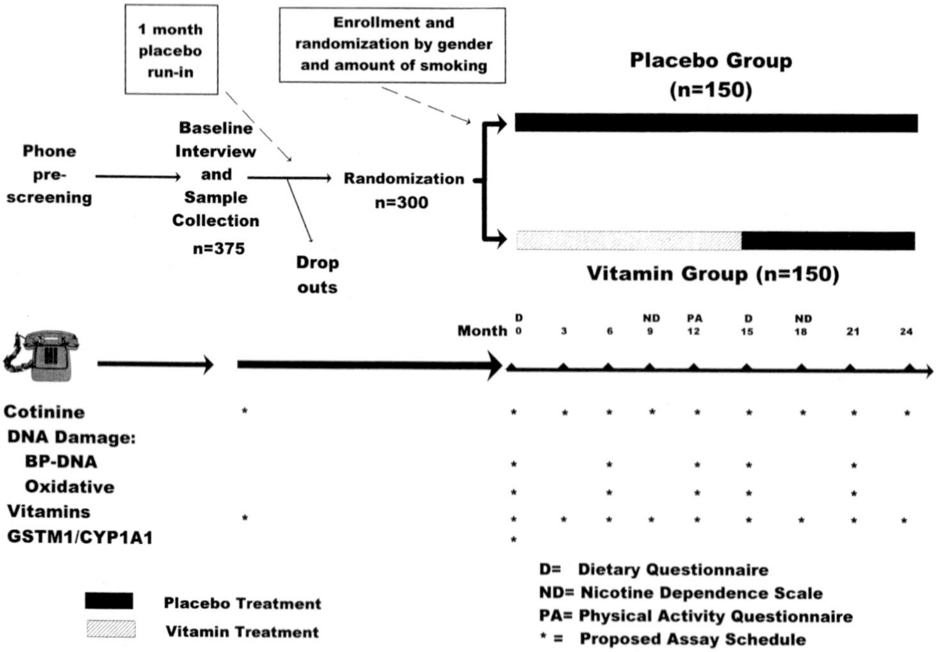

Figure 26.2: Study design.

In this trial, we are measuring two outcome markers of DNA damage (benzo[*a*]pyrene-DNA adducts and 8-hydroxyl-2-deoxyguanosine), plasma vitamin levels including vitamins A, C and E, and carotenoids, and genetic polymorphisms of the GSTM1 and CYP 1A1 genes. Subjects are screened by phone after responding to an advertisement in the local newspaper(s). During the phone screening process they were excluded if they smoked less than 10 cigarettes per day, were currently taking vitamins, have a history or current cancer, under age 18, or are pregnant.

The design calls for enrollment and randomization of a total of 300 smokers (men and women). After a placebo run-in period of one month, the subjects are treated with either a combination of antioxidant vitamins C and E or matching placebos for a 15 months period with nine months of follow up. The subjects return every 3 months for a total of 10 visits over a two-year period.

At each follow up visit his/her exposure and medical history is updated and blood and buccal cells are obtained. Subjects are given results from their previous blood test (lipid profile including triglycerides, HDL, LDL, and total cholesterol). Blood pressure and pulse are reported at each visit. Regular reports describing blood lipid levels are appreciated by the study participants. As an additional incentive, the subjects are financially compensated to return for follow up visits. In studies where "healthy" individuals are asked to participate for multiple years, the financial incentive must be large enough to make it worthwhile, but not so large as to attract those who do not meet the study criteria.

More than half of the required 300 subjects have been enrolled. A small fraction (<1%) was considered ineligible after giving the baseline blood sample due to abnormal liver test results. Loss to follow up is always a concern in longitudinal studies, especially in a trial such as this where the participants are healthy and we rely on incentives to encourage continued participation. We have tried to limit loss to follow up by: 1) placing emphasis on the continuity of the relationship between the patient and the interviewer; 2) the use of frequent calls to make and confirm appointments; and 3) obtaining the names of friends and family members that can be used to trace the subject if they move or if phone contact is unsuccessful, and 4) regular health reports (lipids and blood pressure).

Our protocol calls for measurement of blood cotinine to verify exposure to cigarettes and variability in exposure. Since cotinine can be used to detect passive exposure to cigarette smoke, non-detectable levels of cotinine in the blood of several individuals suggests they were not smoking 10 or more cigarettes per day. The measurement of cotinine in the baseline blood sample can identify subjects who should be excluded, on the basis of non-detectable cotinine.

Compliance with the protocol is measured both by pill counts and by blood vitamin levels. Because the fat-soluble vitamin E has a longer biological half-life than water-soluble vitamin C, we are using the blood vitamin E (α-tocopherol) measurement as the critical micronutrient. In addition, we are measuring the ratio of α-tocopherol to γ-tocopherol, which is altered by supplementation. The α-tocopherol in the supplement competes with γ-tocopherol from the diet, with the net result of an increase in α-tocopherol and a decrease in γ-tocopherol, magnifying the effect of supplementation.

The use of biomarker outcomes in short-term trials will have the advantage of being analyzed not only by the classic intent to treat, where the mean levels of DNA damage are compared between the treatment arms, but also by random effects models (REM). The REM can be used to estimate the slope for the relationship between DNA damage and plasma vitamins over time adjusting for treatment group and other covariates. This repeated measures approach is appropriate for longitudinal data with varying exposures in repeated samplings and for unbalanced data (unequal number of observations at each time point) [7].

If the outcome markers (BP-DNA and oxidative-DNA damage) are modulated by vitamins C and E, we will need to validate the markers in large trials with lung cancer as the endpoint. Incorporation of markers into ongoing trials is an efficient method. For example, a molecular epidemiologic study of biomarkers and lung cancer is being carried out in collaboration with Harvard University and utilizes data and samples collected as part of the Physicians' Health Study (PHS). The study analyzed a panel of biological markers in blood samples collected at baseline in the PHS, as well as markers in the paraffin-embedded tissue from subjects who developed lung cancer during the 13-year follow-up period. The major goal of the project is to investigate the validity of biologic markers that reflect both environmental and genetic factors as indicators of lung cancer risk.

26.6 Conclusions

Molecular mechanisms of susceptibility or protection can be hypothesized from *in vitro* and epidemiologic studies, and promising hypotheses can be tested in a molecular epidemiologic framework. For lung cancer, one mechanism appears to involve both antioxidants and susceptibility genes. The use of early validated markers of DNA damage (e.g., DNA adducts) or intermediate markers (e.g., oncoproteins), in combination with genetic and nutritional markers can help to confirm or refute mechanistic hypotheses. Early molecular markers are especially useful when there is effect modification (e.g., where only those with susceptible genotypes are affected and there is little or no effect in the population as a whole). Although early markers such as DNA damage may not be strongly related to risk, they may be useful in generating information on the biologic response to carcinogenic exposure and chemopreventive agents.

In addition, the early biomarkers can be useful in studies to determine their modulation by a chemoprevention agent, the kinetics of the response to the chemopreventive agent, and the duration of the effect. Since the efficiency of interventions may be increased by studying high risk groups, markers of susceptibility may be used to identify these populations for initial studies.

Intermediate biomarkers (e.g., p53) should also be validated as early response markers or risk markers. This may best be accomplished by a nested-case control design within an ongoing lung cancer chemoprevention trial where cohorts have been optimally sampled for the biomarkers of interest. Intermediate markers with high positive predictive value or attributable proportion may then be used instead of, or in addition to, the cancer endpoint. Given the complexity of carcinogenesis, the combination of early and intermediate biomarkers will improve our understanding of the disease process and our ability to monitor chemoprevention trial efficacy. Use of several markers will increase the likelihood that the biomarkers will reflect the mechanisms of the chemoprevention agent. There is a trade off between the use of early and intermediate markers, with the former being more frequent in healthy populations but less associated with disease; the latter being more infrequent, and offering less opportunity to intervene before disease is established. Although a high PPV would be desirable, in practice it cannot be a rigorous criterion given that it is influenced directly by the prevalence of the disease and therefore, by the study design (e.g., the number of high risk patients or matched controls). The attributable proportion may be a more meaningful statistic, since it reflects the proportion of cases going through the marker on the casual pathway.

In summary, elevated risk of lung cancer has been associated with polymorphisms of genes such as CYP 1A1 and GSTM1, that govern the metabolic activation and detoxification of chemicals. DNA adducts have also been shown to be inversely correlated with vitamin levels in blood. Others have shown that DNA damage can be modulated by treatment with antioxidants [49]. However, it is unclear whether modulation of specific biomarkers (e.g., DNA damage) will reflect a change in incidence of cancer. Therefore, where possible, we must incorporate markers into large-scale Phase III clinical trials to validate them with respect to disease. The use of markers may ultimately reduce sample size and time required to test promising chemopreventive agents.

References

[1] Goodman, G. (1997) The Clinical Evaluation of Cancer Prevention Agents. *Proc. Soc. Exp. Biol. Med.* **216**, 253–259.

[2] Kelloff, G. J; Malone, W. F.; Boone, C. W.; Steele, V. E.; Doody, L. A. (1992) Intermediate biomarkers of precancer and their application in chemoprevention. *J. Cell Biochem.* Suppl. **16G**, 15–21.

[3] Perera, F. P.; Mooney, L. A. (1993) The role of molecular epidemiology in cancer prevention. In: DeVita, V. T. Jr.; Hellman, S.; Rosenberg, S. A. (Eds.) *Cancer prevention.* J. B. Lippincott Co., Philadelphia, p. 1–15.

[4] Schatzkin, A.; Freedman, L. S.; Dorgan, J.; *et al.* (1997) Using and interpreting surrogate endpoints in cancer research. In: Toniolo, P.; Boffetta, P.; Shuker, D. E. G., *et*

al. (Eds.), *Application of biomarkers in cancer epidemiology.* International Agency for Research on Cancer, Lyon, p. 265–271.

[5] Mooney, L. A.; Santella, R. M.; Covey, L.; Jeffrey, A. M.; Bigbee, W.; Randall, M. C.; Cooper, T. B.; Ottman, R.; Tsai, W.-Y.; Wazneh, L.; *et al.* (1995) Decline in DNA damage and other biomarkers in peripheral blood following smoking cessation. *Cancer Epidemiol. Biomarkers Prev.* **4**, 627–634.

[6] Mooney, L. A.; Perera, F. P.; Blaner, W. S.; Karoszka, J.; Frenkel, K. (1997) *Gender differences in autoantibodies to oxidative damage in cigarette smokers.* Conference Proceedings, American Association of Cancer Research.

[7] Laird, N. M.; Ware, J. H. (1982) Random-effects models for longitudinal data. *Biometrics* **38**, 963–974.

[8] Dickey, C.; Santella, R.; Hattis, D.; Tang, D.; Hsu, Y.; Cooper, T.; Young, T.; Perera, F. (1997 Nov 5) Variability in PAH-DNA adduct measurements in peripheral mononuclear cells: Implications for Quantitative cancer risk assessment. *Risk Anal.* **17**, 649–655.

[9] Grinberg Funes, R. A.; Singh, V. N.; Perera, F. P.; Bell, D. A.; Young, T. L.; Dickey, C.; Wang, L. W.; Santella, R. M. (1994) Polycyclic aromatic hydrocarbon-DNA adducts in smokers and their relationship to micronutrient levels and glutathione S-transferase M1 genotype. *Carcinogenesis* **15**, 2449–2454.

[10] Mooney, L. A.; Bell, D. A.; Santella, R. M.; Van Bennekum, A. M.; Ottman, R.; Paik, M.; Blaner, W. S.; Lucier, G. W.; Covey, L.; Young, T. L.; *et al.* (1997) Contribution of genetic and nutritional factors to DNA damage in heavy smokers. *Carcinogenesis* **18**, 503–509.

[11] Hamada, G. S.; Sugimura, H.; Suzuki, I.; Nagura, K.; Kiyokawa, E.; Iwase, T.; Tanaka, M.; Takahashi, T.; Watanabe, S.; Kino, I.; *et al.* (1995) The heme-binding region polymorphism of cytochrome P450 IA1 (CYP IA1), rather than the rsal polymorphism of IIE1 (CYP IIE1), is associated with lung cancer in Rio de Janeiro. *Cancer Epidemiol. Biomarkers Prev.* **4**, 63–67.

[12] Crofts, F.; Taioli, E.; Trachman, J.; Cosma, G. N.; Currie, D.; Toniolo, P.; Garte, S. J. (1994) Functional significance of different human CYP 1A1 genotypes. *Carcinogenesis* **15**, 2961–2963.

[13] Mooney, L. A.; Bell, D. A.; Lucier, G.; Ottman, R.; Santella, R. M.; Tsai, W. Y.; Covey, L.; Young, T. L.; Perera, F. P. (1995) Modulation of DNA adducts in heavy smokers by genetic factors. *Proc. AACR Annual Mtg.* **36**, 711.

[14] Tang, D.; Santella, R. M.; Blackwood, A.; Young, T. L.; Mayer, J.; Jaretzki, A.; Grantham, S.; Carberry, D.; Steinglass, K. M.; Tsai, W. Y.; *et al.* (1995) A case-control molecular epidemiologic study of lung cancer. *Cancer Epidemiol. Biomarkers Prev.* **4**, 341–346.

[15] Tang, D. L.; Rundle, A.; Warburton, D.; Santella, R. M.; Tsai, W.-Y.; Chiamprasert, S.; Zhou, J. Z.; Shao, Q.; Hsu, Y.; Perera, F. (1998) Associations between both genetic and environmental biomarkers and lung cancer: a greater risk of lung cancer in women smokers? *Carcinogenesis* **19**(11), 1949–1953.

[16] Block, G. (1992) The data support a role for antioxidants in reducing cancer risk. *Nutr. Rev.* **50** (7), 207–213.

[17] American Institute for Cancer Research (1997) *Food, nutrition, and the prevention of cancer: A global perspective.* American Institute for Cancer Research, Washington, DC.

[18] Omenn, G. S.; Goodman, G. E.; Thornquist, M. D.; Balmes, J.; Cullen, M. R.; Glass, A.; Keogh, J. P.; Meyskens, F. L.; Valanis, B.; Williams, J. H.; *et al.* (1996 May 2) Effects of a combination of β-carotene and vitamin A on lung cancer and cardiovascular disease. *New Engl. J. Med.* **334** (18), 1150–1155.

[19] Albanes, D.; Heinonen, O. P.; Taylor, P. R.; Virtamo, J.; Edwards, B. K.; Rautalahti, M.; Hartman, A. M.; Palmgren, J.; Freedman, L. S.; Haapakoski J.; *et al.* (1996

Nov 6) α-Tocopherol and β-carotene supplements and lung cancer incidence in the α-tocopherol, β-carotene cancer prevention study: effects of base-line characteristics and study compliance. *J. Natl. Cancer Inst.* **88** (21), 1560–1570.

[20] Hennekens, C. H.; Buring, J. E.; Manson, J. E.; Stampfer, M.; Rosner, B.; Cook, N. R.; Belanger, C.; LaMotte, F.; Gaziano, J. M.; Ridker, P. M.; *et al.* (1996 May) Lack of effect of long-term supplementation with β-carotene on the incidence of malignant neoplasms and cardiovascular disease. *New Engl. J. Med.* **334** (18), 1145–1149.

[21] Patcht, E. R.; Kaseki, H.; Mohammed, J. R.; Cornwell, D. G.; Davis, W. B. (1986) Deficiency of vitamin E in the alveolar fluid of cigarette smokers. Influence on alveolar macrophage cytotoxity. *J. Clin. Invest.* **77**, 789–796.

[22] Frei, B.; Forte, T. M.; Ames, B. N.; Cross, C. E. (1991) Gas phase oxidants of cigarette smoke induce lipid peroxidation and changes in lipoprotein properties in human blood plasma. *Biochem. J.* **277**, 133–138.

[23] Kellerman, G.; Shaw, C. R.; Kellermann, M. L. (1973) Aryl hydrocarbon hydroxylase inducibility and bronchogenic carcinoma. *New Engl. J. Med.* **289**, 934–937.

[24] Kouri, R. E.; McKinney, C. E.; Slomiany, D. J.; Snodgrass, D. R.; Wray, N. P.; McLemore, T. L. (1982) Positive correlation between high aryl hydroxylase activity and primary lung cancer as analyzed in cryopreserved lymphocytes. *Cancer Res.* **45**, 5030–5037.

[25] Nakachi, K.; Imai, K.; Hayashi, S.; Watanabe, J.; Kawajiri, K. (1991) Genetic susceptibility to squamous cell carcinoma of the lung in relation to cigarette smoking dose. *Cancer Res.* **51**, 5177–5180.

[26] Hayashi, S.; Watanabe, J.; Kawajiri, K. (1992) High susceptibility to lung cancer analyzed in terms of combined genotype of P450 IA1 and Mu-class glutathione S-transferase genes. *Jpn. J. Cancer Res.* **83**, 866–870.

[27] Wedlund, P. J.; Kimura, S.; Gonzalez, F. J.; Nebert, D. W. (1994) I462V mutation in the human CYP 1A1 gene: lack of correlation with either the Msp I 1.9kb (M2) allele or CYP 1A1 inducibility in a three-generation family of East Mediterranean descent. *Pharmacogenetics* **4**, 21–26.

[28] Cosma, G.; Crofts, F.; Taioli, E.; Toniolo, P.; Garte, S. (1993) Relationship between genotype and function of the human CYP 1A1 gene. *J. Toxicol. Environ. Health* **40**, 309–316.

[29] Kawajiri, K.; Nakachi, K.; Imai, K.; Watanabe, J.; Hayashi, S. (1993) The CYP 1A1 gene and cancer susceptibility. *Crit. Rev. Oncol./Hematol.* **14**, 77–87.

[30] Kawajiri, K.; Nakachi, K.; Imai, K.; Hayashi, S.; Watanabe, J. (1990) Individual differences in lung cancer susceptibility in relation to polymorphisms of P450 IA1 gene and cigarette dose. *International Symposium of the Princess Takamatsu Cancer Research Fund.* **21**, 55–61.

[31] Ishibe, N.; Wiencke, J. K.; Zuo, Z. F.; McMillan, A.; Spitz, M.; Kelsey, K. (1997 Dec) Susceptibility to lung cancer in light smokers associated with CYP 1A1 polymorphisms in Mexican- and African-Americans. *Cancer Epidemiol. Biomarkers Prev.* **6** (12), 1075–1080.

[32] Taioli, E.; Ford, J.; Trachman, J.; Li, Y.; Demopoulos, R.; Garte, S. (1998 May) Lung cancer risk and CYP 1A1 genotype in African Americans. *Carcinogenesis* **19** (5), 813–817.

[33] Hirvonen, A.; Husgafvel-Pursiainen, K.; Karjalainen, A.; Anttila, S.; Vainio, H. (1992) Point-mutational MspI and Ile-Val polymorphism closely linked in the CYP 1A1 gene: Lack of association with susceptibility to lung cancer in a Finnish study population. *Cancer Epidemiol. Biomarkers Prev.* **1**, 485–489.

[34] Tefre, T.; Ryberg, D.; Haugen, A.; Nebert, D. W.; Skaug, V.; Brogger, A.; Borresen, A.-L. (1991) Human CYP 1A1 (cytochrome P450) gene: Lack of association between the Mspl restriction fragment length polymorphism and incidence of lung cancer in a Norwegian population. *Pharmacogenetics* **1**, 20–25.

[35] Shields, P. G.; Caporaso, N. E.; Falk, R. T.; Sugimura, H.; Trivers, G. E.; Trump, B. F.; Hoover, R. N.; Weston, A.; Harris, C. C. (1993) Lung cancer, race, and a CYP 1A1 genetic polymorphism. *Cancer Epidemiol. Biomarkers Prev.* **2**, 481–485.

[36] Xu, X.; Kelsey, K. T.;Wiencke, J. K.; Wain, J. C.; Christiani, D. C. (1996) Cytochrome P450 CYP 1A1 MspI polymorphism and lung cancer susceptibility. *Cancer Epidemiol. Biomarkers Prev.* **5**, 687–692.

[37] Drakoulis, N.; Cascorbi, I.; Brockmoller, J.; Gross, C. R.; Roots, I. (1994 Feb) Polymorphisms in the human CYP 1A1 gene as susceptibility factors for lung cancer: exon-7 mutation (4889 A to G), and a T to C mutation in the 3′-flanking region. *Clin. Invest.* **72** (3), 240–248.

[38] Seidegard, J.; Pero, R. W.; Markowitz, M. M.; Roush, G.; Miller, D. G.; Beattie, E. J. (1990) Isoenzyme(s) of glutathione transferase (class mu) as a marker for the susceptibility to lung cancer: A follow-up study. *Carcinogenesis* **11** (1), 33–36.

[39] Zhong, S.; Howie, A. F.; Ketterer, B.; Taylor, J.; Hayes, J. D.; Beckett, G. J.; Wathen, C. G.; Wolf, C. R.; Spurr, N. K. (1991) Glutathione S-transferase mu locus: Use of genotyping and phenotyping assays to assess association with lung cancer susceptibility. *Carcinogenesis* **12**, 1533–1537.

[40] Alexandrie, A. K., Sundberg, M. I., Seidegard, J., Tornling, G., Rannug, A. (1994) Genetic susceptibility to lung cancer with special emphasis on CYP 1A1 and GSTM1: A study on host factors in relation to age at onset, gender and histological cancer types. *Carcinogenesis* **15**, 1785–1790.

[41] Bell, D. A.; Thompson, C. L.; Taylor, J.; Clark, G.; Miller, C. R.; Perera, F.; Hsieh, L.; Lucier, G. W. (1992) Genetic monitoring of human polymorphic cancer susceptibility genes by polymerase chain reaction: Application to glutathione transferase. *Environ. Health Persp.* **98**, 113–117.

[42] Ketterer, B. (1988) Protective role of glutathione and glutathione transferases in mutagenesis and carcinogenesis. *Mutat. Res.* **202**, 343–361.

[43] McWilliams, J. E.; Sanderson, B. J. S.; Harris, E. L.; Richert-Boe, K. E.; Henner, W. D. (1995) Glutathione S-transferase M1 (GSTM1) deficiency and lung cancer risk. *Cancer Epidemiol. Biomarkers Prev.* **4**, 589–594.

[44] Harris, C. (1991) Chemical and physical carcinogenesis: Advances and perspectives for the 1990s. *Cancer Res.* **51** (Suppl.), 5023S–5044 S.

[45] Hollstein, M.; Sidransky, D.; Vogelstein, B.; Harris, C. C. (1991) p53 Mutations in human cancers. *Science* **253**, 49–52.

[46] Suzuki, H.; Takahashi, T.; Kuroishi, T.; Suyama, M.; Ariyoshi, Y.; Takahashi, T.; Ueda, R. (1992) p53 Mutations in non-small cell lung cancer in Japan: Association between mutations and smoking. *Cancer Res.* **52**, 734–736.

[47] Kawajiri, K.; Eguchi, H.; Nakachi, K.; Sekiya, T.; Yamamoto, M. (1996) Association of CYP 1A1 germ line polymorphisms with mutations of the p53 gene in lung cancer. *Cancer Res.* **56**, 72–76.

[48] Boone, C. W.; Kelloff, G. J. (1997) Biomarker end-points in cancer chemoprevention trials. In: *Application of biomarkers in cancer epidemiology.* IARC Scientific Publications, Lyon, p. 273–280.

[49] Duthie, S. J.; Ma, A.; Ross, M. A.; Collins, A. R. (1996) Antioxidant supplementation decreases oxidative DNA damage in human lymphocytes. *Cancer Res.* **56**, 1291–1295.

F Diet and Cancer: Evaluation Criteria

27 Approaches Towards Assessment of Carcinogens/Anticarcinogens

Harri Vainio

Cancer mortality in industrialized countries has not undergone a similar decline that has been recorded for cardiovascular mortality over the past three decades. While our knowledge base on identified causal factors for human cancer is quite solid for certain cancers (such as cancer of the lung and bladder), for certain other sites, such as breast and prostate, only little evidence for either modifiable risks or effective intervention has been produced. The main modifiable causes of cancer seem to have a common behavioural pathway: substances that people put in their mouths, including cigarettes, smoking pipes, chewing tobacco and food. Food components are important modifiers of cancer risk. Eating fruits and vegetables appears to carry a cancer preventive effect. Also, when studied in experimental studies, several classes of non-nutritive components of foods could prevent cancer. But when some of these components have been further evaluated in large-scale randomized clinical trials, the results have been disappointing. β-Carotene supplements showed not only no benefit, but increased the risk of lung cancer, as well as cardiovascular and total mortality. This has forced us to rethink the relevance and use of observational data from nutritional epidemiology in evaluating individual components of food. Lycopene and other carotenoids are potential candidates as cancer-preventive agents as well. With the perspective of the β-carotene trials, however, caution is warrented about assuming that such chemicals are safe in pharmacological (supplemental) doses.

28 Diet and Cancer: Public Health Aspects

Martin J. Wiseman

28.1 Abstract

About one quarter of all deaths in England and Wales are due to cancer. Various workers have estimated that dietary factors contribute to the development of around one third of all cancers. The potential importance of diet as a means of avoiding cancer is therefore of major public health importance.

Foods are complex mixtures of chemicals, many of which can have direct or indirect biological effects, both beneficial and adverse. The ultimate impact of foods depends on their composition, the amount consumed, the numbers consuming it; and other aspects of diet and lifestyle which may interact.

Cancer is a complex phenomenon which can take years to develop, resulting from the interaction between the cumulative effects of different exposures over time and the genetic background. It is therefore difficult to test preventive hypotheses directly. Information can be derived from several sources – epidemiological, laboratory, animal and human, including clinical studies. Conclusions can be based only on the totality of evidence. No single methodology is adequate to draw firm conclusions relevant to public health.

A recent report from the UK Committee on Medical Aspects of Food and Nutrition Policy (COMA) has examined the relationship between nutrition and the development of cancer applying this concept. Their analysis was made rigorous by use of a pre-determined scoring system for quality of epidemiological data, and by use of clearly defined terminology. They applied recognised principles to assess the likelihood that observed relationships might be causal. The conclusions so drawn allowed recommendations to be made for the improvement of public health.

The importance of rigorous procedures for analysing diverse data will be discussed, as well as the conclusions and recommendations of the COMA Report.

28.2 Introduction

In the United Kingdom cancers at all sites are the registered cause of around one quarter of all deaths, and the second most common cause of death after cardiovascular diseases. Cancer – its causation, prevention and treatment – is clearly, therefore, of major public health importance. Death is, of course, the ultimate outcome of many cancers, but mortality statistics hide considerable morbidity and suffering. Though it is relatively simple to ascertain deaths due to cancer, the incidence of the disease is more difficult to measure. Nevertheless, improved methodology has allowed the establishment in the UK of regional centres which are successful to a greater or lesser extent in recording the number of incident cancers. The difference in incidence and mortality reflects a number of factors, in particular the natural fatality of the disease, and the success of treatment. For instance non-melanoma skin cancers are relatively common, but almost never fatal. Other cancers may be more treatable than others. Differences in the two may reflect either of these factors, or inaccuracies in registration of incidence. Regional variation in such differences may reflect differences in pattern of disease, or of treatment, or of registration.

Overall numbers of cancers are increasing with time, but this is largely a reflection of the increasing numbers of people surviving to greater age. The peak mortality and incidence of most cancers occurs in the sixth, seventh and eight decades. For some cancers there is a real, age-specific, increase in incidence and mortality. In women, breast cancer has been increasing steadily over the last decade. On the other hand stomach cancer has been declining rapidly over the same period. Cancers are not evenly distributed either regionally – though there is no straightforward pattern – or with socio-economic status. Some cancers tend to be more common in lower socio-economic classes (e.g. stomach, bladder and lung) while others are more common in more affluent groups (e.g. breast, colorectum, prostate and – though numbers are small – melanoma). Overall, however, the incidence of cancers at all sites does not vary substantially by social status – it is the pattern of cancers which varies.

It has been well documented that rates of cancer within populations are not fixed. In addition to the above time trends, rates of cancer have been shown to vary rapidly in populations who migrate to different communities. The pattern of migrant Japanese to the USA changes to that of the host nation within one to two generations. In addition within populations there is a considerable variation. Socio-economic variations are commonly described, but other population groups e.g. vegetarians have been reported to have different cancer patterns than the general population. Together these factors argue for a strong environmental component, as opposed to a fixed heritable one, in the development of cancers. Though clearly at a cellular level a disease of genes, only a small proportion of all cancers is caused by inherited genetic defects. While inherited genetic defects might yet be discovered to account to a greater or lower extent for other cancers, it is certain that the majority are related to environmental exposures accumulated over time, eventually leading to sufficient unrepaired genetic damage to allow the development and progression of a neoplastic tumor.

A few cancers have been clearly linked to specific exposures such as chemicals (e.g. asbestos) or occupations (e.g. scrotal carcinoma in chimney sweeps). However, such clearly defined causal links are the exception, and it can be difficult to identify the exposures which might singly or in combination lead to cancer development. There are a number of reasons for this. First of all cancers usually take decades to develop, and the number of exposures required need not occur in a particular sequence. The influence of environmental exposures may occur therefore, at different times and through different pathways; and for any particular cancer a number of different exposures may be necessary. In order to identify such factors, the relatively crude tool of epidemiology requires long periods of observation; sufficient variation in either incidence or mortality, and in exposures; sufficiently large numbers to exclude the play of chance; and avoidance of bias by methodology or confounding. Such stringency is difficult to achieve and many studies are deficient in one or more respects. Consequently it is necessary to scrutinize the totality of the evidence – epidemiological, clinical and laboratory – to draw conclusions on the possible links between essential exposures and cancer. Amongst environmental exposures, it has been estimated that dietary factors might be responsible for around one third of all cancers in Western populations.

Ascertainment of diet presents particular difficulties. It is often remarkably similar between individuals within populations, but there is considerable within person day to day variation. All methods of estimating diet – from retrospective 24 h recall or food frequency questionnaires to prospective diary or weighed records – are imperfect in one or more respects. In addition there are few validated biomarkers either of dietary exposure or of cancer risk. Furthermore, the relatively fixed nature of dietary patterns offers considerable scope for finding spurious associations from confounding by known or unknown dietary or non-dietary factors. In summary, then, the scientific exploration of the links between nutritional factors and cancers is fraught with difficulty. Nevertheless, increasing public interest in and understanding of health matters has led to a perception – possibly distorted – of the causation of cancer and any possible dietary component.

28.3 Diet and cancer in the UK

It is against this background in the UK that the Committee on Medical Aspects of Food and Nutrition Policy – a committee of independent experts which advises Government on the links between diet, nutrition and health – began a review of nutritional aspects of the development of cancer [1].

Their review was set firmly in the context of the reality of human cancer in the UK, and so was concerned with drawing conclusions which might be relevant to developing public health policy.

28.3.1 The nature of associations

In order to be relevant to public policy, the scientific evidence must reasonably support a causal relationship between the putative risk factor and the disease. With all the imperfections in the evidence base, a number of generally accepted principles can help provide a judgement on the likelihood that any observed association might actually be causal. In his classic text, Bradford Hill expounded these over half a century ago, but the principles stand now as they did then. There are nine features which Hill identified as being important in drawing conclusions about the nature of an association.

Firstly, the *strength* of an association; the greater the increase in risk which accompanies the exposure, the less easy it is to explain an association on the basis of other related – confounding – factors. Secondly, the observation should be observed *consistently* by different observers, in different experimental settings, in different places and at different times. *Specificity* is the third characteristic of an association which can aid the inference of causality, though like all the features, it cannot be taken alone as a critical factor. If the observed exposure is the only one which is linked to the adverse outcome, or if there is only one associated outcome, this lends support to the notion that it might be causative. Fourthly, there should be evidence that the exposure preceded the disease. Cross sectional studies typically cannot identify such a *temporal relation*, and dietary habits which might have changed in response to disease could produce a similar effect in such studies.

A fifth feature is the demonstration of a *biological gradient*. In more common parlance, a dose-response relationship more easily admits a cause-effect relationship. Sixthly, there should be a *biologically plausible* explanation for the observed link – though clearly with incomplete knowledge such a characteristic cannot be demanded. Seventh, the possibility of cause and effect should be *coherent* with overall knowledge of the biology and natural history of the disease, and with trends in its occurrence, and so on.

The possibility of *experimental evidence* to test a specific hypothesis may exist, and in many ways this can be the strongest evidence. But often experimental data are not as generally applicable as would be desired, due to the nature of the intervention, the population studied, the time course or the stage of disease of the intervention. Finally, there is some value in judging data by *analogy* with other known cause-effect relationships. For instance if one substance was known to cause a particular cancer, with considerable and essentially indisputable evidence, a similar link for a related substance might require a different degree of evidence to draw conclusions.

The review was therefore founded on the basis of a systematic review of the epidemiological relationships between specific dietary or nutritional factors and specific cancers. This was supplemented by a review of the experimental evidence – either *in vivo* or *in vitro* – for mechanisms operating in humans which might underlie such associations. Overall conclusions were drawn on the basis of the totality of this evidence.

28.3.2 The epidemiological evidence

The epidemiological review focused on those cancers for which there was some suggestion of a dietary or nutritional link. These included those cancers which are the major contribution to cancer mortality. However, in view of the variability in experimental designs, and in the ascertainment of diet, it was just as important to apply a measure of quality to the assessment of the data. A methodology was therefore developed which scored studies on the basis of this quality, specifically in relation to diet and nutrition. Otherwise good studies which had included a crude measure of diet as a secondary consideration would therefore have not scored well. This is not a criticism of the study, but of its value for the assessment of nutritional links. Two scoring systems were developed – one for cohort studies and one for case control.

The cohort studies were scored on the basis of the dietary assessment; the definition of the cohort; the ascertainment of the outcome; and the analysis employed. Case control studies were scored on dietary assessment, recruitment of subjects, and analysis. The system was validated for within and for between observer variation. Application of this system resulted in categorizing studies as high, intermediate and low scoring. These values were then used in coming to a judgment on the epidemiological data.

The epidemiological review therefore addressed the question of the strength of any association, and its specificity; its consistency (e.g. between cohort and case control or between different studies of the same design); the temporal relation; and the biological gradient. This part of the review also took account of the few, but important, experimental studies in humans with "hard" endpoints of cancer incidence or of mortality.

Again, in order to ensure a systematic approach to the assessment, a defined terminology was used. The evidence for an association was defined in relation to the extent of the evidence – were there none, few or many studies? Was the evidence sufficient or insufficient to draw conclusions? If there was sufficient evidence, was it inconsistent (around half of studies showing an effect), weakly (around half to two thirds of studies in the same direction), moderately (about two thirds to three quarters) or strongly consistent (more than about three quarters of studies in the same direction)? Some value was applied to whether an individual study gave results which reached conventional levels of statistical significance, but weight was also given if a number of "insignificant" studies showed essentially the same effect. In addition, it was felt that where diet was concerned, case control studies might be particularly prone to bias, for instance due to errors in recall influenced by the knowledge of the current illness, and this was a feature of the overall judgement.

28.3.3 Experimental evidence

The generation of experimental models of cancer in laboratory animals is far removed from the occurrence of cancers in humans in the real world. While these studies are valuable in their context, they cannot be extrapolated to the human condition without caution. Apart from the species difference (which can be critical), there are differences in dose, route of exposure, and the experimental conditions which limit their direct relevance to human cancer. Nevertheless, they can provide information regarding possible mechanisms of action, although again any such mechanism cannot be assumed to operate in humans. Similar cautions apply to *in vitro* laboratory data, whether using animal or human material. COMA laid emphasis particularly on data in humans. A consistent terminology was used to define both the *extent* of the evidence (no, little, some, substantial), and whether it was in animals, *in vitro* or in humans; and the *strength* of the evidence (convincing, equivocal, unconvincing or lacking). This approach allowed judgements to be made in a systematic way on whether there might be a biologically plausible explanation for an association.

28.3.4 Drawing overall conclusions

An overall conclusion was only drawn by synthesizing the epidemiological data with the experimental evidence, and making a judgement in the wider context of the coherence of the evidence, and any analogous evidence. In the absence of corroborating evidence from the other category, neither experimental nor epidemiological evidence alone were judged sufficient to draw firm conclusions on causality. The overall evidence for a causal link between a specified dietary or nutritional factor and a cancer was judged to be not enough, weak, moderate or strong. Where there was moderate or strong evidence, recommendations proportionate to the extent of evidence were formulated.

28.4 Diet and cancer – conclusions

In general, the evidence was varied, with studies using different definitions, giving variable degrees of information, different methodologies, and of variable quality. Very few relationships showed a large difference in risk between the highest and lowest exposure, the majority being of the order of 1–2-fold. But for some common cancers at least moderate evidence of risk was found, and though relative risk was not great, because of wide exposure these might be of considerable importance in determining absolute numbers affected, and so of public health significance.

The detailed assessment of the evidence will be presented elsewhere, but COMA found that there was moderate evidence to conclude that:

- avoiding overweight, and especially gain in weight during adult life, would reduce risk of postmenopausal breast cancer and of endometrial cancer
- increasing consumption of fruits and vegetables would reduce risk of cancers of the colorectum and stomach.
- increasing intakes of non starch polysaccharides (NSP – "dietary fibre") would reduce risk of colorectal cancer.
- avoiding high intakes of red and processed meat would reduce risk of colorectal cancer.

COMA recognized the possibility of confounding and covariance in coming to these conclusions, and advised that recommendations arising from this should be implemented in the context of a balanced diet. COMA also advised that high dose purified supplements of β-carotene should be avoided; and that such supplements of other nutrients should be used with caution as they cannot be assumed to be safe.

28.5 Conclusions

Dietary and nutritional factors are thought to account for around one third of cancers in the developed world. There is relatively little evidence for specific diet/nutrition/cancer links to explain this estimate. However a dietary pattern which would be expected to minimize cancer risks can be recommended with at least moderate confidence. This pattern is consistent with dietary advice in respect of risk reduction of other conditions notably cardiovascular disease. Much research needs to be done to enhance the evidence of diet/cancer links, in particular on new methodological approaches.

In the current state of knowledge, statutory regulations to influence diet to reduce cancer risk might be considered disproportionate, but there is room for better coordinated dissemination of information about diet and health, for more education about how to construct healthy diets, and for increasing the choice of and access to a wide variety of foods at affordable prices to all sectors of society. The selection of such elements of public policy is a matter of political decision.

Reference

[1] Department of Health (1998) *Nutritional aspects of the development of cancer. Report of the Working Group on Diet and Cancer of the Committee on Medical Aspects of Food and Nutrition Policy*. London, HMSO (Report on Health and Social Subjects; No. 48).

IV Posters

1 Induction of Unscheduled DNA Synthesis by the Mycotoxin Ochratoxin A in Cultured Urothelial Cells from Pigs and Humans

W. Föllmann, A. Dörrenhaus, A. Flieger, K. Golka and Gisela H. Degen*

Introduction

The mycotoxin ochratoxin A (OTA) is a widespread contaminant of food and feedstuffs. It is known for its nephrotoxic, immunosuppressive, and carcinogenic effects in domestic and laboratory animals. OTA is an etiological factor in Balkan Endemic Nephropathy (BEN) and suspected to cause urinary tract tumors in BEN patients [1]. Recent *in vivo* and *in vitro* studies on the genotoxicity of OTA revealed DNA adducts in the bladder of mice [2a, b], induction of sister chromatid exchanges in primary cultures of porcine urinary bladder epithelial cells (PUBEC) [3] and formation of micronuclei in ovine seminal vesicle cells [4].

Since the transitional epithelium is a potential target for OTA, its effects were studied in cultured normal human urothelial cells (HUC) [5] and PUBEC [6, 7]. The genotoxic potential of OTA was examinated by measuring the induction of unscheduled DNA synthesis (UDS) in serum-free and arginine-deprived cultures by autoradiography [8].

* Institut für Arbeitsphysiologie an der Universität Dortmund, Ardeystr. 67, D-44139 Dortmund, Germany

Materials and methods

HUC were isolated from specimen of ureters of two nephrectomized adults (male 66 year-old; female, 26 year-old) and of two children with ureteropelvic junction stenosis (males, 7 year-old, 9 month-old) and cultured under sterile conditions [5]. PUBEC were isolated from porcine urinary bladders from freshly slaughtered pigs from the slaughterhouse and cultured as HUC. For the performance of the UDS assay the cell proliferation was suppressed by culturing in arginine-deficient Ham's F12-medium for 96 hours. Both culture systems were prepared for autoradiography as described previously [8]. The percentage of cells in repair (CIR) was determined by counting 100–150 cells per treatment group for PUBEC and 50 cells per treatment group for HUC.

Results

In PUBEC incubated with OTA 0.25–5 μM for 24 h, a concentration dependent increase in the number of CIR was observed with a maximum at 1 μM OTA (61±9% of CIR compared to 7±4% in solvent control). Above 1 μM OTA was cytotoxic (Figure 1.1). In HUC derived from two nephrectomized adult patients

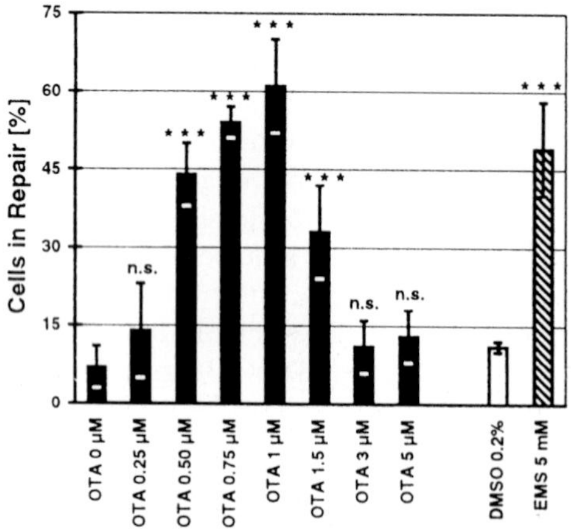

Figure 1.1: DNA repair in primary cultured porcine urinary bladder epithelial cells after *in vitro* incubation with ochratoxin A (OTA), dimethyl sulfoxide (DMSO, solvent control), or ethyl methanesulfonate (EMS, positive control). Differences between treated samples and solvent control (DMSO) are marked with asterices (*p<0.05, **p<0.005, ***p<0.001; Student's t-test; n.s. = not significant).

(male, 66 year-old; female, 26 year-old (Figure 1.2 a, b) and one 7 year-old boy (Figure 1.3 a) UDS induction was observed at concentrations between 0.05–0.75 µM OTA. Morphological signs of cytotoxicity appeared above 0.5 µM OTA and OTA-induced repair decreased. Urothelial cells derived from an infant (9 month-old, boy, Figure 1.3 b) showed a response to ethyl methanesulfonate (5 mM EMS, positive control) but no significant induction of UDS by OTA.

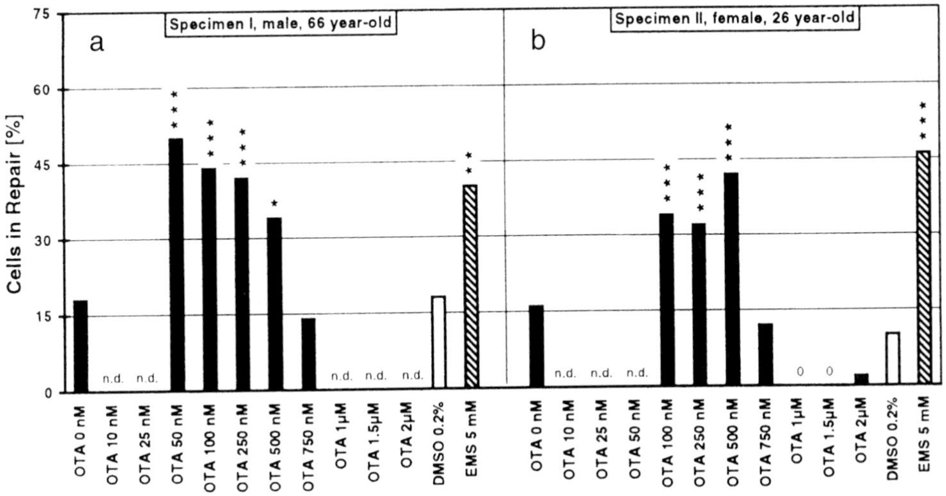

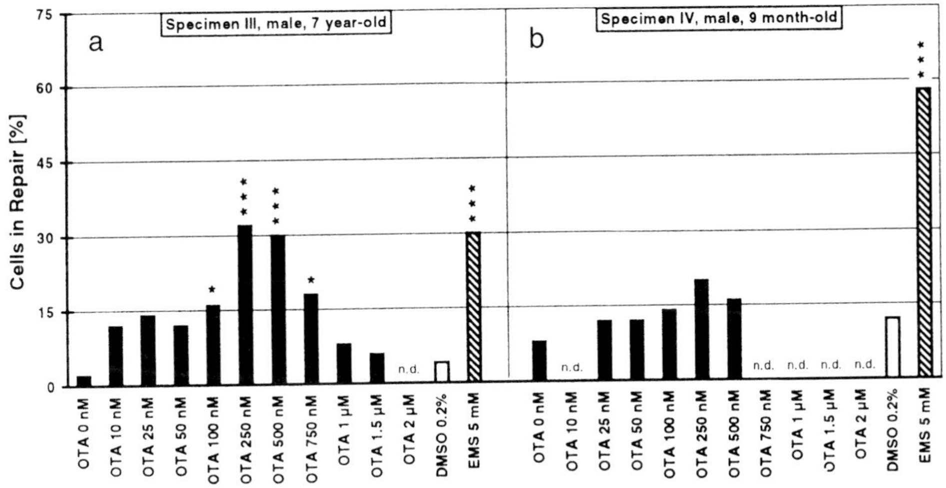

Figures 1.2 a, b and 1.3 a, b: DNA repair in primary cultured urothelial cells of adults and children after *in vitro* incubation with ochratoxin A (OTA), dimethyl sulfoxide (DMSO, solvent control), or ethyl methanesulfoxide (EMS, positive control). Significantly higher numbers of cells in repair in the treated cultures *vs* solvent control are marked with asterices (*$p<0.1$, **$p<0.05$, ***$p<0.01$, Fisher's exact-test), n. d. = not determined.

Conclusions

In both *in vitro* models for the target tissue (PUBEC and HUC) OTA induces DNA-repair (UDS) as a result of DNA-damage. These data as well as previous data [2–4] support the view that OTA is genotoxic.

Interestingly, HUC cultures from four individuals showed a variable UDS response to OTA (Figures 1.2, 1.3). If OTA requires bioactivation in order to affect the DNA, then different activities of metabolizing enzymes in the cultures would affect their response. Further investigations with additional specimen are necessary to clarify this point.

Moreover, it is noteworthy that human cells appeared to be more susceptible to OTA than those from pigs: whereas porcine cells showed a response maximum at 1 μM OTA, maximal induction of DNA repair was observed at much lower concentrations in human cells. This points to a species difference which should be taken into account in a risk assessment of this carcinogenic mycotoxin.

References

[1] IARC Monographs on the evaluation of the carcinogenic risk of chemicals to humans. Vol. 56 (1993) *Some naturally occurring substances: food items and constituents, heterocyclic aromatic amines and mycotoxins*. IARC, Lyon, p. 489–521.

[2a] Pfohl-Leszkowicz, A.; Grosse, Y.; Kane, A.; Creppy, E. E.; Dirheimer, G. (1993) Differential DNA adduct formation and disappearance in three mice tissues after treatment by the mycotoxin ochratoxin A. *Mutat. Res.* **289**, 265–273.

[2b] Obrecht-Pflumio, S.; Grosse, Y.; Pfohl-Leszkowicz, A.; Dirheimer, G. (1996) Protection by indomethacin and aspirin against genotoxicity of ochratoxin A. *Arch. Toxicol.* **70**, 244–248.

[3] Föllmann, W.; Hillebrand, I. E.; Creppy, E. E.; Bolt, H. M. (1995) Sister chromatid exchange frequency in cultured isolated porcine urinary bladder epithelial cells (PUBEC) treated with ochratoxin A and alpha. *Arch. Toxicol.* **69**(4), 280–286.

[4] Degen, G. H.; Gerber, M. M.; Obrecht-Pflumio, S.; Dirheimer, G. (1997) Induction of micronuclei with ochratoxin A in ovine seminal vesicle cell cultures. *Arch. Toxicol.* **71**, 365–371.

[5] Flieger, A.; Reckwitz, Th.; Golka, K.; Schulze, H.; Föllmann, W. (1997) Primary cultures of normal human urothelial cells for toxicological studies. *Naunyn Schmiedeberg's Arch. Pharmacol.* **355**, R141.

[6] Guhe, C.; Föllmann, W. (1994) Growth and characterization of porcine urinary bladder epithelial cells *in vitro*. *Am. J. Physiol.* **266**, F298–F308.

[7] Guhe, C.; Degen, G. H.; Schuhmacher, U.; Kiefer, F.; Föllmann, W. (1996) Drug metabolizing enzyme activities in porcine urinary bladder epithelial cells (PUBEC). *Arch. Toxicol.* **70**, 599–606.

[8] Dörrenhaus, A.; Föllmann, W. (1997) Effects of ochratoxin A on DNA repair in cultures of rat hepatocytes and porcine urinary bladder epithelial cells. *Arch. Toxicol.* **71**, 709–713.

2 *In vitro* Covalent Binding as a Possible Quantitative Indicator for the Genotoxic Potency of Aflatoxin B_1 in Rat and Human

Kerstin Groß-Steinmeyer[1], J. Weymann[1], H. G. Koebe[2] and M. Metzler[3]

Abstract

The mycotoxin aflatoxin B_1 (AFB) naturally occuring in food exhibits genotoxic and carcinogenic effects in animals. The *in vivo* covalent binding of genotoxic chemicals does positively correlate with their carcinogenic potency in animals. Thus, the aim of this study was (i) to compare *in vivo* and *in vitro* covalent binding in the rat using AFB as reference chemical, and (ii) to extrapolate the AFB genotoxicity via *in vitro* covalent binding to humans, because *in vitro* experiments with human biological materials are possible. The results demonstrate a clear comparability between *in vivo* and *in vitro* covalent binding of AFB in rat liver and in rat hepatocytes. Furthermore, the mycotoxin shows less genotoxic effects in human hepatocytes than in rat cells. In conclusion, the findings indicate that humans may be less susceptible to AFB genotoxicity and perhaps to AFB carcinogenicity than rats.

Introduction

Several mycotoxins naturally occuring in food exhibit carcinogenic effects in animals, e. g. aflatoxins, ochratoxin A, griseofulvin and sterigmatocystin [1]. Nevertheless, there is insufficient evidence of human carcinogenicity for these mycotoxins except aflatoxin B_1 (AFB) which shows its genotoxic effects after metabolic activation [2]. Epidemiological studies suggest that AFB is a human carcinogen [3], at least in the presence of hepatitis B virus infection [4]. The *in vivo*-Covalent Binding Index (CBI) of genotoxic chemicals defined as *μmol of bound chemical per mol nucleotide/mmol of administered chemical per kg body weight*

[1] Knoll AG, Drug Metabolism and Pharmacokinetics, P.O. Box 210805, D-67008 Ludwigshafen, Germany
[2] Klinikum Großhadern, Chir. Klinik und Poliklinik, D-81366 München, Germany
[3] University of Karlsruhe, Institute of Food Chemistry, P.O. Box 6980, D-76128 Karlsruhe, Germany

[5] does positively correlate with their carcinogenic potency in animals [6]. Therefore, the first aim of this study was to compare *in vivo* and *in vitro* covalent binding in the rat by using AFB as reference chemical. As *in vitro* experiments with human biological materials are possible, the second subject of this study was the evaluation of a suitable *in vitro*-CBI parameter in order to allow an extrapolation of animal data to those of humans.

Methods

Animal experiments (1 mg/kg BW iv; 2 h; n = 3), liver perfusions (1 mg/kg BW; 2 h; n = 3) and hepatocytes of male Sprague Dawley rats cultured by collagen-film method (1.5 µM; 2,4,6,8 h; n = 3) as well as by collagen-sandwich technique (1.5 µM; 8 h; n = 2) were carried out using the AFB as the [^{14}C]-ring-labelled radio-chemical. Human hepatocytes cultured by collagen-sandwich were also treated with the mycotoxin (1.5 µM; 8 h; n = 3). Cytotoxic effects in hepatocytes were measured as leakage of lactate dehydrogenase (LDH) in the medium. Cytotoxicity of AFB in liver perfusions was determined via bile flow.

The soluble metabolites in the urine of the animal experiments, in the bile and perfusate of the liver perfusions and in the supernatant of the hepatocytes were analyzed by HPLC after C18-purification.

DNA amounts in liver homogenates and hepatocyte lysates were quantified by their fluorescence after reaction with the Hoechst dye No 33258 in order to compare liver equivalents of whole organs with those of cells.

The DNA of whole livers and hepatocytes was purified by a combined proteinase K-SDS-treatment and a subsequent ribonuclease-incubation followed by silica-extraction. Every washing step was carried out as long as the rinsing solutions are free of radioactivity. To determine the covalent binding the isolated DNA was quantified by Hoechst dye No 33258 and known DNA amounts were measured by liquid scintillation counting after chemical hydrolysis by trichloroacetic acid.

Results and discussion

Covalent binding in rat experiments

The modified *in vivo*-CBI parameter CBI′ (*µmol of bound chemical per mol nucleotid/mmol of administered chemical per kg* **liver weight**) meets the requirements for a comparison of *in vitro* and *in vivo* studies by replacement of "body weight" by "liver weight" as a reference point which is determined via DNA quantification in liver homogenates as well as in cell lysates. The CBI′ values of animal experiments, perfused livers and collagen-film cultured rat hepatocytes which all were treated with AFB for 2 hours under comparable dose conditions were 708 ± 74, 630 ± 32 and 736 ± 39, respectively. These findings demonstrate that cultured rat hepatocytes are clearly suitable as an *in vitro* model to quantify the covalent binding level of AFB.

The covalent binding in collagen-film cultured rat hepatocytes treated with 1.5 µM AFB was determined after 2, 4, 6 and 8 hours in order to assess the time dependent alteration of AFB-adduct level in DNA. These results show that at the earliest after 4 hours the bound metabolites have reached a maximum value enduring at an almost constant level with a tendency to decrease after 8 hours. Based on the results of LDH-leakage and analysis of metabolism the findings suggest a steady-state of AFB-DNA-adduct formation and elimination, probably by repair enzymes or spontaneous hydrolysis, after 4 hours.

Covalent binding in rat and human hepatocytes

The CBI′-parameter gives no information about the metabolism rate of the applied chemical. This disadvantage is eliminated by the formation of the *in vitro*-CBI defined as *µmol of bound chemical per mol nucleotid/mmol of* **metabolized** *chemical per kg liver weight.* Due to the unequal metabolism rates of different species this quotient is especially suitable for comparative studies by integrating the internal AFB-exposition within the parameter.

The comparison of rat and human revealed that the CBI′ and *in vitro*-CBI values of the rat are on the average 43-times and 71-times higher, respectively (Table 2.1). The greater difference between rat and human values in the latter case is not surprising: The analysis of AFB and its soluble metabolites in the supernatants of human hepatocytes shows a larger extent of AFB-metabolism than in rat cells (Table 2.1). This means that human hepatocytes could have activated a greater quantity of chemical than rat hepatocytes although this does not result in higher amounts of bound metabolites.

Table 2.1: Comparison of the two CBI parameters in rat and human hepatocytes cultured by collagen-sandwich technique after incubation with AFB (1.5 μM; 8 h).

Species	**Extent of metabolism**	**AFB-CBI′ [%]**	**Equivalent to [%]**	***In vitro*-CBI**	**Equivalent to [%]**
Rat (n = 2)	43	2329	100	5553	100
Human (m; 68 years)	81	62	2,7	76	1,4
Human (f; 48 years)	95	50	2,1	52	0,9
Human (m; 62 years)	50	51	2,2	105	1,9

Extrapolation of animal to human data

Due to the relation between *in vivo* and *in vitro* covalent binding of AFB in the rat as shown in this study, an extrapolation of human *in vitro* genotoxicity to the corresponding *in vivo* data seems to be possible. Consequently, the findings of covalent binding in rat and human hepatocytes indicate that humans may be less susceptible to AFB genotoxicity and perhaps to AFB carcinogenicity than rats.

Limitations

(i) The model was established using the strong DNA-reactive mycotoxin AFB. It is nessecary to examine several other chemicals, in particular moderate and weak genotoxic compounds. Another reason for such a system validation is the influence of hepatocyte culture conditions (collagen-film vs. collagen-sandwich) on the extent of covalent binding. (ii) The biological consequences of DNA-damage are dependent on the size and disposition of the DNA-adducts, their capacity to hydrolyze spontaneously and their repairability. This cannot be assessed completely by binding studies and needs to be minded when extrapolating animal data to those of humans.

Conclusions

The results of the different rat test systems using the strong DNA-reactive mycotoxin AFB show a clear comparability of *in vivo* and *in vitro* covalent binding. Therefore, cultured hepatocytes are an appropriate *in vitro* model to quantify the genotoxicity of AFB.

The *in vitro*-CBI parameter allows the comparison of covalent binding in hepatocytes of different species by integrating the internal exposition of the applied chemical within the quotient. Its evaluation in rat and human hepatocytes revealed a lower genotoxic effect of AFB in human cells. This indicates that humans could be less susceptible to AFB genotoxicity and perhaps to AFB carcinogenicity, too.

It may be possible to assess the genotoxic potency of chemicals via *in vitro* covalent binding in cultured human hepatocytes. This would represent a useful tool to obtain quantitative experimental data for human genotoxicity. Therefore, it could be a little step towards cancer risk assessment of chemicals for humans.

Acknowledgements

The authors would like to thank Andrea Egly, Siegried Krupp and Sabine Nescholta from Knoll AG and Mrs. Kußmaul from Klinikum Großhadern for their technical assistance.

References

[1] Castegnaro, M.; Wild, C. P. (1995) *Nat. Toxins* **3**, 327–331.
[2] Martin, C. N; Garner, R. C. (1978) *Nature* **267**, 863–865.
[3] Peers, F. G.; Linsell, C. A. (1977) *Ann. Nutr. Aliment.* **31**, 1005–1017.
[4] Peers, F. G.; Bosch, X.; Kaldor, J.; Linsell, A.; Pluumen, M. (1987) *Int. J. Cancer* **39**, 545–553.
[5] Lutz, W. K. (1979) *Mutat. Res.* **65**, 289–356.
[6] Lutz, W. K. (1986) *J. Cancer Res. Clin. Oncol.* **112**, 85–91.

3 N-Palmitoyl-HFB_1: A New Metabolite of Hydrolyzed Fumonisin (HFB_1) in Rat Liver Microsomes

Hans-Ulrich Humpf[1], E.-M. Schmelz[2], F. I. Meredith[3] and A. H. Merrill[2]

Fumonisin FB_1 is the predominant member of a family of mycotoxins produced mainly by *Fusarium moniliforme* (Sheldon), which occurs world-wide and grows primarily on yellow corn. Fumonisins are responsible for at least two animal diseases (equine leukoencephalomalacia and porcine pulmonary edema) and are carcinogenic for experimental animals and perhaps humans [1]. Some foods also contain the aminopentol backbone (HFB_1) that is formed upon base hydrolysis of the ester-linked tricarballylic acid (TCA) side chains of FB_1 [2]. Fumonisin FB_1 and, to a lesser extent, HFB_1 inhibit ceramide synthase [3] with either sphinganine or sphingosine as the substrates and with a variety of fatty acyl-coenzyme A's as the co-substrate. Since it was shown that the inhibition is competitive with both, the long-chain base and the fatty-acyl-coenzyme A, fumonisins may be recognized by both substrate binding sites of ceramide synthase. As an immediate consequence of the enzyme inhibition, FB_1 causes sphingosine to accumulate and therefore the sphinganine to sphingosine (Sa/So) ratio increases in tissues and a variety of cell types *in vitro* and *in vivo*. The accumulation of these bioactive compounds initiates a cascade of other intracellular alterations in sphingosine metabolism, which are responsible for a series of cellular effects [4].

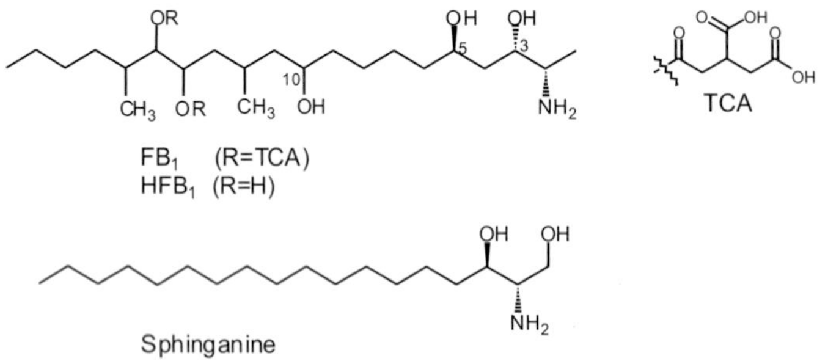

In this study we could show that HFB_1 is not only an inhibitor of the ceramide synthase but also a substrate of the enzyme and is converted to N-palmi-

[1] Lehrstuhl für Lebensmittelchemie, Universität Würzburg, Am Hubland, D-97074 Würzburg, Germany
[2] Department of Biochemistry, Emory University School of Medicine, Atlanta, GA 30322, USA
[3] Toxicology and Mycotoxin Research Unit, USDA, ARS, Athens, GA 30613, USA

toyl-HFB$_1$ in rat liver microsomes (Figure 3.1) [5]. The acylation of FB$_1$ and HFB$_1$ was followed using [1-^{14}C]palmitoyl-CoA as the co-substrate and rat liver microsomes as ceramide synthase source. The assay mixture was incubated with varying concentrations of FB$_1$ and HFB$_1$ (1–10 μM) for 15 min at 37 °C, followed by extraction of the products and TLC analysis. The corresponding reference compounds *N*-pal-FB$_1$ and *N*-pal-HFB$_1$ were synthesized from FB$_1$ and HFB$_1$ using palmitic anhydride as acylating agent. *N*-pal-FB$_1$ and *N*-pal-HFB$_1$ were characterized by MS and NMR spectroscopy [6]. Incubation of rat liver microsomes with 10 μM HFB$_1$ and [1-^{14}C]palmitoyl-CoA produced a radiolabelled product that migrated at the same position on the chromoplate as the reference compound, which indicates that HFB$_1$ is acylated by the ceramide synthase. The formation of *N*-pal-HFB$_1$ by rat liver microsomes could be confirmed by subsequent HPLC-electrospray tandem mass spectrometry analysis of the assay mixture. We were able to detect the protonated molecular ion in the single ion monitoring mode at m/z 645 $[M+H]^+$. The concentration dependence for the formation of *N*-pal-HFB$_1$ fits a simple Michaelis-Menten relationship. Under the assay conditions employed, the apparent K_m was 3.4 μM and V_{max} was 4.0 pmol/min mg microsomal protein for HFB$_1$. Thus, V_{max}/K_m for HFB$_1$ is approximately 1, compared to ca. 100 for the natural substrate sphinganine.

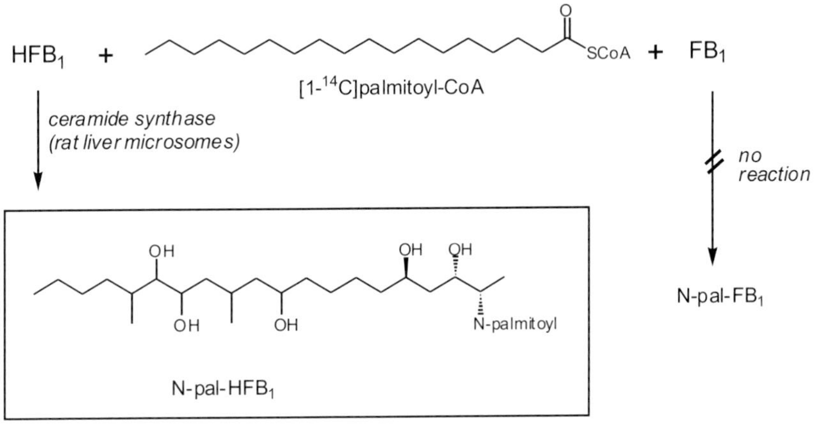

Figure 3.1: Reaction of FB$_1$ and HFB$_1$ with palmitoyl-CoA and formation of *N*-pal-HFB$_1$ using rat liver microsomes as ceramide synthase source.

When 10 μM FB$_1$ was incubated with rat liver microsomes and [1-^{14}C]palmitoyl-CoA, no radiolabel was detected in the region of the *N*-pal-FB$_1$ standard. Therefore, fumonisin FB$_1$ is not acylated by the ceramide synthase (Figure 3.1).

In order to determine whether *N*-pal-HFB$_1$ is cytotoxic, this compound was incubated with HT29 cells, a human colonic cell line that is sensitive to FB$_1$ and HFB$_1$ [7]. The new metabolite *N*-pal-HFB$_1$ was highly toxic for HT29 cells (at least 10-times more toxic than FB$_1$ or HFB$_1$), and all tested concentration of

N-pal-HFB_1 (1–50 μM) caused a significant reduction in the number of viable cells and increased the concentration of free sphinganine.

In conclusion, these studies demonstrate that HFB_1 is converted to an acyl derivative that might play a role in the diseases caused by fumonisins.

Acknowledgement

We thank the Deutsche Forschungsgemeinschaft, Bonn (Hu 730/1–2), and the Josef-Schormüller-Gedächtnisstiftung, Berlin, for financial support.

References

[1] Jackson, L. S.; DeVries, J. W.; Bullerman, L. B. (1996) (Eds.) Fumonisins in food. *Adv. in experimental medicine and biology* Vol. **392**, Plenum Press, New York.

[2] Thakur, R. A.; Smith, J. S. (1996) *J. Agric. Food Chem.* **44**, 1047–1052.

[3] Wang, E.; Norred, W. P.; Bacon, C. W.; Riley, R. T.; Merrill, A. H. (1991) *J. Biol. Chem.* **266**, 14 486–14 490.

[4] Merrill, A. H.; Liotta, D. C.; Riley, R. T. (1996) *Trends Cell Biol.* **6**, 218–223.

[5] Humpf, H.-U.; Schmelz, E. M.; Meredith, F. I.; Vesper, H.; Vales, T. M.; Wang, E.; Menaldino, D. S.; Liotta, D. C.; Merrill, A. H. (1998) *J. Biol. Chem.* **273**, 19 060–19 064.

[6] *N*-pal-HFB_1: ^1H NMR (400 MHz, $CDCl_3$, ppm): $\delta = 0.83$ (d, $J = 7.0$ Hz, 3H), 0.87 (t, $J = 6.6$, 3H), 0.89 (t, $J = 7.4$, 3H), 0.99 (d, $J = 7.0$, 3H), 1.19 (d, $J = 7.0$, 3H), 1.24 (br s, 38H), 1.42 (br s, 6H), 1.8–2.0 (m, 3H), 2.18 (m, 2H), 2.34 (t, $J = 7.4$, 1H), 3.4 (dd, $J = 5.16$, 1H), 3.6–4.1 (m, 5H), 5.75 (d, $J = 8.44$, 1H). ESI-MS: *m/z* 645 $[M+H]^+$.

[7] Schmelz, E. M.; Dombrink-Kurtzman, M. A.; Roberts, P. C.; Kozutsumi, Y.; Kawasaki, T.; Merrill, A. H. (1998) *Tox. Appl. Pharm.* **148**, 252–260.

4 Are Nitroso Bile Acids Involved in Carcinogenesis Following Reflux of Duodenal Juice?

Martin Fein[1], Karl-Hermann Fuchs[1], Stefanie Diem[2] and Markus Herderich[2]

Surgically induced duodenal reflux is carcinogenic in the rat esophagus. So far, no tumor initiating carcinogens have been identified in this model of esophageal adenocarcinoma. One mechanism of carcinogenesis that has been proposed recently relies on the reaction of physiologic bile acids with nitrite to produce carcinogenic *N*-nitroso amides in the presence of bacterial overgrowth [1]. To test this hypothesis, we analyzed duodenal juice for the presence of endogenously formed *N*-nitroso bile acids before and after reflux inducing surgery.

All animals (n = 6) were operated with total gastrectomy and a loop esophagojejunostomy to produce esophageal reflux of duodenal juice. In this model, about 50 % of the animals developed adenocarcinoma after sixteen weeks [2]. Before suturing the anastomosis, duodenal juice (0.05 ml) was aspirated and stored at –80 °C. Standard chow was provided the next day. After 2 weeks (n = 3) and 5 weeks (n = 3), the rats were killed and duodenal juice was aspirated and analyzed within two hours together with the preoperative samples. Commercially available bile acids and chemically synthesized *N*-nitroso derivatives of taurocholic acid (TCA) and glycocholic acid (GCA) were used as reference substances [3]. HPLC-ESI-MS/MS analysis was focused on the detection of molecular ions $[M+H]^+$ of the two major bile acids TCA (*m/z* 516) and GCA (*m/z* 466) and their respective nitroso derivates (*m/z* 545 and *m/z* 495). In addition, selective SRM experiments were performed [4].

All preoperative samples and all samples prepared after 2 and 5 weeks showed the expected pattern of bile derivatives with TCA and GCA as predominating conjugates. However, even SRM experiments failed to demonstrate the presence of any *N*-nitroso taurocholic and glycocholic acids [5]. In addition, other nitroso derivatives could not be detected in any of the samples by neutral loss experiments monitoring the loss of NO (–30 u).

In conclusion, presence of *N*-nitroso bile acids in the duodenal juice could not be demonstrated. Likewise endogenous formation of *N*-nitroso TCA and GCA appears questionable. Therefore, other mechanisms for the initiation of carcinogenesis following reflux of duodenal juice in the rat model have to be taken into consideration.

[1] Chirurgische Universitätsklinik and
[2] Lehrstuhl für Lebensmittelchemie der Universität Würzburg, D-97074 Würzburg, Germany

References

[1] Miwa, K.; *et al.* (1995) *Cancer* **75**, 1426–1432.
[2] Fein, M.; *et al.* (1998) *J. Gastrointest. Surg.* **2**, 260–268.
[3] Dayal, B.; Ertel, N. H. (1997) *Lipids* **32**, 1331–1340.
[4] Selected reaction monitoring SRM: 1.8 mTorr Ar, 15 eV collision energy, scan time 2 s; *m/z* 545 → *m/z* 514, *m/z* 545 → *m/z* 509, *m/z* 495 → *m/z* 464, *m/z* 495 → *m/z* 459.
[5] The limit of detection for the nitroso derivatives was calculated to be 0.1% as compared to the respective unmodified TCA and GCA.

5 The Occurrence of Heterocyclic Aromatic Amines in Various Food Samples Analyzed by High Performance Liquid Chromatography-Electrospray Tandem Mass Spectrometry (HPLC-ESI-MS/MS)

Elke Richling, Dietmar Häring, Markus Herderich and Peter Schreier*

Heterocyclic aromatic amines (HAA) are well-known contaminants in various food. They have been reported to be formed in thermally treated meat and meat products depending on the cooking time and temperature [1–3]. Surprisingly, in non-thermally treated food such as wine PhIP has been identified [4] and even in environmental samples such as rainwater HAA have also been detected [5].

Despite some recently published data about the amounts of HAA in fast-food meat products [6, 7] and commercially available cooked foods [8, 9], there is still information lacking to establish a sufficient analytical base for risk evaluation. With our previously developed sensitive and selective on-line coupling of high performance liquid chromatography and electrospray tandem mass spectrometry (HPLC-ESI-MS/MS) using "selected reaction monitoring" (SRM) we were able to determine HAA in trace levels [10]. With this technique, a number of various foods including processed flavours were analyzed. In this paper, the data obtained in our studies are summarized.

To evaluate the impact of HAA in human daily diet from other sources than meat and meat products, we started to focus on food such as wine [11]. In 24 wines under study HAA were detected in the low ng l^{-1} range. MeIQx and

* Lehrstuhl für Lebensmittelchemie, University of Würzburg, Germany

MeAαC were found to be the most widely distributed HAA in the wines under study. Compared to the quantities of HAA evaluated in meat and meat products, the amounts in wines are 100–1000 fold lower. Our recently performed HPLC-ESI-MS/MS study of commercially available meat products revealed the widely distributed presence of PhIP and MeIQx, ranging from 0.1 to 5.3 ng g^{-1} and 0.1 to 5.2 ng g^{-1}, respectively. Lower amounts were found for 4,8-DiMeIQx and 7,8-DiMeIQx, ranging from 0.2 to 2.0 ng g^{-1} and 0.1 to 0.2 ng g^{-1}, respectively [12].

Our most recently investigations comprised the HPLC-ESI-MS/MS analysis of additional food samples such as musts, soysauces, soups, yeast, gravies, powdered milk, and several processed flavours [13]. Evaluating the data obtained in the studies it can be concluded that an overall occurrence of HAA in daily consumed food exists (Figure 5.1). Highest amounts were found in processed flavours (produced at temperatures <160 °C) and meat products with about <0.1 up to approximately 15 µg kg^{-1} of total HAA. Some foods such as commercially available soups and gravies showed HAA in the low ppb range, too. According to these data the previously calculated daily intake of 100 µg/person/day can be drastically reduced. Moreover, considering the dosage of processed flavours (produced at temperatures <160 °C) in food, they do not contribute to an additional exposure for consumers.

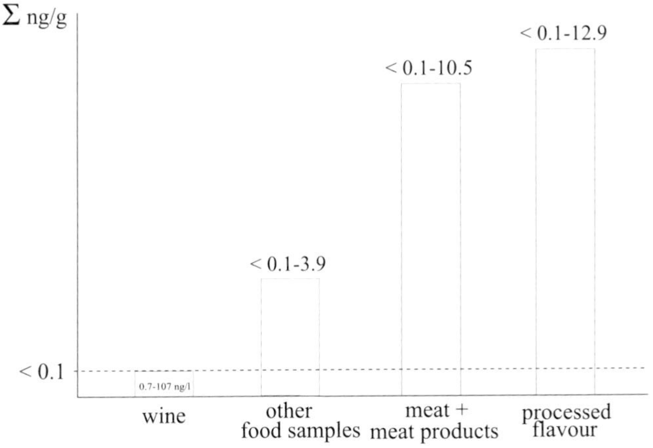

Figure 5.1: Amounts of HAA in food samples determined by HPLC-ESI-MS/MS.

Acknowledgements

The studies were supported by the Forschungskreis der Ernährungsindustrie and the AiF (AiF-FV 10502N). We thank S. Sauer for her skillful assistance.

Abbreviations

MeIQx	2-amino-3,4-dimethyl-3H-imidazo[4,5-*f*]quinoline (77500-04-0)
MeAαC	2-amino-3-methyl-9H-pyrido[2,3-*b*]indole (68006-83-7)
4,8-DiMeIQx	2-amino-3,4,8-trimethyl-3H-imidazo[4,5-*f*]quinoxaline (95896-78-9)
7,8-DiMeIQx	2-amino-3,7,8-trimethyl-3H-imidazo[4,5-*f*]quinoxaline (92180-79-5)
PhIP	2-amino-1-methyl-6-phenyl-imidazo[4,5-*b*]pyridine (105650-23-5)

References

[1] Sugimura, T.; Nagao, M.; Kawachi, T.; Honda, M.; Yahagi, T.; Seino, T.; Sato, S.; Matsukura, N.; Matsushima, T.; Shirai, A.; Sawamura, M.; Matsumoto, H. (1977) In: Hyatt, H. H.; Watson, J.; Winstan, J. A. (Eds.), *Origins of Human Cancer*, Cold Spring Harbour, New York, p. 1561–1577.

[2] Skog, K. (1993) *Food Chem. Toxicol.* **31**, 655–675.

[3] Eisenbrand, G.; Tang, W. (1993) *Toxicology* **84**, 1–82.

[4] Manabe, S.; Suzuki, H.; Wada, O.; Ueki, A. (1993) *Carcinogenesis* **14**, 899–901.

[5] Manabe, S.; Uchino, E.; Wada, O. (1989) *Mutat. Res.* **226**, 215–221.

[6] Knize, M. G.; Sinha, R.; Rothman, N.; Brown, E. D.; Salmon, C. P.; Levander, O. A.; Cunningham, P. L.; Felton, J. S. (1995) *Food Chem. Toxicol.* **33**, 545–551.

[7] Stavric, B.; Matula, T. I.; Klassen, R.; Downie, R. H. (1995) *Food Chem. Toxicol.* **33**, 815–820.

[8] Stavric, B.; Lau, B. P. Y.; Matula, T. I.; Klassen, R.; Lewis, D.; Downie, R. H. (1997) *Food Chem. Toxicol.* **35**, 199–206.

[9] Zoller, O.; Rhyn, P.; Zimmerli, B. (1997) *Lebensmittelchemie* **51**, 6.

[10] Richling, E.; Herderich, M.; Schreier, P. (1996) *Chromatographia* **42**, 7–11.

[11] Richling, E.; Decker, C.; Häring, D.; Herderich, M.; Schreier, P. (1997) *J. Chromatogr. A* **791**, 71–77.

[12] Richling, E.; Häring, D.; Herderich, M.; Schreier, P. (1998) *Chromatographia* **48**, 258–262.

[13] Richling, E. (1998) Doctoral Thesis, University of Würzburg.

6 Determination of HAAs in Samples with Strong Matrix Influences

Marc Vollenbröker and Karl Eichner

Introduction

Heterocyclic aromatic amines (HAAs) like aminoimidazoquinolines, -quinoxalines and -pyridines (IQ-compound) were identified as the main mutagenic compounds in heated meat, fish and their products. Maillard processed food flavours produced by heating a mixture of meat extract, maltodextrins/carbohydrates and flavour enhancers partly show mutagenic activity (Ames his-reversion test). The presence of HAAs may be a possible reason for this observation [1]. Although various methods for the separation and determination of HAAs in meat products by HPLC have been described, their analysis in trace amounts in process flavours heated at high temperatures is difficult because of strong matrix influences during the clean-up procedure as well as during their separation by HPLC. It turned out that not only the recovery rates in spiked samples are very low but also the HPLC column performance and the detection selectivity are not sufficient. Therefore, our goal was to find a method for an effective separation and determination of HAAs in samples with strong matrix influences. Furthermore, this method was also applied for the detection of the HAA PhIP in beer and its ingredients like yeast and malt [2, 3]. Since it was not clear, if PhIP being an ubiquitous environmental contamination was carried over to the raw materials from the environment or if it was formed during the malting and brewing process, we tried to find out a correlation between different processing conditions and the amount of PhIP formed.

Samples

Commercially available processed food flavours produced by heating a mixture of meat extract, maltodextrins/carbohydrates and flavour enhancers were selected. Moreover, maltodextrin and the flavour enhancers inosine-5-monophosphate, hydroxymethylpyrone, guanosine-5-monophosphate were heated in a

Institute of Food Chemistry, University of Münster, Corrensstrasse 45, D-48149 Münster, Germany mixture with meat extract (1 h, 140 °C, aw = 0.52) and the resulting concentrations of HAAs were determined. Furthermore, the influence of creatinine, glucose and phenylalanine on the formation of PhIP in this mixture was investigated.

The analysed beers produced from barley or wheat malt and their ingredients were commercially available. The selected types of beer are the favourites of the german consumers. Following the different brewing processes the possible influences of malting temperature or the brewing yeast species on the amount of PhIP formed were studied. Therefore, beers brewed with light or dark malt as well as beers brewed with *saccharomyces cerevisiae* or *carlsbergensis* were analysed. In addition, malt and yeast as such and a type of beer bottled with yeast were selected.

Clean-up procedure

After extraction of the sample with aqueous HCl the solution is cleaned up by a liquid/liquid-partition with methylene chloride at low pH values. The methylene chloride extract of the alkalized solution is applied on a SI SPE column und eluted with methanolic HCl. After evaporation of the solvent the concentrate is diluted with water, acidified and applied on a SCX SPE column. After washing with diluted HCl and elution with methanolic ammonia the solvent is again evaporated. The concentrate is diluted with water, alkalized and applied on a C18 SPE column. The C18 column is washed with a diluted ammonia solution and eluted with methanolic ammonia. After evaporation of the solvent the IQ-compounds and other HAAs are separated by HPLC.

HPLC separation and detection

The separation was performed on a RP-Select B column (RP-8, 250 × 4 mm) with a gradient of 0.01 *N* triethylammonium phosphate buffer (pH 3) and acetonitrile. A gradient of ammonium acetate buffer (pH 6.0) and methanol was used as a second chromatographic system. Diode-array, fluorescence and electrochemical (ox./red.) detection are used. The standard solution contains the HAAs IQ, MeIQ, MeIQx, 7,8-DiMeIQx, 4,8-DiMeIQx, Norharman, Harman, 4,7,8-TriMeIQx, PhIP, and MeAαC.

Results

The described clean-up procedure renders possible the determination of HAAs in process flavours heated at high temperatures using diode-array, fluorescence and electrochemical detection (recovery rate: 60–70 %). Besides Harman and Norharman the HAAs IQ and MeIQx were detected in some of the analysed process flavours. Contrary to meat extract, in process flavours MeIQx was detected in lower concentrations than IQ.

The added flavour enhancers do not influence the formation of HAAs. An addition of creatinine or glucose slightly increases the concentration of HAAs, an addition of phenylalanine strongly increases the formation of PhIP.

The mutagen and carcinogen PhIP was detected in all beer samples. In accordance with Manabe *et al.* [2] the concentration of PhIP in beer was between 10 and 30 ng/l (recovery rate: 60 %, limit of detection: less than 2 ng/l). Higher amounts as described by Löw [3] were not detected in the analysed samples. There was no significant correlation between the amounts of PhIP detected in beer and the temperatures applied during the malting process; also the yeast species used had no significant influence on the amount of PhIP found in beer, whereas the HAAs Harman and Norharman were detected in concentrated yeast extract (20–80 ng/g). Malt, especially dark malt, contained PhIP in relatively high concentrations (0.01–0.5 ng/g). The different contents of PhIP in light and dark malt did not lead to significantly different concentrations of PhIP in beers brewed with light or dark malt. In Germany the average consumption of beer per annum and head is about 150 litres. According to that the estimated average intake of PhIP per annum and head is about 2 μg. Of course this valuation does not take into account other PhIP sources.

References

[1] Stavric, B.; *et al.* (1997) Mutagenic heterocyclic aromatic amines (HAAs) in processed food flavour samples. *Food Chem. Toxicol.* **35**, 185–197.

[2] Manabe, S.; Suzuki, H.; Wada, O.; Ueki, A. (1993) Detection of the carcinogen 2-amino-1-methyl-6-phenyl-imidazo[4,5-*b*]pyridine (PhIP) in beer and wine. *Carcinogenesis* **14**, 899–901.

[3] Löw, H. (1996) *Heterocyclische Aromatische Amine (HAAs) in Lebensmitteln*, Dissertation TU Graz.

7 Activation of Nitrosamines in Metabolically Proficient Cell Systems to Induce Cytotoxic and Genotoxic Effects

Monika Hofer[1], Christine Janzowski[1], Richard N. Loeppky[2] and Gerhard Eisenbrand[1]

Nitrosamines are potent environmentally relevant carcinogens occuring in foods, cosmetics, industrial products and tobacco smoke. Nitrosamines also can be formed endogenously by nitrosation of amine precursors. To investigate activation pathways, primary rat hepatocytes and transfected V79 cells stably expressing rat CYP 2B1 and human CYP 2E1 [1, 2] were incubated with specific nitrosamines. Activation of nitrosamines was measured comparing cytotoxicity (sulforhodamin B staining) and induction of mutations at the HPRT locus in transfected and parental V79 cells [3]. Induction of DNA single strand breaks (DNA SSB) and formation of the promutagenic O^6-alkylguanine DNA adducts (O^6-alkyl dG) were monitored by alkaline elution and ^{32}P-postlabelling, respectively [4, 5].

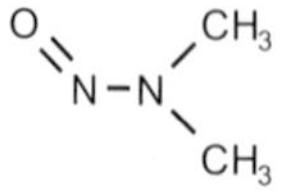

DMN

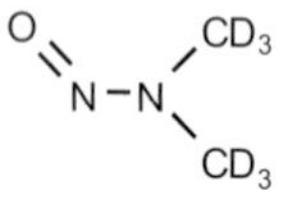

D_6-DMN

Dimethylnitrosamine (DMN) induced strong cytotoxic effects and mutations at the HPRT locus in CYP 2E1 cells at low μmolar concentrations. DMN also induced DNA SSB and formation of O^6-methyl dG adducts in a dose dependent manner in CYP 2E1 transfected cells. The perdeuterated DMN-D_6 clearly was less cytotoxic than its non-deuterated analogue. CYP 2E1 dependent activation was also observed with *N*-nitrosopyrrolidine (NPYR) and *N*-nitrosodiethanolamine (NDELA). NDELA and its β-oxidized metabolite *N*-nitroso-2-hydroxy-morpholine (NHMOR) were identified as CYP 2E1 substrates in cytotoxicity and mutagenicity tests [3]. Comparative cytotoxicity experiments with selectively deuterated NDELA isotopomers (α-D_4-NDELA, β-D_4-NDELA) showed unlabeled NDELA and its β-D_4 isotopomer to exhibit about the same toxicity [6]. In contrast, the α-D_4-NDELA clearly was less active, suggesting an activation mechanism involving α-C-hydroxylation.

[1] Department of Chemistry, Food Chemistry and Environmental Toxicology, University of Kaiserslautern, Erwin-Schrödinger-Str., D-67663 Kaiserslautern, Germany;
[2] Department of Chemistry, University of Missouri, Columbia, MO, 65211, USA

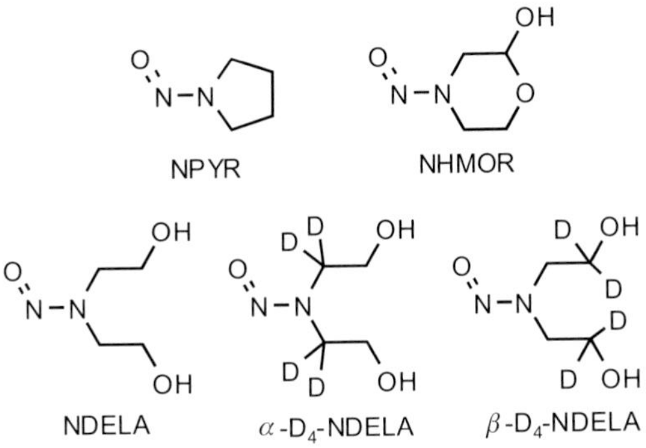

Dibutylnitrosamine (DBN) was activated by CYP 2B1 dependent α-C-hydroxylation, inducing cytotoxic, mutagenic and genotoxic effects and causing O^6-butyl dG formation.

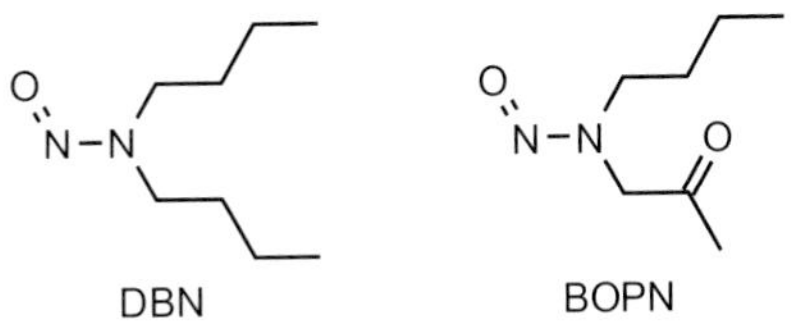

DBN has been shown to be carcinogenic to the urinary bladder. For this organotropic effect, a specific metabolic activation pathway, initiated by ω-oxidation, has been discussed. The main urinary metabolite, an ω-carboxylated nitrosamine, has been found to generate butyl-2-oxopropyl-nitrosamine (BOPN) by mitochondrial β-oxidation. In primary rat hepatocytes cultured on collagen coated flasks, BOPN induced DNA SSB and formation of O^6-methyl dG, its potency being similar to DMN. In CYP transfected V79 cells, BOPN was predominantly toxified by CYP 2B1. To a minor extent it was also a substrate for CYP 2E1. These results strongly support the relevance of 2-oxopropyl-metabolites as key intermediates in nitrosamine related bladder carcinogenesis.

References

[1] Doehmer, J.; Dogra, S.; Friedberg, T.; Monier, S.; Adesnik, M.; Glatt, H.; Oesch, F. (1988) Stable expression of rat cytochrome P450 IIB1 cDNA in Chinese hamster cells (V79) and metabolic activation of aflatoxin B_1. *Proc. Natl. Acad. Sci. USA* **85**, 5769–5773.

[2] Schmalix, W. A.; Barrenscheen, M.; Landsiedel, R.; Janzowski, C.; Eisenbrand, G.; Gonzalez, F.; Eliasson, E.; Ingelmann-Sundberg, M.; Perchermeier, M.; Greim, H.;

Doehmer, J. (1995) Stable expression of human cytochrome P450 2E1 in V79 Chinese hamster cells. *Eur. J. Pharmacol. Environ. Toxicol. Pharmacol. Section* **293**, 123–131.
[3] Janzowski, C.; Scharner, A.; Eisenbrand, G. (1996) Activation of nitrosamines by cytochrome P450 isoenzymes expressed in transfected V79 cells. *Eur. J. Cancer Prev.* **5** *(Suppl 1)*, 154.
[4] Eisenbrand, G.; Hofer, M.; Janzowski, C. (1996) O^6-Alkyl- and cyclic propano-deoxyguanosine adducts in DNA: Formation and cellular consequences. In: Hengstler, J. G.; Oesch, F. (Eds.) *Control mechanisms of carcinogenesis*, Thieme Druck, Meissen, p. 160–192.
[5] Hofer, M.; Janzowski, C.; Eisenbrand, G. (1996) O^6-alkyldeoxyguanosine adduct formation by *N*-nitroso compounds in calf thymus DNA and in metabolically proficient cell systems. *Eur. J. Cancer Prev.* **5** (Suppl. 1), 153.
[6] Loeppky, R. N.; Fuchs, A.; Janzowski, C.; Humberd, C.; Goelzer, P.; Schneider, H.; Eisenbrand, G. (1998) Probing the mechanism of the carcinogenic activation of *N*-nitrosodiethanolamine with deuterium isotope effects: *in vivo* induction of DNA single-strand breaks and related *in vitro* assays. *Chem. Res. Toxicol.* **11**, 1556–1566.

8 2,3,7,8-Tetrachlorodibenzo-*p*-dioxin (TCDD) Suppresses Apoptosis in the Livers of c-*myc* Transgenic Mice and in Rat Hepatocytes in Primary Culture

Wolfgang Wörner[2], Martina Müller[2], Hans-Joachim Schmitz[1] and Dieter Schrenk[1]

A number of promoters of hepatocarcinogenesis in rodents including TCDD have been shown to inhibit apoptosis in preneoplastic foci. In the livers of c-*myc* transgenic mice, increased mitotic activity, karyomegaly and incidence of apoptosis were observed in comparison to the background wild-type strain. After six, nine, and twelve weeks of treatment with 100 ng TCDD/kg × day, a significant reduction of apoptosis was seen in the livers of female c-*myc* transgenic mice. In males, a slight but not statistically significant reduction was observed. The percentage of giant nuclei (karyomegaly) was not affected, however, by TCDD treatment.

In rat hepatocytes in primary culture treated with apoptogenic factors (2-acetylaminofluorene, UV irradiation), 1 nM TCDD inhibited apoptosis measured as internucleosomal DNA fragmentation or as percentage of cells bearing

[1] Food Chemistry and Environmental Toxicology, University of Kaiserslautern, D-67663 Kaiserslautern, Germany
[2] Institute of Toxicology, University of Tübingen, D-72074 Tübingen, Germany

microscopically detectable fragmented and/or condensed chromatin. Concomitantly, the rise in p53 normally observed in hepatocytes after this type of apoptogenic treatment was suppressed by TCDD. Analysis of the phosphorylation pattern of cellular proteins revealed that TCDD stimulated phosphorylation of a variety of proteins including p53 in a concentration-dependent manner. The concentration-response curves for hyperphosphorylation of p53 and Ah receptor (AhR)-regulated induction of CYP 1A-catalyzed 7-ethoxyresorufin-*O*-deethylase (EROD) activity were almost identical indicating the role of the AhR. Hyperphosphorylation of p53 was significantly decreased when the cellular extracts were pretreated with anti-c-*src* antibodies.

These findings suggest that the AhR-associated tyrosine kinase c-*src* which is activated upon ligand binding of the AhR, may hyperphosphorylate a number of transacting cellular proteins including p53. Further research will be aimed at clarifying how tyrosine phosphorylation of p53 and other target proteins of c-*src* affects apoptosis in hepatocytes, and how c-*myc* triggered apoptosis depends on these proteins.

Acknowledgements

The contributions of Dr. A. Buchmann, E. Zabinski (Laboratory of Environmental Toxicology, Institute of Toxicology, University of Tübingen), and S. Vetter (Institute of Toxicology, University of Tübingen) are gratefully acknowledged. C-*myc* transgenic mice were provided by Dr. S. Thorgeirsson (National Cancer Institute, Bethesda, Md., USA). This work was supported by the Deutsche Forschungsgemeinschaft and the Bundesministerium für Forschung und Technologie, Bonn.

References

Wörner, W.; Schrenk, D. (1996) Influence of 2,3,7,8-tetrachlorodibenzo-*p*-dioxin on apoptosis in rat hepatocytes initiated by 2-acetylaminofluorene, UV light or transforming growth factor-β1. *Cancer Res.* **56**, 1272–1278.

Schrenk, D.; Müller, M.; Merlino, G.; Thorgeirsson, S. S. (1997) Interactions of TCDD with signal transduction and neoplastic development in c-*myc* transgenic and TGF-α transgenic mice. *Arch. Toxicol.* Suppl. **19**, 367–375.

9 Metabolism of Benzo[*c*]phenanthrene and Genotoxicity of the Corresponding Dihydrodiols After Activation by Human Cytochrome P450-Enzymes in Comparison to Dibenzo[*a,l*]-pyrene-11,12-dihydrodiol and Benzo[*a*]pyrene-7,8-dihydrodiol

Matthias Baum, Werner Köhl, Steven S. Hecht[1], Shantu Amin[2], F. Peter Guengerich[3] and Gerhard Eisenbrand

Many polycyclic aromatic hydrocarbons (PAH) are potent mutagens. PAHs are activated by cytochrome P450 (CYP 450)-enzymes to dihydrodiol-epoxides as ultimate carcinogens (Figure 9.1). They occur as products of incomplete combustion of organic material and are found as contaminants in food.

The "fjord-region" compound benzo[*c*]phenanthrene (B[*c*]PH) showed only weak carcinogenic activity, but the corresponding fjord-region-dihydrodiol-epoxides were shown to be potent carcinogens in experimental animals [1, 2]. The question therefore needs to be addressed whether metabolic activation of B[*c*]PH occurs in humans. It has been demonstrated earlier that human MCF-7 cells can activate B[*c*]PH [3]. B[*c*]PH-activation has also been demonstrated in V79-cells expressing human CYP 450-enzymes [4].

To elucidate the metabolic activation pathways of the fjord-region-compound B[*c*]PH we incubated the ^3H-labelled compound with human microsomes from liver and lung and also determined the role of individual P450s. Metabolites were identified by HPLC with online-radiodetection. Incubation with liver-microsomes resulted in the formation of 3,4-dihydrodiol as main metabolite, together with 5,6-dihydrodiol and 4-OH B[*c*]PH as minor metabolites (Figure 9.1). Lung microsomes generated mainly 5,6-dihydrodiol with 3,4-dihydrodiol as minor-component.

In liver-microsomes we identified CYP 1A2 and 3A4 as being responsible for the activation by correlating the respective amounts of metabolites formed with CYP 450-isoenzyme activities of the microsomes. In lung microsomes, the known CYP 1A1/1A2-inhibitor α-naphthoflavone completely blocked metabolite

Department of Food Chemistry and Environmental Toxicology, University of Kaiserslautern, Erwin-Schrödinger-Str., D-67663 Kaiserslautern, Germany
[1] Cancer Center, University of Minnesota, Minneapolis, MN 55455, USA
[2] Division of Chemical Carcinogenesis, American Health Foundation, Valhalla, NY 10595, USA
[3] Department of Biochemistry and Center in Molecular Toxicology, Vanderbilt University, School of Medicine, Nashville, Tennessee 37232, USA

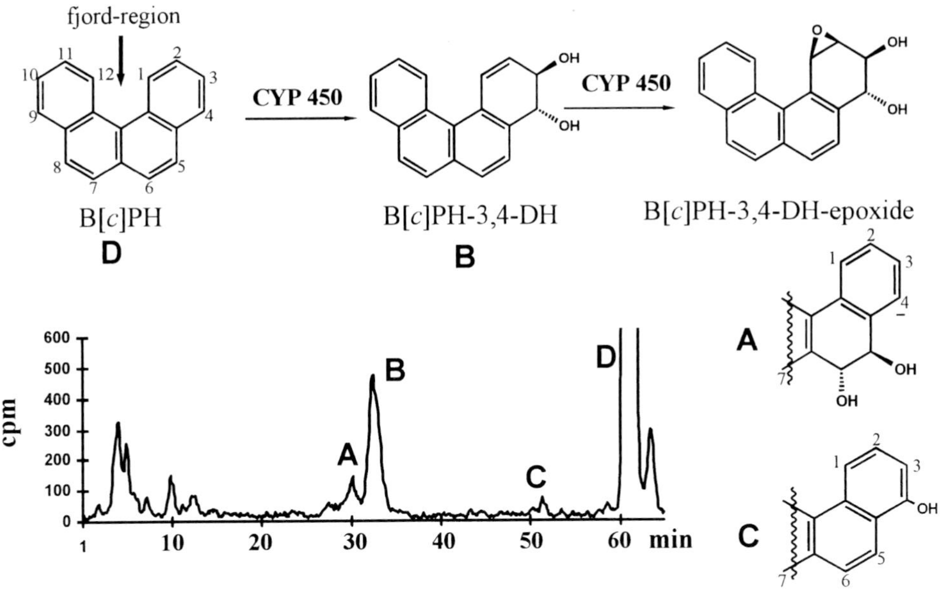

Figure 9.1: Activation of B[c]PH and typical HPLC-chromatogram of an ^{3}H-B[c]Ph-incubation with human liver-microsomes.

formation. In liver-microsomes, formation of B[c]PH-3,4-DH was inhibited. Taken together these findings suggests CYP 1A1 to be of major relevance for metabolite formation in human lung. In human liver CYP 1A2 appears of major importance.

Furthermore, genotoxic and mutagenic potency of several dihydrodiols of fjord-region-compounds were determined. Genotoxic activity was investigated by the *umu*-assay [5] with *Salmonella typhimurium* TA 1535/pSK 1002 containing a *umuC-lacZ* reportergene-construct. Testcompounds were activated with the same set of human liver microsomes that were also used for the B[c]PH-metabolism studies. Dibenzo[*a,l*]pyrene-11,12-dihydrodiol (DB[*a,l*]P-11,12-DH) by far was the most active compound, followed by B[c]PH-3,4-DH which showed activity comparable to benzo[*a*]pyrene-7,8-dihydrodiol (B[*a*]P-7,8-DH). Correlating the observed *umu*-activities with the isoenzyme activities of the microsomes strongly suggested CYP 3A4 to be the relevant isoenzyme for the ultimate metabolic activation of the dihydrodiols. In addition, the known CYP 3A4 inhibitor troleandomycin effectively inhibited *umu*-activities. We therefore used V79-cells expressing human CYP 3A4 together with human NADPH-oxidoreductase [6] to further elucidate mutagenic activities by investigation of HPRT forward mutations. In this system DB[*a,l*]P-11,12-DH induced about 10-fold more mutations than B[*a*]P-7,8-DH or B[c]PH-3,4-DH which is in agreement with the above findings for genotoxicity. In contrast, B[c]PH-5,6-DH had only weak potency in both systems. Our studies demonstrate that B[c]PH can effectively be metabolized by microsomes from human tissues to strongly genotoxic and mutagenic metabolites.

DB[*a,l*]P-11,12-DH, the precursor of the potent carcinogen DB[*a,l*]P-11,12-DH-13,14-epoxide by activation with human liver microsomes became extremely potently mutagenic.

The results mandate a careful analytical search for the occurrence of such compounds in food, consumer products and the environment.

References

[1] Amin, S.; Krzeminski, J.; Rivenson, A.; Kurtzke, C.; Hecht, S. S.; El-Bayoumy, K. (1995) Mammary carcinogenicity in female CD rats of fjord region diol epoxides of benzo[c]phenanthrene, benzo[*g*]chrysene and dibenzo[*a,l*]pyrene. *Carcinogenesis* **16**, 2813–2817.

[2] Hecht, S. S.; El-Bayoumy, K.; Rivenson, A.; Amin, S. (1995) Potent mammary carcinogenicity in female CD rats of a fjord region diol-epoxide of benzo[c]phenanthrene compared to a bay region diol-epoxide of benzo[*a*]pyrene. *Cancer Res.* **54**, 21–24.

[3] Einolf, H. J.; Amin, S.; Yagi, H.; Jerina, D.; Baird, W. M. (1996) Benzo[*c*]phenanthrene is activated to DNA-binding diol epoxides in the human mammary carcinoma cell line MCF-7 but only limited activation occurs in mouse skin. *Carcinogenesis* **17**, 2237–2244.

[4] Seidel, A.; Soballa, V. J.; Raab, G.; Frank, H.; Grimmer, G.; Jakob, J.; Doehmer, J. (1998) Regio- and stereoselectivity in the metabolism of benzo[*c*]phenanthrene mediated by genetically engineered V79 Chinese hamster cells expressing human cytochromes P450. *Environ. Toxicol. Pharmacol.* **5**, 179–196.

[5] Shimada, T.; Oda, Y.; Yamazaki, H.; Mimura, M. M.; Guengerich, F. P. (1994) SOS function tests for studies of chemical carcinogenesis using *Salmonella typhimurium* TA 1535/pSK 1002, NM 2009, and NM 3009. In: Adolph, K. W. (Ed.) *Methods in molecular genetics, Vol. 5. Gene and chromosome analysis.* Academic Press, Orlando, p. 342–354.

[6] Schneider, A.; Schmalix, W. A.; Siruguri, V.; Groene de Els, M.; Horbach, G. J.; Kleingeist, B.; Lang, D.; Böcker, R.; Belloc, C.; Beaune, P.; Greim, H.; Doehmer, J. (1996) Stable expression of human cytochrome P450 3A4 in conjunction with human NADPH-cytochrome P450 oxidoreductase in V79 Chinese hamster cells. *Arch. Biochem. Biophys.* **332**, 295–304.

10 *In vitro* Toxicity and Microsomal Metabolism of the Polycyclic Musk Fixolide and Crysolide

Christine Janzowski, Martin Burkart, Aribert Vetter and Gerhard Eisenbrand

Introduction

Polycyclic musk compounds are widely used synthetic fragrances which substitute nitro musks in cosmetics, toiletry and in detergents [1]. The lipophilic compounds have been found environmentally persistent and to accumulate in fish and other aquatic organisms in concentrations exceeding those of nitro musk flavors [2, 3]. Residues have also been identified in human adipose tissue and in breast milk up to 200 and 600 µg/kg fat, respectively [4–6]. The relevance of this exposure, however, is not clear yet, since only limited information about the toxicity of these fragrances is available.

We investigated two relevant representatives (Figure 10.1) fixolide (AHTN; 7-acetyl-1,1,3,4,4,6-hexamethyltetraline) and crysolide (ADBI; 4-acetyl-1,1-dimethyl-6-*tert*-butylindane) for cytotoxic/genotoxic/mutagenic effectiveness in V79 cells, stably transfected with human CYP 3A4 (h3A4 cells) [7]. Additionally, microsomal metabolism of the compounds was monitored.

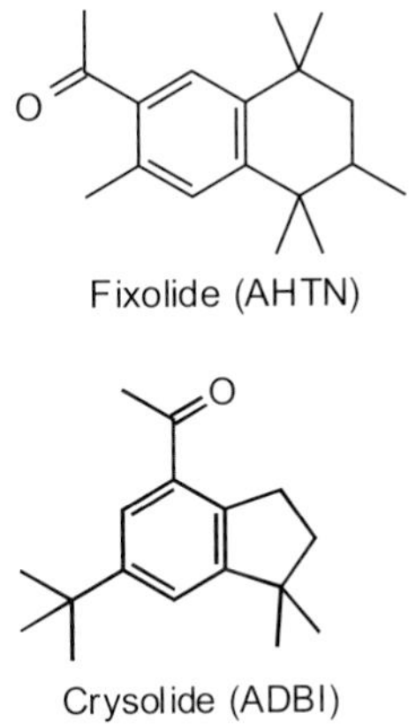

Figure 10.1: Structures of tested compounds.

Food Chemistry and Environmental Toxicology, Dept. of Chemistry, University, Erwin-Schrödinger-Str. 52, D-67663 Kaiserslautern, Germany

Methods

Cytotoxicity was determined after 24 h incubation followed by a 3–4 d postincubation period (until controls approached confluency). Concentration dependent relative viabilities were determined by protein staining with sulforhodamin B [8], IC_{50} were calculated after log-probit transformation. DNA damage was quantified by single cell gel electrophoresis (COMET assay) under alkaline conditions after 1 h incubation of suspension culture in a shaking water bath [9]. Tail intensity of the images was analyzed semi-automatically; viability of cells was controlled by trypan blue exclusion. Mutagenicity testing (HPRT locus) was performed following 24 h incubation [10]. Genotoxicity was screened in *Salmonella typhimurium*, TA1535/pSK1002, carrying a fused gene *umu*C'-'lacZ chimera (*umu*-test, 2 h incubation) in the presence and absence of microsomal activation [11]. Metabolism of the compounds was monitored after incubation with microsomes from Aroclor pretreated male Wistar rats. Parent compounds were quantified by GC/MS (HP 5971; HP5 capillary column) after Extrelut extraction into dichloromethane and concentration in hexane. Losses during clean up were corrected by using an internal standard. Oxidized metabolites were tentatively identified by characteristic mass spectra.

Results and discussion

Crysolide clearly showed an increased toxicity in h3A4 cells (IC_{50}: 73 μM), compared to the parental V79 cells (IC_{50}: 196 μM) (Figure 10.2). Fixolide was more cytotoxic in parental cells (IC_{50}: 42 μM) and only to a minor extent activated by CYP 3A4 (IC_{50}: 31 μM). Both compounds (up to 2 mM) did not induce SOS repair in *S. typhimurium* (*umu*-test) in the presence/absence of Aroclor induced rat liver microsomes. In V79 and h3A4 cells, DNA damage (single cell gel electrophoresis) was not detectable after 1 h incubation with non cytotoxic concentrations (fixolide: 75 μM; crysolide: 100 μM). Similarly, mutations at the HPRT locus were not observed. Microsomal incubation of fixolide and crysolide revealed a characteristic time dependent decrease of initial substrate concentration (100 μM) in the presence of the complete mix (see Figure 10.3).

Moreover, GC/MS analysis showed time and concentration dependent formation of hydroxylated metabolites as detected by a characteristic 16 amu increase of diagnostic fragments. In the case of fixolide, oxygen insertion was confirmed by high resolution GC/MS.

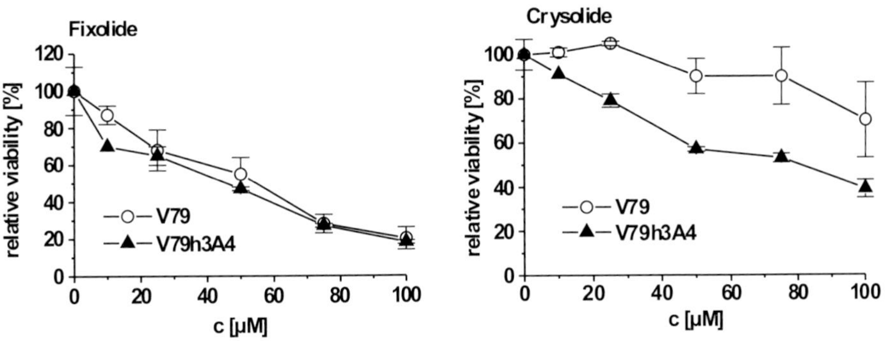

Figure 10.2: Cytotoxicity of fixolide and crysolide in V79 and h3A4 cells (24 h incubation).

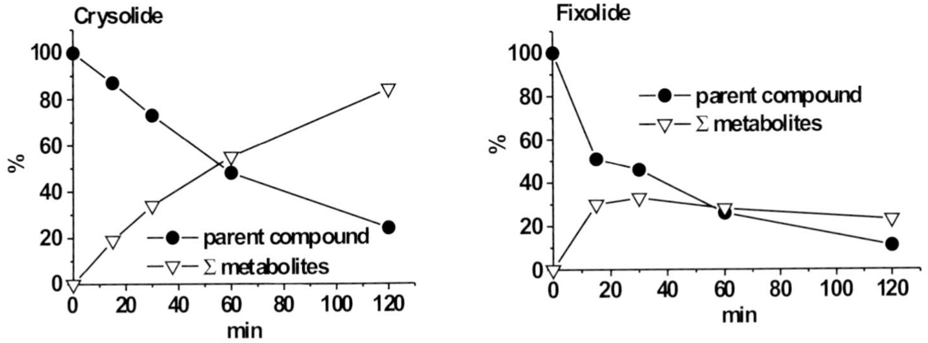

Figure 10.3: Time dependent decrease of parent compound and formation of metabolites at incubation of crysolide and fixolide with Aroclor induced rat liver microsomes.

In conclusion, fixolide and crysolide are not genotoxic or mutagenic in different *in vitro* test systems but show a distinct cytotoxicity, which is increased in the presence of CYP 3A4. Strong evidence for oxidative microsomal metabolism is obtained from incubations with Aroclor induced rat liver microsomes.

Acknowledgments

Thanks are due to C. Thielen for competent assistance, Prof. Dr. J. Doehmer, University of München, for the gift of CYP transfected V79 cells, and Dr. R. Graf, University of Saarbrücken, for high resolution GC/MS analysis.

References

[1] Ohloff, G. (1990) *Riechstoffe und Geruchssinn: Die molekulare Welt der Düfte*, Springer Verlag, Berlin, Heidelberg.

[2] Eschke, H.-D.; Dibowski, H.-J.; Traud, J. (1995) Untersuchungen zum Vorkommen polycyclischer Moschus-Duftstoffe in verschiedenen Umweltkompartimenten. *UWSF-Z. Umweltchem. Ökotox.* **7**, 131–138.

[3] Rimkus, G.; Wolf, M. (1997) Nachweis von polycyclischen Moschus-Duftstoffen in Fisch- und Humanproben. *Lebensmittelchemie* **51**, 81–96.

[4] Eschke, H.-D.; Dibowski, H. J.; Traud, J. (1995) Nachweis und Quantifizierung von polycyclischen Moschus-Duftstoffen mittels Ion-trap GC/MS/MS in Humanfett und Muttermilch. *Dtsch. Lebensm.-Rdsch.* **91**, 375–379.

[5] Müller, S.; Schmid, P.; Schlatter, C. (1996) Occurrence of nitro and non-nitro benzenoid musk compounds in human adipose tissue. *Chemosphere* **33**, 17–28.

[6] Rimkus, G. G.; Wolf, M. (1996) Polycyclic musk fragrances in human adipose tissue and human milk. *Chemosphere* **33**, 2033–2043.

[7] Schneider, A.; Schmalix, W. A.; Siruguri, V.; deGroene, E.; Horbach, G. J.; Kleingeist, B.; Lang, D.; Böcker, R.; Belloc, C.; Beaune, P.; Greim, H.; Doehmer, J. (1996) Stable expression of human cytochrome P450 3A4 in conjunction with NADPH cytochrome P450 oxidoreductase in V79 Chinese hamster cells. *Arch. Biochem. Biophys.* **332**, 295–304.

[8] Skehan, P.; Storeng, R.; Scudiero, D.; Monks, A.; McMahon, J.; Vistica, D.; Warren, J. T.; Bokesch, H.; Kenney, S.; Boyd, M. R. (1990) New colorimetric cytotoxicity assay for anticancer-drug screening. *J. Natl. Cancer Inst.* **82**, 1107–1112.

[9] Singh, N. P.; McCoy, M. T.; Tice R. T.; Schneider E. L. (1988)A simple technique for quantitation of low levels of DNA damage in individual cells. *Exp. Cell Res.* **175**, 184–191.

[10] Bradley, O. M.; Bhuyan, B.; Francis, M. C.; Langenbach,R.; Peterson, A.; Hubermann, E. (1981) Mutagenesis by chemical agents in V79 Chinese hamster cells: a review and analysis of the literature. A report of the Gene-Tox Program. *Mutat. Res.* **87**, 81–142.

[11] Oda, Y.; Nakamura, S.; Oki, I.; Kato, T. (1985) Evaluation of the new system (*umu*-test) for the detection of environmental mutagens and carcinogens. *Mutat. Res.* **147**, 219–229.

11 Special Effects of α,β-Unsaturated Aldehydes in Human Lymphocytes Studied by Premature Chromosome Condensation (PCC)

Eva Ritter, Christiane Meilike, Harry Scherthan and Heinrich Zankl

Introduction

The technique of premature chromosome condensation (PCC) allows the visualization of chromosome lesions in interphase cells. The fusion of interphase target cells with mitotic inducer cells results in rapid condensation of the interphase chromatin into chromosome-like structures after treatment with inactivated Sendai virus. This phenomenon was detected by Johnson and Rao [1] and they called it premature chromosome condensation. Pantelias and Wolff [2] presented a method for the induction of PCC on primary cells in suspension using polyethylene glycol as fusogen and used it for direct analysis of chemically induced clastogenic activity in unstimulated lymphocytes.

In the present study we used this PCC-method to investigate the genotoxic potential of 2-*trans*-butenal (crotonaldehyde, CA), 2-*trans*-hexenal (leaf aldehyde, HX) and 2-*trans*-6-*cis*-nonadienal (ND), which belong to the group of α,β-unsaturated aldehydes. These compounds are of special interest because they naturally occur in many fruits and vegetables. α,β-Unsaturated aldehydes are also used as food additives for flavouring purposes. Hexenal has been identified in about 80 different types of food with the highest natural concentration of 76 ppm in bananas. For crotonaldehyde the highest concentration found was 0.7 ppm, detected in red wine [3].

Mutagenic effects of crotonaldehyde and 2-*trans*-hexenal have already been reported [4]. Covalent interaction with DNA and formation of adducts were also shown [5]. Hexenal caused genotoxic effects on human buccal mucosa cells *in vivo* [6]. We used the PCC-technique to visualize the chromosome-like structures of unstimulated lymphocytes and to monitor directly the initial DNA-damage in comparison to chromosome breaks which were observed by metaphase analysis after PHA-stimulation.

Department of Human Biology and Human Genetics, University of Kaiserslautern, Erwin-Schrödinger-Str. 14, D-67663 Kaiserslautern, Germany

Material and methods

Unstimulated human peripheral blood lymphocytes were treated with different doses of CA (60 μM to 480 μM) and HX (50 μM to 300 μM), in a separate study with ND (20 μM to 80 μM) for 1 hour at 37 °C. HX and ND were dissolved in dimethylsulfoxide (DMSO). The well known mutagens bleomycin (BLM; 10 μg/ml), mitomycin C (MMC; 20 μM) and diethylstilbestrol (DES; 200 μM) were used for control purposes.

PCC analysis

Chinese hamster ovary (CHO) cells were used as mitotic PCC-inducers. Therefore the cells were subcultivated for about two days in medium containing BrdU (2 μM). Then the medium was removed and fresh medium containing BrdU and Colcemid (0.2 μg/ml) was added for the last four hours. The mitotic cells were collected by selective detachment.

For PCC-induction a slightly modified method of Pantelias and Maillie [7] was applied. The induction of DNA-damage was evaluated by counting the G_0-PCC-elements on FPG-stained slides. Cells with more than 46 chromatin-elements were classified as damaged cells.

Additionally the G_0-PCCs were divided into four categories according to their degree of chromosome condensation with category four being for the most extended PCCs.

Fluoresence *in situ* hybridization (FISH)

Fluorescence *in situ* hybridization was used to visualize damage by focusing on a single prematurely condensed chromosome. Therefore a biotin-labelled chromosome 4 specific probe was hybridized to PCC-spreads and detected by FITC-conjugated avidin [8].

Metaphase analysis

For conventional chromosome analysis the blood samples were cultured for 48 hours in RPMI 1640 medium containing 10 μM BrdU. Colcemid was added the last 2 hours. After slide preparation, FPG-staining was carried out [9]. Structural chromosome analysis was performed in first-division metaphases.

Results and discussion

The PCC-experiments revealed that treatment with 2-*trans*-butenal, 2-*trans*-hexenal and 2-*trans*-6-*cis*-nonadienal induced high yields of cells with additional PCC-elements. However, the numbers of aberrant metaphases were only slightly increased (Figure 11.1).

BLM (10 μg/ml) induced in 87.3 %, MMC (20 μM) in 46.3 % and DES (200 μM) in 26.3 % of the cells an increase of PCC-elements.

After treatment with the α,β-unsaturated aldehydes we observed an interesting additional phenomenon: The PCCs of CA-, HX- and ND-treated cells appeared strongly elongated in comparison to the PCCs of untreated cells. All tested α,β-unsaturated aldehydes reduced the condensation in a dose-dependent manner (Figure 11.2).

As visualized by chromosome painting the prematurely condensed chromosome 4 appeared about three-times more elongated than the PCC of an untreated cell. This strong effect of elongation was not observed after BLM-treatment and only slightly visible after incubation with MMC and DES.

These studies confirm the mutagenic potency of the tested α,β-unsaturated aldehydes. The presented data on PCC analysis are in very good agreement with the observation made by others, that the PCC-technique is far more sensitive for detecting mutagenic effects than the analysis of metaphases [10]. This is probably due to the lack of repair in the G_0-PCCs and an exclusion of heavily damaged cells during their progress to mitosis.

The observed elongation of PCCs in cells treated with α,β-unsaturated aldehydes may be caused by alterations of mitotic factors and/or the chromatin structure. Probably the α,β-unsaturated aldehydes interfere with the phosphorylation process of histone H1 and/or histone H3 which are supposed to be involved in PCC-induction being mediated through the activity of the mitosis-promoting factor.

Similar chromosomal elongation was observed after UV-irradiation [11], heat-treatment [12] and exposure to hydrogen peroxide [13]. Our study reveals that other strong chemical mutagens like BLM, MMC and DES do not show this elongating effect. Probably this reaction is restricted to some types of chemical compounds like α,β-unsaturated aldehydes.

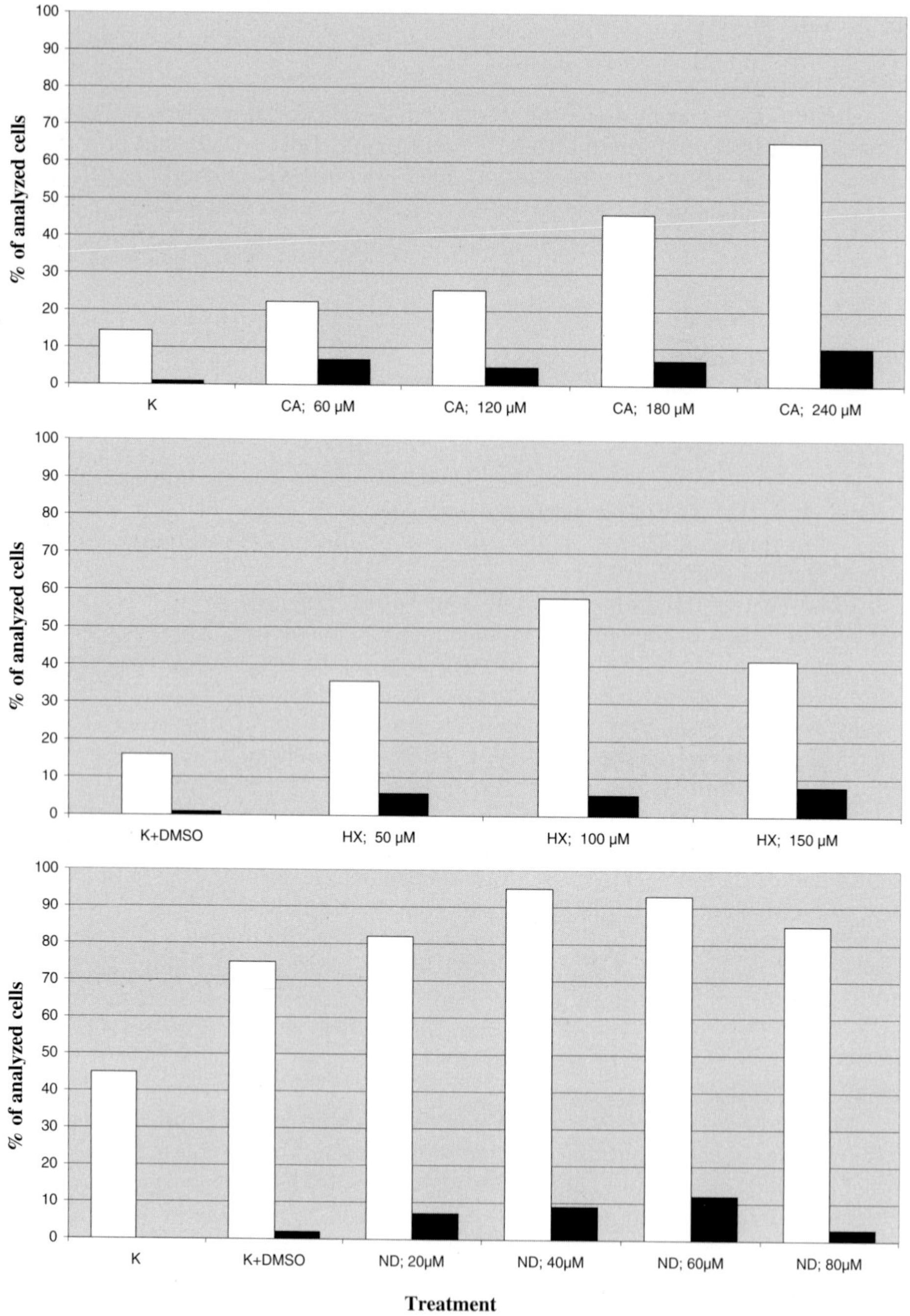

Figure 11.1: Frequency of damaged cells; comparison of PCC and metaphase spreads. Aberrant metaphases without gaps (solid bars), PCC <46 chromatin elements (open bars); K = untreated cells; K+DMSO = cells treated with DMSO.

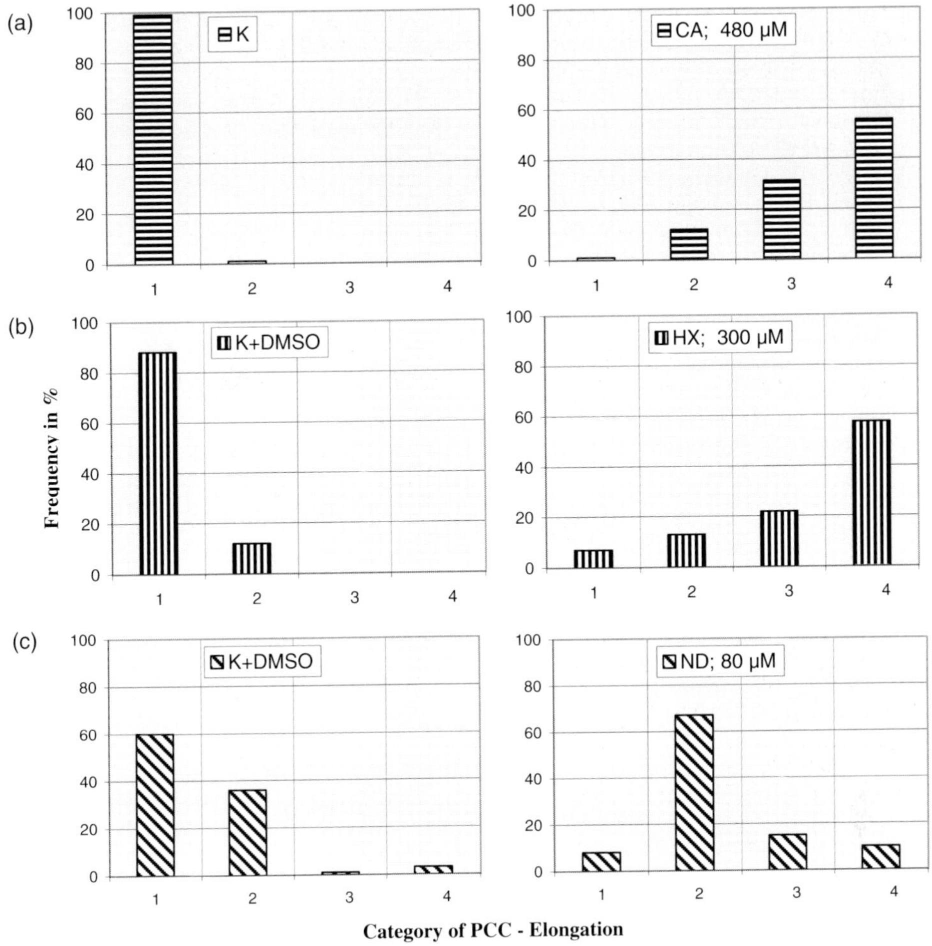

Figure 11.2: Dose response of premature chromosome condensation after treatment with CA (a), HX (b) and ND (c). K = untreated cells; K+DMSO = cells treated with DMSO.

References

[1] Johnson, R. T.; Rao, P. N. (1970) Mammalian cell fusion: Induction of premature chromosome condensation in interphase nuclei. *Nature* **226**, 717–722.

[2] Pantelias, G. E.; Wolff, S. (1985) Cytosine arabinoside is a potent clastogen and does not affect the repair of X-ray-induced chromosome fragments in unstimulated human lymphocytes. *Mutat. Res.* **151**, 65–72.

[3] Feron, V. J.; Til, H. P.; de Vrijer, F.; Woutersen, R. A.; Cassee, F. R.; van Bladeren, P. J. (1991) Aldehydes: occurence, carcinogenic potential, mechanism of action and risk assessment. *Mutat. Res.* **259**, 363–385.

[4] Eder, E.; Scheckenbach, S.; Deininger, C.; Hoffman, C. (1993) The possible role of α,β–unsaturated carbonyl compounds in mutagenesis and carcinogenesis. *Toxicol. Lett.* **67**, 87–103.

[5] Gölzer, P.; Janzowski, C.; Pool-Zobel, B. L.; Eisenbrand, G. (1996) (E)-2-Hexenal-induced DNA damage and formation of cyclic 1,N^2-(1,3-propano)-2′-deoxyguanosine adducts in mammalian cells. *Chem. Res. Toxicol.* **9**, 1207–1213.

[6] Dittberner, U.; Schmetzer, B.; Gölzer, P.; Eisenbrand, G.; Zankl, H. (1997) Genotoxic effects of 2-*trans*-hexenal in human buccal mucosa cells *in vivo*. *Mutat. Res.* **390**, 161–165.

[7] Pantelias, G. E.; Maillie, H. D. (1983) A simple method for premature chromosome condensation induction in primary human and rodent cells using polyethylene glycol. *Somat. Cell Genet.* **9**, 533–547.

[8] Scherthan, H.; Kioschis, P.; Zankl, H. (1990) Application of reflection contrast- and fluorescence microscopy to non-radioactive *in situ* hybridization. *Lab. Practice* **3**, 61–63.

[9] Perry, P.; Wolff, S. (1974) New Giemsa method for the differential staining of sister chromatids. *Nature* **251**, 156–158.

[10] Cornforth, M. N.; Bedford, J. S. (1983) X-ray-induced breakage and rejoining of human interphase chromosomes. *Science* **222**, 1141–1143.

[11] Waldren, C. A.; Johnson, R. T. (1974) Analysis of interphase chromosome damage by means of premature chromosome condensation after X-ray and ultraviolet irradiation. *Proc. Natl. Acad. Sci. USA* **71**, 1137–1141.

[12] Iliakis, G. E.; Pantelias, G. E. (1989) Effects of hyperthermia on chromatin condensation and nucleoli disintegration as visualized by induction of premature chromosome condensation in interphase mammalian cells. *Cancer Res.* **49**, 1254–1260.

[13] Iliakis, G. E.; Pantelias, G. E.; Okayasu, R.; Blakely, W. (1992) Induction by H_2O_2 of DNA and interphase chromosome damage in plateau-phase Chinese hamster ovary cells. *Radiat. Res.* **131**, 192–203.

12 Determination of Apoptosis and Mutagenic Effects of 2-Cyclohexen-1-one in Human Leukemic Cell Lines and Peripheral Blood Lymphocytes

Anja Genzlinger[1], I. Zimmermann[1], C. Janzowski[2] and H. Zankl[1]

Introduction

The α,β-unsaturated cyclic keton 2-cyclohexen-1-one is a widespread carbonyl compound, which was identified in certain foods, tobacco smoke and flavoring materials. It also occurs as environmental pollutant. Furthermore it was found in concentrations up to 15 µM in diet soft drinks sweetened with cyclamate. Acute toxic effects were observed in animal experiments, carcinogenicity was not yet reported. In V79 cells the compound showed cytotoxic and DNA damaging effects [1].

The present study was especially performed to investigate in detail the apoptotic potential of 2-cyclohexen-1-one. Additionally mutagenic effects were studied by cytogenetic methods.

Apoptosis, or programmed cell death, is an active and physiological process in which the cell itself designs and executes the program of its own demise and subsequent disposal. Morphologically apoptosis is characterised by reduction of cell volume, membrane blebbing, chromatin condensation and the formation of membrane vesicles, called apoptotic bodies. Another characteristic event is the activation of endonucleases which preferentially cleave DNA at the internucleosomal sections, resulting in fragments of 180 bp or multiples thereof. In some cell types DNA degradation does not proceed to nucleosomal sized fragments but rather results in 50–300 kbp DNA fragments [2]. Apoptosis can be induced by a variety of stimuli, e.g. glucocorticoids, anticancer drugs, UV irradiation and *anti*-APO-1 antibodies. A variety of flow cytometric methods have been developed for analysis of apoptosis in many cell systems. The easiest and most rapid discrimination of apoptotic cells is based on the analysis of DNA content. Apoptotic cells can be recognised by their low DNA stainability. And in the one parameter histogram they are visualised as sub-G1-peak [3].

In the present study we investigated induction of apoptosis by 2-cyclohexen-1-one by two parameter flow cytometry, a modified DNA/protein staining which discriminates intact and apoptotic cells [4] and also provides infor-

[1] Department of Human Biology and Human Genetics
[2] Department of Food Chemistry and Environmental Toxicology, University of Kaiserslautern, Erwin-Schrödinger-Str. 52, D-67663 Kaiserslautern, Germany

mation about the cell cycle specificity. The kinetics of cell death was also identified by fluorescence microscopy and confirmed by DNA agarose gel electrophoresis.

The mutagenic potential of 2-cyclohexen-1-one was examined by the evaluation of sister-chromatid-exchanges (SCE) and micronuclei frequency in blood lymphocytes. By these two methods a wide range of different mutagenic effects can be detected. Additionally the mitotic index and the proportion of cells from first, second and later mitotic cycles were determined to study the proliferative behaviour of the cells after treatment with 2-cyclohexen-1-one.

Results

The 2-parameter (DNA/protein) flow cytometric analysis revealed a significant dose- and time-related induction of apoptosis by 2-cyclohexen-1-one in the cell lines U937, HL-60 and K562. The differences to control cells became highly significant at 50 μM. In Molt4 cells no apoptosis could be detected by flow cytometry. Human blood lymphocytes already showed marked apoptosis in control cultures and the increase after treatment was only weak. The microscopic study of apoptosis revealed a good correlation with the flow cytometric analysis in U937 and HL-60 cells. Molt4 cells showed at 75 μM also a clear apoptotic cell population, identified by distinct chromatin condensation. In K562 cells however only few apoptotic cells could be microscopically identified. To confirm these results we additionally examined the DNA ladder formation by agarose gel electrophoresis and got positive results in U937, HL-60 and K562 cells. Furthermore it was observed by the flow cytometric analysis that apoptosis preferentially occurred in S-phase of U937 and HL-60 cells and in G1 phase of K562 cells. A significant dose- and time-related increase in mitotic cells was visible too (Figure 12.1).

The mutagenic studies showed significantly increased number of SCE in human blood lymphocytes after treatment with 25 μM 2-cyclohexen-1-one (Figure 11.1). A dose dependent slow down in cell progression was also observed. An increased frequency of micronuclei in lymphocytes could not be detected.

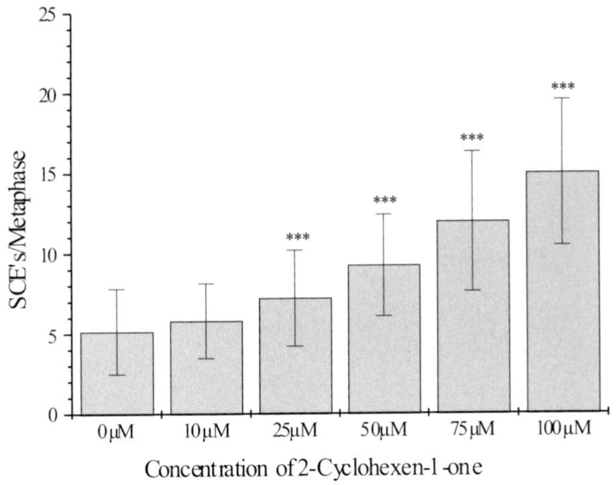

Figure 12.1: Microscopic evaluation of the dose dependent induction of SCE by 2-cyclohexen-1-one in human blood lymphocytes.

Discussion

In the present study 2-cyclohexen-1-one was proven to be a potent inductor of apoptosis in human leukemic cell lines at concentration of 50 µM or higher. However these concentrations are two- to three-times higher than that found in diet soft drinks. Apoptosis was induced preferentially in S and G1 phase and committed a mitotic arrest. The two parameter flow cytometric method for DNA/protein measurement was suitable to discriminate the phases of cell cycle from which apoptosis occurred (Figure 12.2). Apoptosis was also proven by DNA ladder formation in three cell lines. In Molt4 cells however which usually do not show internucleosomal DNA fragmentation [5, 6], apoptosis could only be detected by studying cell morphology. In contrast we could not confirm apoptosis in K562 cells by morphological investigations. These results show that lacking evidence of apoptosis in one study does not prove the absence of apoptosis at all. There are numerous examples in the literature of apoptotic cell death without showing all typical apoptotic features [7]. Therefore application of more than one method, based on different parameters is necessary to detect apoptosis with certainty.

2-Cyclohexen-1-one was also proven to be a mutagenic agent because it increased significantly sister chromatid exchanges in human blood lymphocytes. Additionally the reduced mitotic index and the slow down of cell cycle progression revealed cytotoxic effects. Differences between the cell lines and human

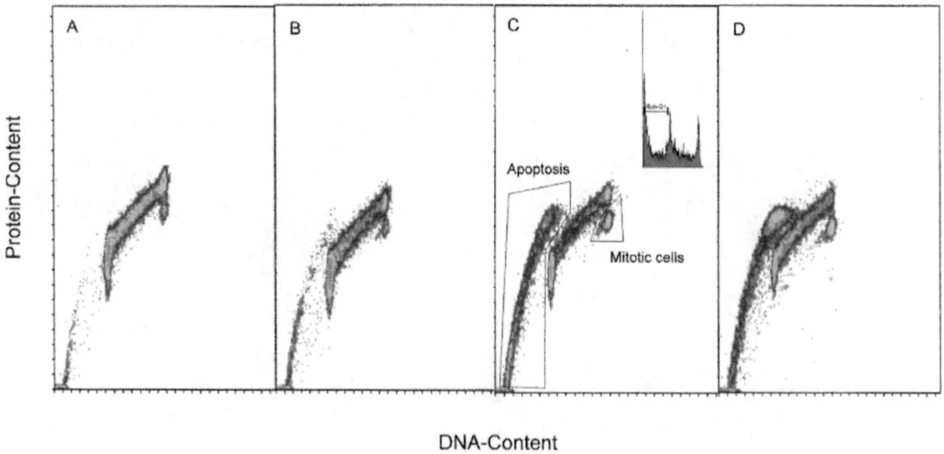

Figure 12.2: Dot plots of the 2-parameter-measurement (DNA/protein) of apoptosis in U937 cells induced by 2-cyclohexen-1-one. A. Control, B. 25 µM/24 h, C. 50 µM/24 h, D. 100 µM/24 h. The additional histogram in C shows, that not all apoptotic cells could be detected with DNA staining alone. The apoptotic region contents apoptotic cells, apoptotic bodies, micronuclei and debris.

blood lymphocytes in induction of apoptosis and mitotic arrest could be due to different patterns of detoxifying enzymes.

Our results indicate therefore that 2-cyclohexen-1-one has not only cytostatic but also apoptotic and mutagenic effects in different human cells. Additional investigations with the related aldehyde *trans*-2-hexenal are in progress.

References

[1] Janzowski, C.; Glaab, V.; Eisenbrand, G. (1997) Cytotoxic and genotoxic effects of 2-cyclohexen-1-one in V79 cells and in *Salmonella typhimurium*. *Proc. Am. Assoc. Cancer Res.* **38** (abstracts only).

[2] Walker, P. R.; Sikorska, M. (1997) New aspects of the mechanism of DNA fragmentation in apoptosis. *Biochem. Cell Biol.* **75**, 287–299.

[3] Nicoletti, I.; Migliorati, G.; Pagliacci, M. C.; Grignani, F.; Riccardi, C. (1991) A rapid and simple method for measuring thymocyte apoptosis by propidium iodide staining and flow cytometry. *J. Immunol. Meth.* **139**, 271–279.

[4] Darzynkiewicz, Z.; Bruno, S.; Del Bino, G.; Gorczyca, W.; Hotz, M. A.; Lassota, P.; Traganos, F. (1992) Features of apoptotic cells measured by flow cytometry. *Cytometry* **13**, 795–808.

[5] Beere, H. M.; Chresta, C. M.; Alejoherberg, A.; Skladanowski, A.; Dive, C.; Larson, A. K.; Hickman, J. A. (1995) Investigation of the mechanism of higher order chromatin fragmentation observed in drug induced apoptosis. *Mol. Pharmacol.* **47**, 986–996.

[6] Zamai, L.; Falcieri, E.; Merhefka, G.; Vitale, M. (1996) Supravital exposure to propidium iodide identifies apoptotic cells in the absence of nucleosomal DNA fragmentation. *Cytometry* **23**, 303–311.

[7] Darzynkiewicz, Z; Juan, G.; Li, X.; Gorczyca, W.; Murakami, T.; Traganos, F. (1997) Cytometry in cell necrobiology. Analysis of apoptosis and accidental cell death (necrosis). *Cytometry* **27**, 1–20.

13 Food Relevant α,β-Unsaturated Carbonyl Compounds: *In vitro* Toxicity, Genotoxic (Mutagenic) Effectiveness and Reactivity Towards Glutathione

Christine Janzowski, V. Glaab, E. Samimi, J. Schlatter[1], B. L. Pool-Zobel[2] and G. Eisenbrand

α,β-Unsaturated carbonyl compounds are widespread in the environment. 2-Alkenals, such as (*E*)-2-hexenal (HEX), (*E*)-2-octenal (OCTA), (2*E*,4*E*)-2,4-hexadienal (HD), (2*E*,6*Z*)-nonadienal (NONA) or cinnamic aldehyde (CA) are generated in food by lipid peroxidation or are added as naturally flavoring additives (see Figure 13.1) [1].

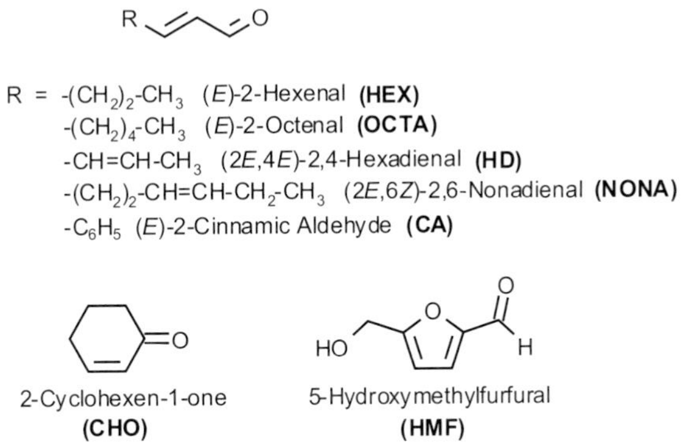

Figure 13.1: Structures of tested compounds.

Dept. of Chemistry, Div. of Food Chemistry and Environmental Toxicology, University, Kaiserslautern, Germany

[1] Swiss Federal Office of Public Health, Zürich, Switzerland

[2] Institute for Nutrition, University, Jena, Germany

Introduction

2-Cyclohexen-1-one (CHO) occurs as a natural ingredient in certain fruits and has also been detected in tobacco smoke. Moreover, it has been identified as a contaminant in cyclamate sweetened soft drinks in concentrations up to 1.4 mg/l [2]. 5-Hydroxymethylfurfural (HMF), a common product in the Maillard reaction, is formed predominantly from sugars as a consequence of thermal food treatment. It occurs in rather high amounts in a wide variety of foods such as dry fruits, juices from dry fruits, bread and caramel products, maximal concentrations exceeding 1 g/kg [3].

Toxicologically, these compounds deserve consideration due to their reactivity towards cellular nucleophiles leading to Michael type adducts. Acute toxic effects, especially dermal toxicity and eye irritation have been observed with 2-alkenals in animal experiments [4]. Crotonaldehyde is carcinogenic, inducing liver tumors after oral application of high concentrations to rats [5]. DNA adducts of 2-alkenals, such as cyclic 1,N^2-propano-dG adducts, have been identified *in vitro* and/or *in vivo* [6]. Mutagenicity has been reported in a limited number of *in vitro* studies [7]. For CHO, formation of dG adducts has also been described [8]. HMF has been investigated *in vitro* and *in vivo* for genotoxic/mutagenic effectiveness, resulting in positive and negative results [9–11]. In addition to direct interaction with critical cellular nucleophiles sulfate conjugation at the allylic hydroxyl group might be relevant [12].

The aim of this study was to establish a structure-activity relationship on toxic/genotoxic/mutagenic effectiveness of these food relevant compounds and to compare their genotoxic potential.

Methods

Viability, DNA damage, mutagenicity and depletion of glutathione (GSH) were comparatively investigated in hamster lung fibroblasts (V79), human colon carcinoma cells (CaCo-2), primary human colon cells and primary rat hepatocytes. Viability (LC_{50}) was monitored by trypan blue exclusion. DNA damage was quantified by single cell gel electrophoresis (COMET assay) under alkaline conditions [13]. After incubation of suspension culture 1 h or 30 min in a shaking water bath, cells were embedded in agarose on microscopic slides, and submitted to lysis. Resulting nucleoids were electrophoresed under alkaline conditions, stained with ethidium bromide and viewed microscopically. Tail intensity of images was quantified by semi-automatic analysis. DNA damaging potency was expressed as DC_{50} (concentration inducing DNA damage in 50 % of cells). Mutagenicity was determined in V79 cells (HPRT locus) after 1 h incubation in tissue culture flasks [7]. GSH depletion, resulting from 1 h incubation of suspended cells, was monitored by photometric analysis of *p*-nitrothiophenol produced from bis(*p*-nitrophenyl)disulfide (DTNB) [14].

Results and discussion

All tested alkenals and CHO induce DNA damage in V79 and CaCo-2 cells after 1 h incubation (Figure 13.2).

For each compound rather similar results were obtained in both cell lines. Within the group of tested compounds, however, DNA-damaging effectiveness varied substantially from NONA (with the highest potency) to CA (weakly active). The same structure-related ranking order was observed for cytotoxic effectiveness, but IC_{50} values varied over a larger range. In primary human colon cells, DNA damage was induced with 400 µM of HEX and 800 µM of CHO. HMF, however, did not exert DNA damage in V79 and CaCo-2 cells up to 80 mM (cytotoxic concentration limit, 75 % viability) and moreover, was not effective in primary rat hepatocytes up to 100 mM (cytotoxic concentration limit). Results on mutagenicity of HEX and CHO in V79 cells are shown in Figure 13.3.

Whereas CHO induced a clear concentration dependent increase of HPRT mutations between 400–1000 µM, with HEX low but significant mutagenic ef-

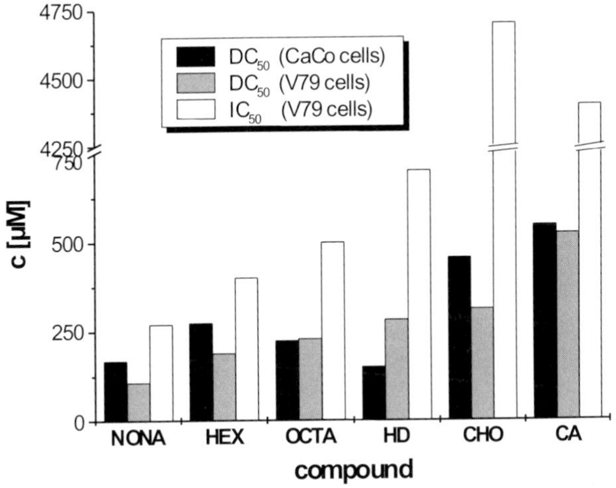

Figure 13.2: DNA damaging potency (DC_{50}) and cytotoxicity (IC_{50}) of 2-alkenals and CHO in V79 and CaCo-2 cells. Incubation time: 1 h.

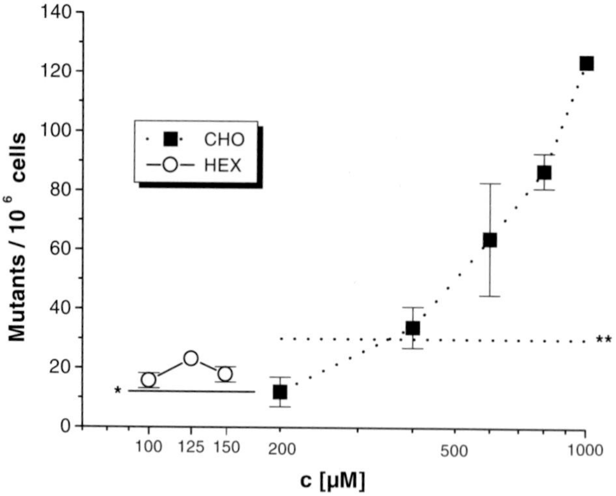

Figure 13.3: Induction of mutations at the HPRT locus in V79 cells. *threshold: 3-fold background, HEX experiments (n = 2); **threshold: 3-fold background, CHO experiments (n = 3, 1000 μM: n = 1); 1 h incubation.

fects were observed between 100–150 μM (mutation rate >3-fold over control). At higher concentrations cell growth in the post-incubation period was drastically inhibited, impairing the assay under standard conditions. With NONA, the most cytotoxic compound, mutations were not detectable under these conditions. HMF was significantly mutagenic after 1 h incubation with 120 mM, inducing mutation rates, 4- to 5-fold over control. Distinct GSH depletion (down to appr. 40 % of control) was observed in V79 cells (GSH content: 21 nmoles/mg protein) following 1 h incubation with HEX (50 μM), CHO (20 μM) as well as with HMF (80 mM).

In conclusion, food relevant 2-alkenals and CHO are genotoxic and mutagenic in mammalian cells. Protective cellular GSH levels were decreased at substantially lower concentrations, than those needed to exert DNA damage and mutagenicity. HMF did not induce DNA damage in V79 and CaCo-2 cells; negative results, obtained in metabolically competent primary hepatocytes, do not support substantial metabolic activation. High HMF concentrations (120 mM), however, induced HPRT mutations in V79 cells, suggesting a weak mutagenic potential of the compound. Compared to HEX, HMF had to be applied at least in 1000-fold higher concentrations to induce similar mutagenic effects. The relevance of the observed GSH depletion has to be further investigated.

Acknowledgments

We thank Dr. J. Schmitz and Prof. Dr. Dr. D. Schrenk for the gift of primary rat hepatocytes. This work in part was supported by Swiss Federal Office of Public Health, Switzerland (Grant No. FE 316980676).

References

[1] Feron, V. J.; Til, H. P.; de Vrijer, F.; Woutersen, R. A.; Cassee, F. R.; van Bladeen, P. J. (1990) Aldehydes: occcurrence, carcinogenic potential, mechanism of action and risk assessment. *Mutat. Res.* **259**, 363–385.

[2] Hahn, H. (1996) Über das Vorkommen von 2-Cyclohexen-1-on in süßstoffhaltigen Erfrischungsgetränken. *Lebensmittelchemie* **50**, 52–54.

[3] Bachmann, S.; Meier, M.; Känzig, A. (1997) 5-Hydroxymethyl-2-furfural (HMF) in Lebensmitteln. *Lebensmittelchemie* **51**, 49–50.

[4] Opdyke, D. L. J. (1975) Monographs of fragrance raw materials. Hexen-2-al. *Food Cosmet. Toxicol.* **13**, 453–454.

[5] Chung, F. L.; Tanaka, T.; Hecht, S. S. (1986) Induction of liver tumors in F344 rats by crotonaldehyde. *Cancer Res.* **46**, 1285–1289.

[6] Chung, F. L.; Young, R.; Hecht, S. S. (1984) Formation of cyclic 1,N^2-propanodeoxyguanosine adducts in DNA upon reaction with acrolein or crotonaldehyde. *Cancer Res.* **44**, 990–995.

[7] Canonero, R.; Maritelli, A.; Marinari, U.; Brambilla, G. (1990) Mutation induction in Chinese hamster lung V79 cells by five alk-2-enals produced by lipid peroxidation. *Mutat. Res.* **244**, 153–156.

[8] Chung, F. L.; Kalpana, R. R.; Hecht, S. S. (1988) A study of α,β-unsaturated carbonyl compounds with desoxyguanosine. *J. Org. Chem.* **58**, 14–17.

[9] Florin, I.; Rutberg, L.; Curvall, M.; Enzell, C. R. (1980) Screening of tobacco smoke constituents for mutagenicity using the Ames' test. *Toxicology* **18**, 219–232.

[10] Shinohara, K.; Kim, E.; Omura, H. (1986) Furans as the mutagens formed by amino-carbonyl reactions. In: Fujimaki, M.; Naniki, M.; Kato, H. (Eds.) *Amino-carbonyl reactions in food and biological systems.* Elsevier, Amsterdam, Netherlands, p. 353–362.

[11] Zhang, X. M.; Chan, C. C.; Stamp, D.; Minkin, S.; Archer, M. C.; Bruce W. R. (1993) Initiation and promotion of colonic aberrant crypt foci in rats by 5-hydroymethyl-2-furaldehyde in thermolyzed sucrose. *Carcinogenesis* **14**, 773–775.

[12] Lee, Y.-C.; Shlyankevich, M.; Jeong, H.-K.; Douglas, J. S.; Surh, Y.-J. (1995) Bioactivation of 5-hydroxymethyl-2-furaldehyde to an electrophilic and mutagenic allylic sulfuric acid ester. *Biochem. Biophys. Res. Commun.* **209**, 996–1002.

[13] Singh, N. P.; McCoy, M. T.; Raymond, R. T.; Schneider, E. L. (1988) A simple technique for quantitation of low levels of DNA damage in individual cells. *Exp. Cell Res.* **175**, 184–191.

[14] Tietze, F. (1969) Enzymatic method for quantitative determination of nanogram amounts of total and oxidized glutathione: Applications to mammalian blood and other tissues. *Anal. Biochem.* **27**, 502–522.

14 Enantioselective Metabolic Activation of Safrole

Robert Landsiedel[1], U. Andrae[2], A. Kuhlow[1], M. Scholtyssek[1] and H. R. Glatt[1]

Safrole (1-allyl-3,4-methylenedioxybenzene, CAS 94-59-7) is a flavour constituent of many spice plants. High concentrations are found in essential oils of sassafras, sweet basil and cinnamon [1, 2]. It is used as flavour compound in perfumes and cosmetics [1].

Safrole is a hepatocarcinogen in mice and rats and exhibits genotoxic effects in various mammalian cells and *in vivo* [3–5 and literature herein]. It is hydroxylated at the benzylic position by cytochromes P450 (reaction scheme in Figure 14.1) [6, 7]. The resulting 1′-hydroxysafrole is a more potent carcinogen than its parent compound [4] and induces mutations in sulfotransferase-proficient *Salmonella* strains [8]. It is a substrate for sulfotransferases to form electrophilic sulfuric acid esters which form DNA adducts [5, 9]. The hydroxysteroid sulfotransferases of humans and rats (hHST and rHSTa) are capable of catalysing this reaction [10]. Studies with brachymorphic mice, which show a reduced rate of synthesis of 5′-phosphoadenosinephosphosulfate (PAPS), the cofactor for sulfotransferases, or using sulfotransferase inhibitors in normal mice and rats indicate that hydroxylation with subsequent sulfonation is the major activation pathway *in vivo* [11]. The activation at the benzylic carbon atom to electrophilic and DNA-binding sulfuric acid esters is a common pathway for many compounds [10]. Safrole was the first example reported in 1973 [4].

The key metabolite of this activation pathway, 1′-hydroxysafrole, is chiral. Cytochromes P450 and sulfotransferases exhibit in many cases a high product and substrate enantioselectivity, respectively [12–14]. These enantioselectivities

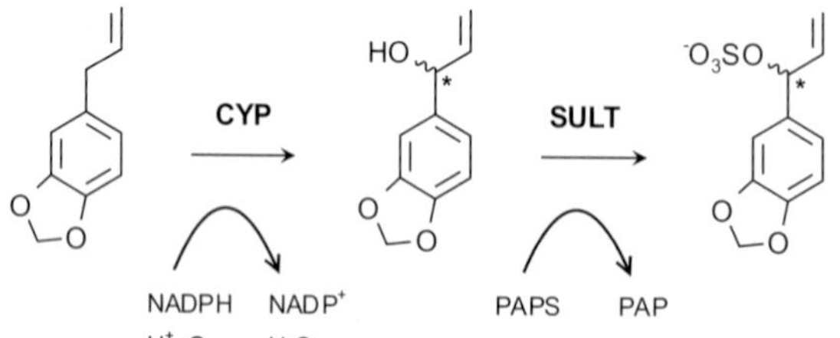

Figure 14.1: Metabolic activation of safrole by cytochromes P450 (CYP) and sulfotransferases (SULT).

[1] German Institute of Human Nutrition, Department of Toxicology, D-14558 Potsdam-Rehbruecke, Germany
[2] GSF, Institute for Toxicology, D-85758 Neuherberg, Germany

may be important factors controlling the metabolic activation. Thus we investigated the product enantioselectivity of cytochromes P450 present in rat liver microsomes and the substrate enantioselectivity of rHSTa.

Safrole was incubated with NADPH-fortified microsomal preparations of rat liver. The formed 1′-hydroxysafrole was separated by HPLC on a cellulose-tris(*N*-3,5-dimethyl-phenylcarbamate) chiral stationary phase (Figure 14.2).

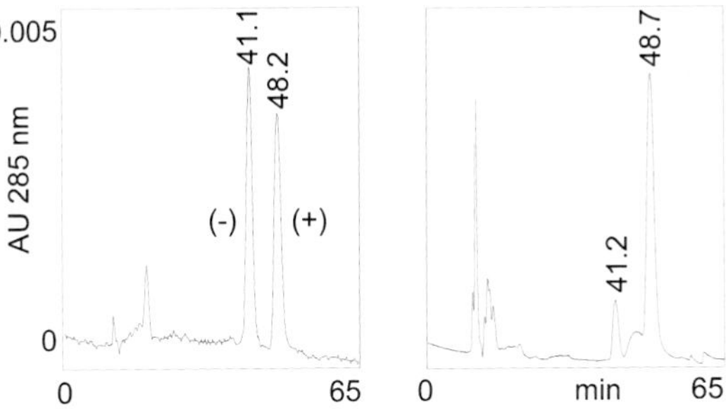

Figure 14.2: Chiral separation of 1′-hydroxysafrole. Racemic standard (left) and metabolites formed from safrole in the presence of NADPH-fortified liver microsomes from Aroclor-treated male rats (right).

Eluting analytes were quantified by their UV absorption at 285 nm and their UV spectra were recorded. Additionally, fractions were collected from the chiral separation and analysed by circular dichroism spectroscopy (Figure 14.3), HPLC-MS (data not shown) and GC-MS (Figure 14.4). Liver microsomes from both male and female rats formed 170±15 pmol 1′-hydroxysafrole/min/mg protein. Pretreatment of male and female rats with Aroclor enhanced this reaction to 1490±180 pmol/min/mg protein. The hydroxylation proceeded enantioselectively. The (+)-enantiomer was preferentially formed. The enantiomeric excess was 0.45±0.10 and 0.60±0.15 for microsomes from untreated and Aroclor-treated animals respectively.

DNA adduct patterns of racemic 1′-hydroxysafrole and the individual enantiomers incubated with rHSTa, its cofactor PAPS and DNA were analysed by

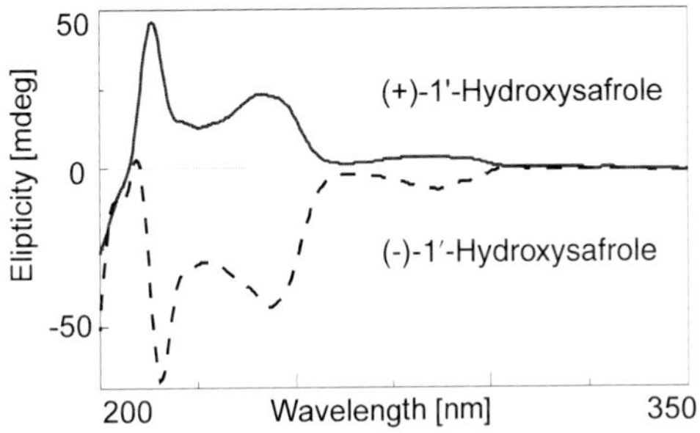

Figure 14.3: CD-spectra of chemically synthesised 1′-hydroxysafrole enantiomers.

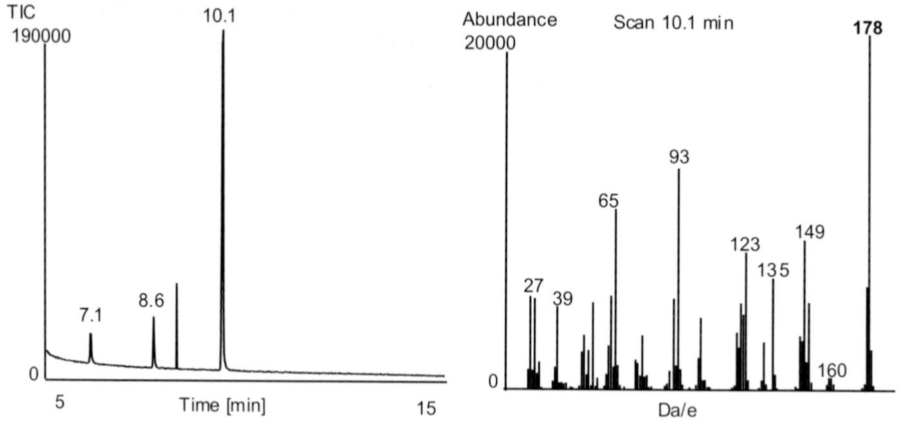

Figure 14.4: GC-MS analysis of peak 2 (retention time 48.7 min, see Figure 14.2) collected from the chiral separation of metabolites of safrole incubated with NADPH-fortified liver microsomes from Aroclor-treated male rats.

^{32}P-postlabelling technique. Both enantiomers formed identical profiles of DNA adducts. These were in good accordance with those published for the livers of safrole-treated rats [5, 9].

The racemate and individual enantiomers of 1′-hydroxysafrole were tested for their genotoxic acitivity in V79-derived cells which express rHSTa [8]. Unscheduled DNA synthesis (UDS), as a measure of DNA repair activity and indirectly of DNA damage, was used as the endpoint of the test. The racemate as well as both individual enantiomers induced UDS in a concentration-dependent manner (Figure 14.5). However, there was a clear enantioselectivity for (–)-1′-hydroxysafrole, which was three-times more effective than its optical antipode.

In conclusion the enantiomer which is the minor product of the first step of the bioactivation of safrole in rat liver microsomes is the more potent genotoxi-

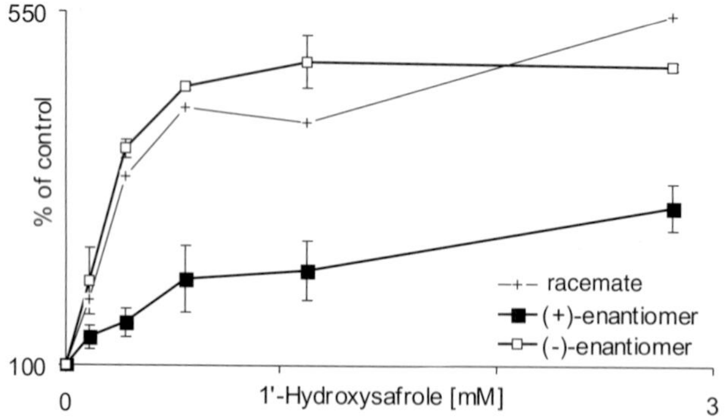

Figure 14.5: Induction of DNA-repair (UDS) by 1′-hydroxysafrole in V79-derived cells expressing rHSTa. Individual enantiomers were tested in two separate experiments (indicated values are means and range) the racemate was tested in a single experiment.

cant under the experimental conditions used (V79 cells engineered for the expression of rat HSTa). In further studies we will investigate the stereochemistry of the metabolic activation by human enzymes.

References

[1] Eisenbrand, G.; Schreier, P. (1995) In: Falbe, J.; Regitz, M. (Eds.). *Römpps Lexikon Lebensmittelchemie*, Thieme Verlag, Stuttgart.

[2] Furia, T.; Bellanca, N. (1975) *Fenaroli's handbook of flavour ingredients*, CRC Press, Cleveland.

[3] Long, E.; Nelson, A.; Fitzhut, O.; Hansen, W. (1963) Liver tumors in rats by feeding safrole. *Arch. Pathol.* **75**, 595.

[4] Borchert, P.; Miller, J.; Miller, E.; Shires, T. (1973) 1'-Hydroxysafrole, a proximate carcinogenic metabolite of safrole in the rat and mouse. *Cancer Res.* **33**, 590.

[5] Daimon, H.; Sawada, S.; Asakura, S.; Sagami, F. (1998) *In vivo* genotoxicity and DNA adduct levels in the liver of rats treated with safrole. *Carcinogenesis* **19**, 141.

[6] Ioannides, C.; Delaforge, M.; Parke, D. (1981) Safrole: its metabolism, carcinogenicity and interactions with cytochrome P450. *Food Cosmet. Toxicol.* **19**, 657.

[7] Miller, J.; Miller, E. (1983) The metabolic activation and nucleic acid adducts of naturally-occuring carcinogens: recent results with ethyl carbamate and spice flavors safrole and estragole. *Br. J. Cancer* **48**, 1.

[8] Glatt, H.; Bartsch, I.; Czich, A.; Seidel, A.; Falany, C. N. (1995) Salmonella strains and mammalian cells genetically engineered for expression of sulfotransferases. *Toxicol. Lett.* **82–83**, 829.

[9] Randerath, K.; Haglund, R.; Phillips, D.; Reddy, M. (1984) ^{32}P-Postlabelling analysis of DNA adducts formed in the livers of animals treated with safrole, estragole and other naturally-occurring alkenylbenzenes. *Carcinogenesis* **5**, 1613–1622.

[10] Glatt, H. (1997) Bioactivation of mutagens via sulfation. *FASEB J.* **11**, 314.

[11] Boberg, E.; Miller, E.; Miller, J.; Poland, A.; Liem, A. (1983) Strong evidence from studies with brachymorphic mice and pentachlorophenol that 1'-sulfoxysafrole is the major ultimate electrophilic and carcinogenic metabolite of 1'-hydroxysafrole in mouse liver. *Cancer Res* **43**, 5163.

[12] Testa, B. (1989) Mechanisms of chiral recognition in xenobiotic metabolism and drug-receptor interactions. *Chirality* **1**, 7.

[13] Landsiedel, R. (1998) Stoffwechsel und Mutagenität benzylischer Verbindungen. Logos Verlag, Berlin.

[14] Banoglu, E.; Duffel, M. (1997) Studies on the interactions of chiral secondary alcohols with rat hydroxysteroid sulfotransferase. *Drug Metab. Dispos.* **25**, 1304.

15 Inhibition of DNA Repair Processes by Inorganic Arsenic

Andrea Hartwig[1], U. D. Gröblinghoff[1], Y. Hiemstra[1] and L. H. F. Mullenders[2]

Introduction

Inorganic arsenic is a long recognized established human carcinogen, causing lung cancer upon inhalation and mainly skin, but also bladder, kidney, lung and liver tumors upon oral exposure. Comparatively high exposures occur at workplaces like copper, zinc and lead smelters, glass works and as a consequence of the application of arsenic-containing pesticides and herbicides. Due to its toxicity, the industrial use in Germany is nowadays restricted to few purposes. Nevertheless, one major problem related to arsenic exposure is the high arsenic content in drinking water in many geographic regions due to natural sources; thus, worldwide there are millions of people chronically exposed to levels of arsenic in drinking water that are high enough to cause severe health problems including cancer [1, 2]. The toxic species is believed to be arsenic(III), which is metabolically methylated to monomethylarsenic acid (MMA) and finally to dimethylarsinic acid (DMA) [3].

Direct and indirect genotoxic effects of arsenic(III)

The mechanism of arsenic-induced carcinogenicity is not clear, which is in part due to the absence of a suitable animal model, but also to the missing mutagenicity in bacterial and mammalian test systems. In contrast, arsenic(III) induces mainly chromatid-type chromosomal aberrations and sister chromatid exchanges. Furthermore, arsenic(III) has been found to increase the mutagenicity and clastogenicity in combination with other DNA damaging agents, including UV-light, benzo[*a*]pyrene, X-rays, alkylating agents and DNA crosslinking com-

[1] Department of Food Chemistry, University of Karlsruhe, D-76128 Karlsruhe, Germany
[2] MGC-Department of Radiation Genetics and Chemical Mutagenesis, Leiden University, Leiden, The Netherlands

pounds in cultured mammalian cells [reviewed in 4, 5]. As possible mechanism, there have been several indications for an interference with DNA repair processes, which has been investigated in more detail in the present study.

Interference with nucleotide excision repair

The major DNA repair pathway involved in the removal of bulky DNA damage induced by environmental and food mutagens is nucleotide excision repair (NER). According to the current understanding, NER is a complex procedure, requiring the coordinated action of about 30 different proteins. It can be roughly subdivided into four different steps, namely the damage recognition followed by the incision at both sides of the lesion in the damaged DNA strand, the repair polymerization and finally the ligation of repair patches [6]. To elucidate the effect of arsenic(III) on NER, we used the alkaline unwinding procedure to quantify the frequency of DNA strand breaks generated during the repair of UV-induced DNA lesions in the presence and absence of specific inhibitors of distinct steps of the repair process. By applying human fibroblasts we observed that two steps of the repair process are inhibited by arsenic(III). At concentrations as low as 2.5 μM the incision frequency was reduced; at higher, cytotoxic concentrations of 20 μM and 50 μM the ligation of the repair patches was inhibited as well (Figure 15.1; for further details see [7]).

During the last years, two NER subpathways have been identified, one removing lesions from the transcribed strand of active genes (transcription-

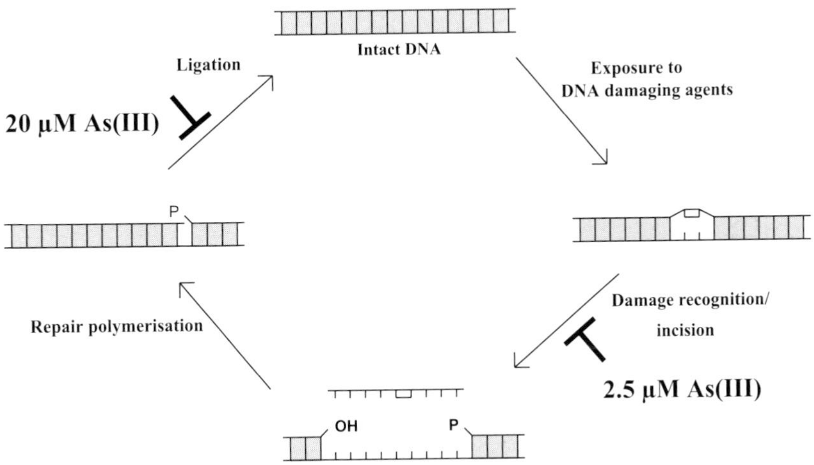

Figure 15.1: Inhibition of distinct steps of nucleotide excision repair by arsenic(III).

coupled repair; TCR) and the other repairing the bulk chromatin (global genome repair, GGR). With respect to UV-induced cyclobutane pyrimidine dimers (CPD), three hierachies of repair can be defined: the comparatively slow removal of lesions from inactive DNA regions, the more accelerated repair of active genes and finally the fast repair of the transcribed strand of active genes [8]. By using partly repair deficient human fibroblasts derived from patients of *xeroderma pigmentosum* group C, where the repair capacity is restricted to TCR [9], and by monitoring the removal of CPD at the gene level by southern blot analysis we observed that arsenic inhibits the repair of CPD in active genes and in inactive DNA regions to about the same extent. Furthermore, TCR is affected as well. While in the absence of arsenite most UV-induced lesions are removed from all parts of the genome within 24 h, the presence of low concentrations of arsenite leads to a persistence of DNA damage in all genomic regions.

Conclusions and perpectives

Taken together, our results provide evidence that arsenite leads to partially repair deficient conditions which may give rise to an accumulation of DNA damage induced by many environmental and food mutagens. As known from patients suffering from the repair deficient disorder *xerodermea pigmentosum*, this may increase the risk of tumor formation. Future studies will have to clarify whether the repair of endogenously induced DNA lesions like those generated by reactive oxygen species is inhibited as well; this would provide an additional hint to understand the arsenic-induced carcinogenicity even in the absence of other DNA-damaging agents.

Acknowledgements

This work was supported by the EC, grant no. EV5V-CT94-0479, and by the Deutsche Forschungsgemeinschaft, grant no. Ha 2372/1-1.

References

[1] IARC (1980) Monographs on the evaluation of the carcinogenic risk of chemicals to humans, Vol. 23: *Some metals and metallic compounds*. IARC, Lyon.
[2] Chiou, H.-Y.; Hsueh, Y.-M.; Liaw, K.-F.; Horng, S.-F.; Chiang, M.-H.; Pu, Y.-S.; Lin, J. S.-N.; Huang, C.-H.; Chen, C.-J. (1995) Incidence of internal cancers and ingested inorganic arsenic: a seven-year follow-up study in Taiwan. *Cancer Res.* **55**, 1296–1300.
[3] Vahter, M.; Marafante, E. (1988) *In vivo* methylation and detoxification of arsenic. In: Craig, P. J.; Glockling, F. (Eds.) *The biological alkylation of heavy elements*. Royal Soc. Chem. Special Publications No. 66, Oxford, UK.
[4] Hartwig, A. (1995) Current aspects in metal genotoxicity. *BioMetals* **8**, 3–11.
[5] Wang, Z.; Rossman, T. G. (1996) The carcinogenicity of arsenic. In: Chang, L.W. (Ed.) *Toxicology of metals*. CRC Press, p. 221–227.
[6] Sancar, A. (1996) DNA excision repair. *Annu. Rev. Biochem.* **65**, 43–81.
[7] Hartwig, A.; Gröblinghoff, U. D., Beyersmann, D.; Natarajan, A. T.; Filon, R.; Mullenders, L. H. F. (1997) Interaction of arsenic(III) with nucleotide excision repair in UV-irradiated human fibroblasts. *Carcinogenesis* **18**, 399–405.
[8] Mullenders, L. H. F.; Vrieling, H.; Venema, J.; van Zeeland, A. A. (1991) Hierarchies of DNA repair in mammalian cells: biological consequences. *Mutat. Res.* **250**, 223–228.
[9] Venema, J.; van Hoffen, A.; Kargagi, V.; Natarajan, A. T.; van Zeeland, A. A.; Mullenders, L. H. F.(1991) Xeroderma pigmentosum complementation group C cells remove pyrimidine dimers selectively from the transcribed strand of active genes. *Mol. Cell Biol.* **11**, 4128–4134.

16 Identification of Antiandrogens with Functional Transactivation *in vitro* Assays and Molecular Modeling Analysis

S. Guth, D. Seng, S. Böhm, B. Mußler and G. Eisenbrand

Over the last years there is an increasing public and scientific discussion about adverse effects to normal endocrine functions induced by xenobiotics. Besides possible endocrine disrupting effects in wildlife also certain hints from epidemiological studies have roused public attention. Male reproductive health is potentially influenced by a large group of anthropogenic compounds. Declining sperm counts, marked increase in testicular cancer rates and other disor-

Dept. of Chemistry, Food Chemistry and Environmental Chemistry, University of Kaiserslautern, Erwin-Schrödinger-Str. 52, D-67663 Kaiserslautern, Germany ders like cryptorchism and hypospadias have been reported in several studies. Primarily attention has been focussed on a possible linkage between exposure to environmental/natural estrogens and adverse male reproductive health. More recently it was found that two agrochemicals [1, 2] with widespread usage elicit rather strong antiandrogenic activity *in vitro* and *in vivo*. The main metabolite of technical DDT in mammalians, *p,p'*-DDE, and a metabolite of the fungicide vinclozolin were identified as androgen receptor antagonists. Although both substances induce the same biological effects they are structurally quite diverse. We became interested into identification of structural elements of such compounds that might be relevant for antiandrogenic efficacy. Therefore we initiated a study on structure-activity relationships assisted by molecular modeling on the basis of CoMFA (Comparative Molecular Field Analysis). The aim is to identify common lead structure elements on the basis of our experimental data and the data of other groups. A structural feature displayed in common antiandrogens like the clinically utilized hydroxyflutamide and casodex shows marked similarity with corresponding elements of metabolites of pesticides vinclozolin and procymidon and also with urea herbicides such as linuron, monolinuron.

Another structural element contained in *p,p'*-DDE shows for instance similarity to some benzophenones widely used as UV-protectors in cosmetics and plastics.

Antiandrogenic potency has been determined mainly in functional transactivation assays until now. Such systems are based on overexpression of the human androgen receptor and cotransfection of steroid hormone controlled promotor-reportergene vectors into cells.

We therefore established a transient *in vitro* transactivation assay in COS-7 cells. The cells were transfected with 10 μg of an androgen receptor expression vector, a MMTV-luciferase reportergene vector and a β-galactosidase control vector using electroporation. The transiently transfected cells were then incubated with the drugs and lysed with a hypotonic buffer. The measured luciferase activity was corrected against the β-galactosidase activity.

Dihydrotestosterone (DHT) induced in a dose dependent manner the expression of the reportergene. *p,p'*-DDE in combination with DHT at a 10–100-fold higher concentration showed a marked reduction of reportergene activity. In addition to this antiandrogenic activity it exhibited also weak androgenic activity at high concentrations (Figure 16.1). Oxybenzone, a structural analogue of *p,p'*-DDE, acted in this test system also as a partial agonist/antagonist (Figure 16.2).

Linuron, a common used herbicide and structural analogue of hydroxyflutamide was also able to reduce the reportergene activity induced by DHT in a dose dependent manner (data not shown).

Preliminary results support the theoretical determined structural homology of oxybenzone to *p,p'*-DDE and of linuron to the vinclozolin metabolite M2 or to hydroxyflutamide. With the help of these lead structures it is possible to identify and characterize further structural analogues by on-line structural database search (Beilstein-crossfire). This test principle offers advantages in the identification of potential endocrine disruptors or new antiandrogenic drugs.

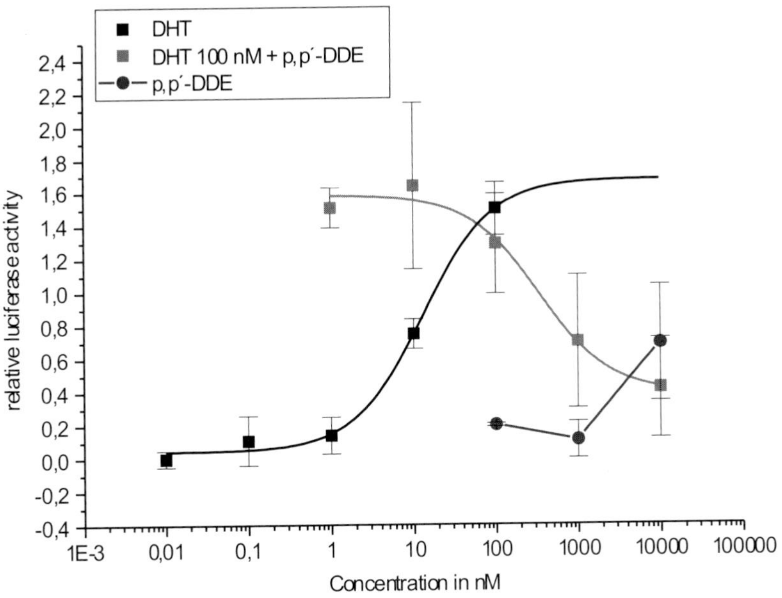

Figure 16.1: Dose dependent induction and inhibition of the androgen controlled reportergene luciferase by DHT and *p,p'*-DDE or a mixture of 100 nM DHT and increasing concentrations of *p,p'*-DDE respectively.

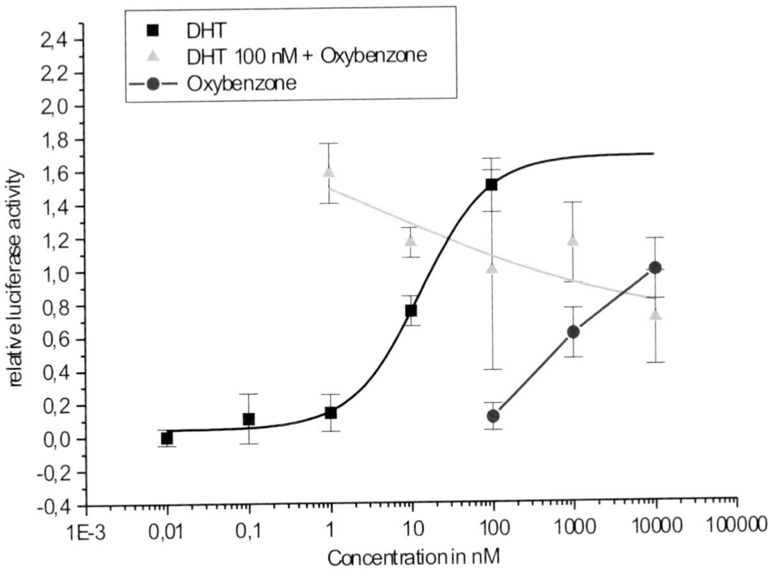

Figure 16.2: Dose dependent induction and inhibition of the androgen controlled reportergene luciferase by DHT and oxybenzone or a mixture of DHT 100 nM and increasing concentrations of oxybenzone respectively.

It is hoped that the combination of functional *in vitro* assays with molecular modeling analysis will bring about new insights into relevant structural elements of antiandrogens and might be helpful to identify new compounds of potential significance by advanced library search.

References

[1] Kelce, W. R.; Stone, C. R.; Lows, S. C.; Kemppainen, J. A.; Wilson, E. (1995) Persistent DDT metabolite *p,p'*-DDE is a potent androgen receptor antagonist. *Nature* **375**, 581–585.

[2] Wong, C.; Kelce, W. R.; Madhabananda, S.; Wilson, E. (1995) Androgen receptor antagonist versus agonist activities of the fungicide vinclozolin relative to hydroxyflutamide. *J. Biol. Chem.* **270** (34), 19 998–20 003.

17 Oxidative Metabolism and Genotoxicity of Enterolactone and Enterodiol

Eric Jacobs and Manfred Metzler

Abstract

We have studied the hitherto unknown oxidative metabolites of the mammalian lignans enterolactone (ENL) and enterodiol (END) in rat and human liver microsomes *in vitro* and in the urine of humans *in vivo* after the ingestion of flaxseed for five days. With hepatic microsomes of Aroclor-pretreated rats, ENL gave rise to twelve monohydroxylated metabolites, six of which had the additional hydroxy group at the aromatic and six at the aliphatic moiety of the molecule. Human liver microsomes generated a very similar profile of metabolites. The six aromatic hydroxylation products of ENL were also found in the urine of male and female humans and were identified with the help of synthesized reference compounds. END yielded three aromatic and four aliphatic hydroxy-

Institut für Lebensmittelchemie, Universität Karlsruhe, P.O. Box 6980, D-76128 Karlsruhe, Germany lated metabolites *in vitro*. The three metabolites generated by aromatic hydroxylation were also found in human urine and identified using synthetic standards. The genotoxic potential of ENL and END was investigated at various endpoints in cultured Chinese hamster V79 cells *in vitro*. Both compounds were devoid of aneugenic, clastogenic or mutagenic activity at concentrations up to 100 μM.

Introduction

The mammalian lignans ENL and END are weak estrogens formed by intestinal bacteria from plant lignans ingested with various types of food [1, 2]. They can reach concentrations in body fluids more than 10,000-times higher than those of endogenous steroidal estrogens. Epidemiological and biochemical studies suggest beneficial effects for mammalian lignans, especially with respect to hormone-dependent types of cancer [3]. However, the structural similarity of the lignans with known carcinogens such as the synthetic estrogen diethylstilbestrol raises the possibility of negative health effects. Thus the aim of our study was to clarify the oxidative metabolism and the genotoxic potential of ENL and END.

Results

To study the oxidative metabolism of ENL and END, microsomal incubations with hepatic microsomes of Aroclor-treated male Wistar rats were carried out in a final volume of 2 ml 0.1 M phosphate buffer pH 7.4 containing 25 μM test compound, 2.5% DMSO and 1 mg microsomal protein. Analysis of the extracted metabolites by HPLC/DAD and GC/MS revealed two groups of monohydroxylated metabolites for both lignans. The hydroxylation can take place at the aromatic or the aliphatic moiety of the molecule (Figure 17.1). Whereas ENL yielded six aromatic and six aliphatic hydroxylated metabolites, END gave rise to three aromatic and four aliphatic hydroxylation products. Analogous incubations with human microsomes showed qualitatively similar elution profiles in HPLC and GC/MS with a lower amount of aromatic hydroxylated metabolites.

The six products of aromatic hydroxylation were identified with the help of synthesized reference compounds. They carry the additional hydroxy group in

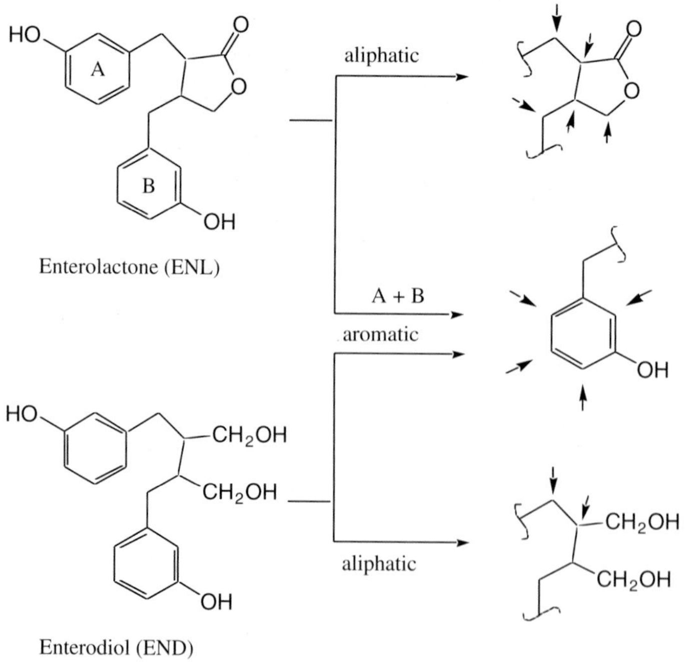

Figure 17.1: Possible positions of ENL and END for the metabolic introduction of an additional hydroxy group.

para- or ortho-position to the parent phenolic groups of ENL or END, as is shown for ENL in Figure 17.2.

The oxidative *in vivo* metabolites of ENL and END were investigated in the urine of male and female humans after the daily ingestion of 16 g flaxseed over a period of five days. A cleanup procedure which was a modification of the method of Adlercreutz [4] was used prior to GC/MS analysis of the urine (Figure 17.3). Six aromatic hydroxylation products of ENL (peaks 6, 7, 8, 9, 11, 12) and three of END (peaks 4 a, 4 b, 5) were detected. These proved to be the same aromatic hydroxylation products as those formed *in vitro*.

The genotoxic potential of ENL and END was studied at different endpoints *in vitro*. Both lignans neither affected microtubule assembly under cell-free conditions nor did they induce mitotic arrest, micronuclei or HPRT mutations in cultured Chinese hamster V79 cells at concentrations up to 100 μM [5].

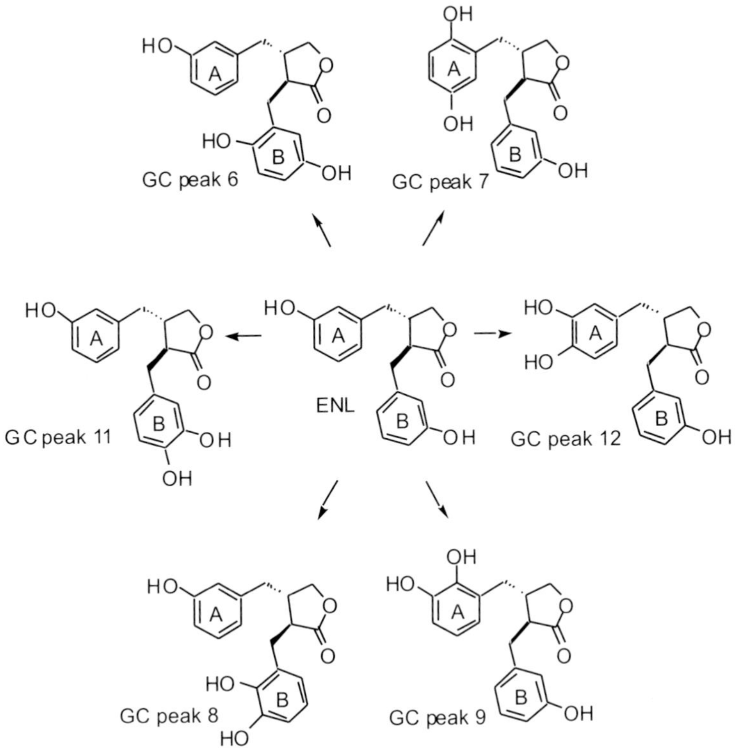

Figure 17.2: Identified metabolites of ENL. The GC peak numbers refer to Figure 17.3.

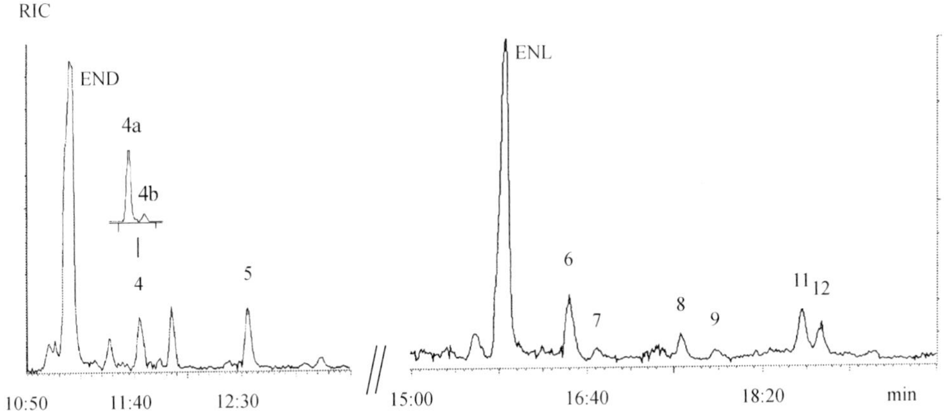

Figure 17.3: GC/MS analysis of the urine of a male volunteer after flaxseed supplementation for five days.

Conclusion

Our study has shown that ENL and END undergo oxidative metabolism *in vitro* and *in vivo*. Both lignans are devoid of aneugenic, clastogenic or mutagenic activity at various *in vitro* endpoints. The genotoxic potential of the oxidative metabolites remains to be clarified.

Acknowledgement

This study was supported by the Deutsche Forschungsgemeinschaft (Grant Me 574/9-2).

References

[1] Setchell, K. D. R.; Axelson, M.; Sjövall, J.; Gustafsson, B. E. (1982) Origin of lignans in mammals and identification of a precursor from plants. *Nature* **298**, 659.
[2] Thompson, L. U.; Robb, P.; Serraino, M.; Cheung, F. (1991) Mammalian lignan production from various foods. *Nutr. Cancer* **16**, 43.
[3] Adlercreutz, H. (1995) Phytoestrogens: epidemiology and a possible role in cancer protection. *Environ. Health Perspect.* **103** (Suppl. 7), 103.
[4] Adlercreutz, H.; Honjo, H.; Higashi, A.; Fotsis, T.; Hämälainen, E.; Hasegawa, T.; Okada, H. (1991) Urinary excretion of lignans and isoflavonoid phytoestrogens in Japanese men and women consuming a traditional Japanese diet. *Am. J. Clin. Nutr.* **54**, 1093.
[5] Kulling, S. E.; Jacobs, E.; Pfeiffer, E.; Metzler, M. (1998) Studies on the genotoxicity of the mammalian lignans enterolactone and enterodiol and their metabolic precursors at various endpoints *in vitro*. *Mutat. Res.* **416**, 115.

18 Genotoxic Potential of Isoflavone and Coumestane Phytoestrogens

S. E. Kulling, Leane Lehmann, S. Mayer and M. Metzler

Abstract

Our study revealed striking differences between phytoestrogens with closely related structures concerning the type and strength of *in vitro* genotoxicity. Coumestrol caused DNA strand breaks, structural chromosomal aberrations, CREST-negative micronuclei and HPRT mutations in cultured Chinese hamster V79 cells. The same profile of genetic damage was noted for genistein, whereas its 4′-methyl ether biochanin A disrupted the cytoplasmic microtubule complex and the mitotic spindle of V79 cells. Daidzein was devoid of genotoxicity at the endpoints used in our study.

Introduction

Phytoestrogens are endogenous constituents of edible plants, and are therefore part of our daily food. In spite of the widespread occurrence of phytoestrogens and the high exposure through certain diets, only few studies have dealt with the question whether these compounds have genotoxic potential. Genistein (GEN), daidzein (DAI), biochanin A (BCA) and coumestrol (COUM) (Figure 18.1) were not mutagenic in *Salmonella typhimurium* strains TA 98, TA 100, and TA 1598 in the absence or presence of rat liver supernatant for metabolic activation [1]. Genistein was reported to cause DNA strand breaks due to its inhibition of topoisomerase II [2]. Coumestrol was found to induce micronuclei in cultured human chorionic villi cells [3].

We have investigated the effects of the isoflavone phytoestrogens GEN, DAI and BCA as well as the coumestane phytoestrogen COUM at various endpoints for genetic damage in cultured Chinese hamster V79 cells.

Institut für Lebensmittelchemie, Universität Karlsruhe, P.O. Box 6980, D-76128 Karlsruhe, Germany

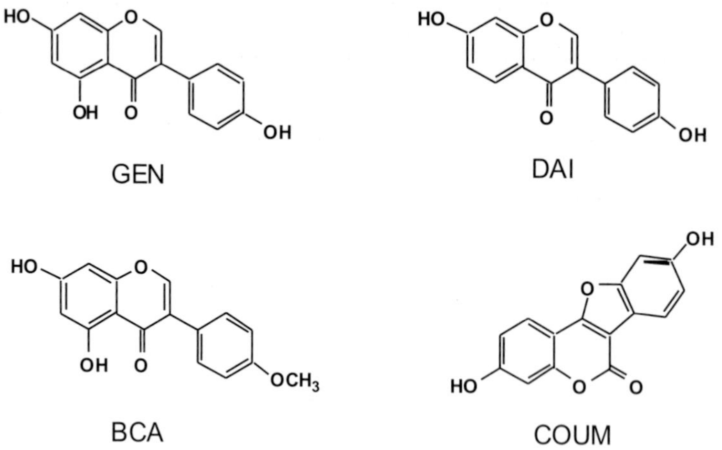

Figure 18.1: Molecular structures of the phytoestrogens GEN, DAI, BCA and COUM.

Results

Neither COUM, GEN nor DAI affected the cytoplasmic microtubule complex (CMTC) or the mitotic spindle, implying that they are not aneuploidogenic (Table 18.1). However, COUM and GEN but not DAI were strongly clastogenic, inducing DNA strand breaks and micronuclei containing acentric chromosomal fragments as shown with CREST antikinetochore antibodies (CREST-negative micronuclei). Moreover, COUM was a clear inducer of gene mutations at the HPRT locus in V79 cells, whereas GEN was only marginally active and DAI inactive at this endpoint. COUM and GEN also induced structural chromosomal aberrations in cultured peripheral human lymphocytes.

In contrast to the other phytoestrogens, BCA exhibited aneuploidogenic potential in V79 cells: it caused mitotic arrest and disrupted the CMTC as well as the mitotic spindle; it also induced micronuclei containing whole chromosomes (CREST-positive micronuclei) (Table 18.1 and Figure 18.2).

A combination of BCA and GEN caused an increased number of CREST-positive and decreased number of CREST-negative micronuclei in comparison with either BCA or GEN alone.

Table 18.1: Genotoxic effects of various phytoestrogens (CMTC, cytoplasmic microtubule complex; MN, micronuclei; n.d., not determined).

Assay and observed effect	DAI 50–100 μM	GEN 25 μM	COUM 35–100 μM	BCA 25 μM
Tubulin staining in V79 cells				
Mitotic spindle disturbances	–	–	–	+++
Disruption of CMTC	–	–	–	+
Mitotic arrest assay in V79 cells				
Delay or stop of cells in mitosis	–	–	–	+++
Micronucleus assay in V79 cells				
Induction of CREST-positive MN	–	–	–	++
Induction of CREST-negative MN	–	+++	++	–
HPRT mutation assay in V79 cells				
Mutations at the HPRT locus	–	(+)	++	n.d.
Alkaline elution assay in V79 cells				
Induction of DNA single strand breaks	–	n.d.	+	n.d.
Neutral elution assay in V79 cells				
Induction of DNA double strand breaks	–	n.d.	+	n.d.
Cytogenetic studies in cultured human lymphocytes				
Induction of structural chromosomal aberrations	–	+++	++	n.d.

Conclusion

The results of this study clearly show that some but not all phytoestrogens exhibit genotoxic potential. Despite their structural similarity, individual phytoestrogens vary with respect to the type and strength of genotoxicity.

Acknowledgement

Supported by the Deutsche Forschungsgemeinschaft (Grant Me 574/9-2).

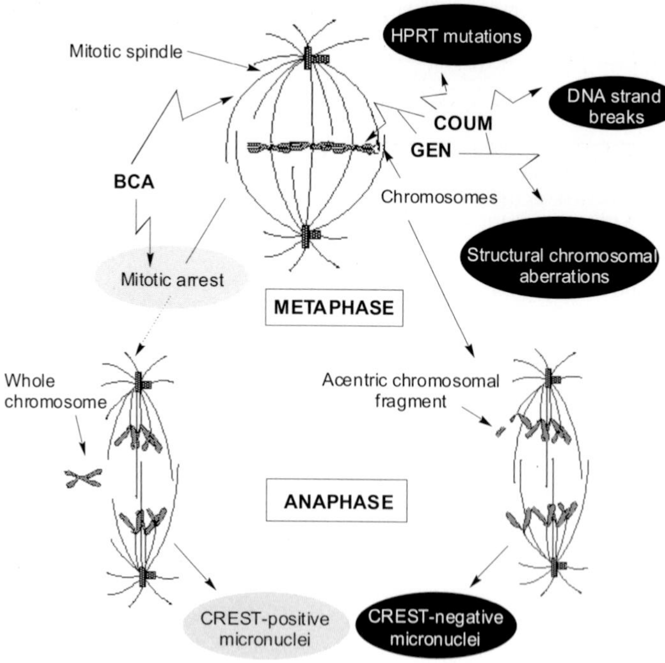

Figure 18.2: Aneugenic (gray circle) and clastogenic (black circle) effects of various phytoestrogens.

References

[1] Bartholomew, R. M.; Ryan, D. S. (1980) Lack of mutagenicity of some phytoestrogens in the Salmonella/mammalian microsome assay. *Mutat. Res.* **78**, 317.

[2] Kiguchi, K.; Constantinou, A. I.; Huberman, E. (1990) Genistein-induced cell differentiation and protein-linked DNA strand breakage in human melanoma cells. *Cancer Commun.* **2**, 271.

[3] Schuler, M.; Huber, K.; Zankl, H.; Metzler, M. (1996) Induction of micronucleation, spindle disturbance, and mitotic arrest in human chorionic villi cells by 17-estradiol, diethylstilbestrol, and coumestrol. In: Li, J. J.; Li, S. A.; Gustafsson, J.-A.; Nandi, S.; Sekely, L. I. (Eds.) *Hormonal carcinogenesis II*. Springer-Verlag, New York, p. 458.

19 Study on Mechanisms by Which Carrot and Tomato Juice May Reduce Oxidative DNA-Damages *in vivo*: *in vitro* Effects of All-*trans*-β-Carotene and Lycopene

B. L. Pool-Zobel[1], Michael Glei[1], B. Spänkuch[2] and G. Rechkemmer[2]

Introduction

A previous dietary intervention study with 23 healthy, non smoking male subjects had shown that the consumption of vegetable juices significantly reduces oxidative DNA damage in peripheral lymphocytes. This reduction may be due to scavenging of reactive oxygen species (ROS) and to induction of chemopreventive enzymes such as glutathione *S*-transferase (GST), which also may deactivate ROS.

Aim

It was the aim of this study to determine whether major carotinoid ingredients of the protective juices (namely lycopene of tomato juice and all-*trans*-β-carotene of carrot juice) are responsible for the *in vivo* effectiveness of the intervention study. For this, the compounds were assessed *in vitro* in human lymphocytes for their capacity to reduce oxidative DNA damage, modulate repair of endogenous and bleomycine-induced damage, and to induce GST-π-expression.

[1] Institute for Nutrition and Environment, Dornburger Str. 25, D-07743 Jena, Germany
[2] Federal Research Centre for Nutrition, Engesserstr. 20, D-76131 Karlsruhe, Germany

Material and methods

Peripheral lymphocytes were isolated from healthy volunteers by gradient centrifugation. They were incubated with or without 2 μM of water soluble carotenoids (kindly provided by L. E. Schlipalius, Betatene Corporation, Australia) and worked up in the COMET assay using the repair specific enzyme endonuclease III to reveal oxidised DNA bases. Proteins were isolated from aliquots of the treated lymphocytes and analysed by ELISA to determine GSTπ. Modulation of DNA repair was assessed by inducing damage with bleomycin and observing kinetics of damage-persistence in the presence of the carotenoids.

Results

All-*trans*-β-carotene leads to a reduction of DNA breaks and oxidised DNA bases in lymphocytes from 6 subjects who had refrained from eating carotenoid-rich foods for 2 weeks. Also GSTπ protein levels were increased from 3.9 ± 1.5 to 7.1 ± 2.2 ng GST$\pi/10^6$ cells (means$\pm$SEM, n=9). In subjects on normal diets oxidised DNA bases were again reduced by all-*trans*-β-carotene. Similar studies with lycopene are on going. Finally, bleomycin-induced DNA damage was less pronounced in lymphocytes incubated with all-*trans*-β-carotene than in untreated or in lycopene-treated cells.

Conclusions

The potential of all-*trans*-β-carotene to reduce oxidative DNA damage in cells and to affect cellular metabolism (GST-repair modulation) is probably an important mechanism by which carrot juice consumption can protect *in vivo*.

20 Carotenoids and Vitamin D: Micronutrients Capable to Stimulate Gap Junctional Communication

Wilhelm Stahl, S. Nicolai, A. Clairmont and H. Sies

Introduction

The inhibitory effects of retinoids and carotenoids on neoplastic transformation initiated by carcinogens correlates with their ability to stimulate gap junctional communication (GJC) [1]. Gap junctions allow the direct exchange of molecules with a molecular weight <1000 Da and are involved in the intercellular exchange of nutrients and signalling compounds such as cAMP or diacylglycerol. There is growing evidence that gap junctions play a role in growth control, cell differentiation, and the regulation of morphogenesis. Several tumor promotors have been demonstrated to inhibit GJC.

In epidemiological studies high carotenoid serum levels were associated with a reduced risk for several types of cancer. Lycopene intake was related to a lower risk of prostate cancer [2]. Effects on GJC have been discussed as biochemical mechanisms of protective properties of carotenoids, apart from their antioxidant acitivity. Since also non-provitamin A carotenoids (lycopene or canthaxanthin) are capable of inducing GJC, it has been suggested that their effects are independent of retinoid signalling pathways. However, non-provitamin A carotenoids might be metabolized or decompose to yield non-vitamin A retinoids with biological activities in regulatory processes. 4-Oxo-retinoic acid has been identified as a ligand of the nuclear retinoic acid receptor β (RAR-β).

An interaction of retinoic acid and vitamin D signalling pathways has been described depending on the formation of heterodimers between the retinoic X receptor (RXR) and the vitamin D receptor (VDR). This prompted us to study also the effect of cholecalciferol on GJC.

Institut für Physiologische Chemie I, Heinrich-Heine-Universität Düsseldorf, P.O. Box. 101007, D-40001 Düsseldorf, Germany

Materials and methods

Chemicals: Lucifer yellow and vitamin D were purchased from Sigma (Deisenhofen, Germany); carotenoids were obtained from Hoffmann-La Roche (Basel, Switzerland); all other chemicals were from Merck, (Darmstadt, Germany). The other carotenoids were synthesized as described by Devasagayam *et al.* [3].

Cells and Culture conditions: Murine embryo fibroblasts C3H/10T (ATCC No. CCL 226) were cultured with fibroblast growth medium (Promo Cell, Heidelberg, Germany) supplemented with 10% FCS. When confluence was reached FCS was decreased to 3% and the cells were treated with the indicated concentrations of the test compounds. The medium with the respective compounds was exchanged every 3 days. The final concentration of solvents (tetrahydrofuran or acetone) in the culture medium was 0.5% [4, 5].

GJC-Assay: GJC was assayed by microinjection of the fluorescent dye Lucifer yellow CH (10% in 0.33 M LiCl) into cells of confluent cultures by means of a microinjector and micromanipulator (Eppendorf, Hamburg, Germany). The number of fluorescent neighbors of the injected cells was scored 5 min after injection and serves as an index of GJC.

Results and discussion

Induction of GJC by carotenoids

β-Carotene, echinenone, canthaxanthin, cryptoxanthin, and 4-hydroxy-β-carotene efficiently induce GJC [6]. The analog of β-carotene, retro-dehydro-β-carotene is as effective as the parent compound. Echinenone and 4-OH-β-carotene exhibit about similar activity. Cryptoxanthin is somewhat less active than the corresponding 4-OH-β-carotene.

No stimulatory effects were found for capsorubin or violerythrin. A direct comparison of five- and six-membred ring carotenoids can be taken from dinorcanthaxanthin and canthaxanthin. The six-membered ring carotenoid is about twice as active as its five-membered ring analog. The effects could be related to the intact carotenoids or their decomposition products.

Two synthetic novel compounds, C22-polyene-tetrone-diacetal and C28-polyene-tetrone were tested for their activity in the gap junctional assay in comparison to canthaxanthin as a positive control [7]. The C22-carotenoid induced GJC at concentrations of 10^{-5} M; the induction was in the range of canthaxanthin. Lower levels (10^{-6} M) of the carotenoid were not active. No inductory effects on GJC were observed with C28-polyene-tetrone.

The effect of C22-polyene-tetrone-diacetal correlated with the inhibition of methylcholanthrene induced transformation of the cells. At 10^{-5} M the C22-compound completely inhibited carcinogen-induced transformation when added during the post-initiation phase of carcinogenesis. Comparable results were obtained with canthaxanthin.

The demonstration that synthetic carotenoids possess biological activities comparable to one of the most active natural compounds suggests that rational synthesis of carotenoids with improved pharmacological properties should be possible.

Induction of GJC by 4-oxo-retinoic acid generated from canthaxanthin

Both isomers of 4-oxo-retinoic acid (all-*trans* and 13-*cis*) enhanced GJC in C3H/10T 1/2 cells, accompanied by an increase of the expression of connexin 43 *m*-RNA [4]. Decomposition fractions of canthaxanthin were isolated by preparative HPLC and shown to be active in the cell communication assay. Two of the decomposition products were identified as all-*trans*- and 13-*cis*-4-oxo-retinoic acid. Therefore, it is concluded that the biological activity of canthaxanthin is at least in part due to the formation of active decomposition products such as 4-oxo-retinoic acid. To obtain a significant induction of cell-cell-communication in 10T 1/2 cells, only 0.1% of the 10 µM incubation mixture with canthaxanthin needs to be converted to 4-oxo-retinoic acid, which is active at even 10 nM levels.

It is interesting to note that the non-provitamin A carotenoid canthaxanthin decomposes to yield the non-vitamin A retinoid 4-oxo-retinoic acid which exhibits regulatory activity. Thus, other carotenoids might serve as precursors for retinoid derivatives with interesting biochemical properties.

Induction of GJC by vitamin D

Vitamin D3 (cholecalciferol) induces GJC in 10T 1/2 cells at concentrations between 0.01 and 1.0 µM [5], as assayed by the dye transfer method. The extent of induction is similar to that obtained with canthaxanthin or retinoic acid at 10 and 1 µM, respectively. Vitamin D2 also induces GJC. At elevated concentrations of vitamin D3 (5 µM) there is a suppression of GJC, reversible upon exposure to all-*trans* retinoic acid (1 µM) after removal of vitamin D3 from the medium. Conversely, communication between cells prestimulated with all-*trans* retinoic acid (1 µM) rapidly decreases when the retinoid is replaced by vitamin D3 (5 µM). The inductory effects of vitamin D at low concentrations are rather slow as compared to the rapid inhibition of GJC at high vitamin D levels. This implies two different vitamin D dependent regulatory mechanisms of GJC, one being activatory, possibly by regulation of connexin gene expression. The other mechanism is inhibitory, likely operating through changes in cellular calcium concentration.

The physiologically active metabolite of vitamin D, 1,25-dihydroxy vitamin D, stimulates GJC at concentrations of 10^{-8} to 10^{-11} M [8]. In fibroblasts devoid of a functional vitamin D receptor (VD-R) there is no effect on GJC. Parallel to the increase in GJC there is a VD-R-dependent increase in connexin 43 protein and connexin 43 *m*RNA levels.

The data demonstrate a role of vitamin D in the regulation of GJC. This novel property of Vitamin D may contribute to the antiproliferative effects of vitamin D exhibited in some types of cancer.

References

[1] Pung, A.; Rundhaug, J. E.; Yoshizawa, C. N.; Bertram, J. S. (1988) *Carcinogenesis* **9**, 1533–1539.

[2] Giovannucci, E.; Ascherio, A.; Rimm, E. B.; Stampfer, M. J.; Colditz, G. A.; Willett, W. C. (1995) *J. Natl. Cancer Inst.* **87**, 1767–1776.

[3] Devasagayam, T. P. A.; Werner, T.; Ippendorf, H.; Martin, H.-D.; Sies, H. (1992) *Photochem. Photobiol.* **55**, 511–514.

[4] Hanusch, M.; Stahl, W.; Schulz, W.; Sies, H. (1995) *Arch. Biochem. Biophys.* **317**, 423–428.

[5] Stahl, W.; Nicolai ,S.; Hanusch, M.; Sies, H. (1994) *FEBS Lett.* **352**, 1–3.

[6] Stahl. W.; Nicolai, S.; Briviba, K.; Hanusch, M.; Broszeit, G.; Peters, M.; Martin, H.-D.; Sies, H. (1997) *Carcinogenesis* **18**, 89–92.

[7] Pung, A.; Franke, A.; Zhang, L.-X.; Ippendorf, H.; Martin, H.-D.; Sies, H.; Bertram, J. S. (1993) *Carcinogenesis* **14**, 1001–1005.

[8] Clairmont, A.; Tessmann, D.; Stock, A.; Nicolai, S.; Stahl, W.; Sies, H. (1996) *Carcinogenesis* **17**, 389–1391.

21 Protective Effects of Vitamins C and E in Smokers Monitored by Ascorbyl Radical Formation in Plasma and the Frequency of Micronuclei in Lymphocytes

Matthias Schneider[a], Kerstin Diemer[b], Karin Engelhart[a], Heinrich Zankl[b], Wolfgang E. Trommer[c] and Hans K. Biesalski[a]

Oxidation induced by cigarette smoke is widely thought to be associated with an increased risk of several diseases including lung cancer and atherosclerosis. The gas phase of cigarette smoke contains numerous compounds of free radicals which have been shown to be involved in the oxidation process. Epidemiologic studies suggest a protective role for the antioxidant compounds β-carotene, ascorbic acid and α-tocopherol. In this work we studied the effects of vitamins C and E intake on the signal intensity of the ascorbyl radical used as a marker of oxidative stress in human plasma. Furthermore, we investigated mutagenic effects using the micronuclei test. Twelve smokers and non-smokers followed a 14 days supplementation with 1000 mg ascorbic acid daily during the first week and 1000 mg ascorbic acid and 335.5 mg α-tocopherol daily during the second week. Measurement of these antioxidants showed lower plasma concentrations in smokers than in non-smokers. After 7 days intake of vitamin C the group of smokers had a significant increase in the EPR signal intensity ($p<0.05$) and plasma ascorbic acid ($p<0.05$). Non-smokers had no such increases. Further treatment for 7 days with vitamin E revealed no effects on ascorbate radical formation in both groups, but a significant ($p<0.0001$) decline was observed in non-smokers after washout of the supplements. The micronuclei test performed by the cytokinesis-block method as an indicator of mutagenic effects showed significantly higher values of micronuclei in smokers compared to non-smokers. During supplementation with vitamins C and E a reduction of the micronuclei frequently was evident in smokers and non-smokers. However, the decrease was only significant in smokers ($p<0.05$) by administration of both vitamins.

[a] Institut für Biologische Chemie und Ernährungswissenschaften, Universität Hohenheim, D-70593 Stuttgart, Germany
[b] Fachbereich Biologie/Abteilung Humanbiologie der Universität Kaiserslautern, Germany
[c] Fachbereich Chemie/Biochemie der Universität Kaiserslautern, D-67653 Kaiserslautern, Germany

22 Anticarcinogenic Effects of Genotoxic Carcinogens Based on Cell-Cycle Delay, Calculated for the Two-Stage Clonal Expansion Model of Carcinogenesis

Annette Kopp-Schneider[1] and W. K. Lutz[2]

Dose-response relationships for tumor induction in animal bioassays for carcinogenicity are often sublinear (convex) at the low-dose end. In some cases, the background tumor incidence even appears to be decreased at the lowest dose levels. Recent observations indicate that DNA damage can result in a delay of the cell cycle. In view of the understanding that cell proliferation constitutes a risk factor for carcinogenesis the question was addressed to what extent a decrease in the rate of cell turnover could result in a reduction of the spontaneous tumor incidence or balance out a concomitant increase in the rate of mutation.

Calculations were based on the exact solution [1–3] for the two-stage clonal expansion model of carcinogenesis [4, 5]. Assuming a constant number of 10^6 normal stem cells, rates of 10^{-7}/day for both transitions from normal to premalignant and from premalignant to malignant cells, and rates of cell division and cell death for premalignant cells of 0.5 and 0.49/day (= background rates of "cell turnover"), a 2-year cumulative tumor incidence of 10.5 % was calculated.

These parameters were used as a starting point to investigate the effect of changes in mutation and cell turnover rates on the cumulative 2-year tumor incidence. In all investigations, the ratio of death to birth rate was kept constant at 98 %. In a first approach, the combined effect of increasing mutation rates and decreasing cell turnover rates was analyzed and the "point of equivalence" was determined, i. e. the percent decrease in rates of cell turnover required to compensate the given increase in mutation rates:

Table 22.1: Cumulative 2-year tumor incidence as a function of an increase in mutation rates and a decrease in cell turnover rates.

Decrease in birth/death rates [%]	Increase in mutation rates [%] 0	3	10	30	100
0	10.5	11.0	12.3	16.0	30.4
3	9.4	9.8	11.0	14.5	28.1
10	7.1	7.5	8.4	11.2	22.7
30	3.0	3.2	3.7	5.0	11.1
Point of equivalence [%]	–	1.5	4.5	12.0	31.5

[1] Biostatistics Unit, German Cancer Research Center, D-69120 Heidelberg
[2] Department of Toxicology, University of Würzburg, D-97078 Würzburg

Levels of reduction of the rates of cell proliferation up to 46 % have been observed *in vivo* [6], indicating that the scenario is compatible with life.

In view of the fact that high doses of genotoxic carcinogens can result in cytotoxicity and regenerative hyperplasia, J-shaped dose responses for the rates of cell birth and death were modeled by shifted quadratic functions for the respective rates. Combinations with linearly dose-dependent mutation rates generated, under certain conditions, J-shaped dose-response curves also for the 2-year cumulative tumor incidence.

The dose in a J-shaped dose-response curve at which the tumor incidence reverts to the background value could be considered to represent a "threshold dose". As an example, for a 30 % increase in mutation rates and a 10 % decrease in cell turnover rates (both at dose 1), the "threshold dose" can be calculated to be 0.8 dose units. Table 22.2 shows the threshold values for all situations considered in Figure 22.1.

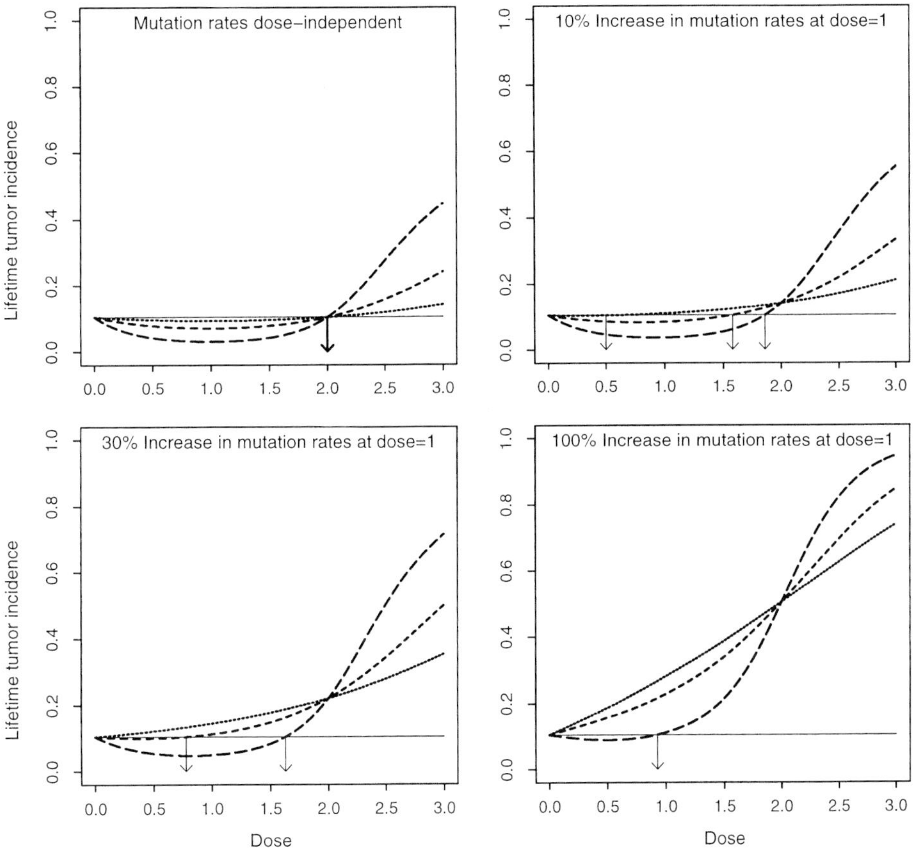

Figure 22.1: Shapes of dose-response curves generated by a superposition of a dose-linear increase in the rates of mutation (0, 10 %, 30 % and 100 % increase at dose 1 in the four charts respectively) with a parabolic shape of the dose-response rates of cell turnover. Percent decrease at dose 1: ······, –3 %, - - - -, –10 %, – – –, –30 %.

Table 22.2: Threshold doses calculated for a superposition of a dose-linear increase in the mutation rates with a parabolic shape of the dose-response rates of cell turnover.

Decrease in birth/death rates [%]	Increase in mutation rates [%] 0	10	30	100
3	2.0	0.5	–	–
10	2.0	1.6	0.8	–
30	2.0	1.9	1.6	0.9

The superposition of a number of dose-response relationships which govern the process of carcinogenicity allows to explain mechanistically many different shapes of curves, including linear-sublinear shapes ("convex" in mathematical terms), J-shapes, and thresholds [7]. This approach might reconcile opposing views on thresholds on a biologically plausible mechanistic basis and show a way for the estimation of threshold doses.

Although the model used for the analysis is very simple, the results are expected to carry over to more realistic models which take into account that the process of carcinogenesis involves more than two steps, that model parameters might be cell type- and age-dependent and that the mutation rates are more complicated functions of DNA damage, repair and replication.

References

[1] Kendall, D. G. (1960) Birth-and-death processes, and the theory of carcinogenesis. *Biometrika* **47**, 316–330.

[2] Moolgavkar, S. H.; Venzon, D. J. (1979). Two-event model for carcinogenesis: Incidence curves for childhood and adult tumors. *Math. Biosci.* **47**, 55–77.

[3] Moolgavkar, S. H.; Knudson, A. G. (1981) Mutation and cancer: A model for human carcinogenesis. *J. Natl. Cancer Inst.* **66**, 1037–1052.

[4] Kopp-Schneider, A.; Portier, C. J.; Sherman, C. D. (1994) The exact formula for tumor incidence in the two-stage model. *Risk Anal.* **14**, 1079–1080.

[5] Zheng, Q. (1994). On the exact hazard and survival functions of the MVK stochastic carcinogenesis model. *Risk Anal.* **14**, 1081–1084.

[6] Lutz, U.; Lugli, S.; Bitsch, A.; Schlatter, J.; Lutz, W. K. (1997) Dose response for the stimulation of cell division by caffeic acid in forestomach and kidney of male F344 rat. *Fund. Appl. Toxicol.* **39**, 131–137.

[7] Lutz, W. K. (1998) Dose-response relationships in chemical carcinogenesis: superposition of different mechanisms of action, resulting in linear-sublinear curves, practical thresholds, J-shapes. *Mutat. Res.* **405**, 117–124.

23 Occurrence of Emodin, Chrysophanol and Physcion in Vegetables, Herbs and Liquors. Genotoxicity and Antigenotoxicity of the Anthraquinones and of the Whole Plants

Stefan O. Müller, Marko Schmitt[1], Wolfgang Dekant, Helga Stopper, Josef Schlatter[2], Peter Schreier[1] and Werner K. Lutz

Introduction

Hydroxyanthraquinones are the active principles in a large number of plant-derived drugs, such as laxatives. Furthermore, their occurrence in some plants has been described [1]. In many *fungi imperfecti* and some toadstools, anthraquinone derivatives have been discussed as the toxic metabolites and coloring matters [1]. Most of the hydroxyanthraquinones are present as pharmacologically inactive glycosides in plant extracts but are thought to be activated by glycosidic cleavage *in vivo* by microorganisms in the intestinal flora [2]. The concentration of anthraquinones in food plants can be constitutive or induced by infection. In addition, a contamination of foods by enzyme preparations from fungi used in food production has been discussed [3].

The genotoxic effects of the 1,8-dihydroxyanthraquinones are still under debate. Recently, we have demonstrated unequivocally a genotoxicity of the anthraquinones emodin, aloe-emodin and danthron in mammalian cells [4]. Here, we present data on the presence of emodin, chrysophanol and physcion in vegetables, herbs and liquors, and additional data on the genotoxic effects of the anthraquinones, using the micronucleus test, the single-cell gel electrophoresis assay (COMET assay) and the mutation assay in mouse lymphoma L5178Y tk+/– cells. In the micronucleus test, the vegetables were also tested as a whole, in order to investigate putative antigenotoxic effects.

Department of Toxicology, University of Würzburg, Versbacher Str. 9, D-97078 Würzburg, Germany

[1] Chair of Food Chemistry, University of Würzburg, Am Hubland, D-97074 Würzburg, Germany

[2] Swiss Federal Office of Public Health, Food Sciences Division, c/o Institute of Veterinary Pharmacology, CH-8057 Zürich, Switzerland

Materials and methods

Lyophilisates: washed vegetable was lyophilized and powdered. The powder was suspended in DMSO at a concentration of 25 mg/ml.

Extracts: washed vegetable was homogenized in a blender and lyophilized. The powder was suspended in bicarbonate/acetonitrile and treated with ultrasound. The suspension was extracted with chloroform, either directly or after filtration. Chloroform was evaporated and the residue resuspended in DMSO.

We used RP-HPLC analysis for the quantitative determination of the content of the anthraquinones emodin, chrysophanol and physcion. Because LC-MS-analysis was not more sensitive than RP-HPLC, LC-MS analysis was only performed to verify anthraquinone occurrence in selected materials. Additionally, we applied a GC-MS method to quantify the contents of chrysophanol and physcion in selected vegetables.

Methods were performed as described in [5].

Results and discussion

Occurrence of anthraquinones in vegetables, herbs and herbal-flavoured liquids

We have shown for the first time the presence of anthraquinones in a number of vegetables. Physcion predominated, with maximum contents in beans (32 mg/kg fresh weight), followed by lettuce (5.8) and peas (3.6). Anthraquinones were not detected in all analyzed herbs. However, in some herbs emodin also occurred in high concentration. Garden sorrel showed an exceptional high content of emodin (261 mg/kg). Watercress also contained a significant amount of all three anthraquinones, higher than the other herbs. As liquors contain herbal ingredients, we investigated several common and often consumed herbal-flavoured liquids. In 7 of 11 analyzed herbal-flavoured liquors all three anthraquinones were detected in a ranking chrysophanol > emodin > physcion. The concentrations of the individual compounds varied between 0.02 mg/kg and 3.7 mg/kg. The anthraquinones in liquors might be of interest only in the case of an abuse.

Although the content of anthraquinones was low, a maximum possible intake of up to 4 mg anthraquinone in a common 100 g serving of beans may be

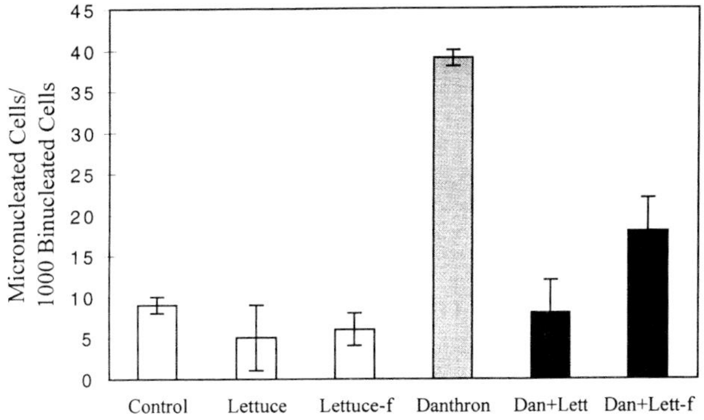

Figure 23.1: Cytokinesis-block micronucleus test in mouse lymphoma cells with lettuce extracts (250 µg/ml, Lettuce/Lett = unfiltered and Lettuce-f/Lett-f = 0.2 µm filtered), and with danthron (Dan, 7.5 µg/ml), in the presence and absence of the extracts. All incubations were performed with 6 µg/ml cytochalasin B; control: 1% DMSO. Mean and standard deviation is given (n = 3). Decreases of micronuclei frequencies of the coincubation (danthron + lettuce) compared to danthron alone were significant according to Students' t-test (Dan + Lett $p<0.001$; Dan + Lett-f $p<0.01$).

deduced from the maximum determined concentrations. For laxative drugs, the maximum therapeutical daily dose is 30 mg of anthraquinone glycoside. For senna, a maximum of up to 4 mg anthraquinones per day is used. Therefore, the dose of total anthraquinones that could be present in 100 g beans is comparable to the maximum daily dose of anthraquinones taken up with the use of a senna laxative. Because of the observed ranking chrysophanol > emodin > physcion, also seen in rhubarb root extracts, the use of this root extract as a part of the herbal ingredients of the spirits can be assumed. However, for a toxicological assessment of liquors, the anthraquinone content is considered irrelevant.

Genotoxicity of anthraquinones in *Salmonella typhimurium* and mouse lymphoma cells

Emodin, chrysophanol and physcion showed no mutagenicity in the Ames test with TA100 and TA2638 with or without metabolic activation using a S9-mix from Aroclor 1254-pretreated male rats. Mutagenic activity in eukaryotic cells was measured by the induction of TFT-resistance in mouse L5178Y cells. Emodin induced a moderate increase in relative mutation frequency (>2 fold over control at 105 µM). In a similar concentration range, chrysophanol and physcion did not meet the criterion of a 2-fold mutation frequency when compared to the solvent control. Toxicity as shown by a reduction in the cloning efficiency was low in the applied dose range for emodin and chrysophanol. For physcion, a reduction of the cloning efficiency to less than 50% of the control value was observed at the higher concentrations tested. With respect to the induction of micronuclei in the same cell line, we have previously shown a dose-dependent increase for emodin [4]. Chrysophanol and physcion were much less potent and the effect was not clearly concentration dependent. In the COMET assay performed in a similar dose range, none of the three anthraquinones showed an effect, in the dose range tested.

It should be mentioned here that chrysophanol can be metabolized by a cytochrome P450-dependent reaction [6] to the known genotoxin aloe-emodin. Therefore, a genotoxification pathway relevant for the situation *in vivo*, may be deduced for chrysophanol.

Previously published data from mammalian mutation assays for anthraquinones by other investigators were not consistently positive (e.g. [7]). However, the earlier data related to the hemizygous hprt locus, a locus known to be less sensitive. In the present study, we used the highly sensitive heterozygous thymidine kinase (tk) locus in the L5178Y cell line.

Genotoxicity and antigenotoxicity of the vegetables

To assess whether the vegetables as a whole contain additional genotoxic constituents, the lyophilisates of lettuce, beans and peas were tested in the micronucleus test in mouse lymphoma cells, alone and in combination with the known genotoxic (and carcinogenic) anthraquinone danthron. None of the vegetable lyophilisates increased the frequency of micronuclei. Lettuce extracts (of the unfiltered and of the 0.2 μm-filtered suspension, described in section 23.2 (Materials and Methods) did not induce higher micronucleus frequencies as compared to the control (Figure 23.1, white bars). The positive control danthron induced a high frequency of micronuclei. When co-incubated with danthron, both types of lettuce extracts produced a marked reduction of the micronucleus frequency. The extract of the unfiltered suspension abolished the micronucleus induction completely, whereas the extract of the filtered suspension inhibited the genotoxicity of danthron to a somewhat lesser degree (Figure 23.1).

Although three common vegetables have been shown to contain anthraquinones, data on their genotoxicity alone should not be used to derive a human health risk. The vegetables as a whole contain numerous other constituents some of which might also have protective effects [8, 9]. Chloroform-extracts of unfiltered, suspended lettuce lyophilisate had a strong protective effect. Filtration of the suspension reduced the protective effect.

In conclusion, together with the strong epidemiological evidence for a cancer-protective effect of vegetable consumption, it could be postulated that the mutagenicity from substances taken up with vegetables is more than compensated by protective effects. For the evaluation of a putative human health risk from dietary mutagens, an assessment should not be based exclusively on measured concentrations of mutagens.

References

[1] Thomson, R. H. (1986) *Naturally Occurring Quinones III. Recent Advances*. Chapman and Hall, London.

[2] Hattori, M.; Akao, T.; Kobashi, K.; Namba, T. (1993) Cleavages of the *O*- and *C*-glucosyl bonds of anthrone and 10,10′-bianthrone derivatives by human intestinal bacteria. *Pharmacology* **47** (suppl. 1), 125–133.

[3] Gross, M.; Levy, R.; Toepke, H. (1984) Zum Vorkommen und zur Analytik des Mycotoxines Emodin. *Nahrung* **1**, 31–44.

[4] Müller, S. O.; Eckert, I.; Lutz, W. K.; Stopper, H. (1996) Genotoxicity of the laxative drug components emodin, aloe-emodin and danthron in mammalian cells: topoisomerase II mediated? *Mutat. Res.* **371**, 165–173.

[5] Müller, S. O.; Schmitt, M.; Dekant, W.; Stopper, H.; Schlatter, J.; Schreier, P.; Lutz, W. K. (1999) Occurrence of emodin, chrysophanol and physcion in vegetables, herbs and liquors. Genotoxicity and anti-genotoxicity of the anthraquinones and of the whole plants. *Food Chem. Toxicol.* **37**, 481–491.

[6] Müller, S. O.; Stopper, H.; Dekant, W. (1998) Biotransformation of the anthraquinones emodin and chrysophanol by cytochrome P450 enzymes: Bioactivation to genotoxic metabolites. *Drug Metabolism and Disposition* **26**, 540–546.

[7] Westendorf, J.; Marquardt, H.; Poginsky, B.; Dominiak, M.; Schmidt, J.; Marquardt, H. (1990) Genotoxicity of naturally occurring hydroxyanthraquinones. *Mutat. Res.* **240**, 1–12.

[8] Ames, B. N. (1983) Dietary carcinogens and anticarcinogens. *Science* **221**, 1256–1264.

[9] Mitscher, L. A.; Telikepalli, H.; McGhee, E.; Shankel, D. M. (1996) Natural antimutagenic agents. *Mutat. Res.* **350**, 143–152.

24 Degradation of Flavonoids by the Human Intestinal Bacteria *Eubacterium Ramulus* and *Enterococcus Casseliflavus*

R. Simmering, Heiko Schneider and M. Blaut*

Flavonoids are present in many plants and in food derived thereof. Owing to their pharmacological properties there is a growing interest in their metabolism. It is known that flavonoids are subject to degradation by intestinal bacteria, but there is a lack of information on the bacteria and the degradational pathways involved. Two types of bacteria capable of degrading the very common flavonoid

* German Institute of Human Nutrition (DIfE), Department of Gastrointestinal Microbiology, Arthur-Scheunert-Allee 114–116, D-14558 Potsdam-Rehbrücke, Germany

quercetin-3-glucoside (isoquercitrin) were isolated from human feces and subjected to further examinations.

Enterococcus casseliflavus utilized only the glucose moiety of isoquercitrin, but did not degrade the aglycon any further. In contrast, *Eubacterium ramulus* also degraded the aglycon producing acetate, butyrate, ethanol, 3,4-dihydroxyphenylacetic acid and H_2, 3,4-dihydroxybenzaldehyde, and phloroglucinol as minor products.

A 16S rRNA-targeted oligonucleotide probe (Eram997) specific for *E. ramulus* was designed and validated to examine the occurrence of this bacterial species in human fecal samples. Using *in situ* hybridization we detected up to 10^9 cells/g fecal dry mass.

The ability of *Eubacterium ramulus* and *Enterococcus casseliflavus* to degrade isoquercitrin *in vivo* was tested in gnotobiotic rats. No isoquercitrin but quercetin was detected in feces and urine of germfree rats. In rats monoassociated with *E. casseliflavus* larger amounts of quercetin were detected in feces and urine. In rats monoassociated with *E. ramulus* or diassociated with *E. casseliflavus* and *E. ramulus* 3,4-dihydroxyphenylacetic acid was found in feces and urine.

25 Distribution Kinetics of a Flavonoid Extract in the Gastrointestinum of Rats

Holger Pforte, Jürg Hempel and Gisela Jacobasch*

Introduction

Flavonoids are polyphenolic plant secondary metabolites that may exert anti-atherosclerotic and anti-carcinogenic activities in mammals. To be biologically effective, dietary flavonoids must be bioavailable. Deglycosylation of flavonoids to the aglycone form by the intestinal microflora was regarded as prerequisite to absorption for many years. The digestive absorption in the form of glycosides could be proven only recently [1].

* German Institute of Human Nutrition, Department of Food Chemistry and Preventive Nutrition, Arthur-Scheunert-Allee 114–116, D-14558 Bergholz-Rehbrücke, Germany

To get a better insight into flavonoid absorption we investigated the gastrointestinal fate of a glycosides mixture of flavones and flavonoles from a parsley cell culture that was extracted, purified and characterized as reported [2].

Methods

- Oral administration by gavage of 200 μl flavonoid glycosides extract (corresponding to 0.942 mg flavonoid aglycones) from parsley cell culture to female Wistar rats, sacrifice of the animals after 1, 1.5, 2, 4 or 12 hours respectively and taking of stomach, small intestine, cecum, colon and their contents, liver and kidneys.
- Organs and their contents were lyophilized. Organs were degreased with heptane and Soxhlet-extracted with methanol for 4 hours under reduced pressure at about 45 °C. Organ contents were extracted with 70 % methanol in an ultrasound bath.
- Separation of the extracted phenolic compounds by RP-HPLC with acetonitrile gradient in diluted acetic acid and detection by DAD.
- Quantification of flavonoids by relating the respective sum of peak areas at $\lambda = 354$ nm to the amount administered.

Results

- Administered flavonoid glycosides are partly deglycosylated already within the stomach contents of rats. Quercetin aglycone (Q) is detectable at 1 hour after gavage and with progressing hydrolysis additionally the aglycones apigenin (Apig), chrysoeriol (Chrys) and isorhamnetin (Isorh) 4 hours later.
- Within the stomach walls the flavonoid concentration increases up to 4 hours (Figure 25.1). At that time mainly the aglycones Q, Apig, Chrys and Isorh are found accompanied by only traces of glycosides.
- At 1 hour the extract from the small intestine wall shows a flavonoid pattern deviating from the starting material as well as from the stomach contents at the same time. Apart from some glycosides Q is detectable.

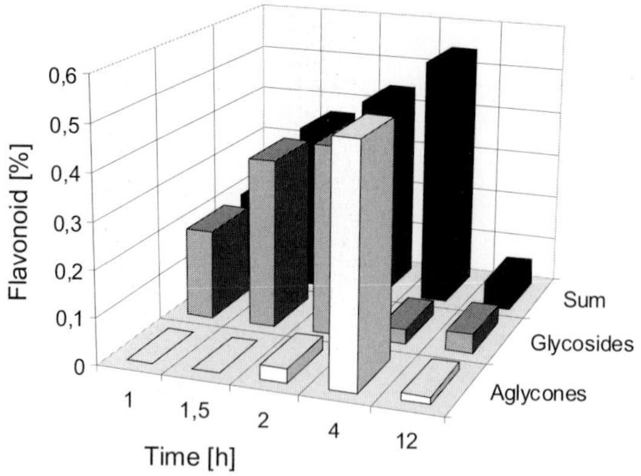

Figure 25.1: Recovery [%] of administered flavonoids in stomach wall.

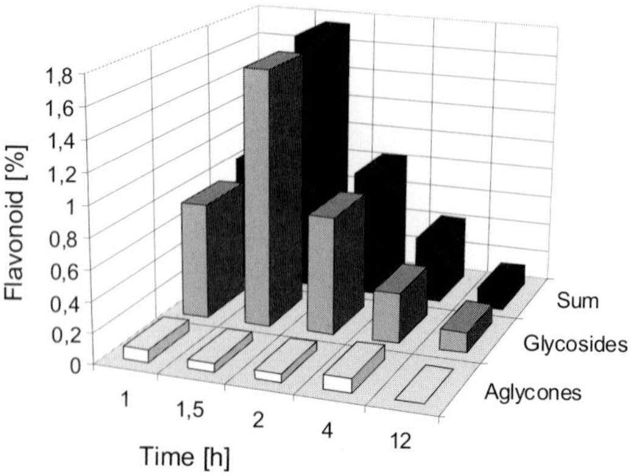

Figure 25.2: Recovery [%] of administered flavonoids in small intestine wall.

- Maximum concentration of flavonoids in the small intestine wall is detected after 1.5 hours. The aglycones level remains constant up to 4 hours (Figure 25.2).
- After 4 hours the aglycones Q, Apig, Chrys and Isorh are detected in the cecum contents. These aglycones are accompanied by some glycosides with 1 major peak in the cecum wall.
- After 4 hours flavonoids are found in the colon wall at a very low level.
- The amount of total flavonoids in the gastrointestinal tract is nearly zero after 12 hours.
- Flavonoid glycosides are accumulated in the liver with higher values after 1.5 hours and 12 hours (Figure 25.3).
- There were no flavonoids detected in kidneys and urine.

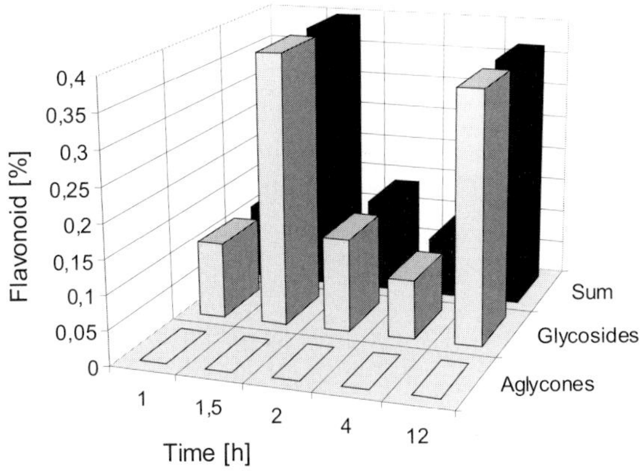

Figure 25.3: Recovery [%] of administered flavonoids in liver.

Conclusion

Results show the beginning deglycosylation of dietary flavonoid glycosides already in the stomach. The uptake of flavonoids in the form of aglycones or glycosides differs between the investigated parts of the gastrointestinal tract.

Acknowledgement

The study was supported by the Innovationskolleg INK 26/A1–1 from the Deutsche Forschungsgemeinschaft.

References

[1] Paganga, G.; Rice-Evans, C. A. (1997) *FEBS Letters* **401**, 78–82.
[2] Hempel, J.; Pforte, H.; Raab, B.; Engst, W.; Böhm, H.; Jacobasch, G. *Nahrung/Food* **43**, 201–204.

26 Flavones and Inhibition of Tumor Cell Growth: New Aspects on the Mechanism of Action

Ellen Niederberger, *Susanne Meiers*, A. Genzlinger, H. Zankl, W. C. Tang, G. Eisenbrand and D. Marko

Flavones, isolated by us from the roots of *Scutellaria baicalensis*, a major plant constituent of popular herbal medicine used in Asia (Japan, China) for the prevention of chronic diseases and malignancies, comprise baicalein, wogonin and skullcapflavon II, as well as the respective glucuronides baicalin and wogonoside. These compounds were tested for their growth inhibitory properties on human tumor cell lines and their mechanism of action.

Baicalin, baicalein, skullcapflavon II and wogonoside inhibited the growth of human tumor cell lines (e.g. LXFL 529L, a large cell lung tumor cell line) in the micromolar range, whereas wogonin showed no inhibitory effect up to 100 μM. Baicalein has been reported earlier to inhibit DNA-topoisomerase II by stabilising the covalent topoisomerase-DNA-intermediate in a ternary complex. Our studies show that in the same concentration range also topoisomerase I is inhibited by stabilisation of the otherwise cleavable complex. In contrast, the respective glucuronide baicalin exhibited no inhibitory properties on topoisomerase I and II. In accordance with these results, baicalein was found to induce DNA strandbreaks as detected in the COMET assay (single cell gel electrophoresis), whereas no strandbreaks were found after treatment with baicalin.

Furthermore, baicalein has been reported to inhibit the activity of cAMP degrading phosphodiesterases (PDE). Our own results confirmed the PDE inhibitory activity and we could show that baicalein is able to inhibit the cAMP-specific isoenzyme family PDE4 (IC_{50} = 10 μM). The respective glucuronide baicalin is only a weak PDE4 inhibitor. However, when whole cells (LXFL 529L) were treated with baicalein, no inhibition of the intracellular PDE activity was detectable and the intracellular cAMP level remained unchanged.

Flavones have also been reported to inhibit receptor tyrosine kinases. We found, that baicalin, wogonin, skullcapflavon II and wogonoside showed only weak inhibitory properties on the tyrosine kinase activity of the epidermal growth factor receptor (EGFR) with IC_{50} values exceeding 60 μM. In contrast, baicalein was found to be a potent inhibitor of the EGF receptor with an IC_{50} of 1.1 ± 0.5 μM. In A431 cells overexpressing the EGFR, Baicalein was significantly more active than in other cell lines. The result indicates, that EGFR inhibition plays an important role in the biological activity of baicalein. However, although baicalein is a potent inhibitor of the isolated EGFR, the dose- and time-depen-

Dept. of Chemistry, Division of Food Chemistry and Environmental Toxicology, and Dept. of Biology, Human Biology and Human Genetics, University of Kaiserslautern, Erwin-Schroedinger-Str. 52, D-67663 Kaiserslautern, Germany, e-mail: marko@rhrk.uni-kl.de dent induction of DNA strandbreaks by the compound indicates that the inhibition of topoisomerases I and II substantially contributes to its mechanism of action. Flow cytometry and morphological studies of LXFL 529L cells showed an arrest of the cells in the G_2/M-phase of the cell cycle together with marked induction of apoptosis. The induction of these events can result from both, inhibition of topoisomerases and of the EGF receptor.

27 Food Derived Flavonoids That Affect Proliferation, Differentiation and Apoptosis in Human Colon Carcinoma Cells (HT-29) and Their Mode of Action

Uwe Wenzel, Sabine Kuntz, Stefan Storcksdieck, Ulrike Jambor de Sousa and Hannelore Daniel*

Introduction

Dietary flavonoids, that belong to the flavone, flavanone or isoflavone subgroups (Figure 27.1), are discussed in relation to cancer incidence, especially of breast and colorectal cancers [1, 2]. In spite of the indirect epidemiological evidence for their anticancer activity, informations on the mechanisms by which selected flavonoids might contribute to cancer prevention at the cellular level are sparse. Proliferation, differentiation and apoptosis are parameters known to be disregulated during the development of colorectal cancers [3, 4]. Therefore, by using human colon cancer cells (HT-29), we investigated the influence of flavonoids on these parameters and investigated how they exert their effects on a molecular basis.

* Institut für Ernährungswissenschaft, Technische Universität München, Hochfeldweg 2, D-85350 Freising-Weihenstephan

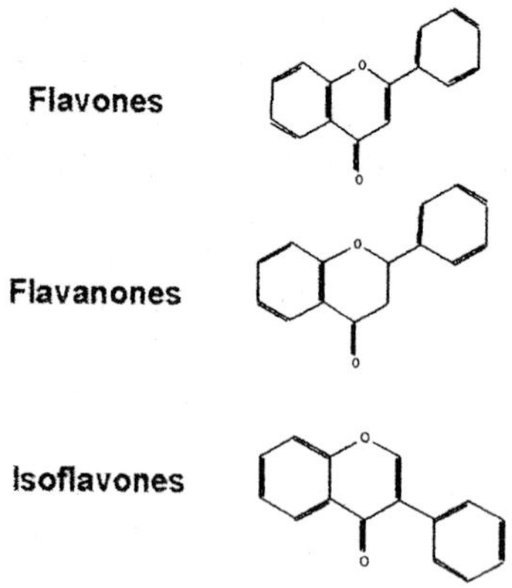

Figure 27.1: Basic structures of flavonoids.

Methods

Cell culture: HT-29 cells were cultured in RPMI-1640 media (Gibco) and passaged at 70 % confluency. Medium was supplemented with 10 % FCS, 2 mM Glutamine, 100 U/ml Penicillin and 100 μg/ml Streptomycin. Cultures were maintained in a humidified atmosphere of 95 % air and 5 % CO_2 and were passaged at preconfluent densities.

Cell proliferation- and cytotoxicity-assays: Cells were seeded at a density of 2.5×10^3 per well onto a 24 well plate and allowed to adhere for 24 h. Cells were then incubated for 72 h in the absence or the presence of flavonoids. Total cell counts were determined by Sytox (Molecular Probes) that becomes fluorescent after DNA binding. Therefore cells were lyssed by 1 % Triton-X 100 and cell numbers were determined based on a calibration curve. Cytotoxicity was assessed by Sytox-fluorescence after 5×10^4 cells adhered for 3 h and subsequent exposure of the flavonoids for 3 h.

Differentiation-assay: After having reached 40 % confluency on 25 cm^2-culture flasks cells were incubated for 72 h with the test compounds. Harvested cells were pelleted and alkaline phosphatase activity as a marker for differentiation was determined by the release of fluorescein from the substrate fluorescein-diphosphate.

Intracellular Ca^{2+} ($[Ca^{2+}]_i$) measurements: Cell monolayers were loaded with the Ca^{2+}-sensitive fluorescent dye preincubating the cells with 5 μM of Fura-2-AM for 30 min at 37 °C. The buffers alone or containing the effectors were applied to the monolayers subsequently and the fluorescence of Fura-2 was measured at 515 nm after excitation at 340 nm and 390 nm using a microtiter plate reader. Ca_i^{2+} was calculated based on the formula of Grynkiewicz *et al.* [5].

Apoptosis-assays: 80–90 % confluent cell monolayers grown on 6 well-plates were incubated for 72 h with the test compounds and after lysis the cytosol fraction was prepared by ultracentifugation. Cytosolic Apopain-activity determined by hydrolysis of fluorogenic CPP32/apopain substrate served as an early marker of apoptosis.

RT-PCR form RNA of HT-29 cells: First strand cDNA synthesis from isolated RNA of HT-29 cells was accomplished with an Oligo-dT probe. PCR amplification of the products was achieved by addition of gene-specific back and forward primers, respectively. Amplification was done with 35 cycles (95 °C denaturation for 1 min, 55 °C hybridization for 2 min, 72 °C extensions for 2 min). The first strand cDNA amounts used were adjusted to yield a PCR product whose densitometric signal was in a linear range when RNA from untreated control cells was used. RT-PCR products were separated on a 1% agrose gel and visualized by ethidium bromide.

Results

Flavone, three OH-substituted flavones (fisetin, baicalein, quercetin), the flavanone naringin and the isoflavone genistein were tested for their ability to affect cell proliferation, differentiation and apoptosis. All flavonoids tested inhibited proliferation of HT-29 in a dose dependent manner, with EC_{50} values between 20 μM (flavone) and 170 μM (naringin) (Figure 27.2). The ability of flavone to promote differentiation and to induce apoptosis were found to be more pronounced than those of the anti-tumor agent camptothecin and the other flavonoids tested (Figure 27.3). Measurements with Fura-2 revealed a slow increase in $[Ca^{2+}]_i$ caused by flavone (Figure 27.4A) that was not due to a loss of plasma membrane integrity (Figure 27.4B). Moreover, long term exposure to flavone is required for a sustained increase in $[Ca^{2+}]_i$ (Figure 27.4A) and for induction of apoptosis (Figure 27.5).

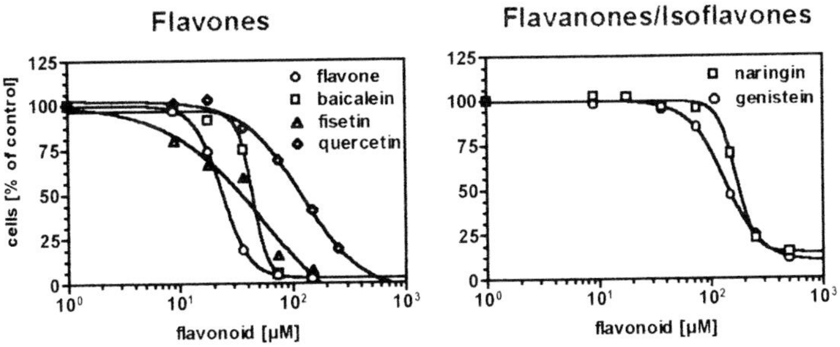

Figure 27.2: Dose-dependent effects of flavonoids on proliferation of HT-29 cells.

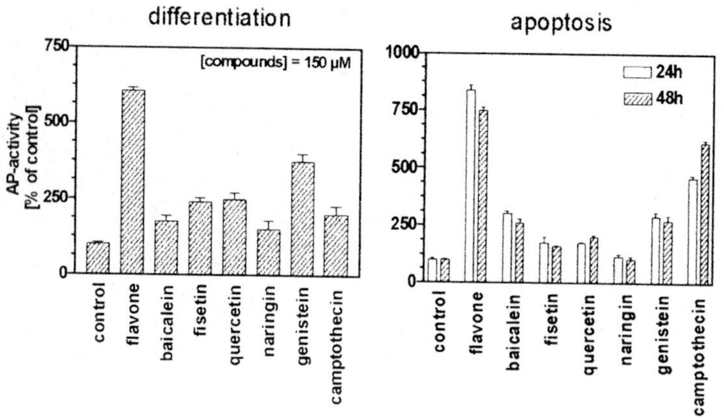

Figure 27.3: Differentiation and apoptosis of HT-29 cells induced by selected flavonoids.

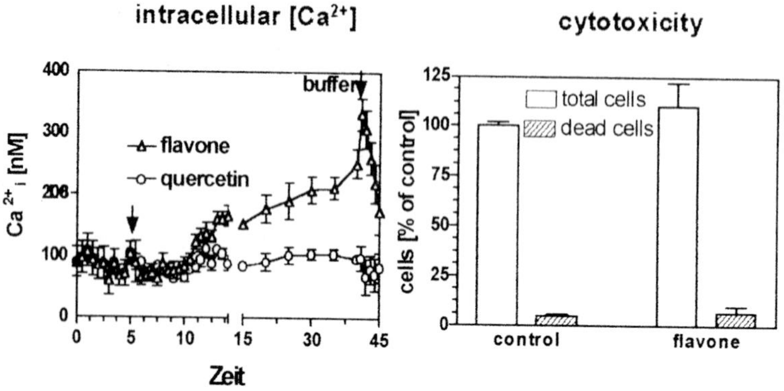

Figure 27.4: Effects of flavone on $[Ca^{2+}]_i$ and on cellular cytotoxicity.

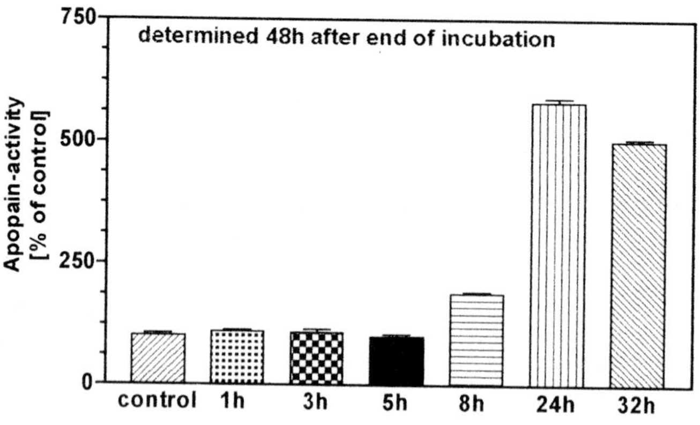

Figure 27.5: Flavone effects on apopain activity as a function of exposure time.

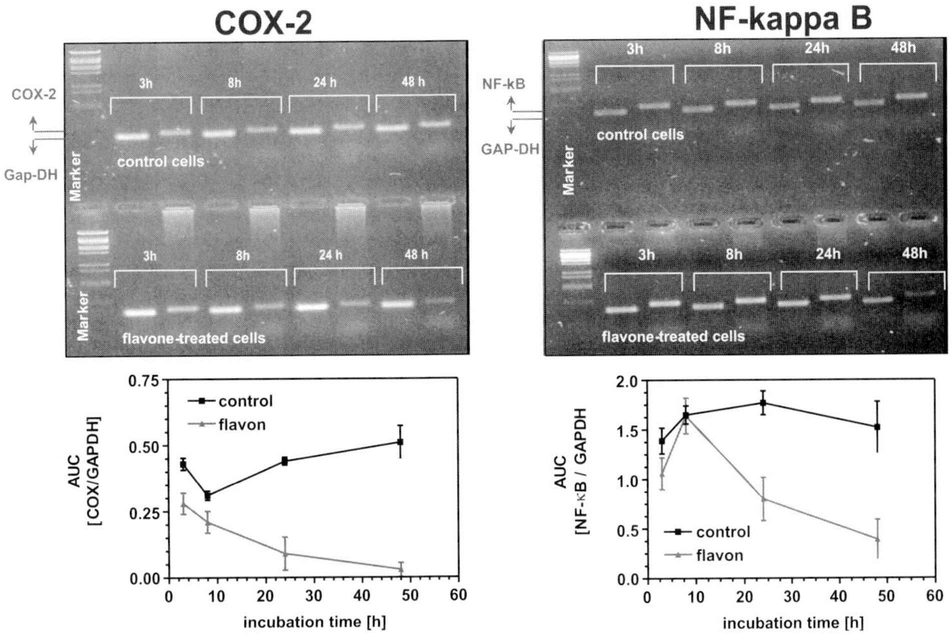

Figure 27.6: Effects of flavone on gene expression.

Flavone most likely exerts its effects by an reduction in expression of COX-2 and NF-kappa B mRNA (Figure 27.6).

Discussion

Several flavonoids were tested with regard to their ability to affect proliferation, differentiation and apoptosis in HT-29 cells. Flavone was most potent in promoting differentiation and induction of apoptosis. The effects of flavone were associated with a slow and sustained increase in $[Ca^{2+}]_i$ that as a second messenger could affect gene expression leading to the observed phenomena. From the genes affected, COX-2 and NF-kappa B play a crucial role in the prevention of apoptosis and cell cycle arrest [6, 7]. Moreover, COX-2 is overexpressed in most cases of colon carcinomas [8]. Therefore, the observed pronounced suppression of COX-2 and NK-kappa B mRNA-expression by flavone exposure may provide a useful strategy for colon cancer prevention and/or cancer therapy.

References

[1] Witte, J. S. *et al.* (1996) Relation of vegetable, fruit, and grain consumption to colorectal adenomatous polyps. *Am. J. Epidemiol.* **144**, 1015–1027.

[2] Messina, M.; Barnes, S. (1992) The role of soy products in reducing risk of cancer. *J. Natl. Cancer Inst.* **83**, 541–546.

[3] Scheppach, W. *et al.* (1995) Role of short-chain fatty acids in the prevention of colorectal cancer. *Eur. J. Cancer* **31**, 1077–1080.

[4] Potten, C. S. (1997) Epithelial cell growth and differentiation. *Am. J. Physiol.* **36**, G253–G257.

[5] Grynkiewicz, G. *et al.* (1985) A new generation of Ca^{2+} indicators with greatly improved fluorescence properties. *J. Biol. Chem.* **260**, 3440–3450.

[6] Sheng, H. *et al.* (1998) Modulation of apoptosis and Bcl-2 expression by prostaglandin E2 in human colon cancer cells. *Cancer Res.* **58**, 362–366.

[7] Van Antwerp, D. J. *et al.* (1998) Inhibition of TNF-induced apoptosis by NF-B. *Trends Cell Biol.* **8**, 107–111.

[8] Reddy, B. S. *et al.* (1996) Evaluation of cyclooxygenase-2 inhibitor for potential chemopreventive properties in colon carcinogenesis. *Cancer Res.* **56**, 4566–4569.

List of Participants and Contributors on the Symposium of the DFG-Senate Commission on Food Safety (SKLM) "Carcinogenic/Anticarcinogenic Factors in Food: Novel Concepts?"

Dr. Christine B. Ambrosone
Division of Molecular Epidemiology
(CBA) National Center for Toxicological Research
3900 NCTR Road
Jefferson, AR 72079
USA
Tel.: 001 870 543 7528
FAX: 001 870 543 7773

Dr. Lenore Arab
School of Public Health
University of North Carolina
Chapel Hill, NC 27599-7400
USA
Tel.: 001 919 966 7450
FAX: 001 919 966 2089

Prof. Dr. Helmut Bartsch
Deutsches Krebsforschungszentrum
Division of Toxicology and Cancer Risk Factors
Im Neuenheimer Feld 280
D-69120 Heidelberg
Germany
Tel.: 0049 6221 423300
FAX: 0049 6221 423359

Dr. Sheila A. Bingham
Dunn Clinical Nutrition Center
Medical Research Council
Hills Road
Cambridge CB2 2DH
Great Britain
Tel.: 0044 1223 415695
FAX: 0044 1223 413763

Dr. Anthony T. Diplock †
University of London
United Medical and Dental School
Guys' Hospital
Guys' Campus: London Bridge
London SE1 9RT
Great Britain
Tel.: 0044 171 9554621
FAX: 0044 171 403 7195

Prof. Dr. Guy Dirheimer
Institut de Biologie Moléculaire et Cellulaire du C.N.R.S.
15, rue René Descartes
F-67084 Strasbourg
France
Tel.: 0033 8841 7056
FAX: 0033 8860 2218

Dr. Bernd Epe
Universität Mainz, Institut für Pharmazie
Abt. für Pharmakologie und Toxikologie
Staudinger Weg 5
D-55099 Mainz
Germany
Tel.: 0049 6131 394309
FAX: 0049 6131 395521

Dr. Montserrat García-Closas
Nutritional Epidemiology Branch
Division of Cancer Epidemiology and Genetics
6130 Executive BLVD. MSC7374
Bethesda, MD 20892 7374
USA
Tel.: 001 301 496 6426
FAX: 001 301 402 0916

Dr. Ronald W. Hart
National Center for Toxicological Research/FDA
3900 NCTR Road
Jefferson, AR 72079-9502
USA
Tel.: 001 870 543 7116
FAX: 001 870 543 7576 (543 7332)

Dr. Stephen S. Hecht
University of Minnesota Cancer Center
Box 806 – UMHC
420 Delaware St.
S.E. Minneapolis, MN 55455-0392
USA
Tel.: 001 612 624 7604
FAX: 001 612 626 5135

Prof. Dr. Werner K. Lutz
Department of Toxicology
University of Würzburg
Versbacher Str. 9
D-97078 Würzburg
Germany
Tel.: 0049 931 201 5402
FAX: 0049 931 201 3446

Dr. Margareth McCredie
Department of Preventive and Social Medicine
University of Otago
P. O. Box 913
Dunedin
New Zealand
Tel.: 00 643 479 7201
FAX: 00 643 479 7298

Dr. LaVerne Mooney
Columbia University School of Public Health
Division of Environmental Health Sciences
60 Haven Avenue B109
New York, NY 10032
Tel.: 001 212 304 7281
FAX: 001 212 544–1943

Dr. Robert W. Owen
Deutsches Krebsforschungszentrum
Toxikologie und Krebsrisikofaktoren
Im Neuenheimer Feld 280
D-69120 Heidelberg
Germany
Tel.: 0049 6221 423 338 (3316)
FAX: 0049 6221 423359

Dr. Wolfgang Pfau
Department of Toxicology and Environmental Medicine of the Fraunhofer Society
Hamburg University Medical School
Grindelallee 117
D-20146 Hamburg
Germany
Tel.: 0049 40 4123 5905
FAX: 0049 40 42838 5316

Frau Prof. Dr. Beatrix L. Pool-Zobel
Institut für Ernährung und Umwelt
Friedrich Schiller Universität
Dornburgerstr. 25
D-07743 Jena
Germany
Tel.: 00493641949671
FAX: 00493641949672

Dr. John D. Potter
Fred Hutchinson Cancer Research Center
1100 Fairview Ave. N., MP 702
Seattle, WA 98109-1024
USA
Tel.: 0012066674683
FAX: 0012066675977

Prof. Dr. Gerhard Rechkemmer
Federal Research Center for Nutrition
Institute of Nutritional Physiology
D-76344 Eggenstein-Leopoldshafen
Germany
Tel.: 00497247823600
FAX: 0049724722820

Prof. Dr. Helmut K. Seitz
Department of Medicine
Salem Medical Center
Zeppelinstr. 11–33
D-69121 Heidelberg
Germany
Tel.: 00496221483200
FAX: 00496221483494

Dr. David E. G. Shuker
MRC Toxicology Unit
Hodgkin Building, University of Leicester
PO Box 138, Lancaster Road
Leicester LE1 9HN
Great Britain
Tel.: 00441162525573
FAX: 00441162525616

Prof. Dr. Wolfgang Scheppach
Klinikum der bayrischen Julius-Maximilians-Universität Würzburg
Luitpoldkrankenhaus
Josef Schneider Straße 2
D-97080 Würzburg
Germany
Tel.: 00499312013183
FAX: 00499312013534

Dr. Takuji Tanaka
Kanazawa Medical University
1-1 Daigaku, Uchinada
Ishikawa 920-0293
Japan
Tel.: 0081 76 286 2211 (3613)
FAX: 0081 76 286 6926

Dr. Steven R. Tannenbaum
Massachusetts Institute of Technology
Department of Chemistry and Whitaker College (Room 56–311)
Cambridge, MA 02139-4307
USA
Tel.: 001 617 253 3729
FAX: 001 617 258 8676

Dr. Lilian U. Thompson
Department of Nutritional Sciences
Faculty of Medicine
University of Toronto
150 College Street
Toronto, Ontario, M5S 3E2
Canada
Tel.: 001 416 978 3523
FAX: 001 416 978 5882

Prof. Dr. Harri Vainio
Division of Health Risk Assessment
Institute of Environmental Medicine at Karolinska Institute
P.O. Box 210
Doktorsringen 16 C
S-171 77 Stockholm
Sweden
Tel.: 0046 8728 75 30
FAX: 0046 8 34 38 49

Prof. Dr. Henk J. van Kranen
Department of Carcinogenesis, Mutagenesis and Genetics
Laboratory of Health Effects Research
National Institute of Public Health and the Environment
Bilthoven
The Netherlands
Tel.: 0031 30 274 2182
FAX: 0031 30 274 4446

Prof. Dr. Martin J. Wiseman
Burson-Marsteller
24–28 Bloomsbury Way
London WC1A 2PX
Great Britain
Tel.: 0044 171 831 6262
FAX: 0044 171 430 1033

Dr. Chung S. Yang
Laboratory for Cancer Research
College of Pharmacy
Rutgers University
Piscataway, NJ 08854-8020
USA
Tel.: 001 908 445 5360
FAX: 001 732 445 0687